Physical Metallurgy

WILEY SERIES ON THE SCIENCE AND TECHNOLOGY OF MATERIALS

J. H. Hollomon, Advisory Editor

Physical Metallurgy
   Bruce Chalmers

Ferrites
   J. Smit and H. P. J. Wijn

Zone Melting
   William G. Pfann

The Metallurgy of Vanadium
   William Rostoker

Introduction to Solid State Physics, Second Edition
   Charles Kittel

New York
London

Physical Metallurgy

by
Bruce Chalmers
GORDON MCKAY PROFESSOR OF METALLURGY
DIVISION OF ENGINEERING AND
APPLIED PHYSICS
HARVARD UNIVERSITY

John Wiley & Sons, Inc.
Chapman & Hall, Limited

TN690
.C513

Copyright © 1959 by John Wiley & Sons, Inc.

All Rights Reserved

This book or any part thereof must not be reproduced in any form without the written permission of the publisher.

Library of Congress Catalog Card Number: 59-14983
Printed in the United States of America

Preface

The purpose of this book is to present, in logical and systematic order, the ideas and concepts that have led to and resulted from the great advance in the understanding of the behavior of metals and alloys that has occurred in the last few decades. The underlying theme is that the behavior of a metal is uniquely determined by its properties and the environment (stress, temperature, etc.) to which it is subjected; that the properties are uniquely determined by the structure in the most general sense of the term; and that the structure is uniquely determined by the chemical composition and the history of the material. It cannot be claimed that all the links in this chain are complete in detail; but they are sufficiently complete for us to understand all the relevant phenomena qualitatively, and some of them quantitatively.

The method of presentation that has been selected is to develop a physical understanding of the various processes and to compare the experimental information with the theoretical expectation sufficiently often for the application to real situations to become an integral part of the whole. As the main emphasis is on the development of physical concepts that can be applied to real cases, it was not appropriate to include a large amount of quantitative detail. It is, in any case, instructive to look up the detailed properties of specific materials in the appropriate handbooks.

It is hoped that this book will be suitable as a first textbook for students of physical metallurgy whose preparation includes some chemistry and physics, not only because some knowledge of chemistry and physics is required, but also because the whole approach and point of view may be too unfamiliar to students who have not previously studied the physical sciences. It is also hoped that it may be useful as an outline of the most modern concepts for metallurgists whose training antedated the development of the present point of view. No attempt has been made to achieve the kind of completeness that has characterized many textbooks in the past; the main reason for this is twofold. In the first place, it is necessary to establish a framework of understanding before very many details can be usefully absorbed, and secondly, the student should be compelled to go to a variety of sources to follow the interests that are aroused. This subject cannot be mastered from any one book, because it is being developed by many different people with many different points of view. My own views may color some of the topics that are discussed herein; the student should be on his guard against accepting any theory or statement as being final.

Some problems for the student to attempt are presented in the Appendix. There are three ways in which problems may be useful: 1. to test the student's ability to make quantitative use of available information, 2. to develop the student's ability to use the literature effectively, and 3. to test the extent to which concepts are understood by the student. The problems are of various types, many of them contributing to all three of these processes. They are of various degrees of difficulty and vary widely in the time that they will require, some taking only a few minutes if the point is understood, others demanding several hours of literature research and writing. It is not expected that any one class will cover the whole range of problems suggested here.

It may be useful to the student to discuss briefly the literature that is relevant. This may be done by following the history of an unspecified theory or piece of information. The item originates in a laboratory or in the mind of a research worker; it is normally published in a *paper* that appears in one of the scientific or technical journals.* At least 12,000 metallurgical papers are published each year, and it is clearly impossible even to see all of them. All relevant papers are

---

* The journals devoted specifically to metallurgy are *Acta Metallurgica, Journals of Metals, Transactions of the American Society for Metals, Journal of the Institute of Metals, Journal of the Iron and Steel Institute* (*London*), the English translations of the Russian *Physics of Metals and Metallography,* and many others in various languages. Papers concerned with metals also appear in many other periodicals.

abstracted, and the abstracts are published in specialized abstract journals.*

Of increasing importance, in view of the growth of the subject, are the review articles written by experts and usually consisting of a critical discussion of the present state of knowledge in a specialized field. Such review articles are published in various forms, including the American Society for Metals Annual Seminar volumes and *Progress in Metal Physics*. These "secondary publications" usually contain exhaustive bibliographies from which earlier work can be found. More comprehensive surveys of special fields also appear in the form of monographs. Quantitative information that is of importance is collected from time to time in handbooks, of which the *Metals Handbook* and the *Metals Reference Book* are the most useful to metallurgists.

Because of the availability of review articles and monographs on most of the subjects discussed in this book, references to original papers are not given; instead each chapter is followed by some suggestions for further reading; it is most desirable that a student should "follow up" some detailed topics, in order to acquire a technique for studying the literature, to learn to appreciate the way in which the present position has been built up, and to see the processes by which it will advance.

Experimental techniques are not discussed in this book; they are many and varied, ranging from the preparation of metal for examination under the microscope, X-ray diffraction, and temperature measurement, which have been very adequately described in specialized books, to nuclear resonance and radiation damage, which are relatively new and have not been so thoroughly exploited or discussed. The "standard" techniques must be learned in the laboratory, as an integral part of any course in physical metallurgy; they can be learned as by-products of experiments in which phenomena are measured or observed.

This book is written primarily as a textbook for a course in which the student makes his first serious contact with physical metallurgy, and it is intended to provide the basis on which more advanced and detailed studies can subsequently be based. It is my opinion that the introduction of physical metallurgy as a scientific, rather than a descriptive, subject can be achieved at any desired stage of the student's

* Abstracts of metallurgical papers are published in English in *Metals Review* (U.S.A.), *Metallurgical Abstracts* (U.K.), and the translations of the Russian *Metallurgical Abstracts*.

career, but that much is gained if this is delayed until the basic sciences have been built up to a fairly advanced level. It is hoped that this book will be suitable in level and content for juniors, seniors, or graduate students who are embarking on the study of physical metallurgy with the intention of making a career in this challenging and rewarding field.

I would like to take this opportunity to express my thanks to the many friends and colleagues who have helped me directly or indirectly in the preparation of this book and to the many students, past and present, who have taught me so much.

Those who supplied illustrations are mentioned elsewhere; those who supplied ideas and criticism are too numerous to mention individually; Richard S. Davis and Kenneth A. Jackson, however, have been very stimulating colleagues during the period of planning and writing, and their contribution, though intangible, is very significant.

Much material from many sources has been used; the decision not to include references to original papers has in some cases led to the absence of credit for material where authorship is identifiable. This applies particularly to illustrations taken from "review" and "seminar" compendiums. Reference to these sources will always identify the original author. It is to be hoped that the serious reader will pursue his studies at least that far.

I am also indebted to the following for permission to use copyright material: Academic Press, Inc.; *Acta Metallurgica;* Addison-Wesley Press, Inc.; American Institute of Mining, Metallurgical, and Petroleum Engineers, Inc.; American Society for Metals; Butterworths Scientific Publications; National Research Council; The Institute of Metals; The International Nickel Company, Inc.; Interscience Publishers, Inc.; McGraw-Hill Book Company, Inc.; Pergamon Press, Ltd.; Philosophical Library; Prentice-Hall, Inc.; D. Van Nostrand Company, Inc.; and John Wiley & Sons, Inc.

BRUCE CHALMERS

Cambridge, Mass.
October, 1959

Contents

1 The Structure of the Atom   1

1.1 The Nucleus   1
1.2 The Electrons   2
      The Four Quantum Numbers   2
      Electron Configuration of the Elements   3
      Electron Configuration and Chemical Properties   3
1.3 The Periodic Table   6
1.4 The "Excited State"   8
1.5 Ionization Potentials   8
1.6 Further Reading   9

2 Aggregates of Atoms   10

2.1 The Stability of a Structure   10
2.2 The Electrons in an Aggregate   11
      Ionic or Heteropolar Bonding   12
      Covalent or Homopolar Bonding   13
      Metallic Bonding   14
      Van der Waals Forces   14
      Definition of "Metal"   14
2.3 Minimum Free Energy Condition   15
      Equilibrium Between Phases: Kinetic Considerations   17
      Equilibrium Between Phases: Thermodynamic Considerations   20
2.4 Structures of Crystalline Solids   21

2.5 Equilibrium Crystal Structures of the Elements   22
    The 8 − N Rule   28
    Structures of the Metals   30
    Sizes of Atoms in Metals   31
2.6 Structure of Liquids and Vapors   34
2.7 Alloys   36
    Phase Diagrams   36
    Ternary Diagrams   40
    Substitutional Solid Solutions   45
    Ordering and Clustering   45
    Short Range Order   46
    Long Range Order   46
    Types of Ordered Structure   47
    The Long-Range-Order Parameter   50
    Antiphase Domains   51
    Thermodynamic Considerations   52
    Nonstoichiometric Alloys   52
    Interstitial Solid Solutions   53
    Limit of Solubility   53
2.8 Equilibrium Between Solid and Liquid Alloys   54
    Kinetic Considerations   54
    Thermodynamic Considerations   59
2.9 Intermediate Phases   63
    The Energy of Electrons in a Crystal   64
    Brillouin Zones   64
    Hume-Rothery or Electron Phases   67
    Ionic and Covalent Compounds   69
    Ionic Compounds   70
    Covalent Compounds   71
    Interstitial Compounds   73
    Laves Phases   73
    Heats of Formation of Intermediate Phases   74
2.10 Further Reading   76

3 Structure-Insensitive Properties   77

3.1 Classification of Properties   77
3.2 Structure-Insensitive Properties   78
    Density of the Elements   78
    Change of Volume on Melting   78
    Density of Alloys   80
    Interstitial Solid Solutions   80

Elastic Properties 81
Elastic Constants 81
Elastic Properties of Polycrystalline Aggregates 82
Single-Phase Alloys 82
Polyphase Alloys 83
3.4 Melting Point 84
Pure Metals 84
Alloys 85
3.5 Thermal Expansion 85
Variation of Interatomic Distance with Temperature 85
Relationship Between Thermal Expansion and Melting Point 86
Abnormal Thermal Expansion 87
Thermal Expansion of Uranium 89
Invar 89
3.6 Magnetic Properties 89
3.7 Electrical Properties 90
3.8 Further Reading 91

4 Imperfections in Crystals 92

4.1 Types of Imperfection 92
4.2 Point Imperfections 93
Lattice Vacancies 93
Interstitial Atoms 94
Substitutional Atoms 95
Complex Point Defects 95
4.3 Line Imperfections 95
Dislocations 95
Energy of a Dislocation 100
Motion of Dislocations 101
Complete and Partial Dislocations 103
Segregation at Dislocations 106
Evidence for the Existence of Dislocations 107
Etching 107
Decoration 107
Electron Microscopy 107
X-Ray Diffraction 109
Dislocation Content of Crystals 109
4.4 Surface Imperfections 110
Crystal Boundaries 113
"Equilibrium" Disposition of Crystal Boundaries 121

CONTENTS

      Subboundaries   124
      Twin Boundaries   125
      Interphase Interfaces   128
      Solid-Liquid Interfaces   130
      Solid-Vapor Interfaces   133
      Ferromagnetic Domain Walls   136
4.5 Further Reading   138

5   Structure-Sensitive Properties   **139**

5.1 Stress-Strain-Time Relationships   139
5.2 Plastic Deformation   139
      Yield Stress for the Perfect Single Crystal   140
      Dislocation Theory of Plastic Deformation   144
      The Frank-Read Source   146
      Slip Systems   149
      The Stress-Strain Curve   149
      Nonhomogeneous Deformation Processes   157
      Twinning   158
      Deformation Bonds   159
      Effect of Size   161
      Effect of Environment   162
      Effect of Temperature   162
      Plastic Properties of Solid Solutions   163
      The Cottrell Mechanism   166
      The Fisher Mechanism   167
      The Suzuki Mechanism   168
      Experimental Evidence   168
      Hardening by a Second Phase   172
      Effects of Crystal Boundaries   174
      The Single Boundary   174
      Polycrystalline Aggregates   177
      Effect of Crystal Size on Plastic Properties   179
      Substructure Hardening   180
5.3 Creep   181
      The Influence of Time on Plastic Deformation   181
      Creep   184
      Effects of Stress and Temperature on Creep   185
      Effect of Structure on Creep   186
5.4 Anelasticity   191
5.5 Fracture   194
      Theoretical Strength   194

Types of Fracture   195
Criterion for Crack Propagation   196
Origin of Cracks   201
The Fracture Curve   202
Effect of Temperature on Fracture   205
Effect of Time; Delayed Fracture   206
Hydrogen Embrittlement   206
Fatigue   207
Characteristic Appearances of Fractures   210
5.6   Mechanical Properties   213
Mechanical Testing   213
The Tensile Test   214
Hardness Testing   220
Impact Testing   221
Special Mechanical Tests   222
Nondestructive Testing   223
5.7   Magnetic Properties   223
Magnetically Soft Materials   225
Magnetically Hard Materials   228
5.8   Further Reading   230

6   Change of State   231

6.1   Control of Shape and Structure   231
6.2   Solidification   231
The Process of Freezing   232
The Influence of Crystallography   232
"Smooth" and "Dendritic" Freezing   234
6.3   Nucleation   235
Qualitative Discussion   235
Calculation of Critical Size   238
Distribution of Embryo Sizes   240
Heterogeneous Nucleation   241
Nucleation of Freezing   243
Homogeneous Nucleation   243
Heterogeneous Nucleation   246
Surface Nucleation   246
6.4   Imperfections Resulting from Freezing   248
Dislocations   248
6.5   Freezing of Liquids Containing More than One Component   249
Redistribution of Solute   250

CONTENTS

      Zone Refining   255
      Constitutional Supercooling   258
      Nucleation in Supercooled Melt   261
6.6 Two-Phase Alloys   262
      Liquid of Eutectic Composition   263
      Eutectic System, Noneutectic Composition   269
      Peritectic System   271
6.7 Structure Resulting from Solidification   272
      Size, Shape, and Orientation of Crystals   272
6.8 Porosity   277
      Effects of Shrinkage   277
      Effects of Gas Evolution   280
      Interaction of Shrinkage and Gas Evolution   280
6.9 Segregation   281
      Normal Segregation   282
      Coring   282
      Interdendritic Segregation   283
      Inverse Segregation   283
      Gravity Segregation   283
6.10 Heat Flow Considerations   284
      Pure Metals   285
      Alloys   285
6.11 Applications of Solidification   285
      Ingots   286
      Continuous Casting   286
      Arc Melting   287
      Vacuum Melting   288
      Castings   288
      Sand Mold Casting   289
      Investment Casting   289
      Shell Molding   289
      Permanent Mold Casting   289
      Special-Purpose Casting Operations   290
      Joining   290
      Coating   290
6.12 Electroplating   291
6.13 Growth of Crystals from Vapor   292
6.14 Melting   295
6.15 Solution   295
      Solution in Liquid Metals   296
      No Intermediate Phase   296
      With Intermediate Phase   297

CONTENTS

Solution in Aqueous Media   298
Electrochemical Potential   298
Inhibitors   301
Sacrificial Protection   302
Protection by "Insulation"   302
Types of Corrosion   302
Stress Corrosion   304
Corrosion Fatigue   304
Effect of Temperature on Corrosion   306
Electropolishing   306
Chemical Polishing   306
6.16 Further Reading   306

7 Deformation, Radiation Damage, and Recovery Processes   307

7.1 The Structure of Deformed Metals   307
Dislocation Content   308
Diffraction Effects   308
Shapes of Crystals   309
Two-Phase Alloys   309
Preferred Orientation, Deformation Texture   311
Work Hardening   312
7.2 Radiation "Damage"   313
Types of "Damage"   314
Ionization Effects   314
Collision Effects   314
Fission Effects   317
Effects of Radiation on Properties   317
Without Fission   317
Effects Produced by Irradiation Accompanied by Fission   321
7.3 Recovery Processes   321
Annealing Out of Point Defects   322
Recovery (Polygonization)   322
Residual Stresses and Stress Relief   326
Growth of Uranium during Cyclic Temperature Changes   327
Recrystallization   328
Kinetics of Recrystallization   328
Mechanisms of Recrystallization   330
Recrystallization Temperature   330

Recrystallization Textures 332
Recrystallization Twins 334
Stored Energy 335
Properties of Recrystallized Metals 336
Grain Growth 337
Normal Grain Growth 337
Secondary Recrystallization 339
Factors that Affect Grain Growth 340
Abnormal Grain Growth 341
Grain Shapes in Two-Phase Alloys 341
7.4 The Shaping of Metals 342
Deformation Processes 343
Cold Deformation and Hot Deformation 343
Types of Processes 344
Properties of Metals Fabricated by Deformation 347
Hot-Deformation Processes 348
Cold-Deformation Processes 349
Cutting Processes 349
Machining 349
Abrasion and Wear 351
Polishing and Burnishing 352
7.5 Solid-State Welding 352
Powder Metallurgy 354
Other Solid-State Bonding Techniques 355
7.6 Further Reading 356

8 Solid-State Transformations 357

8.1 Classification of Solid-State Transformations 357
8.2 Diffusionless, or Martensitic, Transformations 359
General 359
Crystallography of Diffusionless Transformations 359
Reversibility of Diffusionless Transformations 360
Effect of Plastic Deformation 361
The Martensitic Transformation in Steel 361
The Structure of Martensite 362
The Morphology of Martensite 364
The Properties of Martensite 365
Volume Change in Martensite Transformation 366
Nucleation and Growth of Martensite 366
Stabilization of Austenite 367
Diffusionless Transformations in Titanium Alloys 368

CONTENTS

8.3 Diffusion 369
    Diffusion in Crystals 370
    The Unit Process of Diffusion 370
    Jump Frequency 370
    The Diffusion Coefficient 371
    Discussion of Diffusion Mechanisms 377
    Kirkendall Effect 378
    Effect of Irradiation on Diffusion 380
    Dislocation Diffusion 380
    Grain Boundary Diffusion 380
    Surface Diffusion 384
    Diffusion in Liquid Metals 384
8.4 Precipitation and Solution 385
    Definition 385
    Reversible and Irreversible Precipitation 385
    Nucleation of Precipitate 386
    Distribution of Nucleation Sites 388
    Growth of Precipitates 388
    Kinetics 388
    Morphology 390
    Effect of Quenching Rate 391
    Effect of Strain on Precipitation 393
    Crystal Structures of Precipitates 393
    Effect of Precipitation on Properties 394
    Binary Aluminum-Copper Alloys 394
    Other Aluminum Alloys 396
    Beryllium Copper 398
    Creep Resistant Alloys 399
    Tempering of Martensite 399
    High-Speed Steels 400
    Precipitation of Solute on Dislocations 401
    The Yield Point 401
    Strain Aging and Quench Aging 402
    "Precipitation" of Gases in Metals 403
    Graphitization 404
    Solution of Precipitates 405
    Spheroidization 405
    Diffusion Controlled "Allotropic" Transformation 406
    Heat Treatment and Properties of Titanium Alloys 410
8.5 The Eutectoid Transformation 415
    Pearlite 415
    Kinetics of Austenite-Pearlite Transformation 418

Properties of Pearlite   420
Metallography of Pearlite   422
The Bainite Transformation   425
Influence of Alloying Elements on T.T.T. Curve   426
Effect of Alloying Elements on Equilibrium Conditions   426
Effect of Alloying Elements on Kinetics of Transformation   427
Continuous Cooling Curves   427
Quenching   432
Hardenability   434
Effect of Austenite Grain Size   434
Calculation of Hardenability   435
Advantages of High Hardenability   435
Martempering   437
Austempering   437
Austenitic Steels   437
Annealing of Steel   438
8.6 Order-Disorder Changes   438
Effect of Ordering on Properties   439
Density   440
Mechanical Properties   440
Electrical Properties   440
Magnetic Properties   441
8.7 Reaction with Environment   441
Oxidation   441
Thermodynamic Considerations   441
Combustion of Metals   441
Rate of Oxidation   444
Effect of Temperature   445
Oxidation of Alloys   445
Carburization and Decarburization   447
Decarburization   447
Carburization   447
8.8 Further Reading   448

Appendix: Problems   **449**

Index   **453**

1

The Structure of the Atom

1.1 The Nucleus

All atoms may be regarded as built up of various numbers of protons, neutrons, and electrons; the relevant properties of these three fundamental particles are given in Table 1.1.

Each atom has a nucleus, consisting of protons and neutrons concentrated in a volume that is much smaller than that of the atom, and extranuclear electrons equal in number to the protons in the nucleus. This number, the atomic number Z, characterizes the chemical element to which the atom belongs.

Since the mass of the electron is only about 1/1800 that of the proton, the mass of the atom can be regarded as being concentrated in the nucleus; the nucleus has a radius of the order of 10^{-12} cm, while the radius of the atom is 10^{-8} to 10^{-7} cm.

The electrostatic charge on a proton is equal in magnitude and opposite in sign to that on an electron; the atom as whole, therefore, is electrically neutral. When this state of electrostatic neutrality is disturbed by adding or removing electrons, the atom becomes an ion.

The nucleus of an atom contains, in addition to the protons, a roughly

Table 1.1. **Properties of the Fundamental Particles**

	Charge (esu)	Mass (grams)
Proton	4.802×10^{-10}	1.672×10^{-24}
Neutron	0	1.675×10^{-24}
Electron	4.802×10^{-10}	9.107×10^{-28}

equal number of neutrons, which contribute to its mass but not to its charge. The different isotopes of an element correspond to different numbers of neutrons in the nucleus, the number of protons being constant (Z). The mass of the atom depends on the total number (N) of protons and neutrons, but does not equal N times the mass of the proton (or neutron) because the energy of association of the protons and neutrons in the nucleus corresponds, by Einstein's relationship between mass and energy, to a decrease of mass. The chemical "atomic weight" is the weighted average of the atomic masses of the various isotopes that are present. The variations in properties and behavior between the different isotopes of an element can be neglected for the purposes of this book.

1.2 The Electrons

There are various ways of describing the location of the extranuclear electrons. They may be described as rotating about the nucleus in circular or elliptical orbits, in which they can remain without any loss of energy. Only those orbits that are characterized by discrete integral quantum numbers are permissible. The alternative description of the location of an electron is in terms of a wave equation, which specifies the probability of the electron being at any point in a system which consists of the atom under consideration and all other atoms associated with it.

The four quantum numbers. For most purposes in the present treatment it is convenient to regard the electrons as moving orbitally around their nuclei. The state of an electron is defined by four quantum numbers, as follows:

The first quantum number n arises from the postulate of Bohr that the only circular orbits in which an electron could remain without radiating energy are those in which the angular momentum of the electron is an integral multiple of $h/2\pi$, where h is Planck's constant. The possible orbits, therefore, have angular momenta of $h/2\pi$, $2h/2\pi$, $3h/2\pi$, $\cdots$; these orbits are characterized by $n = 1$, $n = 2$, $n = 3$, $\cdots$. The radius of the orbit is proportional to n^2 for a given element, but it also depends on the charge on the nucleus (Z).

The necessity for a second quantum number arises from the fact that a series of elliptical orbits have approximately the same energy as a circular orbit, but different angular momenta. The second quantum number l represents the fact that the elliptical orbits can only have eccentricities that correspond to integral values of l; or that the angular momentum, as well as the energy, can only change from one stationary state to another. The second quantum number l must be integral and must be less than n. $l = n - 1$ corresponds to the special case of a circular orbit.

The third quantum number m_l may be considered to correspond to the component of the angular momentum along the axis of an externally imposed magnetic field. m_l can have any integral value from $+l$ to $-l$, including zero. The fourth quantum number m_s can only have values $+\frac{1}{2}$ and $-\frac{1}{2}$ and represents the component of the magnetic moment due to the spin of the electron.

The "state" of an electron is completely defined by its four quantum numbers, and, according to the "exclusion principle," there can only be one electron in any one "state" in a system. At this point we are considering single, isolated atoms in which one atom constitutes the whole system. The complications that arise when the atoms are close enough together to form parts of the same system constitute the main theme of this book.

In an atom in its normal state, the electrons occupy the lowest permissible energy states, and therefore the disposition of the electrons in the isolated neutral atom of an element depends only on its atomic number. Difficulties arise in specifying this disposition in the heavier elements because it is not always possible to predict which of several quantum states should have the lowest energy.

Electron configuration of the elements. It is convenient to adopt the spectroscopic notation for describing the quantum state of the electrons; the value of n is followed by s, p, d, or f, according to whether $l = 0, 1, 2,$ or 3, and the number of such electrons is added as a superscript; for example, $(1s)^2$ corresponds to an atom containing two electrons for which $n = 1$, $l = 0$.

The total permissible number of electrons with a given combination of n and l follows from the rules given above; for example, if $n = 1$, $l = 0$, $m = 0$, $m_l = 0$, $m_s = \pm\frac{1}{2}$, a total of two electrons; for $n = 2$, $l = 0$ or 1, $m_l = 0$ when $l = 0$, $m_l = 0$ or ± 1 when $l = 1$, $m_s = \pm\frac{1}{2}$, allowing a total of $(2s)^2 + (2p)^6 = 8$, etc.

Table 1.2 shows the electron configurations of the elements.

Electron configuration and chemical properties. There are several features of this table that are relevant. It will be seen that the elements lithium 3, sodium 11, potassium 19, rubidium 37, cesium 55, and francium 87 represent the stages in the development at which the first electron appears in the quantum "shells" for which $n = 2, 3, 4, 5, 6,$ and 7 respectively. These elements, the alkali metals, have closely analogous chemical properties; the common factor in their electron configuration is the single electron in an otherwise unoccupied shell. The elements

Table 1.2. Electron Configurations of the Elements *

Element and Atomic Number	$n=$ 1 $l=$ 0	2 0	2 1	3 0	3 1	3 2	4 0	4 1	4 2	4 3
1 H	1									
2 He	2									
3 Li	2	1								
4 Be	2	2								
5 B	2	2	1							
6 C	2	2	2							
7 N	2	2	3							
8 O	2	2	4							
9 F	2	2	5							
10 Ne	2	2	6							
11 Na	2	2	6	1						
12 Mg	2	2	6	2						
13 Al	2	2	6	2	1					
14 Si	2	2	6	2	2					
15 P	2	2	6	2	3					
16 S	2	2	6	2	4					
17 Cl	2	2	6	2	5					
18 A	2	2	6	2	6					
19 K	2	2	6	2	6		1			
20 Ca	2	2	6	2	6		2			
21 Sc	2	2	6	2	6	1	2			
22 Ti	2	2	6	2	6	2	2			
23 V	2	2	6	2	6	3	2			
24 Cr	2	2	6	2	6	5	1			
25 Mn	2	2	6	2	6	5	2			
26 Fe	2	2	6	2	6	6	2			
27 Co	2	2	6	2	6	7	2			
28 Ni	2	2	6	2	6	8	2			
29 Cu	2	2	6	2	6	10	1			
30 Zn	2	2	6	2	6	10	2			
31 Ga	2	2	6	2	6	10	2	1		
32 Ge	2	2	6	2	6	10	2	2		
33 As	2	2	6	2	6	10	2	3		
34 Se	2	2	6	2	6	10	2	4		
35 Br	2	2	6	2	6	10	2	5		
36 Kr	2	2	6	2	6	10	2	6		

* From *The Structure of Metals and Alloys*, W. Hume-Rothery and G. V. Raynor, Institute of Metals Monograph and Report Series No. 1, 3rd ed., London, 1954, pp. 14–16.

Table 1.2 (continued)

Element and Atomic Number.	$n=$ 1 $l=$ —	2 —	3 —	4 0	1	2	3	5 0	1	2	6 0
37 Rb	2	8	18	2	6			1			
38 Sr	2	8	18	2	6			2			
39 Y	2	8	18	2	6	1		2			
40 Zr	2	8	18	2	6	2		2			
41 Nb	2	8	18	2	6	4		1			
42 Mo	2	8	18	2	6	5		1			
43 Tc	2	8	18	2	6	6		1			
44 Ru	2	8	18	2	6	7		1			
45 Rh	2	8	18	2	6	8		1			
46 Pd	2	8	18	2	6	10		—			
47 Ag	2	8	18	2	6	10		1			
48 Cd	2	8	18	2	6	10		2			
49 In	2	8	18	2	6	10		2	1		
50 Sn	2	8	18	2	6	10		2	2		
51 Sb	2	8	18	2	6	10		2	3		
52 Te	2	8	18	2	6	10		2	4		
53 I	2	8	18	2	6	10		2	5		
54 Xe	2	8	18	2	6	10		2	6		
55 Cs	2	8	18	2	6	10		2	6		1
56 Ba	2	8	18	2	6	10		2	6		2
57 La	2	8	18	2	6	10		2	6	1	2
58 Ce	2	8	18	2	6	10	2	2	6		2
59 Pr	2	8	18	2	6	10	3	2	6		2
60 Nd	2	8	18	2	6	10	4	2	6		2
61 Pm	2	8	18	2	6	10	5	2	6		2
62 Sm	2	8	18	2	6	10	6	2	6		2
63 Eu	2	8	18	2	6	10	7	2	6		2
64 Gd	2	8	18	2	6	10	7	2	6	1	2
65 Tb	2	8	18	2	6	10	8	2	6	1	2
66 Dy	2	8	18	2	6	10	10	2	6		2
67 Ho	2	8	18	2	6	10	11	2	6		2
68 Er	2	8	18	2	6	10	12	2	6		2
69 Tm	2	8	18	2	6	10	13	2	6		2
70 Yb	2	8	18	2	6	10	14	2	6		2
71 Lu	2	8	18	2	6	10	14	2	6	1	2
72 Hf	2	8	18	2	6	10	14	2	6	2	2

immediately preceding the alkali metals are helium, neon, argon, krypton, xenon, and radon, the inert gases that are characterized by their lack of chemical reactivity. Here the common factor is the disposition of the electrons in filled shells or subshells. Another group of elements that have much in common in their chemical behavior are the halogens fluorine, chlorine, bromine, iodine, and astatine, each of which has one electron less than the number required to fill the outermost occupied

Table 1.2 (continued)

Element and Atomic Number.	\multicolumn{11}{c}{Principal and Secondary Quantum Numbers.}
$n =$	1	2	3	4	5				6			7
$l =$	—	—	—	—	0	1	2	3	0	1	2	0
73 Ta	2	8	18	32	2	6	3		2			
74 W	2	8	18	32	2	6	4		2			
75 Re	2	8	18	32	2	6	5		2			
76 Os	2	8	18	32	2	6	6		2			
77 Ir	2	8	18	32	2	6	7		2			
78 Pt	2	8	18	32	2	6	8		2			
79 Au	2	8	18	32	2	6	10		1			
80 Hg	2	8	18	32	2	6	10		2			
81 Tl	2	8	18	32	2	6	10		2	1		
82 Pb	2	8	18	32	2	6	10		2	2		
83 Bi	2	8	18	32	2	6	10		2	3		
84 Po	2	8	18	32	2	6	10		2	4		
85 At	2	8	18	32	2	6	10		2	5		
86 Rn	2	8	18	32	2	6	10		2	6		
87 Fr	2	8	18	32	2	6	10		2	6		1
88 Ra	2	8	18	32	2	6	10		2	6		2
89 Ac	2	8	18	32	2	6	10		2	6	1	2

The exact electronic configuration of the later elements is uncertain.

90 Th .
91 Pa .
92 U .
93 Np .
94 Pu .
95 Am .
96 Cm .

shell. These and other similarities of properties have led to the conclusion that the chemical properties of the elements are intimately associated with the electrons that have the highest value of n.

1.3 The Periodic Table

The resulting regularities in the properties are emphasized by the periodic table of the elements, one form of which is given in Table 1.3.

The periodic table is complicated by the fact that, for example, 4s electrons appear, in potassium and calcium, before the 3d levels have been filled, but the 3d level is filled to its maximum of ten electrons (zinc) before any electrons enter the 4p level. The number of electrons in the 4s level is therefore either one or two from potassium to zinc, and the elements in this series do not fall into the sequence, as regards properties, that would be expected from the regularity of the periodic table from hydrogen to calcium. The elements 21 to 30 (scandium to zinc) are called the "first transition group."

A similar sequence of electron configurations is found from elements 39 to 48, where the $5p$ level does not start to fill until the $4d$ level contains ten electrons. This provides the second transition group of elements.

Table 1.3. Periodic Table of the Elements *

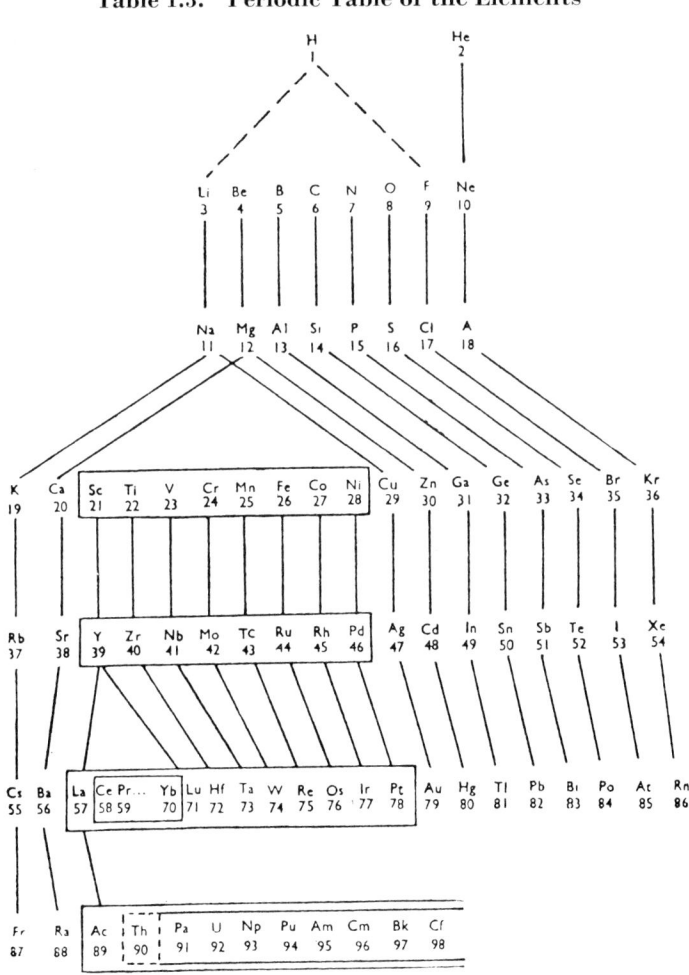

* From *Structure of Metals*, Hume-Rothery and Raynor, p. 6.

The 4f level remains unfilled until after the 6p level contains two electrons; the elements 57 (lanthanum) to 70 (ytterbium) correspond to the filling of the 4f level to 14 electrons, while the $n = 5$ and $n = 6$ shells remain practically constant. These elements are the rare earths (or lanthanides) which, as would be expected, have very similar chemical properties corresponding to their similar outer electron configurations.

The transuranic elements (93 $\cdots$) are believed to be another group analogous to the lanthanides, corresponding to the addition of electrons to the 5f level.

1.4 The "Excited State"

Our consideration of the structure of individual atoms, as distinct from aggregates of more than one atom, will conclude with a discussion of the conditions under which an atom may be in a state that is different from the normal. The normal state is that in which the electrons are distributed in the least energetic levels, i.e., so that there is no unoccupied level that has a lower energy than any occupied level.

The absorption of energy by an atom may correspond to an increase in the energy of one or more of its electrons, which can either be raised to higher (unoccupied) levels or ejected entirely from the atom. In either case this leaves a "lower" level unoccupied, and the atom is then in an "excited state," from which it can return to normal by the emission of energy and the return of electrons to their normal levels.

The levels involved in these processes are similar to those previously defined, using, in general, higher values of the quantum numbers than are occupied in the normal atom. The energy is emitted as electromagnetic radiation (X-ray, ultraviolet, or visible) of frequency ν that is related to the amount of energy E being radiated by the relationship $E = h\nu$, where h is Planck's constant.

1.5 Ionization Potentials

The energy that is required to remove an electron from an atom is called the ionization potential; a sequence of ionization potentials corresponds to the removal of successive electrons, the atom becoming positively charged with one, two, three, etc., units. An ionization potential depends upon the quantum state of the electron that is removed, and upon the charge of the atom or ion from which it is removed; it requires more work to remove an electron from a positive ion than from a neutral atom. As an example, the potentials, in volts, required for the removal of the first five electrons from carbon are: 11.2, 24.7, 47.6, 64.2, and 390. The large increase between the fourth and fifth values occurs because the first four electrons are from $n = 2$ levels while the fifth comes from the $n = 1$ level. This is an illustration of the stability of a filled shell.

1.6 Further Reading

More detailed discussion of the electron configuration and its interpretation in terms of quantum mechanics can be found in many specialized books; *Atomic Theory for Students of Metallurgy* by W. Hume-Rothery, Institute of Metals Monograph and Report Series No. 3, 2nd ed., London, 1952, presents this material in relation to the study of metals.

2

Aggregates of Atoms

2.1 The Stability of a Structure

There are an infinite variety of ways in which a number of atoms could be arranged in relation to each other; most of these would spontaneously change into more stable groupings. The purpose of the present discussion is to consider the criteria that determine the most stable way of grouping any collection of atoms. A stable arrangement is one towards which any other arrangement tends to change spontaneously.

It is necessary to examine the characteristics of a change that tends to occur spontaneously. A spontaneous change is one that occurs without external work being done on the system and which reduces its capacity to do external work, and therefore its free energy.

The free energy F of a system may be expressed in the form

$$F = U - TS + pv$$

where U, the internal energy, is the amount by which the energy of the system is less than it would be if the atoms were all separated from each other, T is the temperature, S the entropy, p the pressure, and v the volume.

For condensed systems, i.e., solids and liquids, in which the volume is changed very little by moderate changes in pressure, the term pv is relatively unimportant, and attention will therefore be concentrated on the internal energy, the temperature, and the entropy. At low temperatures, the value of U will have a predominant influence on the stability of an arrangement of atoms. We will therefore first consider the influence of aggregation on the internal energy U of a collection of atoms. The implications of T and S will be discussed subsequently.

2.2 The Electrons in an Aggregate

The permissible energy levels for the electrons of an isolated atom have been discussed in Chapter 1. If we now consider an aggregate of atoms, however, we find that the energies of the electrons are changed, and that electrostatic forces arise between the component parts of the system. The system, in which there can be only one electron with a given set of quantum states, must now be considered to include all the atoms in the aggregate; the energies of the electrons of each atom are modified by the proximity of the other atoms. Each individual level that exists for the single atom is replaced by a band of levels. Each level in a band is different from each other level, but since there is a level for each atom, they are extremely close together for the number of atoms in any experimental sample. It is therefore permissible to consider a band as consisting of a continuous series of levels. It is necessary, however, to distinguish between the "inner" electrons, which are virtually unaffected by the proximity of other atoms, and the "outer" electrons, whose energies are changed by significant amounts. The constancy of the inner levels is demonstrated by the fact that the characteristic X-ray frequencies are almost unaffected by the state of aggregation of the atoms. It is this fact that permits the identification of chemical elements in a composite sample by fluorescence analysis.

The interaction between atoms can lead to their aggregation only if it lowers their energy, and we must therefore consider how the energy of an atom may be reduced by the proximity of other atoms. The question is whether the proximity of other atoms allows electrons to find lower energy levels than are available in the isolated atoms. If so, the lowest energy state of the "collection" of atoms is that in which the electrons are so redistributed.

There are two ways of approaching the problem of the energy of the electrons in an aggregate of atoms. The electrons, and the energy levels, may be considered to be localized; an electron is at any time either moving freely or else associated with a particular atom. Alternatively, the electron may be described in terms of a wave equation, which has a finite value everywhere in the aggregate; the energy levels of the individual atoms are no longer considered, but are now properties of the aggregate as a whole. Which approach to select depends upon the problem under consideration; neither is able at the present time to make precise predictions, but each gives a useful physical picture in appropriate circumstances.

Ionic or heteropolar bonding. The simplest case is that in which an electron from one type of atom finds a lower energy level in a second type of atom, with the result that the aggregate now consists of positive and negative ions instead of neutral atoms. For example, the energy of the single 2s electron of sodium is relatively high; that is to say, it can be removed from the atom, to form an ion, with a small expenditure of work. This is typical of the elements which have a small number of electrons in the outermost shell (electropositive elements); by comparison, it is very difficult to remove electrons from atoms of the inert gases, because the number of electrons is just sufficient to form a closed or complete outer shell. It also follows that an atom, such as chlorine, with an almost complete outer shell, will accept an available electron very readily (electronegative element). There is a greater *decrease* in energy when an electron is added to a chlorine atom than the *increase* corresponding to the detachment of an electron from a sodium atom, and the result is the transfer of electrons from the electropositive to the electronegative atoms. An aggregate of sodium and chlorine atoms therefore quickly becomes an aggregate of positive sodium ions and negative chlorine ions. The energy of such an aggregate is further decreased if the nearest neighbors of each ion are ions of the opposite sign, as a result of the electrostatic attraction between opposite charges. A typical arrangement of ions is shown, in two dimensions, in Fig. 2.1. Each ion has, in the plane of the diagram, four nearest neighbors of the opposite sign, and is surrounded, at a greater distance, by a ring of four ions of its own sign.

The array of atoms is in fact three-dimensional, being composed of layers similar to that of Fig. 2.1, superimposed so that a line normal to the plane of the diagram passes through positive and negative ions alternately. Figure 2.1 therefore represents a section of the array in any one of three mutually perpendicular planes. Each ion has 6

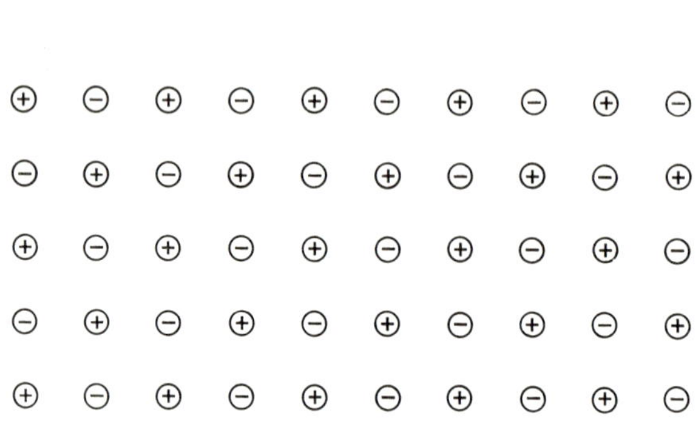

Fig. 2.1. Section of ionic crystal.

nearest neighbors of opposite sign and 12 next nearest neighbors of similar sign. Because the distance between unlike ions is less than the distance between like ions, the electrostatic forces of attraction between the former are greater than those of repulsion between the latter; the pattern would shrink indefinitely if each ion could be completely replaced by a point charge, as was implicitly assumed above. An additional force, however, must be taken into account: this is the electrostatic repulsion that occurs between the outer electron shells of two ions if they approach each other, even if the ions are of opposite sign. The forces of attraction and repulsion just balance when the ions are at a definite distance apart.

It will be recognized that no positive ion in this arrangement is uniquely associated with an individual negative ion, and it follows that the concept of the *molecule* does not apply in this type of structure, which constitutes a *crystal*. It should be pointed out that the term "crystal" describes an array of atoms arranged in a pattern that results from the repeated translation, by fixed amounts in suitable co-ordinate directions, of a unit pattern, the "unit cell."

It is desirable to distinguish between the "crystal lattice," which is an array of points, each of which has an identical environment of such points, and a "crystal," which may be regarded as a structure consisting of an identical *"group"* of atoms similarly related to each lattice point.

Covalent or homopolar bonding. A second way in which the energy of atoms can be lowered by aggregation is for electrons to be shared between atoms which each have partly filled outer shells. Thus, two atoms of chlorine, each with one vacancy in the $3p$ level, may each share one electron with each other; similarly, an atom of germanium has four electrons in the $n = 4$ shells; it can share each of these electrons with a neighboring atom, at the same time sharing one of the electrons of each of the four neighboring atoms. Thus each atom has eight shared electrons, which provides the low energy grouping of eight electrons. This type of bond is described as covalent or homopolar. It will be seen that electron sharing is only possible if the atoms are so close to each other that the electrons can pass from an "orbit" of one atom to an "orbit" of the other without becoming completely detached from either, or if the wave functions representing the electron densities overlap sufficiently. The minimum energy is achieved when the atoms are close together; this may be interpreted in terms of the electrostatic attraction between the shared electrons and the positive ion that is formed when the electrons move away from an atom. In the case of covalent bonding, either molecules or crystals may be formed. Two atoms of chlorine share two electrons to form a chlorine molecule; in germanium, each atom surrounds itself by four neighbors, with each of which it shares two electrons. Thus each germanium atom is at the center of a tetrahedron of germanium atoms. This configuration can be extended indefinitely as a three-dimensional pattern, the diamond-cubic crystal structure. Covalent bonding also occurs between dissimilar atoms, as in the case of the chemical compound InSb; there are, however, many compounds in which the bonding is somewhere between ionic and covalent.

Metallic bonding. In both ionic and covalent bonding it is permissible to regard the electron energy levels as unchanged by the proximity of other atoms; however, this does not apply in cases in which electrons are easily detached and are not accommodated in nearly filled shells. In such cases it is necessary to consider the electron energy levels outside the closed shell as characteristic of the aggregate rather than of the atom. Each energy *level* is spread into an energy band as a result of the interaction; the band must in principle be regarded as consisting of discrete levels, sufficient in number to accommodate all the electrons. The bands belong to the aggregate as a whole, and the electrons therefore are no longer associated with particular atoms (this only applies to the valence electrons). The electron energies are lower when the electrons are in energy bands than when they are in isolated atom levels. The energy of an aggregate is therefore less than that of the isolated atoms, and we can again superimpose the decrease in energy due to the electrostatic attraction between the electrons and the positive ions. The extent to which the ions are pulled together by these attractive forces is limited by the mutual repulsion of the ions themselves and of their closed electron shells.

The "metallic" bond can only exist when positive ions can be formed; however, it is not necessary to make a clear distinction between metallic and covalent bonding; it is in some ways more appropriate to regard the metallic bond as an extension of the covalent bond into situations in which the electrons frequently change the identity of the ions between which they are shared.

van der Waals forces. The fourth source of bonding energy arises from the fact that neutral atoms, if in close proximity, attract each other as a result of the redistribution in space of the negative charge of the electrons. This is the "van der Waals" type of bond. It is much weaker than the other types of bond, but it has some significance in certain solids.

Definition of "metal." The four types of bonding have been described above in terms of electrons and orbits. They may also be described in terms of wave functions; this would provide a more rigorous and quantitative account of the reduction of energy resulting from the aggregation of atoms; it would also show that the four types of bonding are not as distinct and mutually exclusive as the foregoing paragraphs would suggest. It is probably more realistic to consider the bond in most cases as a mixture of the various types.

The term *metal*, as used hereafter, describes an aggregate of atoms in which the bonding is such that electrons can move in response to an applied electric field, even when that field is quite small, and in which the number of electrons that can be so moved is comparable with the number of atoms. This description is not a rigorous definition of a metal, but it invokes one of the most characteristic properties of metals, and, by implication, the corresponding type of bonding. It does not imply that a given element is or is not a metal, but does imply that a given crystal has this characteristic. There are, however, borderline cases, such as germanium, which at ordinary temperatures is a semiconductor, while at high temperatures it is more metallic in character.

2.3 Minimum Free Energy Condition

The purpose of the following discussion is to explore the conditions that determine what structure an aggregate of atoms will have. It is clear that a particular structure is not a unique property of a particular collection of atoms, because the same substance can exist in alternative forms, such as solid and liquid. It is evident that the temperature is one of the conditions that influences the structure. It follows that the energy of an aggregate alone does not determine the stability of its structure. The temperature of a substance corresponds to the mean kinetic energy of thermal agitation of the atoms of that substance; the value of the mean kinetic energy is $\frac{1}{2}kT$ for each degree of freedom, where k is Boltzmann's constant and T is the absolute temperature. In a gas, this thermal energy corresponds to the free movement of atoms or molecules through distances that are large compared with their own size, interrupted by collisions with other atoms or molecules. In a solid or a liquid, the atoms can only move through very short distances, and the thermal motion of each atom is looked upon as a vibration, with an assignable mean frequency but continually changing in direction and amplitude. As with any other vibrating system, the energy alternates between kinetic and potential. The thermal energy of the atoms in a solid is related to the temperature in exactly the same way as in a gas; however, the energy of the atoms, which in a gas is kinetic energy of translation, is distributed between kinetic energy and potential energy, the latter arising from the forces of repulsion that exist when the atoms are at and near the end points of their vibration. The kinetic energy of an atom is a maximum when it passes through the mid-point of its vibration and the potential energy is then zero. The kinetic energy at that time is equal to kT. When the atom is at the end point of its vibration, its kinetic energy is zero, and its potential energy equal to kT. The *average* values of the kinetic energy and of the potential energy are each $\tfrac{1}{2}kT$. The actual values of the kinetic energy at any instant are distributed according to the Boltzmann distribution. This leads to the result that the fraction p of atoms having a kinetic energy in excess of any particular value Q is given by $p = \exp(-Q/kT)$.

This is represented diagrammatically in Fig. 2.2. The temperature therefore determines what proportion of the atoms, at any moment, have an energy greater than any specified value Q. An atom in a crystal is normally in a state of thermal vibration about a minimum energy position; it must therefore have a higher energy, Q, in order to move its mean position to some other point.

The temperature therefore determines the probability that an atom has enough energy to change its position; this is represented in Fig. 2.3a in which two adjacent energy minima, A and B, are shown; C is the "activated state" through which it must pass in going from A to B or B to A. An atom at A can only move to B if it momentarily has an energy Q, and if certain other conditions are fulfilled. The same considerations will still apply if point B corresponds to a higher energy than A, as in Fig. 2.3b, and it follows that the atoms may not always be in the lowest energy positions, although, in the case of Fig. 2.3b, the probability of an atom returning to A is higher than that of its

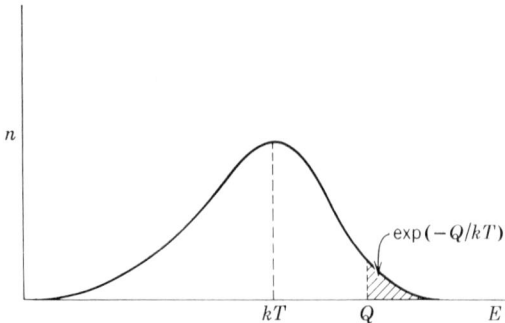

Fig. 2.2. Distribution of thermal energy.

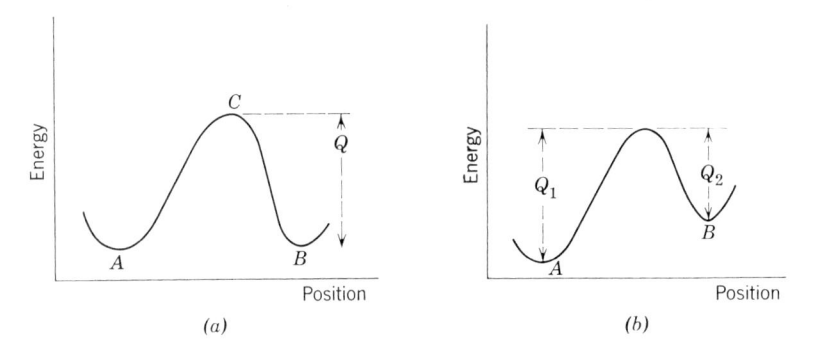

Fig. 2.3. Activated process.

transit from A to B. The atom will therefore spend more time at A than at B, i.e., in its lower energy position, but it will spend some time at B.

A generalization of this argument shows that the minimum energy criterion alone does not determine the arrangement of the atoms. It is opposed by the tendency of thermal agitation to put the atoms into higher energy positions. The result of this conflict between opposing tendencies depends on the temperature.

It is necessary to assume, or to conclude from observation, that only a finite number of different arrangements of atoms are possible; each different and distinct arrangement is called a *phase*. A phase has the same structure everywhere and is separated from any other phase by a surface, the *interface*. The discussion of which phase is stable under what conditions can be initiated by a consideration of the processes that take place at the interface between two phases which are in contact with each other; one phase may grow at the expense of the other, or else they are in equilibrium with each other.

Equilibrium between phases: kinetic considerations. The physical processes that govern the relative stabilities of various alternative arrangements of phases can be discussed entirely in terms of the surface of contact of the two phases as follows: Consider two possible arrangements, or phases, A and B of a collection of similar atoms (Fig. 2.4). In order to determine which of these is the stable form, we will assume that they are in contact at an interface PQ. Let us consider the conditions that determine whether A increases at the expense of B, or vice versa, or neither. This will depend upon whether more atoms move from state A to state B, or from B to A in any unit of time. The two phases are in equilibrium if the two rates are equal. In order to

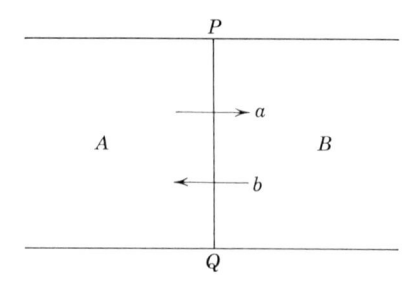

Fig. 2.4. Atom movements at an interface.

visualize the situation, it may be helpful to regard A as a crystalline solid and B as a liquid, but the following considerations are quite general, and can be applied to any two possible states of aggregation of similar atoms.

It will be assumed that phase A has a lower internal energy per atom than phase B, and that atoms may leave state A and take up state B, and vice versa.

An atom in one phase at the interface will move into and become part of the other phase if it satisfies several conditions simultaneously, as follows: it must have a high enough energy to release it temporarily from its immediate surroundings (this is zero if the phase is a vapor or gas); it must move in the appropriate direction to reach the other phase; and it must reach a site in which it becomes a part of the phase. The rate at which this happens depends upon the following parameters:

1. The number of atoms per unit area of interface.

2. The number of times per second that each atom has an energy in excess of the required value.

3. The probability that the atom is moving in an appropriate direction.

4. The probability that the atom arrives at a point where it can be "accepted" by the other phase.

These may be expressed in detail as follows:

1. The number of atoms per unit area of interface d_A or d_B is a function of the density of the phase, and of the arrangement of the atoms. In general it is slightly higher for crystalline phases than for liquids, and much lower for vapors or gases.

2. The probability that an atom has thermal energy Q when it is part of a population of atoms at a temperature T is given by $p = \exp(-Q/kT)$, where Q is the thermal energy required to raise the atom to the energy level at which it can move from A to B or from

AGGREGATES OF ATOMS

B to A. It will require different amounts of *thermal* energy for these two processes because the atoms in states A and B start with different "nonthermal" or internal energies. The values of the two activation energies will be represented by Q_a and Q_b for the transitions from A to B and B to A respectively. If the atom undergoes ν_A or ν_B vibrations per second, it will have the required energy $\nu_A \exp(-Q_a/kT)$ and $\nu_B \exp(-Q_a/kT)$ times per second.

3. The probability that an atom has a sufficient component of motion normal to the interface will be represented by the factors G_a and G_b. These factors depend on the curvature of the interface but this aspect will be neglected at this point. (See p. 237.)

4. When an atom has sufficient energy and a suitable direction of motion, it may cross the interface and become part of the other phase, or it may return as a result of an elastic collision. The probability of its "acceptance" by the other phase is the "accommodation coefficient" A_a or A_b. It is practically unity for evaporation (transfer to a vapor phase), considerably lower for condensation of vapor to liquid because of the much higher density of the liquid, and lower still for freezing (transfer from liquid to solid) because an atom can only be accepted by a crystal when it impinges upon it a point that is very close to a position characteristic of the crystal.

The rate of transfer from A to B is

$$R_a = d_A A_a G_a \nu_A \exp(-Q_a/kT)$$

or

$$R_a = B_a \exp(-Q_a/kT)$$

where B_a is the product of d_A, A_a, G_a, and ν_A, and is the number of transitions per second per unit area of interface that would be made if all the atoms had sufficient energy all the time.

The rate of transfer from B to A is, similarly,

$$R_b = B_b \exp(-Q_b/kT)$$

The two phases are in equilibrium (i.e., can exist in contact with each other indefinitely) at the temperature T_E at which $R_a = R_b$. At T_E,

$$B_a \exp(-Q_a/kT_E) = B_b \exp(-Q_b/kT_E)$$

or

$$\frac{B_a}{B_b} = \exp\left(\frac{Q_a - Q_b}{kT_E}\right) \quad \text{or} \quad \ln\frac{B_a}{B_b} = \frac{Q_a - Q_b}{kT_E}$$

$Q_a - Q_b$ is the difference in energy, per atom, of the two phases, i.e., the latent heat or the heat of transformation. The equilibrium temperature is given by $T_E = [(Q_a - Q_b)/k] \ln(B_b/B_a)$, which consists of the "energy" term $Q_a - Q_b$ and the probability term $\ln(B_b/B_a)$. It also follows that, at temperatures above T_E, R_a is greater than R_b, and so the whole material would very soon be in the form of phase B, while at temperatures below T_E, phase A is the stable form.

It is therefore established that two phases of different energy can be in equilibrium, at the temperature at which the energy difference is just balanced by the difference in the geometrical characteristics (B) of the two arrangements of atoms. It also follows that the phase that is in equilibrium below such a temperature is the phase of lower internal energy per atom.

Equilibrium between phases: thermodynamic considerations. An alternative way of expressing the condition for equilibrium is as follows: The "activated state" is the lowest possible energy for the transition from B to A and for the transition from A to B; it follows that the difference between Q_a and Q_b is equal to the difference between the "nonthermal" energies U_A and U_B, per atom, of the two states A and B. Therefore,

$$U_A - U_B = kT_E \ln \frac{B_a}{B_b}$$

or

$$U_A - kT_E \ln B_a = U_B - kT_E \ln B_b$$

The parameters d, A, G, and ν correspond to physical properties that relate to the geometry of the phase and its vibration frequency, and they jointly control the probability of an interface transition. This is evidently related to the entropy S of a phase, which can be interpreted as the logarithm of the probability of the particular arrangement of atoms, or of other arrangements energetically indistinguishable from it. In terms of entropy, the condition for equilibrium is $U_A - T_E S_A = U_B - T_E S_B$, which corresponds to the kinetic criterion derived above if $\Delta S = k \ln(B_a/B_b)$.

The effect of pressure has been ignored in the above treatment; it is usually unimportant in physical metallurgy. It can, however, contribute to the equilibrium condition if the densities of the phases are different. The free energy expression then becomes

$$U_A - T_E S_A + pV_A = U_B - T_E S_B + pv_B$$

In the kinetic expression this would be taken into account by adding to the work done in reaching the activated state a component equal to the pressure times the change in volume per atom. This is on the basis that an atom must increase the volume it occupies in order to reach the activated state, and that it must do work in so doing. This increases the value of Q by an amount that depends on the volume per atom of the phase concerned.

The kinetic considerations outlined above give, for phases containing a single type of atom, the net rate at which one phase would change into another, and, as a special case, the conditions for equilibrium. The latter coincides with the thermodynamic condition for equilibrium of a one-component system (i.e., containing only one kind of atom).

The condition for equilibrium is that the free energy of the system has the lowest possible value; alternatively, if any other phase were present, and in contact with the equilibrium phase or phases, it would tend to transform into the equilibrium phase or phases.

The equilibrium condition, therefore, defines the direction of any change that occurs spontaneously, i.e., without external work being done on the system; any changes, if they occur, are towards the equilibrium state.

It should also be observed that, as would be expected from the kinetic considerations, a phase does not necessarily transform into a more stable phase unless there is an interface between the two phases. The conditions required for creating an interface are usually more severe, and so a less stable phase may remain in existence because of the difficulty of nucleating the more stable phase. This is discussed in Chapter 6.

A reaction does not necessarily occur at a finite rate even if an interface exists, because the temperature may be so low that kT is small compared with Q, in which case $\exp(-Q/kT)$ would be very small and the reaction would be vanishingly slow.

2.4 Structures of Crystalline Solids

It follows from the considerations outlined above that the lowest energy arrangement of atoms is not necessarily the stable arrangement at any given temperature. At a sufficiently high temperature, a higher energy arrangement which also has a higher entropy is stable. It is found, however, that there are only three general types of arrangements of atoms that are stable, namely crystalline, liquid, and vapor (or gas). The crystalline arrangement of atoms or ions has already been mentioned; it is the lowest energy type of structure. It will be discussed in more detail later.

The salient features of the crystalline form of matter are the extension, over great numbers of atoms, of the regularity of the arrangement; the consistency, in a given crystal, of the distance between the mean positions of neighboring atoms; and the constancy, in a given crystal, of the number of atoms at these distances. These features are all results of the minimum energy aspect of the stability of the phase.

2.5 Equilibrium Crystal Structures of the Elements

Each element has at least one crystal structure that is the equilibrium form of that element in a range of temperatures below its melting point. These structures are shown in Table 2.1. It will be seen

Table 2.1. Crystal Structures of the Elements *

Element	Structure †	Temperature Range
Li	B.C.C. F.C.C. C.P. Hex $c/a = 1.563$	Formed by cold work at low temperatures
Na	B.C.C. F.C.C.	Formed by cold work at low temperatures
K	B.C.C.	
Rb	B.C.C.	
Cs	B.C.C.	
Cu	F.C.C.	
Ag	F.C.C.	
Au	F.C.C.	
Be	C.P. Hex $c/a = 1.568$	
Mg	C.P. Hex $c/a = 1.623$	
Ca	F.C.C. C.P. Hex $c/a = 1.638$	$<440°C$ $>440°C$
Sr	F.C.C.	
Ba	B.C.C.	
Zn	C.P. Hex $c/a = 1.856$	
Cd	C.P. Hex $c/a = 1.886$	
Hg	Simple rhombohedral	
Al	F.C.C.	
Sc	F.C.C. C.P. Hex $c/a = 1.585$	
Y	C.P. Hex $c/a = 1.588$	
La (α)	C.P. Hex $c/a = 1.613$	$<350°C$
La (β)	F.C.C.	$>350°C$

Table 2.1 (continued)

Element	Structure †	Temperature Range
Ga	Orthorhombic	
In	F.C. Tetragonal $c/a = 1.075$	
Tl (α)	C.P. Hex $c/a = 1.598$	$<234°C$
Tl (β)	B.C.C.	$>234°C$
Ti (α)	C.P. Hex $c/a = 1.587$	$<882°C$
Ti (β)	B.C.C.	$>882°C$
Zr (α)	C.P. Hex $c/a = 1.592$	$<852°C$
Zr (β)	B.C.C.	$>852°C$
Hf (α)	C.P. Hex $c/a = 1.587$	$<1950°C$
Hf (β)	B.C.C.	$>1950°C$
Th	F.C.C.	
C (diamond)	Diamond cube	
C (graphite)	Graphite	
Si	Diamond cube	
Ge	Diamond cube	
Sn (gray)	Diamond cube	$<16°C$
Sn (white)	Tetragonal $c/a = 0.546$	$>16°C$
Pb	F.C.C.	
V	B.C.C.	
Nb	B.C.C.	
Ta	B.C.C.	
Pa	B.C. Tetragonal $c/a = 0.825$	
P (black)	Orthorhombic	
As	Rhombohedral	
Sb	Rhombohedral	
Bi	Rhombohedral	
Cr	B.C.C.	
Mo	B.C.C.	
W (α)	B.C.C.	
W (β)	Cubic	
U (α)	Orthorhombic	$<662°C$
U (β)	Tetragonal	$662°C–770°C$
U (γ)	B.C.C.	$>770°C$
Se	Hex	
Te	Hex	
Po	Probably simple cubic	

Table 2.1 (continued)

Element	Structure †	Temperature Range
Mn (α)	Cubic	<718°C
Mn (β)	Cubic	718°C–1100°C
Tc	C.P. Hex $c/a = 1.604$	
Re	C.P. Hex $c/a = 1.614$	
Fe (α)	B.C.C.	<906°C
Fe (γ)	F.C.C.	906°C–1401°C
Fe (δ)	B.C.C.	>1401°C
Co (α)	C.P. Hex $c/a = 1.623$	<1120°C
Co (β)	F.C.C.	>1120°C
Ni	F.C.C.	
Ru	C.P. Hex $c/a = 1.582$	
Rh	F.C.C.	
Pd	F.C.C.	
Os	C.P. Hex $c/a = 1.579$	
Ir	F.C.C.	
Pt	F.C.C.	
Ce (α)	C.P. Hex $c/a = 1.62$	
Ce (β)	F.C.C.	
Pr (α)	C.P. Hex $c/a = 1.613$	
Pr (β)	F.C.C.	
Nd	C.P. Hex $c/a = 1.613$	
Eu	B.C.C.	
Gd	C.P. Hex $c/a = 1.587$	
Tb	C.P. Hex $c/a = 1.580$	
Dy	C.P. Hex $c/a = 1.579$	
Ho	C.P. Hex $c/a = 1.580$	
Er	C.P. Hex $c/a = 1.582$	
Tm	C.P. Hex $c/a = 1.580$	
Yb	F.C.C.	
Lu	C.P. Hex $c/a = 1.584$	

* From *The Structure of Metals and Alloys*, W. Hume-Rothery and G. V. Raynor, Institute of Metals Monograph and Report Series No. 1, 3rd ed., London, 1954, pp. 87–95.

† B.C.C. = body-centered cube; F.C.C. = face-centered cube; C.P. Hex = close-packed hexagonal.

AGGREGATES OF ATOMS

from this table that there are a number of elements which can exist in two or more allotropic or polymorphic forms.

The most extreme example of polymorphism is that of plutonium, which exists in six different structures. The temperature ranges of the six solid phases, and some of their properties, are shown in Table 2.2.

Table 2.2. Allotropic Forms of Plutonium

Phase	Temperature Range (°C)	Crystal Structure	Density (grams/cc)	Coefficient of Linear Expansion (/°C × 10^6)	Change of Volume or Transformation (%)
α	<120	Simple monoclinic	19.8	50.8	α → β 8.9
β	120–210	Unknown (complex)	17.6	38.0	β → γ 2.4
γ	210–300	F.C. orthorhombic	17.2	a −20 b 40 c 84	γ → δ 6.7
δ	300–450	F.C.C.	15.9	−8.6	δ → δ′ −0.4
δ′	450–470	B.C. tetragonal	16.0	a 305 c −659	δ′ → ε −3.0
ε	470–640	B.C.C.	16.5	30	

A crystal structure can be described in two distinct ways. The first method is to define the structure in terms of its "unit cell," of which the shape (cubic, tetragonal, etc.) and the dimensions (length of side

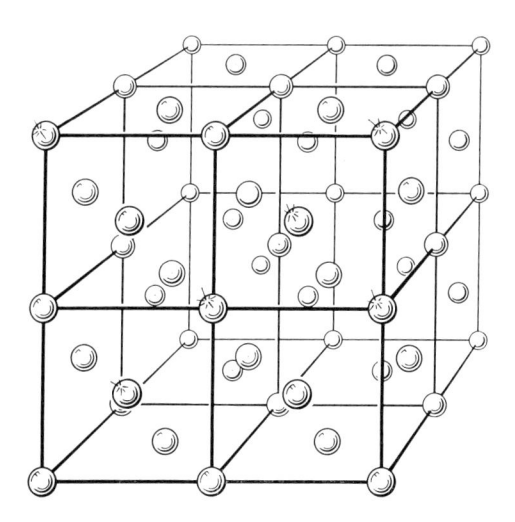

Fig. 2.5. Positions of atoms in face-centered cubic crystal.

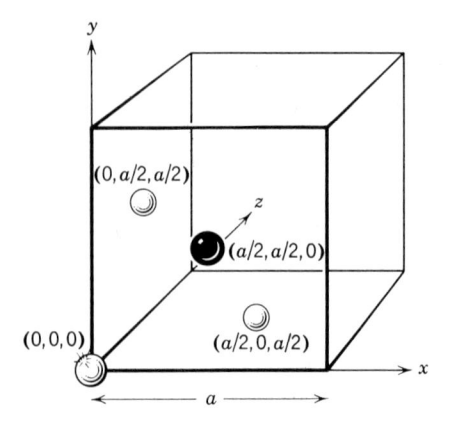

Fig. 2.6. Unit cell of face-centered cubic crystal.

of cube, etc.) are specified. The numbers and positions of the atoms in relation to the unit cell must also be specified. This is usually expressed in a co-ordinate system of which the axes are edges of the unit cell, and the units are its dimensions.

For example, the face-centered cubic structure is one in which the unit cell is a cube, and atoms are located at each corner and at the center of each face of an array of cubes. This is illustrated in Fig. 2.5. The structure is specified by a cubic unit cell of size a, with atoms at (0, 0, 0), (½, ½, 0), (0, ½, ½), and (½, 0, ½), as shown in Fig. 2.6. All the remaining atom sites in the crystal are obtained by translating the unit cell by units of a in the three co-ordinate directions.

The alternative method of describing a structure is in terms of the "packing" of the atoms. The face-centered cubic structure can be built up as follows: Let a layer of atoms (considered as hard spheres) be arranged in a plane so that they are as closely packed as possible. This

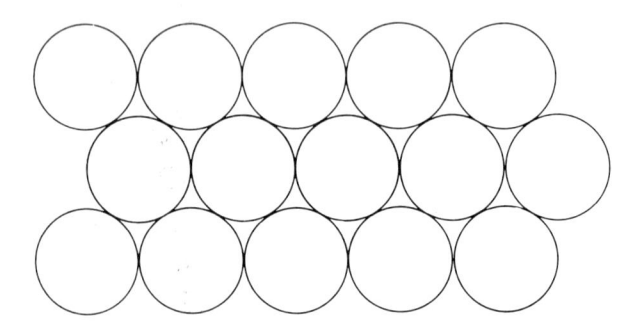

Fig. 2.7. Close-packed layer of atoms.

Fig. 2.8. Positions of atoms in second layer.

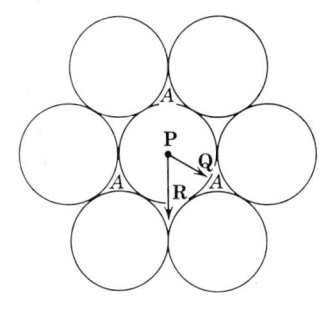

is the arrangement in which each atom has six others in contact with it. It is shown in Fig. 2.7. Now let a succession of such layers be stacked one upon the other, so that the layers are as close together as possible. The center of each atom will be directly above a point which is equidistant from the centers of three atoms in the layer immediately below. Points A in Fig. 2.8 indicates a set of such positions. Only one half of the similar positions are occupied, and there are therefore two possible ways of choosing the location of each layer with respect to the one below it. A useful way of representing these two relative positions is by the symbols △ and ▽. The symbol △ is here used to represent a position of a layer such that each layer is displaced, relative to the one below it, by the vector **PQ** in Fig. 2.8. The symbol ▽ corresponds to the displacement by the vector **PR** of Fig. 2.8.

In building a crystal structure, a sequence of such layers is required. A relationship represented by △ or ▽ must be chosen for each successive layer. The sequence that exists in actual crystals is either △△△△△, ⋯, or △▽△▽△, ⋯. In each case, each atom has 12

Fig. 2.9. Close-packed plane in unit cube of face-centered cubic structure.

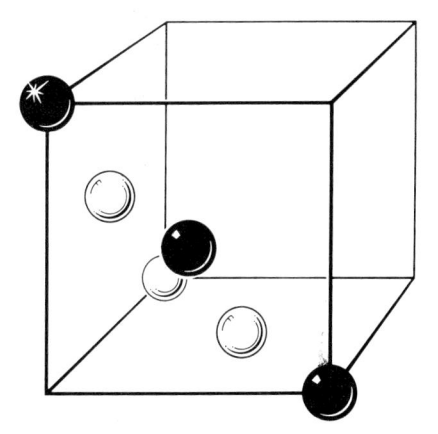

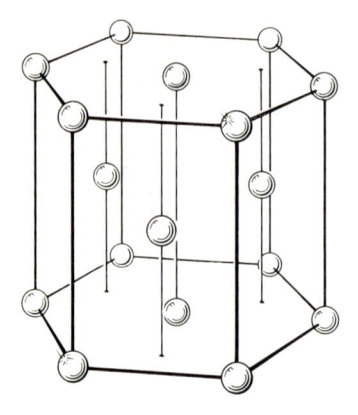

Fig. 2.10. Close-packed hexagonal structure.

nearest neighbors at the same distance from it, and the only difference is in the relative position of second nearest and more distant neighbors. The structure △△△△△, ⋯, is identical with the face-centered cubic structure as described above. Figure 2.9 shows the relationship between the close-packed planes and the unit cell of the structure. The six atoms whose positions are indicated lie on a close-packed plane. The alternative arrangement △▽△▽△, ⋯, corresponds to the close-packed hexagonal structure shown in Fig. 2.10. Here the basal plane of the hexagonal cell is the close-packed plane. An alternative description of the stacking sequence is to represent each of the three possible positions of a plane by the letters A, B, and C. The face-centered cubic structure corresponds to the sequence $ABCABCAB$, ⋯, while the close-packed hexagonal structure is $ABABAB$, ⋯.

A third structure commonly encountered in metals is the body-centered cubic structure, represented in Fig. 2.11. There are several ways of arriving at this structure by the stacking of layers, but none of these is particularly useful for understanding the structure.

The 8 − N rule. The face-centered cubic and the close-packed hexagonal structures may be looked upon as fulfilling a demand for maximum co-ordination number: in many of the elements the requirement is for a co-ordination number that allows covalent bonding with all the near neighbors. The number of nearest neighbors must in such cases be equal to the number of electrons that are to be shared with neighboring atoms in order to produce filled outer electron shells. If N is the group number of an element (the number of electrons in the outer shell), then each atom must have $8 - N$ neighbors each of which contributes a shared electron and receives a shared electron. Each atom then has an outer shell of eight electrons, some of which

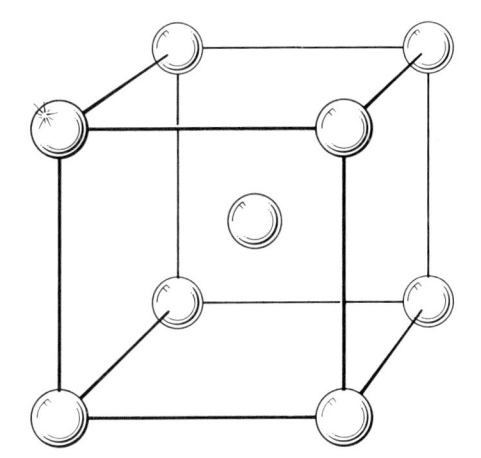

Fig. 2.11. Body-centered cubic structure.

are shared with neighboring atoms. This rule, the $8 - N$ rule, only applies to atoms in groups 4 to 7 because the elements in groups 1 to 3 and the transitional elements have insufficient electrons to share them covalently. The electrons are shared metallically in such cases. Many of the elements in groups 4, 5, 6, and 7 conform to the $8 - N$ rule as shown in Table 2.3.

Table 2.3. Some Elements with Structures that Conform to the $8 - N$ Rule

N	4	5	6	7
Co-ordination Number	4	3	2	1
	C			
	Si			
	Ge	As	$Se(\alpha)$	
	Sn (gray)	Sb	Te	I
		Bi		

It will be seen from Table 2.1 that there are some elements that do not have typical metallic structures (face-centered cubic, body-centered cubic, or close-packed hexagonal) and do not follow the $8 - N$ rule. In some cases the structures are more complicated, but sometimes they are only slightly distorted forms of the simpler structures. Mercury and indium are examples.

Structures of the metals. The metallically bonded elements are mainly body-centered cubic, face-centered cubic, or close-packed hexagonal. The face-centered cubic structure corresponds to the close packing of spheres, as does the ideal close-packed hexagonal structure (i.e., having an axial ratio of 1.63). In all real cases of the close-packed hexagonal structure, however, the ratio departs from this value and the structure corresponds to the close packing of spheroids. In each of these structures there are 12 near neighbors. The body-centered cubic structure does not derive from the close packing of spheres; each atom has eight near neighbors. It appears that the choice of the crystal structure depends on the interplay of the various factors that determine the free energy; it is not at present possible to explain the particular structures of the metallic elements, although some consistencies are seen if Table 2.1 is examined.

It is noteworthy that all the alkali metals (group 1) have the same structure: body-centered cubic at ordinary temperatures; it has, however, been shown recently that several of them have a stable face-centered cubic form at low temperatures. This indicates that the face-centered cubic form has the lower energy, but that the entropy of the body-centered phase is higher than that of the face-centered. The other elements with a single valency electron and filled inner levels (copper, silver, and gold) are face-centered cubic at all temperatures.

Metals with two valency electrons are not quite so consistent; beryllium, magnesium, zinc, and cadmium are close-packed hexagonal, although the packing is that of spheroids rather than of spheres. Calcium has two forms: face-centered cubic below 450°, and close-packed hexagonal above that temperature. Strontium and barium (body-centered cubic) and mercury (rhombohedral) illustrate the tendency for the heavier elements to depart from the more regular behavior of the lighter ones.

In the case of the transition metals, it is not quite so clear what the valency is, because it is probable that electrons that should otherwise be free to take their part in the general electron population may find lower energy levels in the otherwise unfilled lower levels of their own atoms. Nevertheless, most of the transition metals have the typical face-centered cubic, body-centered cubic, or close-packed hexagonal structures; in many cases there are two or more forms, with appropriate transition temperatures. It should be noted that iron is body-centered cubic both above and below the range of stability of its face-centered cubic modification. It is thought that its ferromagnetic properties at lower temperatures contribute to the stability of the body-centered cubic phase.

It will also be observed that a given element, in addition to having alternative crystal structures with essentially the same type of bonding, but with different energies and entropies, may also have alternative forms with different types of bonding, and therefore very different characteristics. The best example of this is tin, which is stable as a metal (free electron bonding) at temperatures above 16°C, while at lower temperatures the stable form is gray tin, which has a covalent bonded structure obeying the $8 - N$ rule, and is nonmetallic in its properties. Another example is carbon.

It follows that the description of a substance as a metal does not refer to the type or types of atom that are present; it refers to the type of bonding between the atoms. This is very closely related to the electrical conductivity, because in a typical metal the electrons can accept small increments of energy, and are therefore able to move freely. This also accounts for the typical appearance of metals, which is due to the fact that the electrons are free to receive and reradiate light without much loss of energy; hence the high specular reflectivity of metals.

Sizes of atoms in metals. It is convenient for many purposes to regard an atom in a metal as having a definite size, which may be defined by the distance between its center and that of its neighbor. This distance is that at which the various forces acting on the atom are in equilibrium.

In a metal, the forces can be considered as (*a*) the attractive forces between electrons and positive ions, (*b*) the repulsion between the complete electron shells of the positive ions, and (*c*) the repulsion between the positive ions as a result of their similar positive changes.

The forces of attraction can be represented as in Fig. 2.12, in which two positive ions are each considered to be attracted by a number of electrons, whose instantaneous positions are represented. Other positive ions are ignored. It is apparent that all the forces of attraction will have components that tend to pull the positive ions towards each other. The resultant force of attraction will reach a maximum when the ions are rather close together, and will decrease when they are separated. Figure 2.13 shows the type of variation that would be expected. The force of repulsion between the ions would also be expected to increase with decreasing distance, and to increase very rapidly as the electron shells begin to impinge upon each other. The force of repulsion and the resultant force are shown in Fig. 2.13. The resultant force is zero at some distance E; at shorter distances, the net force is one of repulsion, while at greater distances, the force is one of attraction. Extension into a three-dimensional array changes

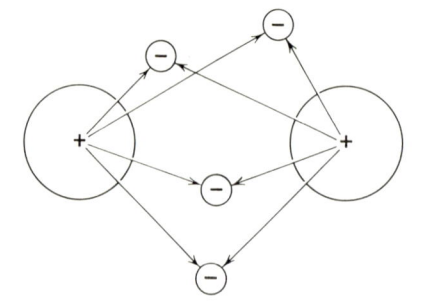

Fig. 2.12. Forces of attraction between electrons and positive ions.

the detail but not the principle of this representation of the forces. The example chosen for Fig. 2.13a is one in which the resultant force of attraction is so large that the atoms are essentially in contact; i.e., the rapidly rising repulsion between the electron shells determines their distance apart. In Fig. 2.13b, a case is illustrated in which the forces of attraction are weaker and the electron shells are not in contact. The former case is typified by copper, and the latter by sodium. The interatomic distance so defined is simply related to the length of side a of the unit cell of the structure; in the face-centered cubic structure, for example, the interatomic distance is $(a\sqrt{2})/2$. The closest distance of approach, and in some cases the second closest distance, are shown in Fig. 2.14, from which it will be seen that there is a gradual increase of size with atomic number superimposed on a periodic varia-

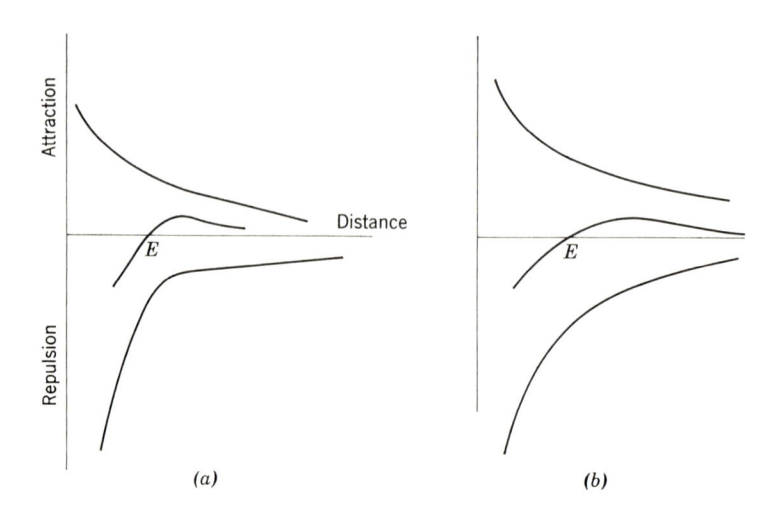

Fig. 2.13. Forces between atoms. (a) "Hard" atom. (b) "Soft" atom.

AGGREGATES OF ATOMS

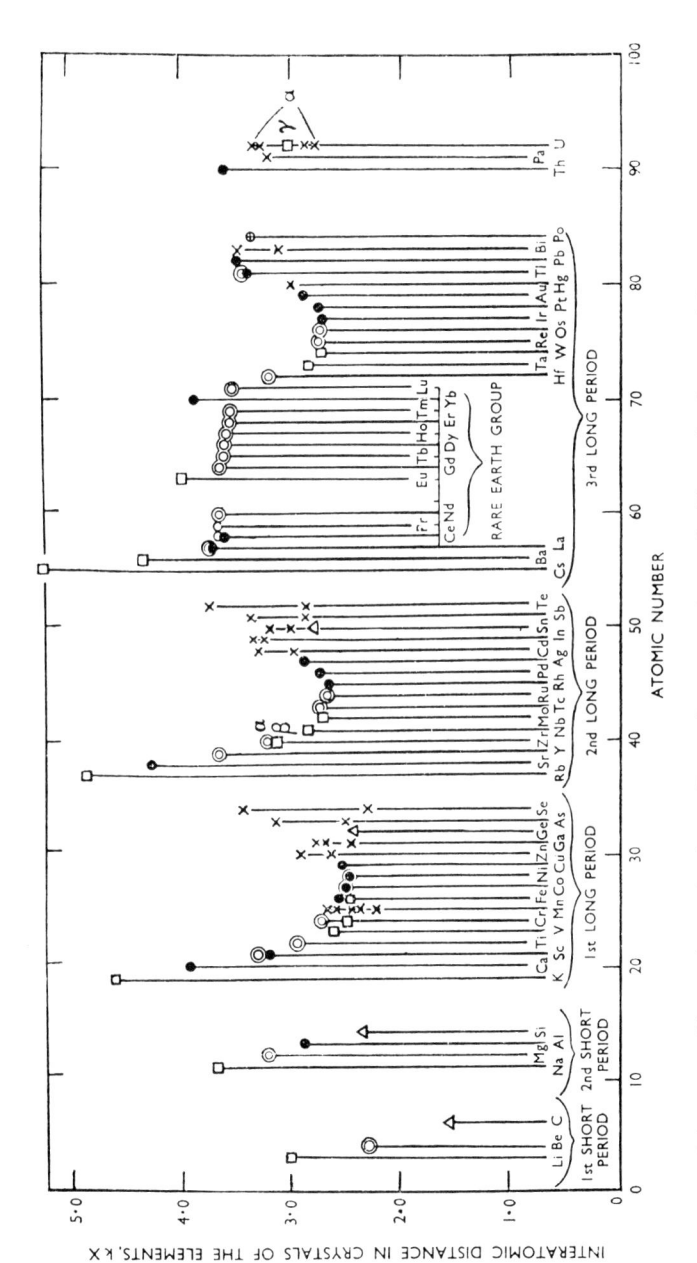

Fig. 2.14. Chart of atomic sizes. From *Structure of Metals*, Hume-Rothery and Raynor, p. 62.

tion that reproduces the periodicity of the periodic table. It will also be observed that most of the metals fall within a rather narrow range of sizes.

The radius of an atom (or ion) as determined for a particular crystal structure is not a real characteristic of that atom, because when the same atom appears in different crystal structures it displays different radii. An empirical correction, due to Goldschmidt, requires that if the radius of an atom were a when it was in a structure of co-ordination number 12, its radius would be $0.97a$, $0.96a$, and $0.88a$ if it were in a structure of co-ordination number 8, 6, or 4 respectively. This can be explained qualitatively on the basis that there is more room for the atom to be "squeezed" out of its spherical shape when it has a lower co-ordination number.

2.6 Structure of Liquids and Vapors

The structure of liquid metals can be described in terms of the tendency, already pointed out in connection with crystals, for neighboring atoms to conform to a very well defined distance apart. The same electrostatic forces that are responsible for the interatomic distance in crystals are operative in liquids, and it is, accordingly, found that the least interatomic distance in a liquid is often very similar to that of the crystal of the same metal. The *number* of atoms at this distance is, however, different. As has been seen, many metals crystallize with a structure in which each atom is surrounded by 12 (i.e., has a co-ordination number of 12) nearest neighbors, each of which is similarly surrounded, etc.

The number of atoms at approximately the "nearest neighbor" distance is about 11 for many metals which have a co-ordination number 12 in the solid; some metals which crystallize in the body-centered cubic structure (co-ordination number 8) appear to have about 7 nearest neighbors in the liquid. It is evident that the structure of the liquid is much less orderly than that of the crystal. There are two general theories of the structure of liquids; one is that the atoms or molecules are truly random as regards separation and direction of line of centers of neighbors; the only limitation to these two parameters would be the number of atoms per unit volume. The alternative view, which appears more acceptable for metals, is that shown in two-dimensional schematic form in Fig. 2.15, in which a shows the crystal, and b the corresponding liquid.

It will be evident that the density of the liquid should be rather less than that of the crystal and that, whereas there is long range order

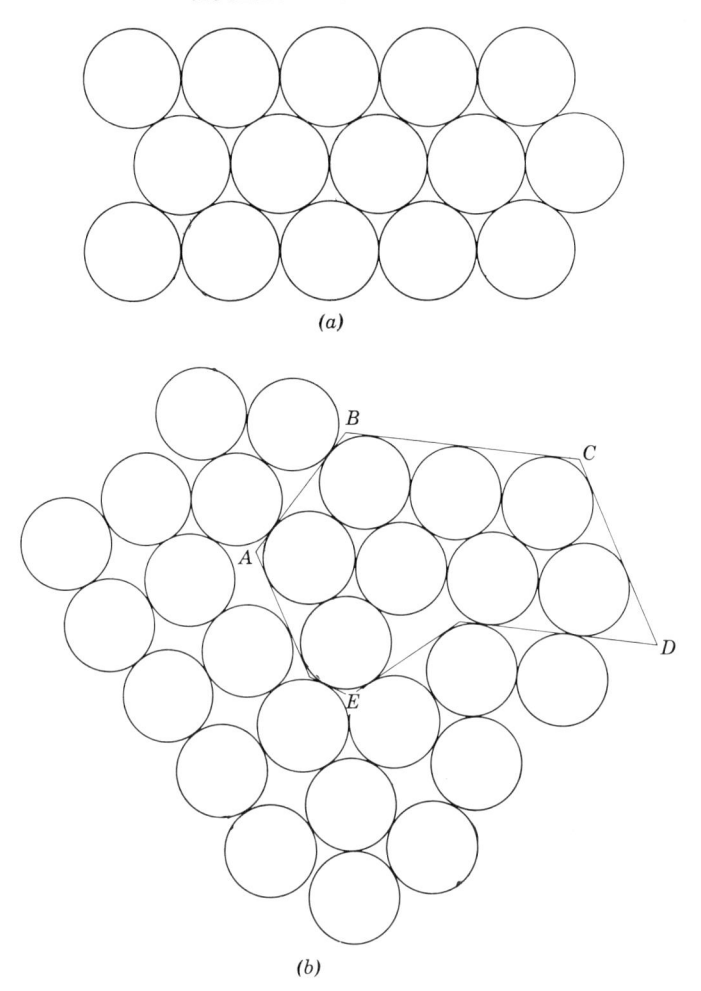

Fig. 2.15. Schematic diagrams of structures of (a) crystal and (b) liquid.

in the crystal, there is none in the liquid in the sense that a knowledge of the positions of the atoms in one region gives no information as to the positions of the atoms in any other region, except to a limited extent about the immediately adjacent atoms.

It will also be seen that most of the atoms are related to most of their immediate neighbors exactly similarly in the crystal and in the liquid; the area denoted by $ABCDE$ in the liquid (Fig. 2.15b) is in fact identical in its arrangement with an equal number of atoms in the crystal. The latent heat of fusion, correspondingly, is small (4–5%

of the latent heat of vaporization for most metals); the energy of the liquid is, however, higher than that of the crystal by this amount because the atoms are not all in the lowest energy relationship with the maximum number of neighbors, as is the case for the crystal. The energy of the vapor or gas is much higher, because the atoms or molecules are separated so that for almost all the time there is no interaction, and therefore no reduction of energy due to aggregation.

2.7 Alloys

One of the major problems of physical metallurgy is to understand and predict the structures and compositions of the phases that will exist in equilibrium in an alloy of a given composition at a given temperature.

We must therefore next consider the equilibrium structures when more than one type of atom is present. The presence of a second element N may change the stable form, at a given temperature T, of a solid metal M in any of three ways:

1. It may form a liquid solution at temperature T.
2. It may modify the structure of M in either of two ways: (a) N atoms may be substituted for some of the M atoms, leaving the structure essentially the same; this is a substitutional solid solution of N in M; or (b) some atoms of N may occupy sites, not normally occupied by M in the crystal P, without otherwise changing the structure; this is an interstitial solid solution of N in M. Or,
3. An entirely different crystal structure from that of M may be formed; such a structure constitutes an *intermediate phase*, which depends for its stability on the presence of both M and N atoms. There are various types of intermediate phase, each corresponding to a low energy structure; each type of intermediate phase depends on a different reason for the reduction in energy.

Phase diagrams. Before considering in detail the conditions which determine the composition and temperature ranges in which phases can exist, it is desirable to outline the conventional methods for exhibiting the results of experiments or computations which provide such data. When an element is under consideration, the equilibrium structure is a function only of the temperature and pressure; for most purposes in physical metallurgy we can ignore the effects of pressure, and we may regard the equilibrium structure as being determined only by the temperature. All the relevant information can be conveyed by a series of temperature ranges, and the structures existing therein.

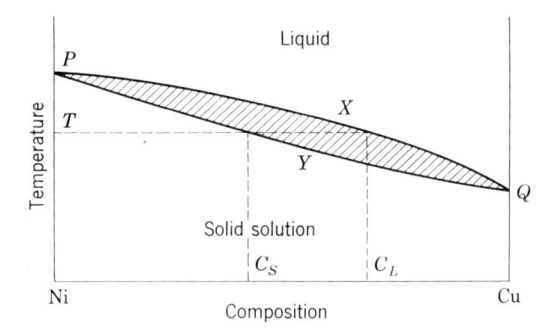

Fig. 2.16. Binary phase diagram: continuous solid solutions.

When more than one type of atom is present, however, the problem is not so simple, because the composition is a second variable. The equilibrium structure therefore depends on temperature and composition, and the limits of existence of a particular structure can be most conveniently represented on a diagram in which temperature is one variable, and, in the case of a binary system, composition is the other.

A phase diagram or equilibrium diagram is a map showing which phase or phases exist in equilibrium for any combination of temperature and composition.

One of the simplest binary phase diagrams is that of the copper-nickel system, shown in Fig. 2.16. There are only two phases in this system: the liquid phase and the solid phase. The line PXQ is the *upper* limit of the temperature range in which the solid phase exists. The line PYQ is the *lower* limit of the liquid range. The solid and liquid fields overlap, and there are, therefore, combinations of temperature and composition which correspond to the existence, in equilibrium, of solid and liquid. The two-phase region of the diagram is shown hatched in Fig. 2.16.

In order that two phases may be in equilibrium, they must be at the same temperature. At temperature T in Fig. 2.16, solid and liquid would be in equilibrium if their compositions were C_S and C_L respectively, and they would both exist if the over-all composition of these alloys were between C_S and C_L. The lines PXQ and PYQ are known as the *liquidus* and *solidus* respectively. The relative amounts of the two phases are given by the "lever rule" as follows: In a two-phase field (Fig. 2.17) the amounts L and S of the two phases of compositions C_L and C_S are given by $Lp = Sq$, where p and q are the differences between the over-all composition and the compositions C_S and C_L.

All binary phase diagrams consist of single-phase fields separated

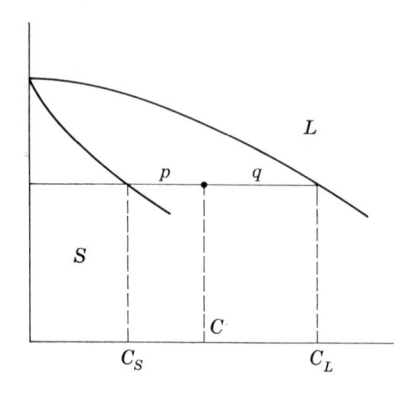

Fig. 2.17. The lever rule.

by two-phase fields except at *points* where the two phases are in equilibrium and have the same composition. This only occurs for pure substances or for compounds or when the phase boundaries are horizontal.

Three-phase equilibrium occurs at points in the phase diagram at which the line where the single-phase field of phase A is limited by its equilibrium with phase B intersects with the line where phase A is limited by its equilibrium with phase C. At this point, A is in equilibrium with both B and C. This is shown in Fig. 2.18. Figure 2.19 shows the phase diagram for the system aluminum-silicon. At point E the liquid phase is in equilibrium with two solid phases. Point E is described as a eutectic point.

Other three-phase equilibria are illustrated in Fig. 2.20, in which portions of phase diagrams are shown. In each case, there are two phases above and one below the equilibrium temperature, or vice versa, and one or more of the phases may be liquid. The nomenclature depends on both of these aspects, as shown in Table 2.4.

Many phase diagrams are complicated by the presence of numerous

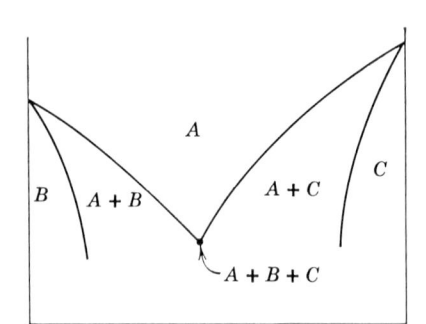

Fig. 2.18. Equilibrium of three phases in a binary system.

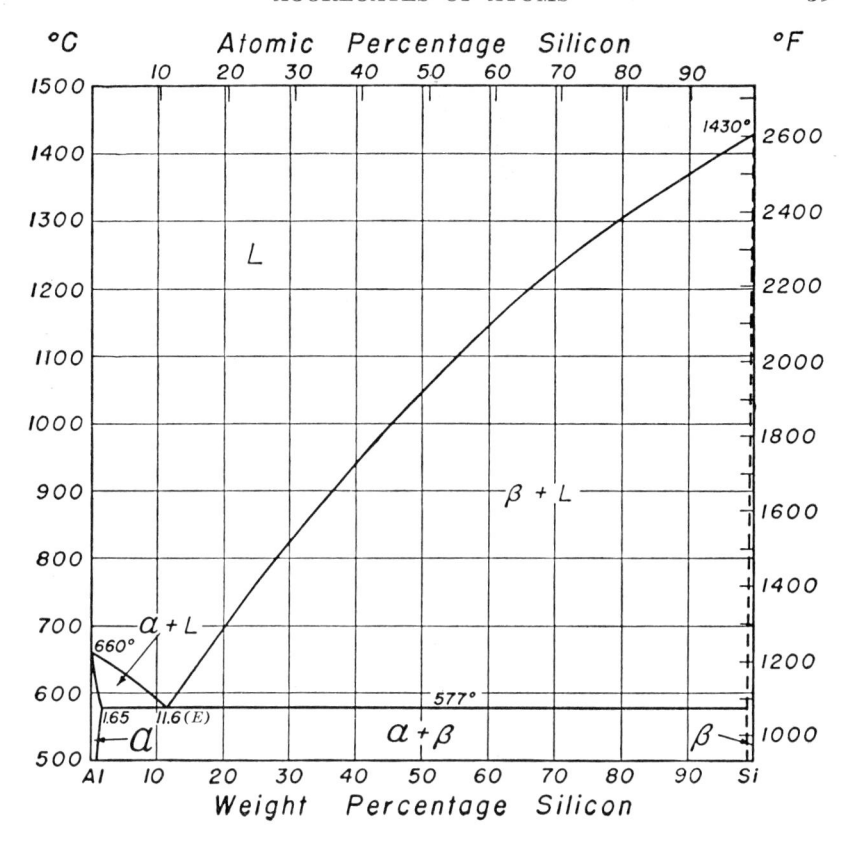

Fig. 2.19. Phase diagram for the aluminum-silicon system. From *Metals Handbook*, American Society for Metals, Cleveland, 1948, p. 1166.

phases and the corresponding three-phase equilibrium points; but they can always be resolved into single-phase fields which touch only at points, and two-phase fields which meet at three-phase equilibrium points.

Examples of more complicated phase diagrams are given in Fig. 2.21.

Many published phase diagrams are probably not very precise; occasionally diagrams are found to be qualitatively impossible, because they do not satisfy the following three criteria:

1. The Gibbs phase rule, which states that $P + F = C + 1$ in a system in which temperature and composition are the only variables that are significant. P is the number of phases in equilibrium with each

40 PHYSICAL METALLURGY

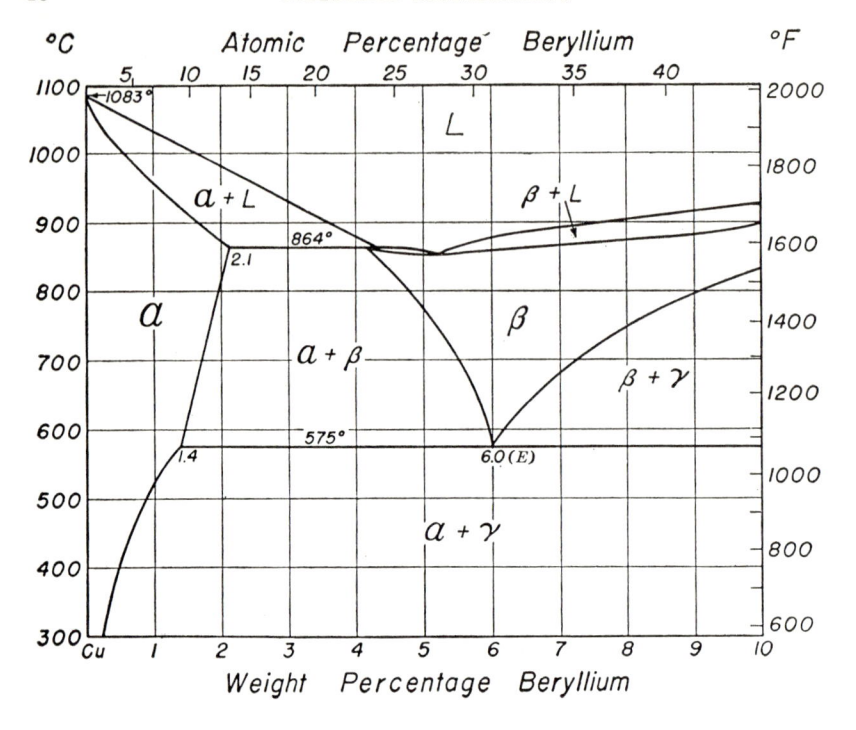

Fig. 2.20a. Part of the copper-beryllium system. From *Metals Handbook*, ASM, p. 1176.

other, F is the number of degrees of freedom, and C is the number of components.

The equilibrium between two phases in a binary system therefore has 1 degree of freedom; this means that equilibrium can exist over a range of temperatures, but the compositions are not independently variable.

2. The rule, which follows from the phase rule, that single-phase fields must be separated by two-phase fields except at *points* of contact (in binary system).

3. The rule that the continuation, through points of intersection, of single-phase field boundaries must enter two-phase fields.

Ternary diagrams. The discussion so far has been limited to binary systems; the same considerations apply to ternary and more complex systems as to binary systems, and the same types of phases exist under equivalent conditions. The main new problems that arise are those of visualization and representation, since two parameters are

AGGREGATES OF ATOMS

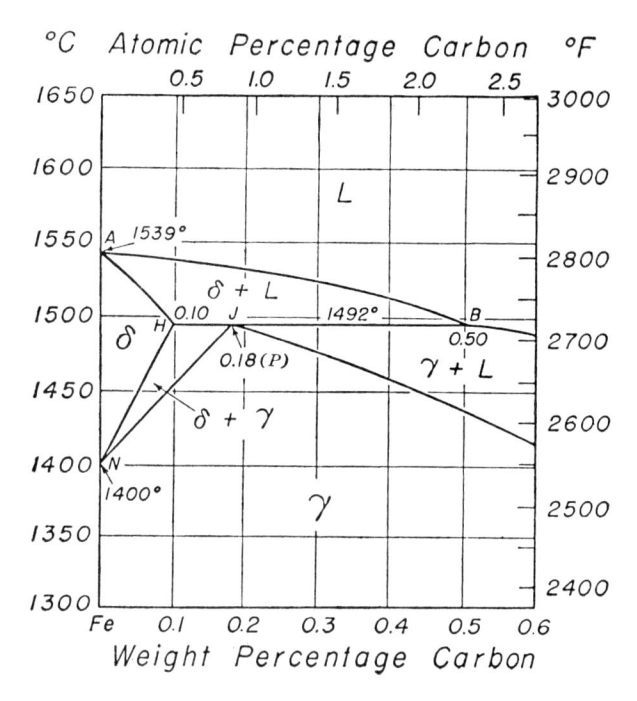

Fig. 2.20b. Part of the diagram for the iron-carbon system, showing the peritectic point P. From *Metals Handbook*, ASM, p. 1181.

Table 2.4. Types of Three-Phase Equilibria in Binary Systems

	A	B	C	
Single phase at higher temperature	Liquid Liquid Solid Solid	Solid Liquid Solid Solid	Solid Solid Solid Liquid	Eutectic Monotectic Eutectoid
Single phase at lower temperature	Solid Solid Solid Liquid	Solid Solid Liquid Liquid	Solid Liquid Liquid Solid	Peritectoid Peritectic

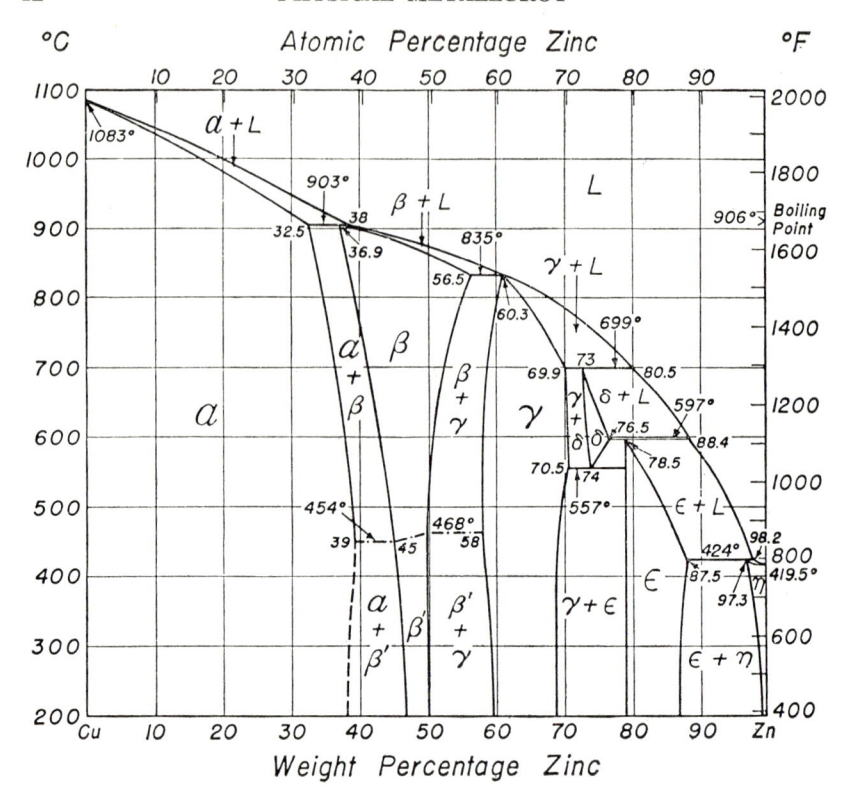

Fig. 2.21a. Copper-zinc diagram. From *Metals Handbook*, ASM, p. 1206.

required to describe the composition of a ternary system, or $(n-1)$ for an n component system; the two-dimensional phase diagram of a binary system (temperature-composition) becomes a three-dimensional diagram for a ternary system, and cannot be completely represented geometrically for more complex cases. The lines separating single- and two-phase regions of a binary diagram become surfaces in a ternary system, and the points representing three-phase equilibria in the binary diagram become lines in a ternary diagram; four-phase equilibria are possible in ternary systems and are defined by points.

A ternary system can be partially represented in two dimensions by a series of isothermal sections of the ternary diagram, as illustrated in Fig. 2.22a. Compositions are represented by points within (or on) the equilateral triangle ABC, and temperature by distances normal to this plane. The isothermal section $A'B'C'$ divides into solid, liquid, and two-phase regions. In order to represent the equilibrium of a particular solid composition with the appropriate liquid composition, it is necessary to use *tie lines*, such as TT' in Fig. 2.22b. The case represented in this diagram is a particularly simple one, as no intermediate phases are included; there are, however, many ternary diagrams in which ternary intermediate phases may be found. An example is shown in Fig. 2.23.

For some purposes, it is more useful to take the projection, on the base triangle, of the liquidus or solidus or solvus surfaces, showing the "contour" lines for various temperatures. An example is given in

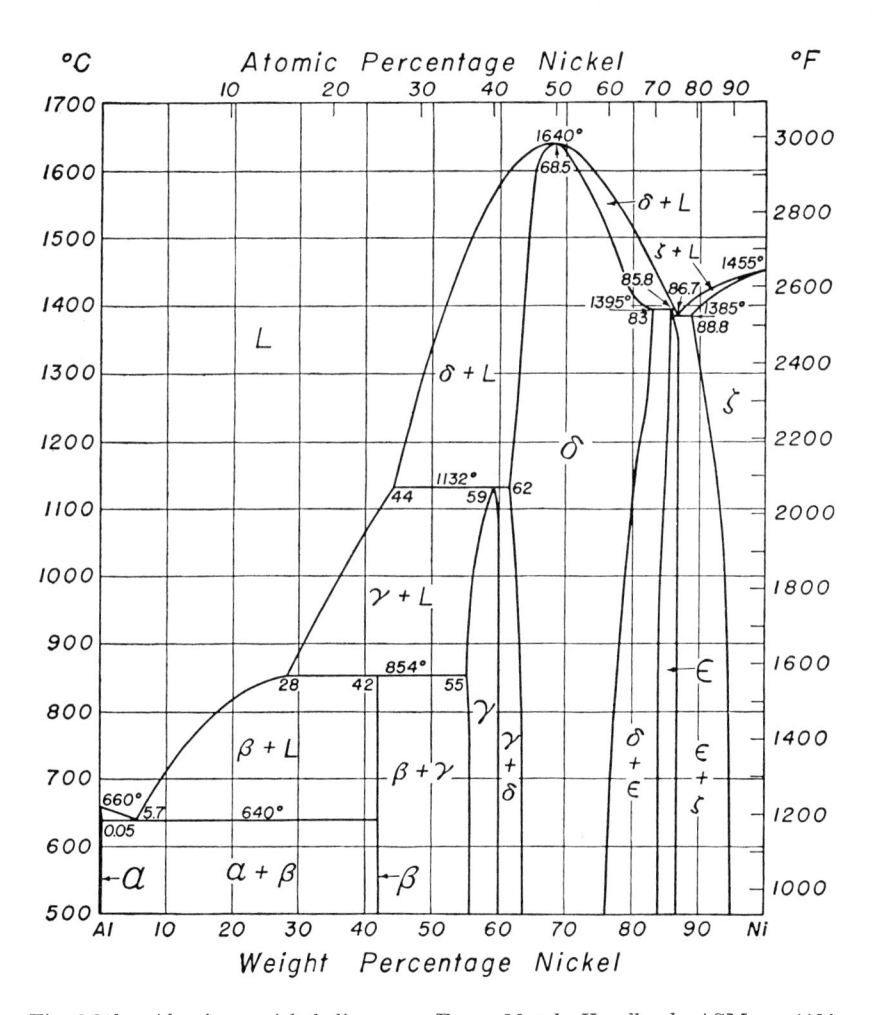

Fig. 2.21b. Aluminum-nickel diagram. From *Metals Handbook*, ASM, p. 1164.

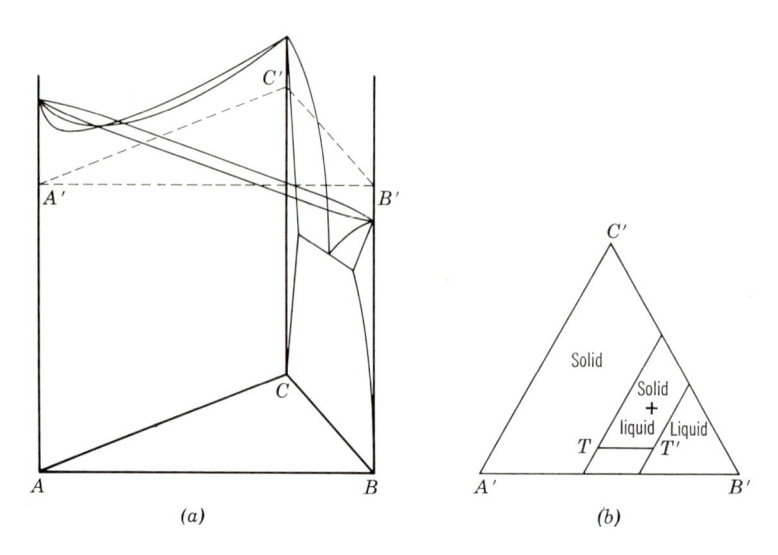

Fig. 2.22. Simple ternary system.

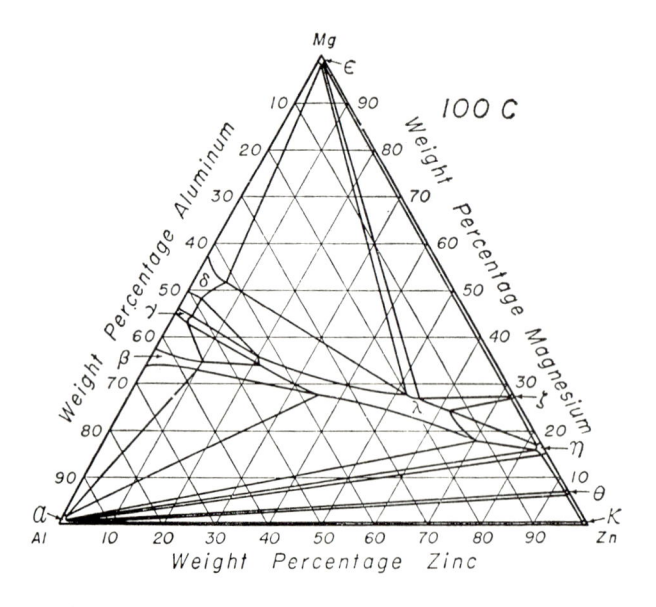

Fig. 2.23. Isothermal section at 100°C of the aluminum-magnesium-zinc system. From *Metals Handbook*, ASM, p. 1247.

AGGREGATES OF ATOMS

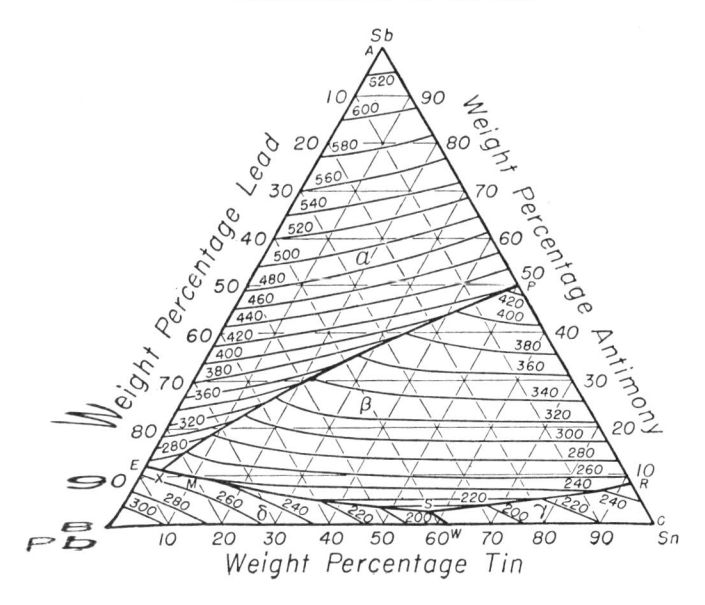

Fig. 2.24. Liquidus surface for lead-antimony-tin system. From *Metals Handbook*, ASM, p. 1267.

Fig. 2.24. Point x is a ternary eutectic point corresponding to the equilibrium of the liquid with three solid phases.

A third useful section of a ternary diagram is the "pseudo-binary" section, in which the concentration of one component is held constant, and the proportions of the other two are varied, as in a binary system.

Substitutional solid solutions

ORDERING AND CLUSTERING. If the structure of a solid solution is characteristic of the element A, and if some atoms of B are substituted for atoms of A in this structure, A is called the solvent and B the solute. The distribution of B atoms on the lattice sites depends on whether there is a preference, energetically, for B atoms to have A atoms as nearest neighbors, or to have B atoms as nearest neighbors. These two conditions lead to *ordering* and *clustering* respectively. The magnitude of ordering or clustering is defined by the departure of nearest neighbor numbers from those predicted as a result of pure chance. It is seldom, if ever, found that the distribution is exactly random. The tendency towards ordering is a result of a lower energy for the ordered than for the disordered arrangement of the atoms. This may be due to a decrease in elastic strain energy or it may be connected with the energies of the electrons; an ordered array, on the other

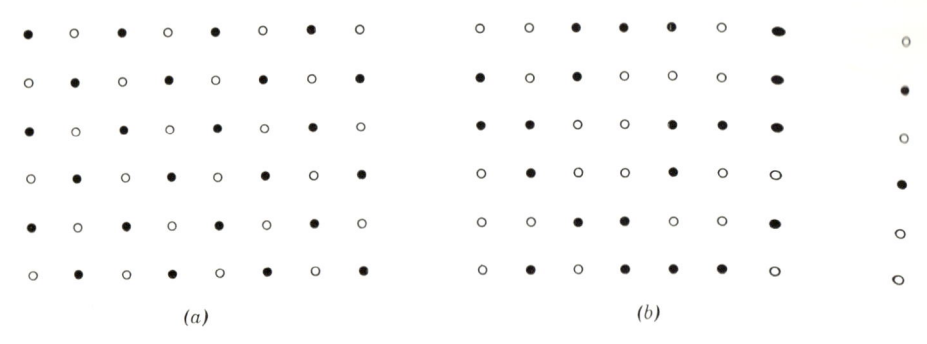

Fig. 2.27. (a) Ordered and (b) disordered states of a simple cubic crystal.

Body-Centered Cubic, 1:1. An example of this is FeAl, in which the iron atoms are considered to be at the points $(0, 0, 0)$ of the unit cell; the aluminum atoms are then at the points $(½, ½, ½)$. This structure is represented in Fig. 2.28.

Face-Centered Cubic, 1:1. In this structure, characteristic of CuAu, the gold atoms may be regarded as occupying the cube corners $(0, 0, 0)$ and the top and bottom face center positions $(½, ½, 0)$, while the copper atoms occupy the side face centers $(0, ½, ½)$ and $(½, 0, ½)$. This is shown in Fig. 2.29, from which it will be seen that it is equivalent to alternate $(0, 0, 1)$ planes containing only gold and only copper atoms. In this case the unit cell is no longer cubic, but has become

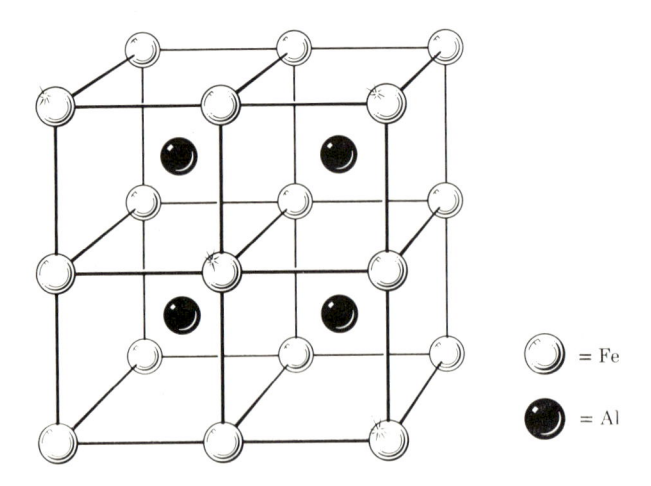

Fig. 2.28. Ordered structure of FeAl.

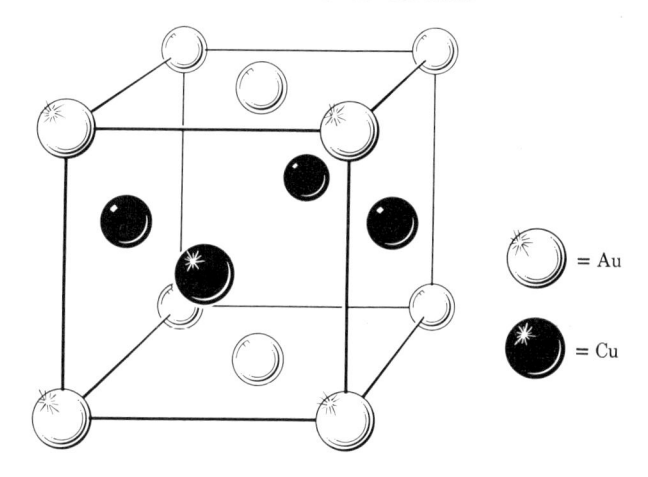

Fig. 2.29. Ordered structure of CuAu.

tetragonal with the "base" dimensions slightly in excess of the height.

Body-Centered Cubic, 3:1. The unit cell for the ordered form of Fe$_3$Al consists of four body-centered cubic cells, the corners of which are all occupied by iron atoms, while body center positions are occupied alternately by aluminum atoms and iron atoms. The structure is shown in Fig. 2.30.

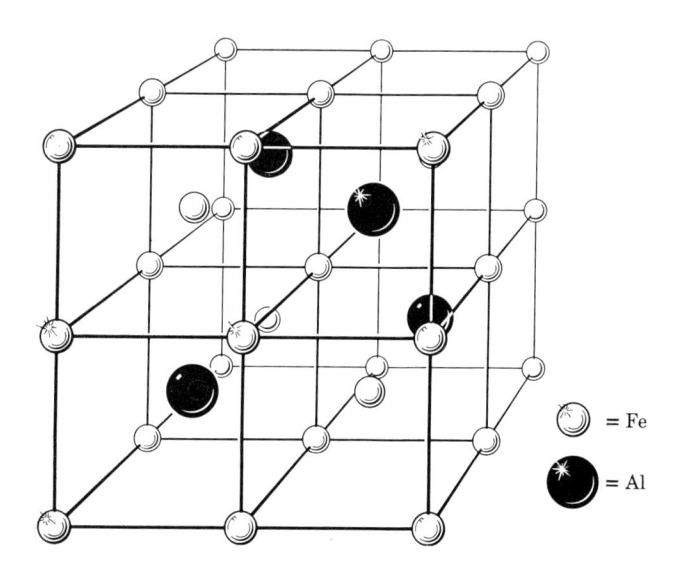

Fig. 2.30. Ordered structure of Fe$_3$Al.

Face-Centered Cubic, 1:3. This is typified by AuCu$_3$, in which the gold atoms occupy the cell corners $(0, 0, 0)$ and the copper atoms are at the centers of the faces $(0, \frac{1}{2}, \frac{1}{2})$, $(\frac{1}{2}, 0, \frac{1}{2})$, and $(\frac{1}{2}, \frac{1}{2}, 0)$.

Close-Packed Hexagonal, 3:1. Mg$_3$Cd forms an ordered structure, in which the cadmium atoms occupy alternate sites in alternate lines in each "layer" of the structure. A typical layer is shown in Fig. 2.31. The stacking of these "layers" is the same as for any other close-packed hexagonal structure. The alternative (face-centered cubic) packing of these layers would give the AuCu$_3$ type of structure.

Ternary Structures. Ordered structures are not confined to binary systems, but are sometimes found in ternary alloys; Cu$_2$MnAl is an example, in which the three types of atoms are distributed regularly on a body-centered cubic structure of the FeAl type, but in which the aluminum and the manganese atoms share the body center planes, instead of iron and aluminum sharing them. The structure consists then of alternate layers, each of which is a square array, which contain only copper atoms, and an alternation of manganese and aluminum atoms.

Defect Lattices. In some systems ordering is only achieved when some of the sites are vacant. Ni$_2$Al$_3$ is an example, in which the FeAl structure is modified by the addition of more aluminum. The vacancies also assume ordered positions and this decreases the symmetry of the structure. A considerable number of ordered systems which include vacancies have been found. They can be regarded as ternary systems in which the vacancies constitute one component.

THE LONG-RANGE-ORDER PARAMETER. The ordered structures described above are examples of complete ordering; however, as will be shown later, it is often found that the driving force for ordering decreases as ordering proceeds and it is consequently necessary to consider partly ordered alloys. It is assumed here that the composition is exactly that required for complete ordering.

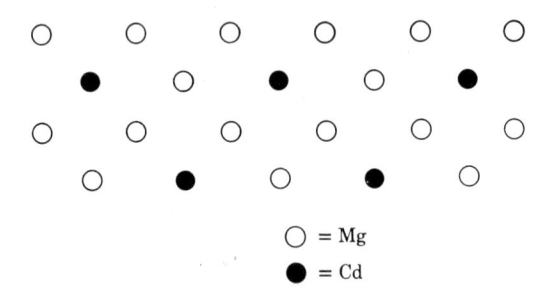

Fig. 2.31. Ordered structure of Mg$_3$Cd.

AGGREGATES OF ATOMS

The extent of short range ordering is represented by the short-range-order parameter discussed above; the degree of long range order is better expressed in terms of the probability that the atom sites are occupied by the species that would be present if the crystal were completely ordered. In a completely ordered structure, it is possible to assign a type of atom to each site in the crystal.

The departure from this criterion of order increases as the crystal becomes more disordered, but it is possible to destroy long range order completely and still to have some short range order.

A long-range-order parameter S may be defined as follows: Consider a crystal containing $2N$ sites, and consisting of N atoms of A and N atoms of B. Let each of the $2N$ sites be labeled a or b, according to which type of atom it would contain if order were complete. For complete order, each a site contains an A atom; for complete disorder (random distribution) half of the a sites contain A atoms. Let the number of a sites occupied by A atoms be n. For complete order $n = N$; for disorder $n = N/2$. For a degree of order to be specified, let $n = M$. Then the degree of order S is defined by

$$S = \frac{M - (N/2)}{N - (N/2)}$$

This varies from zero to unity for the range from complete disorder to complete order.

ANTIPHASE DOMAINS. A difficulty with this type of parameter is as follows: Consider a crystal, represented in two dimensions in Fig. 2.32, in which each half is completely ordered, but the two ordered arrangements are "out of phase" with each other. The two parts are said to be antiphase domains.

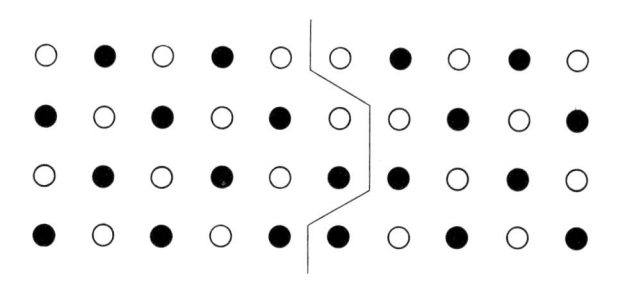

Fig. 2.32. Antiphase domains.

The crystal obviously has a high degree of order and yet the parameter S would be zero, because half of the atoms would be displaced from their proper lattice points for either way of choosing the a sites and the b sites. If the antiphase domains are very large, the state of order within each can be represented by S; but if they are small, the significance of such a parameter becomes less evident.

THERMODYNAMIC CONSIDERATIONS. It is possible, in principle, to determine whether the completely ordered or the completely disordered state of a particular alloy should have the lower free energy, and it might be inferred that the alloy would, in its equilibrium condition, be completely ordered below that temperature and completely disordered above it. This conclusion would be incorrect, because both the energy and entropy depend in a complicated way on the values of both the long-range-order and the short-range-order parameters. As a result of this, there is a critical temperature Tc above which there is *no* long range order, while there is a range of temperatures below Tc in which the long-range-order parameter is less than unity. There is some doubt as to whether the transformation is ever a really abrupt one (like melting, and characterized by a latent heat) or a continuous one with a high specific heat over a finite range of temperatures.

NONSTOICHIOMETRIC ALLOYS. When an alloy that can undergo ordering has a composition that is not stoichiometrically exact, it appears that the ordering reaction involves some diffusion and the subdivision of the alloy into distinct phases; this is because, below the ordering temperature, the free energy can be reduced by allowing part of the material to be exactly at the ordering composition; this of course increases the entropy of the system but the energy of ordering is sufficient to compensate for it. Thus it is not unreasonable to regard the transformation as a heterogeneous two-phase transformation, modified by the continuity of the ordering reaction. The phase diagram can be considered as consisting of single-phase regions separated by two-

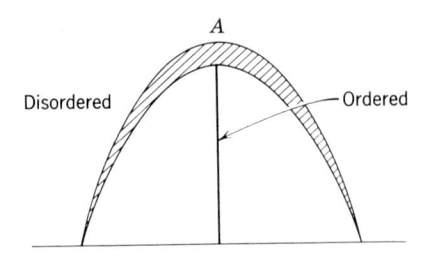

Fig. 2.33. Phase boundaries between ordered and disordered states.

AGGREGATES OF ATOMS

phase regions except for the region around A (in Fig. 2.33), in which the two phases are not separated by the usual phase boundary.

Interstitial solid solutions. Solute atoms in interstitial sites always produce some increase in energy and they never interact to cause clustering. The solubilities are usually low and the distance between solute atoms is so large that no ordering tendency is apparent. Interstitial solid solutions only occur when the solute atom is small compared with the solvent atom; for example, carbon dissolves interstitially in iron, and hydrogen dissolves interstitially in aluminum.

Limit of solubility. There are a few cases in which two metals are soluble in each other in all proportions. An example is the system gold-silver; conditions for unlimited solubility are that the two metals have the same crystal structure (e.g., both face-centered cubic) and that the sizes of the atoms be very similar: a difference of size of about 15% is sufficient to prevent complete solubility. An additional condition is that an intermediate phase must not form, as this would in itself limit the solubility. The criteria for the formation of intermediate phase will be discussed below; at this point it is sufficient to observe that they depend upon differences between the atoms, either in size or in electronic characteristics.

When solubility is not complete, its extent is limited either by disparity in size, or crystal structure, or by the intervention of a second phase; the examples in the next paragraph illustrate these points.

The system silver-copper (Fig. 2.26) is one in which both elements have the face-centered cubic structure; the diameters of the atoms are 2.55 A for copper and 2.88 A for silver. Electronically they are rather similar (see p. 4), and it may be concluded that the difference in atomic size is the reason for limited solubility; the situation is similar for the system sodium-potassium (Fig. 2.34). On the other hand, on the basis of size and crystal structure, it might be expected that aluminum (atomic diameter 2.85 A) and silver (atomic diameter 2.88 A) should be mutually soluble. Figure 2.35 shows that this is not so, and that the intervention of two intermediate phases is the reason.

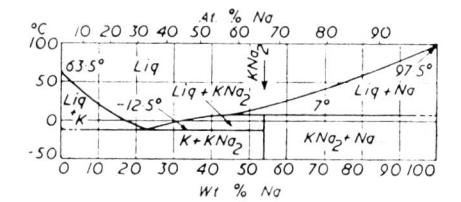

Fig. 2.34. Sodium-potassium system. From *Metals Reference Book*, Vol. I, 2nd ed., C. J. Smithells, Ed., Interscience Publishers, New York, 1955, p. 405.

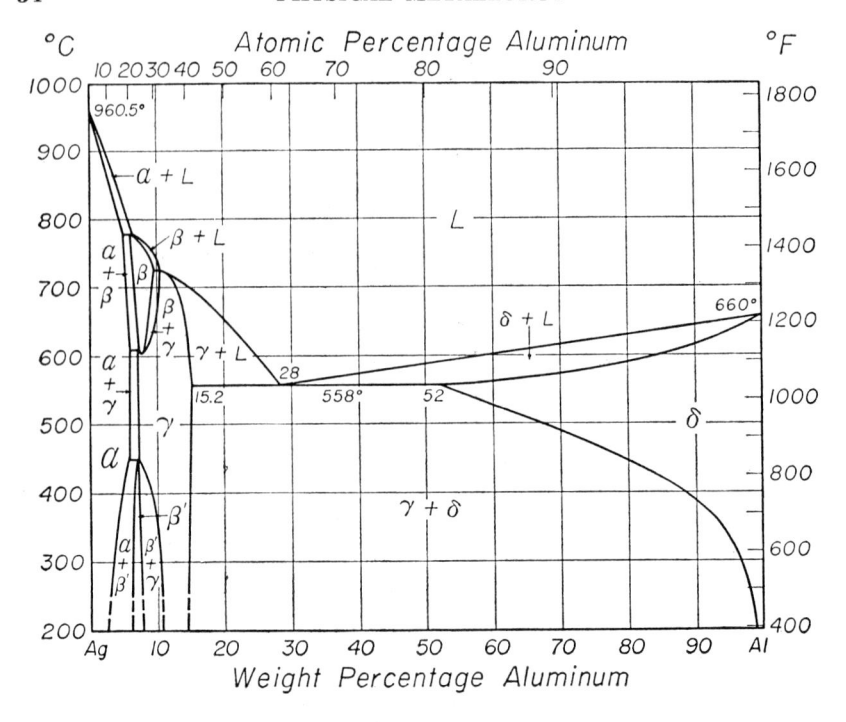

Fig. 2.35. Silver-aluminum system. From *Metals Handbook*, ASM, p. 1146.

2.8 Equilibrium Between Solid and Liquid Alloys

Kinetic considerations. It is instructive to consider the equilibrium between solid solutions and liquid phases from the point of view of the atom movements at an interface. Let us discuss the equilibrium of a crystalline phase consisting of a solvent A and solute B, the composition of which is represented by $C_S{}^A$ or $C_S{}^B$, with a liquid phase of composition $C_L{}^A$ or $C_L{}^B$. It is not assumed that the compositions of the solid and the liquid are the same. We will assume that atoms of A and B in the solid require activation energies $Q_S{}^A$ and $Q_S{}^B$ respectively to cross into the liquid, and that atoms of A and B in the liquid require $Q_L{}^A$ and $Q_L{}^B$ to cross into the solid. We will also assume that there is a probability constant B_M (see p. 19) for an atom with sufficient energy to cross from solid to liquid, and a different constant B_F for the liquid to solid transition. The condition for equilibrium between solid and liquid is that equality of rates in the two directions must be achieved both by the A atoms and by the B atoms. These rates are given by the

AGGREGATES OF ATOMS

product of the concentration of the atom type, the probability constant, and the energy term, i.e., $CB \exp(-Q/kT)$. Thus

$$C_S^A B_M \exp\left(-\frac{Q_S^A}{kT_E'}\right) = C_L^A B_F \exp\left(-\frac{Q_L^A}{kT_E'}\right) \quad \text{for the atom species } A$$

and

$$C_S^B B_M \exp\left(-\frac{Q_S^B}{kT_E'}\right) = C_L^B B_F \exp\left(-\frac{Q_L^B}{kT_E'}\right) \quad \text{for the atom species } B$$

where T_E' is the temperature at which this equilibrium condition is satisfied. The simplest assumptions are that the two kinds of atoms are interchangeable in the liquid, and therefore $Q_L^A = Q_L^B$, and that the energy of each B atom in the solid is greater than the energy of the A atoms by an amount that does not depend on the concentration. This is equivalent to assuming an ideal solution. In this case Q_S^B is constant, and less than Q_S^A. The equilibrium conditions now become

$$\frac{C_S^A}{C_L^A} \cdot \frac{C_L^B}{C_S^B} = \exp\left(\frac{Q_S^A - Q_S^B}{kT_E'}\right)$$

For small concentrations of B (for which C_S^A and C_L^A are close to unity) this becomes

$$\frac{C_S^B}{C_L^B} = \exp\left(\frac{Q_S^B - Q_S^A}{kT_E'}\right)$$

C_S^B/C_L^B for small concentrations is the distribution coefficient k. Qualitatively, this means that if the B atom can only be accommodated in the A crystal with considerable energy, it will be present in the solid at much lower concentration than in the liquid. We can also consider the effect of the solute on the equilibrium temperature T_E'; since

$$C_S^A B_M \exp\left(-\frac{Q_S^A}{kT_E'}\right) = C_L^A B_F \exp\left(-\frac{Q_L^A}{kT_E'}\right)$$

and

$$B_M \exp\left(-\frac{Q_S^A}{kT_E}\right) = B_F \exp\left(-\frac{Q_L^A}{kT_E}\right)$$

it follows that

$$\frac{C_L^A}{C_S^A} = \frac{(Q_S^A - Q_L^A)(T_E - T_E')}{T_E T_E'}$$

or

$$\ln \frac{C_L^A}{C_S^A} = \frac{L \,\Delta T}{T_E T_E'}$$

or
$$\ln\left(\frac{I - C_L{}^B}{I - C_S{}^B}\right) = \frac{L\,\Delta T}{T_E T_{E'}} \quad \text{(for small values of } \Delta T\text{)}$$

or
$$-C_L{}^B + C_S{}^B = \frac{L\,\Delta T}{T_E{}^2}$$

or
$$\frac{C_S{}^B}{\Delta T} - \frac{C_L{}^B}{\Delta T} = \frac{L}{T_E{}^2}$$

This is equivalent to a form of van't Hoff's equation, and gives the difference in slope of the liquidus and solidus for small concentrations in terms of the melting point of the pure solvent and the latent heat of fusion. This relationship shows that, when a solute is added to a pure metal, the compositions of the solid and the liquid that are in equilibrium with each other are related to the latent heat of fusion and the melting point of the solvent metal and to the amount by which the equilibrium temperature differs from the melting point of the pure metal.

The change in equilibrium temperature with increasing concentration may be described in physical terms as follows; assuming that, as discussed above, the solute has a lower activation energy for melting than the solvent, the distribution coefficient k will be less than unity; thus $C_L{}^A > C_S{}^A$. Consider now the rates of melting and of freezing *of the solvent* in such a system. The rate of each process is that for the pure metal reduced by a factor that is equal to its concentration, $C_L{}^A$ or $C_S{}^A$. Thus the rate for freezing is reduced by a larger factor than the rate of melting. Figure 2.36 represents these rates; full lines show the rates for the pure solvent, and broken lines for the liquid and solid solutions.

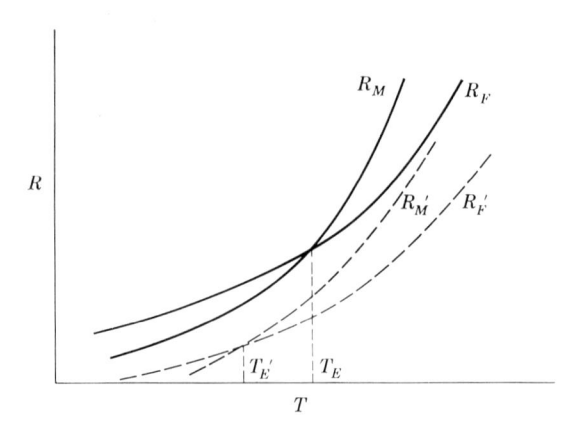

Fig. 2.36. Effect of solutes on interface processes.

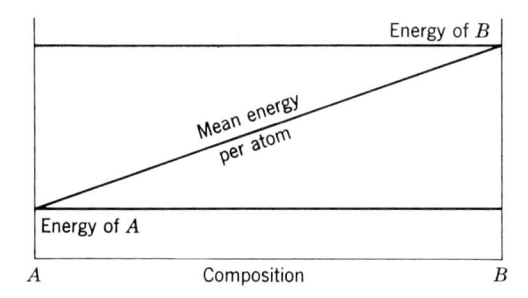

Fig. 2.37. Ideal case of energy of atoms in solid solution.

The rate of freezing is shown as decreased by a greater amount than the rate of melting. It is seen that T_E' must be lower than T_E.

The discussion so far has assumed that the energy per atom of B is independent of the concentration; this would lead to a continuous series of solid solutions of B in A (or A in B) from pure A to pure B. The energies are represented in Fig. 2.37. This case is that of the ideal solution. The point of view presented above, if developed for this case without the limitation of small concentrations, leads to the liquidus and solidus shown in Fig. 2.38. A number of metallic systems resemble this ideal case (e.g., gold-silver, copper-nickel, etc.).

In many other cases, however, the structures of the pure A and B phases are different. This corresponds to the case in which the crystal with the structure characteristic of A, but composed of B, would have a higher free energy than the normal crystal of B, and vice versa. Atoms of A in solid solution in B, and atoms of B in solid solution in A, have lower activation energies for melting than their solvent atoms. Hence, the two equilibrium conditions are no longer continuous with each other (as in Fig. 2.38), but are now as shown in Fig. 2.39. Point

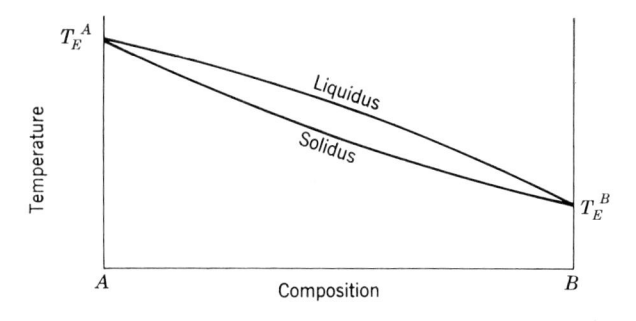

Fig. 2.38. Phase diagram for ideal solution.

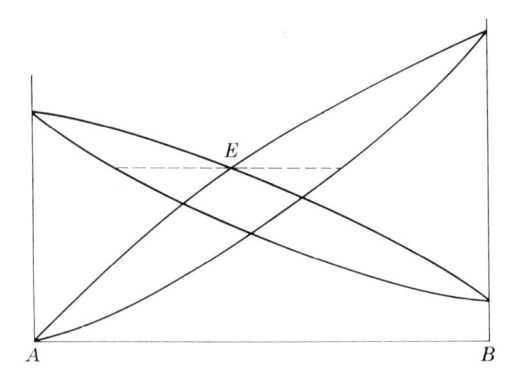

Fig. 2.39. Phase diagram for incompatible structures.

E corresponds to equilibrium between the liquid and two solid phases.

The detailed shapes of the liquidus and solidus curves are never exactly as predicted from the assumption of ideality. Departure from this condition corresponds to a dependence of the activation energy for melting on the concentration; but the qualitative concept expressed above is still valid in such cases.

The foregoing discussion for the equilibrium of two phases is not in principle limited to binary solid-liquid equilibria. With suitable modifications it will represent the equilibrium between solid phases, and it can be extended to systems containing more than two atomic species.

One important consequence of the equilibrium concept can be readily seen from kinetic considerations. Consider two phases α and β, each of which is in equilibrium with a third phase γ. Then equilibrium may be expressed as

$$R^A_{\alpha \to \gamma} = R^A_{\gamma \to \alpha}$$

and

$$R^A_{\beta \to \gamma} = R^A_{\gamma \to \beta}$$

It follows from these two expressions that $R^A_{\alpha \to \beta} = R^A_{\beta \to \alpha}$ if the values of both A and G (see p. 19) are characteristic of one phase, rather than of the transition. Clearly, the accommodation coefficient is a property of the phase to which the atom is moving, and the geometric factor G is a property of the phase from which the atom comes. Thus it follows that if two phases are each in equilibrium with a third phase, these are in equilibrium with each other, since the equilibrium relationship given above applies to each component of the system.

Thermodynamic considerations.

A qualitative explanation for the existence and shape of the two-phase regions can also be given in terms of thermodynamic considerations. Quantitative predictions cannot in general be made because it is not possible at present to predict the energies or the entropies of the various phases. The general problem is as follows: Consider two metals A and B, and let us attempt to predict the free energies of all likely structures containing A and B. The first possibility is that of substitution of A for B or of B for A in the stable crystal of B or of A. The initial effect of substitution is to increase the entropy, because a crystal of A containing a few atoms of B has more different possible arrangements than a crystal of pure A. If the strain energy is not too large (i.e., if the atoms are of nearly the same size), then the energy is not increased greatly and the free energy follows the curve PQ (Fig. 2.40).

If A and B have the same crystal structure, it is possible for this substitution to continue until a crystal of B is reached, and the curve then takes the form PQR. If A and B do not have the same crystal structure, then B has an alternative structure with a lower free energy S; a free energy curve for the substitution of A into B would then have the form STU. The equilibrium condition of lowest free energy in such a system is found as follows: A common tangent XY is drawn to the two free energy curves. To the left of X, the single phase A, with the appropriate amount of substituted B, is the lowest possible free energy condition; similarly to the right of Y. Between X and Y, however, the free energy is lower if the material is divided into appropriate proportions of X and Y, rather than having a composition between these values, corresponding to point T, for example, on the free energy curve for B. The free energy of the mixture of X and Y corresponds to point V on the common tangent. The amounts of X and Y are in the

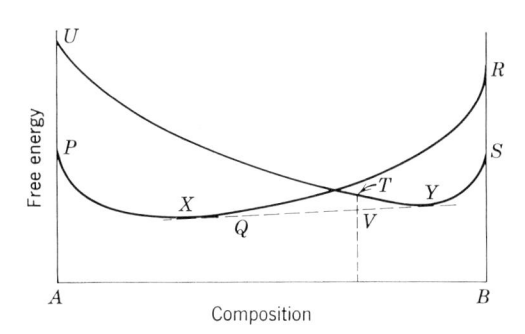

Fig. 2.40. Free energy curves for incompatible structures.

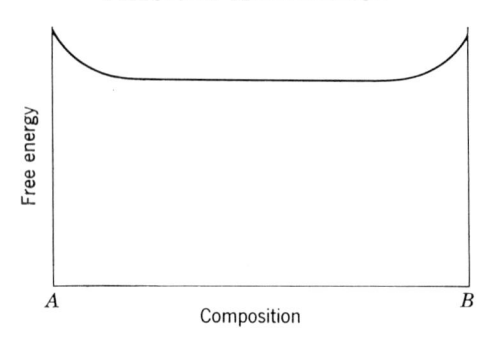

Fig. 2.41. Free energy curve for liquid solutions.

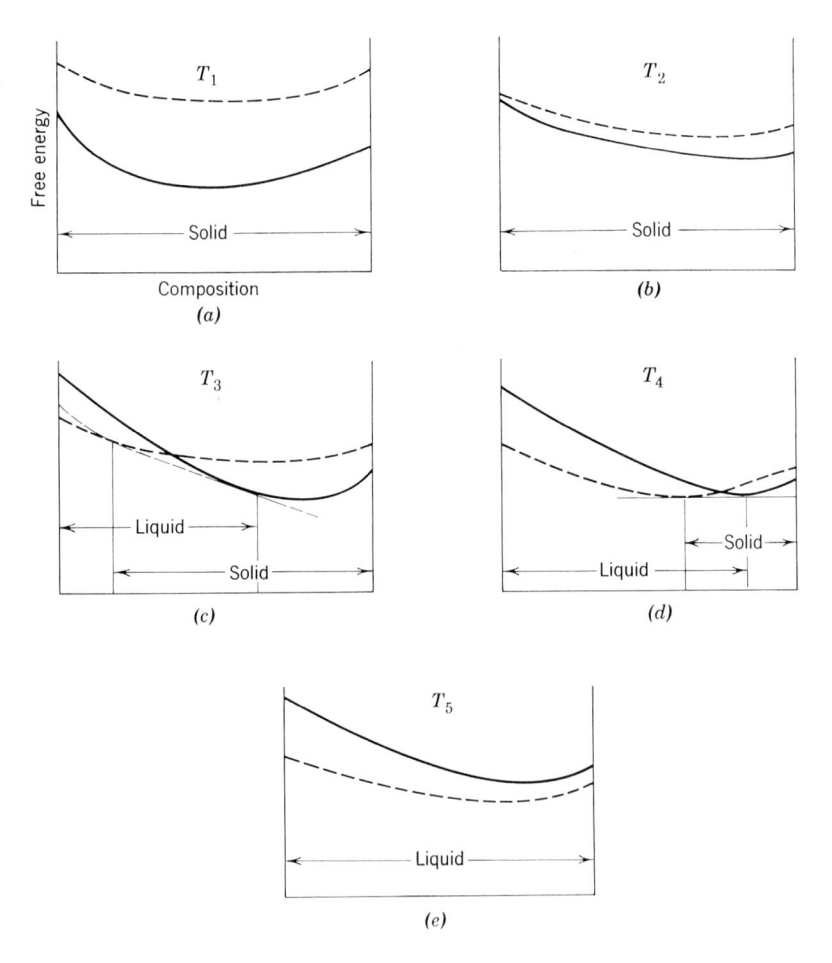

Fig. 2.42. Free energy curves for solid and liquid phases of simple binary system.

ratio of VY and VX. It also follows that the substances with compositions X and Y are in equilibrium with each other, since any change of either composition would increase the free energy of the system.

There are other possible phases, besides those based on the crystal of A and the crystal of B, that must be considered; the liquid phase is the most obvious. It will be assumed that the effect of mixing different metals in molten form is to contribute very little to the energy, but that the entropy is still increased for the same reasons as for the crystal. The liquid, therefore, usually has a single free energy curve and in such cases there is only one liquid phase; the free energy curve is of the form shown in Fig. 2.41. At low temperatures, this curve lies entirely above those for the crystal phases and it has no significance. However, the effect of raising the temperature is to increase the relative importance of the entropy term of the free energy, and so the free energy is reduced as the temperature rises. The entropy is greater for the liquid than for the crystal, and so the free energy curve moves downward faster for the liquid than for the crystal as the temperature is raised. A series of free energy curves at successively higher temperatures is shown in Fig. 2.42. In this example, the solid phase has the same structure at all compositions; its free energy is represented by the full line, while the free energy for the liquid is represented by the broken line. In each case the equilibrium phase (or phases) is indicated along the composition axis.

The phase diagram for this system is shown in Fig. 2.43. The limits of the single-phase region in this diagram are the same as the points of contact of the common tangents in Fig. 2.42. The two-phase region is the composition range that is spanned by the common tangent.

If the two substances A and B have different crystal structures, there

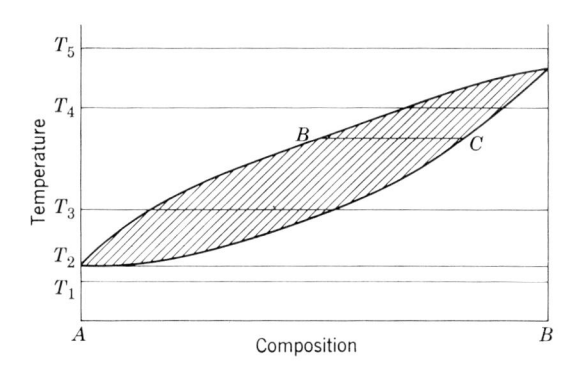

Fig. 2.43. Phase diagram derived from the curves of Fig. 2.42.

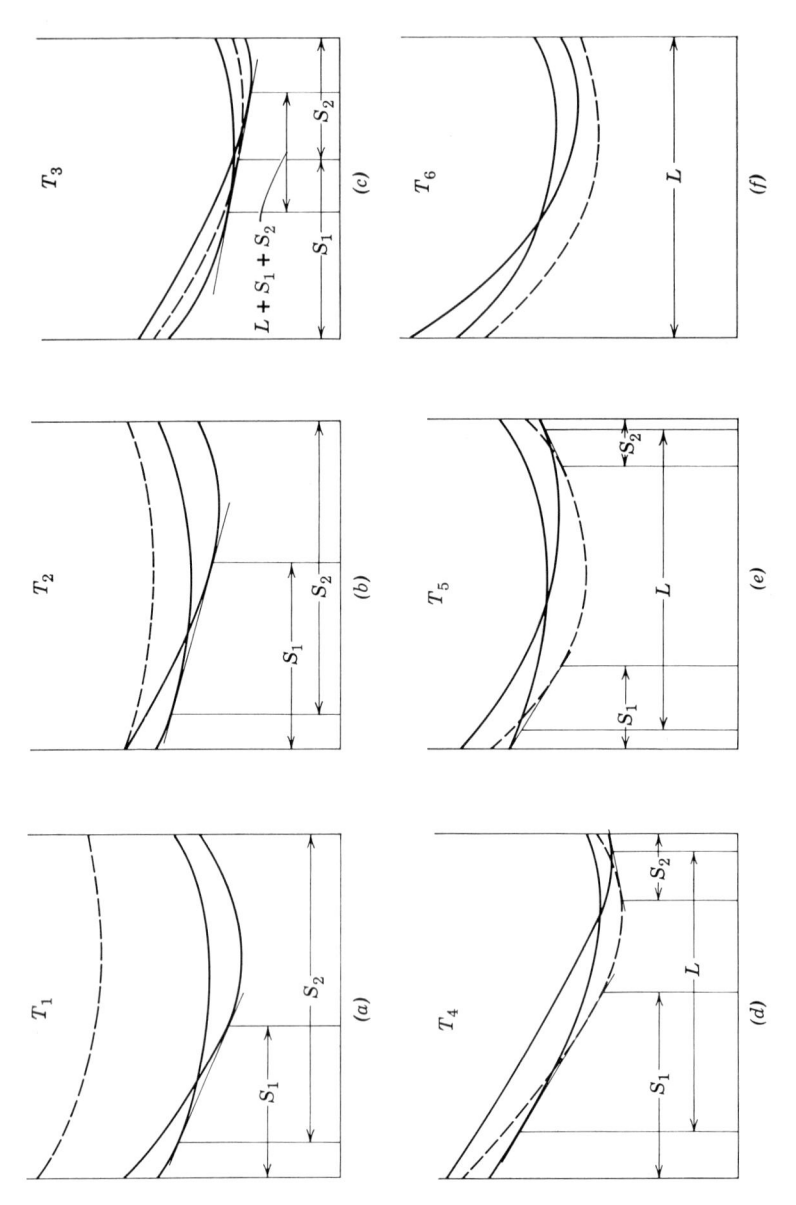

Fig. 2.44. Free energy curves for binary eutectic system.

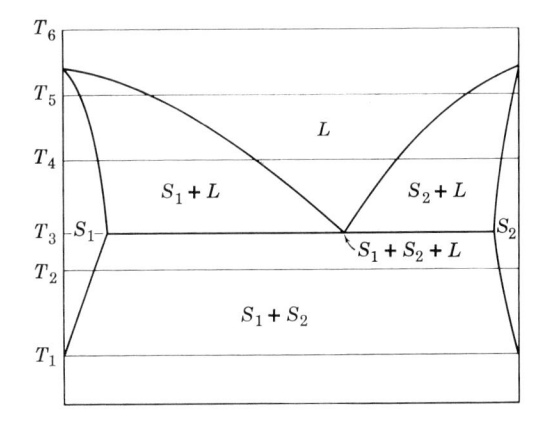

Fig. 2.45. Phase diagram corresponding to free energy curves of Fig. 2.44.

will, as pointed out above, be two separate free energy curves for the two solid phases. Figure 2.44 shows a series of schematic free energy curves for such a system. The phase diagram derived from these free energy curves is given in Fig. 2.45.

The thermodynamic criterion for the equilibrium between two phases in a single component system is, as discussed earlier, that the free energies of the two phases are equal. This criterion does not apply in systems containing more than one component. This will be evident from Figs. 2.40 and 2.42. The condition for equilibrium is that the free energy of one phase shall increase and that of the other decrease *by an equal amount* if a small quantity of any component is moved from one phase to the other. This is expressed more formally by stating that the molar free energies of each component must be equal in the two phases. It should be emphasized that these considerations apply to any two-phase equilibrium, and can be generalized for systems with more than two components.

2.9 Intermediate Phases

The existence of an intermediate phase indicates that a lower free energy is made possible by a distribution of atoms that limits the solid solubility and uses some of the solute atoms and some of the solvent atoms to build a new phase. This may occur for either of two reasons: the free energy of the solid solution may increase so drastically with concentration that virtually any other phase will be possible, or else a new phase of particularly low free energy may be formed. The former situation arises when the two kinds of atoms are very different in size (more than about 15%); the elastic strain energy required to accommodate more than a very few such solute atoms becomes very high. The second situation arises when an intermediate phase of low free energy can form.

An intermediate phase has, therefore, a crystal structure that depends for its stability on the participation of more than one kind of atom. The necessary reduction of energy, compared with the solid solutions, may be caused by (a) more favorable electron energy levels, (b) ionic or covalent bonding between the different atomic species, or (c) geometrical relationships between the sizes of the atoms. It should not be concluded that only one of these causes can operate at a time, and it is more realistic to regard many of the intermediate phases as existing as a result of the combined effect of two or more.

The energy of electrons in a crystal. We will consider first the phases which are essentially metallic in character, but which owe their low energy, and therefore their stability, to the occurrence of a structure that will accommodate a particular number of electrons.

It is necessary to consider briefly the way in which the electron energy depends on the structure and composition in order to appreciate the significance of the electron content of an alloy. It is instructive for this purpose to look at the electrons in terms of the wave concept, and to consider the electron energies in relation to the crystal structure. This provides one of the criteria for the energy of a crystal structure.

BRILLOUIN ZONES. The availability of low energy levels for the electrons is a function of the detailed packing of the atoms, because the limits of the energy bands in which electrons can exist are determined by the geometrical relationships between the positions of the atoms.

A descriptive, but not rigorous, explanation is as follows: The first consideration is the existence of energy levels in a large assembly of atoms. It can be shown that the number of energy levels available for valency electrons is not infinite, but is proportional to the number of atoms in the system. The number of energy levels within a given small range of energies is denoted by $N(E)$ and this is proportional to $E^{1/2}$, so that the relationship between $N(E)$ and E is parabolic, as shown in Fig. 2.46. Since the electrons tend to take the lowest available levels, they will all have energies less than some value denoted by E_{max}.

If we were concerned with a uniform isotropic continuum, the value of E_{max} would be entirely independent of the direction in which the electron was moving; however, the crystal is not such a medium and the next concept to be considered is the effect of the periodic arrangement of the positive ions on the availability of energy levels; this

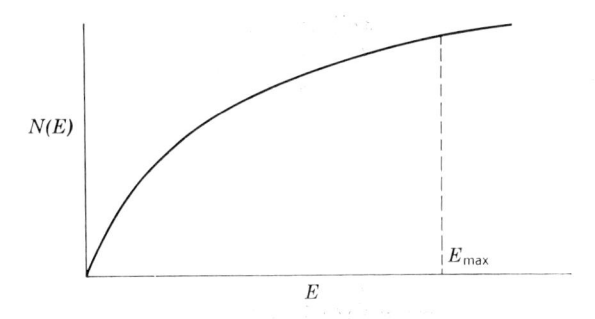

Fig. 2.46. Density of states curve for continuum.

necessitates the consideration of the direction of movement of electrons. Hence we can no longer consider only the energy, which is a scalar quantity, but we must also consider the momentum, which is a vector quantity. The wavelength λ that is associated with a free electron is given by $1/\lambda = mv/h$, where m and v are the mass and velocity of the electron and h is Planck's constant. The speed and direction of motion of an electron can, therefore, be represented by a vector **k** which is in magnitude proportional to $1/\lambda$. For theoretical reasons it is convenient to use $2\pi/\lambda$ as the value for the wave vector **k**.

The motion of any electron can be represented by a point of which the distance and direction from an origin correspond to the magnitude and direction of **k**. The co-ordinate system in which the motion of electrons is represented in this way is called "k space."

An electron could travel in any direction and with any velocity in a truly continuous medium; a metal crystal, however, has a periodic distribution of positive charges (the metal ions) in it, and these interact with the moving electrons in the same way that the periodic lattice of a crystal interacts with X rays; it diffracts when the Bragg condition, $n\lambda = 2d \sin \theta$, is satisfied. This means that for any value of λ there are certain directions in which the electron cannot travel; it also means that certain points in k space correspond to momentum vectors which cannot be achieved, and the corresponding energy levels are accordingly not available. Electrons must therefore have either appreciably less energy or appreciably more energy than these values.

The points in k space which correspond to the Bragg condition lie on planes, the geometry of which depends upon the crystal structure. The planes enclose finite volumes, which contain finite numbers of energy levels. They are known as Brillouin zones. The first Brillouin zone is the most significant one. If all the electrons can be accommodated in levels that are well within the first Brillouin zone, then the highest energy electrons (those corresponding to E_{max}) will lie on a sphere in k space. This surface, whether it is spherical or not, is called the Fermi surface. If the surface of this sphere approaches the first Brillouin zone it will be distorted somewhat, because the most energetic electrons have lower energies if they approach the Bragg condition than if they do not. Hence states are closer together when the Fermi surface is close to a Brillouin zone.

As a result, there are more electrons with energies just below the value at which the Fermi surface just touches the Brillouin zone than there would be at the same energies if the Brillouin zone were not there, i.e., if the electrons were moving in a uniform field instead of in a periodic field. The "density of states" curve therefore follows the broken curve (Fig. 2.47) from A to B, at which point the Fermi surface touches the Brillouin zone, and the number of possible energy states is reduced from B onward because electrons cannot move in particular directions with energies corresponding to values just above E_B. The number of available states therefore decreases sharply above E_B. There are several important consequences of the fact that electrons cannot move with any energy in any direction in a crystal; if the number of electrons is just sufficient to fill a Brillouin zone, which does not overlap with the next zone, it is possible to increase the energy of any electron only by giving it a sufficient increment of energy to lift it into the next Brillouin zone. This is equivalent to stating that such a crystal is an insulator if the increment is large.

If the energy gap between successive zones is small, and if the first zone is filled, electrons can be moved into the second zone by thermal energy. In this case the electrical conductivity increases with rise of temperature, and the crystal is a semiconductor. The conductivity of

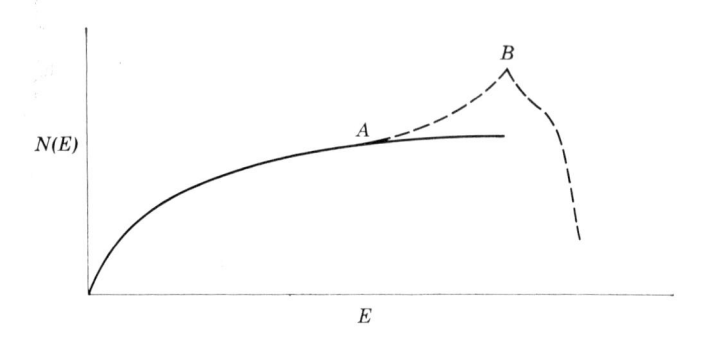

Fig. 2.47. Effect of crystal structure on density of states curve.

a semiconductor can be increased either by increasing the number of electrons, so that some are forced to occupy the lowest levels of the second zone, or by reducing the number of electrons, so that there are some unoccupied electron energy levels in the first zone, or by raising the temperature sufficiently to cause some electrons to move from one zone to the next. The two conditions which are brought about by substituting atoms of higher or of lower valency for a very small proportion of those that would otherwise be present, are referred to as n-type and p-type conductivity respectively.

On the other hand, in metallic conductors the electrons do not fill the first Brillouin zone and can therefore receive small increments of energy, or else the first and second Brillouin zones overlap in some places so that electrons can move into the second zone without having to overcome an energy discontinuity.

These crystals therefore can permit the movement of electrons under small potential differences. The resistivity in this case increases with temperature because of the increased scattering of the electron waves, at "non-Bragg" angles, as a result of increasing thermal movement of the atoms.

HUME-ROTHERY OR ELECTRON PHASES. Another important consequence of the maxima of the $N(E)$ versus E curve is that the crystal structure having the lowest energy may be the one which allows the electrons a more favorable range of energy levels. This is one of the criteria that contribute to the selection of the crystal structure with the lowest energy. In the first place, the criterion may provide an acceptable explanation of the $8 - N$ rule, in terms of electron energies, as an alternative to the shared electron bond point of view discussed above. For example, the elements of group 4 (carbon, silicon, germanium, and tin) each have four "available" electrons, and a favorable structure would be one which just accommodated all four electrons in the first Brillouin zone. The diamond-cubic structure has a first Brillouin zone which contains four electrons per atom, and is therefore just filled. Any alternative structure would require some of the electrons to occupy a second zone, and to correspond to a higher total electron energy. It is probable that the other "$8 - N$" structures could be accounted for in similar terms.

In each of these cases, the number of electrons per atom is characteristic of the element selected. However, the number of electrons can be changed by the substitution of atoms of different valencies, and in some cases this accounts for the stability of an intermediate phase.

The body-centered cubic or β phase of the copper-zinc system is an example of the type of intermediate phase that owes its stability to a crystal structure that accommodates the electrons in relatively low energy states. Copper is a metal in which each atom contributes one electron to the population of electrons that take part in the bonding process. These electrons are easily contained in the first Brillouin zone of the face-centered cubic structure; however, if the number of electrons is increased by substituting elements of higher valency, such as zinc (2), aluminum (3), tin (4), etc., for copper, the average number of electrons per atom is increased. When this reaches a value of about 1.4 electrons per atom, the face-centered cubic structure can no longer accommodate further electrons without considerable increase in energy, whereas a body-centered cubic structure has lower energy if it contains about 1.5 electrons per atom. This follows from the fact that the density of states curve B for the body-centered cubic crystal is higher than for the face-centered cubic structure A in this region (Fig. 2.48). The assumption implicit in this approach is that the zinc and copper atoms only differ in their valency, and this is clearly not correct; however, there are a substantial number of alloy systems that have intermediate phases at compositions close to those which correspond to $3/2$ electrons per atom. Many of these phases have the body-centered cubic structure, and they are metallic in character, as would be expected from the fact that the electrons are not associated with individual atoms.

Analogous phases are also found with $21/13$ electrons per atom and $7/4$ electrons per atom; the striking feature of these phases is the fact that the actual atomic proportions vary widely, but the number of electrons per atom is constant. Some examples are shown in Table 2.5. It is necessary to assume that the transition elements (iron, nickel, cobalt, etc.) contribute no electrons.

The phases described above, known as Hume-Rothery, or electron, phases, do not depend for their stability on the two types of atoms

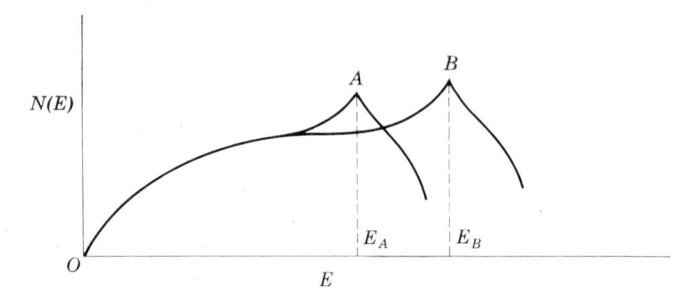

Fig. 2.48. Density of states curves for face-centered cubic (A) and body-centered cubic (B) structures.

AGGREGATES OF ATOMS

Table 2.5. Some Hume-Rothery or Electron Phases *

Electrons per Atom

$3/2$			$21/13$	$7/4$
B.C.C. Structure	Cubic Manganese Structure	C.P. Hex Structure	γ Brass Structure	C.P. Hex Structure
CuBe	Cu$_5$Si	Cu$_3$Ga	Cu$_5$Zn$_8$	CuZn$_3$
CuZn	AgHg	Cu$_5$Ge	Cu$_5$Cd$_8$	Cu$_3$Sn
Cu$_3$Al	Ag$_3$Al	AgZn	Cu$_5$Hg$_8$	Ag$_5$Al$_3$
Cu$_3$Ga	Au$_3$Al	Ag$_3$Al	Cu$_9$Al$_4$	
Cu$_3$In	CoZn$_3$	Ag$_3$Ga	Cu$_{31}$Si$_8$	
Cu$_5$Sn		Ag$_3$In	Ag$_9$In$_4$	
AgMg		Ag$_5$Sn	Mn$_5$Zn$_{21}$	
Ag$_3$Al		Ag$_7$Sb	Fe$_5$Zn$_{21}$	
FeAl		Au$_3$In	Na$_{31}$Pb$_8$	
CoAl		Au$_5$Sn		
PdIn				

* From *Structure of Metals*, Hume-Rothery and Raynor, p. 197.

occupying specified sites in the crystal, although they may be ordered at low enough temperatures; from this point of view they are solid solutions, and they do not have any of the characteristics of chemical compounds. They can exist over a substantial range of compositions, which in many cases does not quite include the simple atomic ratio. The phase diagram for the copper-zinc system (Fig. 2.21a) shows all three Hume-Rothery phases; the β phase at composition about CuZn (3:2), the γ phase at about Cu$_5$Zn$_8$ (21:13), and the δ phase at about CuZn$_3$ (7:4).

It has also been shown that analogous ternary phases exist at similar electron-to-atom relationships.

There are many other cases in which a change in electron concentration is associated with a change in crystal structure. It should be pointed out that another solution to the problem of nonavailability of low energy electron levels is the defect lattice, in which the effective number of electrons per atom is reduced by some atom sites of the crystal structure remaining vacant; the number of electrons per unit cell of the crystal would be the criterion in this case.

Ionic and covalent compounds. The Hume-Rothery phases may be either ordered or disordered; they do not depend for their stability on the occupation of particular sites by atoms of one species; the position of the atom can be interchanged without destroying stability, and in this respect these phases can be regarded as solid solutions.

All the other types of intermediate phase depend upon the two (or more) atomic species occupying nonrandom sites in the crystal. They fall into two main classes: 1. those that depend, at least partly, on electron interaction, either ionic or covalent, between the two species of atom, and 2. those that owe their stability to geometrical relationships controlled by the relative sizes of the atoms.

IONIC COMPOUNDS. The former class can be regarded as "compounds" in the chemical sense of the term and are frequently formed between metals and the elements of groups 4, 5, and 6 that can behave electronegatively by acquiring electrons. Examples are Mg_2Si, Mg_2Sn, Mg_3As_2, and $MgSe$, in all of which magnesium may be regarded as having its normal valency of 2, while Si, Sn, As, and Se have their normal

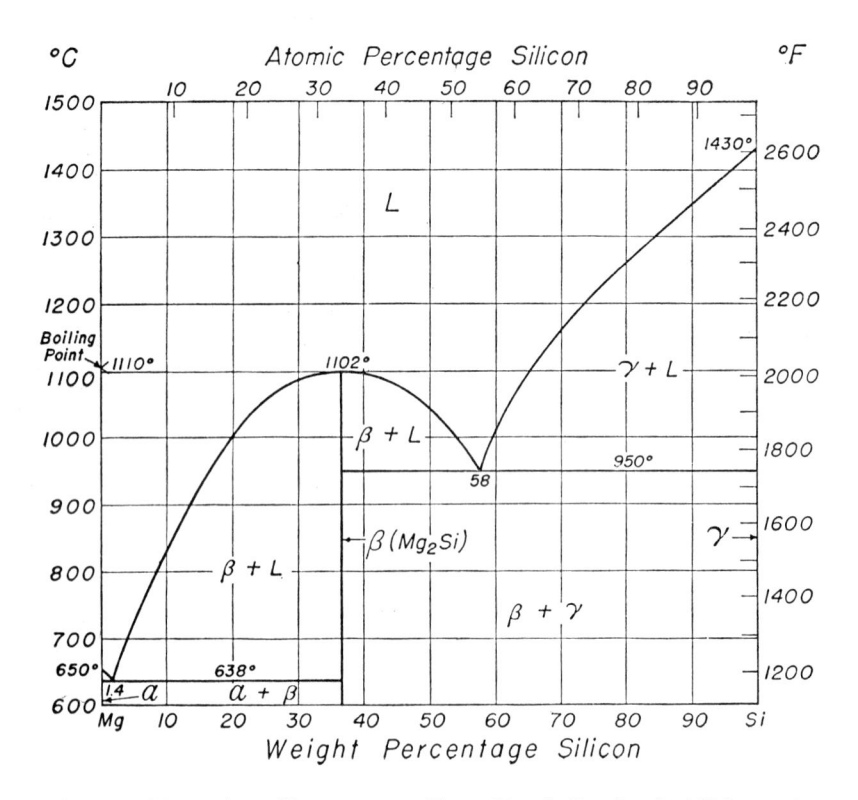

Fig. 2.49. Magnesium-silicon system. From *Metals Handbook*, ASM, p. 1226.

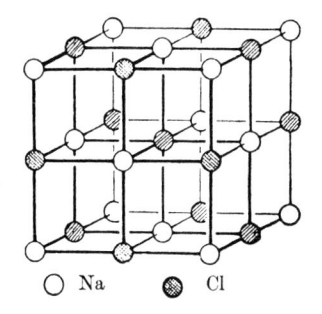

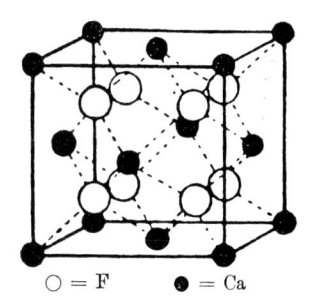

Fig. 2.50. Sodium chloride and calcium fluoride structures. From *Structure of Metals*, Hume-Rothery and Raynor, Figs. 116 and 117, p. 180.

negative valencies of 4, 4, 3, and 2 respectively. It is found that ionic compounds of this type are formed only when the metal is strongly electropositive, or the other element strongly electronegative, as is the case for the lighter elements of a group. Thus zinc, which is less electropositive than magnesium but also has a valency of 2, does not form a compound with tin, while it does so with antimony. The phase diagram for the magnesium-silicon system is shown in Fig. 2.49. It will be observed that Mg_2Si does not have an extended composition range. A number of ternary compounds of this type are known; an example is LiMgSb. The ionic compounds take up either the sodium chloride structure, when the two atomic species are present in equal numbers, or the calcium fluoride structure, when the atomic ratio is 2:1. The two structures are shown in Fig. 2.50.

COVALENT COMPOUNDS. There are a number of intermediate phases in which the bonding is predominantly covalent, and in which there are four electrons per atom. The zinc blende structure and the wurtzite structure, which are cubic and hexagonal respectively, satisfy the condition that each atom has four nearest neighbors of the other species. The former structure is similar to that of diamond. Some examples of compounds of this type are given in Table 2.6. Some of these compounds are semiconductors, as would be expected from the analogy of the structure, electron-atom ratio, and Brillouin zone to germanium and silicon.

A very important group of intermetallic compounds are those with the so-called "nickel arsenide" structure. They form a link between the essentially nonmetallic (ionic or covalent) types and those which are definitely metallic. The nickel arsenide structure is hexagonal, as shown in Fig. 2.51, and the axial ratio varies between the extreme values

Table 2.6. Some Compounds with Zinc Blende and Wurtzite Structures *

Zinc Blende Structure		Wurtzite Structure
BeS	AlAs	ZnS
ZnS	BeTe	CdS
HgS	CdTe	MgTe
AlP	AlSb	CdSe
ZnP	GaSb	AlN
BeSe	ZnSb	InN

* From *Structure of Metals*, Hume-Rothery and Raynor, p. 184.

of 1.21 for CuSn to 1.75 for TiSe. The structure is based on a close-packed hexagonal lattice of the anions, with the cations lying in interstitial sites of the anion lattice, forming "sheets" of cations between the anions. This would result in equal numbers of anions and cations. However, some extra cations can be accommodated in smaller interstitial holes, or some may be missing, and so the numbers of the two types of atoms need not be equal, as they would be in the ideal case. In these structures, the anion is a "metalloid," such as sulfur, selenium, tellurium, tin, or antimony, and the cation is a transitional metal, such as chromium, nickel, iron, etc.

In the extreme case, where the metalloid is replaced by metals of groups 3, 4, or 6, it is possible for the excess transitional metal to reach the level at which the composition corresponds to A_2B. In the latter case the compound is strongly metallic in character, whereas in the case in which there is less than one metal atom to each metalloid, the character is essentially ionic. The compositions of some compounds of the nickel arsenide type are given in Table 2.7.

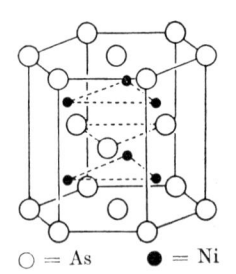

Fig. 2.51. Nickel arsenide structure. From *Structure of Metals*, Hume-Rothery and Raynor, Fig. 122, p. 188.

○ = As   ● = Ni

Table 2.7. Some Compounds of the Nickel Arsenide Type *

TiS$_2$	Fe$_2$Ge	Ni$_3$Sn$_2$
VS	Fe$_3$Sn$_2$	NiAs
CrSb	FeSe	Cu$_4$In$_3$
CrSe	Co$_2$Ge	Cu$_6$Sn$_5$
Mn$_7$Ge$_4$	Co$_3$Sn$_2$	
Mn$_2$Sn	CoSb	
MnTe	Ni$_2$Ga	

* Data from *Structure of Metals*, Hume-Rothery and Raynor, p. 191.

Interstitial compounds. An important class of compounds are those between some metals and hydrogen, carbon, nitrogen, or boron. The characteristic features of these compounds are their high stability and their nonmetallic character. They are all formed when a small atom forms a compound with a larger one, and the structures seem to be derived from face-centered and body-centered cubic structures by the distortions necessary to insert the smaller atoms into the interstitial sites of the structure. The high degree of stability is apparently a result of electron sharing between the smaller atoms and the metallic crystal. Interstitial compound phases are not formed unless the radius of the smaller atom is less than about two-thirds that of the metallic one. Some examples of interstitial compounds are given in Table 2.8.

Table 2.8. Some Interstitial Compounds *

Compound	Melting Point (°C)	Compound	Melting Point (°C)
TiC	3150	NbC	3500
TiN	2940	NbN	2200
VC	2200	HfC	3900
ZrC	3530	TaC	3900
ZrN	2980	TaN	3087

* From *Structure of Metals*, Hume-Rothery and Raynor, p. 220.

Laves phases. When the metallic bond is responsible for the aggregation of atoms of a single species, the resulting structure is usually a very simple one, namely, face-centered cubic, body-centered cubic, or close-packed hexagonal; but when two different species of atoms are present, the energy may be lower when a structure is formed that depends on the size relationship of the two kinds of atom. In particular, many phases are formed at compositions corresponding to AB_2, in which the radius of the A component is about 20% larger than that of the B component. These phases, known as Laves phases, do not appear to depend on electrochemical or electronic relationships between the two kinds of atoms, and they are metallic in character. There are three closely related structures, in each of which the smaller atom lies at the corners of an array of tetrahedrons, within each of which is a larger atom. The difference between the three structures lies in the relative positions and orientations of the tetrahedrons, and therefore in the relationship between the positions of the A atoms. The three structures are described as the $MgCu_2$, $MgZn_2$, and $MgNi_2$ types respectively. Table 2.9 shows some of the Laves phases that have been identified.

Table 2.9. **Some Compounds of the Laves Types** *

$MgCu_2$ Type	$MgZn_2$ Type	$MgNi_2$ Type
$AgBe_2$	$BaMg_2$	$ReBe_2$
$BiAn_2$	$CuAg_2$	FeB_2
$CuAl_2$	$CuBe_2$	$MoBe_2$
$CeAl_2$	KNa_2	$TaCo_2$
$CeFe_2$	$MoBe_2$	WBe_2
$GdFe_2$	$NbFe_2$	$ZrFe_2$
KBi_2	$ReBe_2$	
$ZnAl_2$	$TaFe_2$	
$NaAu_2$	UNi_2	

* From *Structure of Metals*, Hume-Rothery and Raynor, p. 228.

In spite of the advances that have been made in understanding the nature of the various kinds of phases that can be formed in an alloy, it is still not possible to predict from first principles the phase diagram for a particular alloy system. This is because the free energy differences that may determine whether a particular phase is stable are small compared with the free energies of the phases themselves, if the free energy of the isolated atoms is taken as a datum, as it must be for calculations of free energies of crystalline phases.

Although detailed predictions cannot be made, the considerations outlined above give a useful guide of what might be expected in a given alloy system.

Heats of formation of intermediate phases. Some indication of the decrease in internal energy that accompanies the formation of the various types of intermediate phases may be gained from a study of their heats of formation. Table 2.10 gives some values for different

Table 2.10. Heats of Formation of Some Intermediate Phases *

A_{298} in kcal/gram-atom

Phase	ΔH	Phase	ΔH
Hume-Rothery Phases			
CuZn	2.5	AgCd	1.3
Cu_5Zn_8	2.9	Ag_5Cd_8	1.4
$CuZn_3$	2.0	$AgCd_3$	1.2
Cu_9Al_4	5.4	AgZn	1.7
Cu_3Zn_8	1.4	Ag_5Zn_8	1.9
Cu_3Zn	1.8	$AgZn_3$	1.3
AlFe	6.1	AlNi	16.0
Ionic Phases			
Mg_2Si	19.0	Mg_2Sn	18.3
Covalent Phases			
BeS	55.9	AlSb	23.0
ZnSe	34.0	InSb	7.8
ZnTe	28.8	HgTe	3.0
Nickel Arsenide Type Phases			
CoSe	10.0	NiTe	9.0
CoTe	9.0	NiSb	15.8
CoSb	10.0		
Interstitial Compounds			
TiC	43.9	NbN	59.0
NiN	80.4	NbC	33.7
ZnN	82.2	Fe_3C	−5.4
Laves Phases			
$AuPb_2$	1.5		
$CuMg_2$	21.3		
$CuCd_2$	30.0		

* Data from *Metallurgical Thermochemistry*, O. Kubaschewski and E. Ll. Evans, 2nd ed., John Wiley & Sons, New York, 1956.

types of phases. These are typical examples, selected from the very incomplete data that are available.

The most striking features of these data are the low values of the heats of formation of the Hume-Rothery or electron phases, and the very high values of some of the interstitial phases. The negative value for Fe_3C corresponds to the fact that this phase is not stable. Its existence is a result of its formation, under suitable conditions, from an even less stable supersaturated solution of carbon in iron.

2.10 Further Reading

A. H. Cottrell, *Theoretical Structural Metallurgy*, Edward Arnold and Co., London, 1948.

W. Hume-Rothery and G. V. Raynor, *The Structure of Metals and Alloys*, Institute of Metals Monograph and Report Series No. 1, 3rd ed., London, 1954.

C. S. Barrett, *Structure of Metals*, McGraw-Hill Book Co., New York, 1952.

F. N. Rhines, *Phase Diagrams in Metallurgy*, McGraw-Hill Book Co., New York, 1956.

G. V. Raynor, "Progress in the Theory of Alloys," Chapter 1 in *Progress in Metal Physics 1*, Bruce Chalmers, Ed., Butterworths Scientific Publications, London, 1949.

3

Structure-Insensitive Properties

3.1 Classification of Properties

The ideal crystalline phase consists of an array of atoms each of which has as its mean position a point reached by repeated integral translations of the occupied points in the unit cell of the structure. The atomic positions so defined are mean positions because the atoms are always in a state of thermal agitation which causes them to vibrate about their mean positions. Many of the properties of such an array of atoms can be calculated or estimated; it is found that, whereas some of the properties of real crystals can be accounted for, at least approximately, on this basis, there are other properties that deviate so seriously that an entirely different basis of prediction is required. It is believed that the latter type of property, described as structure-sensitive, depends upon departures of the structure from perfection, while the former (structure-insensitive) is characteristic of the perfect crystal, modified to a minor extent by the imperfections that account for the structure-sensitive properties.

Various properties of engineering significance are classified in this way in Table 3.1.

An important distinction between the two classes of properties is that, whereas the structure-insensitive properties are well-defined properties of a phase, the structure-sensitive properties are dependent not only on the composition and crystal structure of the material but also on structural details that depend upon the previous history of the sample. Thus the structure-sensitive properties are properties of a particular sample of a material, while the structure-insensitive properties relate to the material. Different samples of the same material have essentially identical structure-insensitive properties, but the structure-sensitive properties are identical only when the previous treatment has been equivalent.

Table 3.1. **Structure-Sensitive and Structure-Insensitive Properties**

	Structure-Insensitive	Structure-Sensitive
Mechanical	Density Elastic moduli	Fracture strength Plasticity
Thermal	Thermal expansion Melting point Thermal conductivity Specific heat Emissivity	
Electrical	Resistivity (metallic) Electrochemical potential Thermoelectric properties	Resistivity (semiconductor and at low temperatures)
Magnetic	Paramagnetic and diamagnetic properties	Ferromagnetic properties (including magnetostriction)
Optical	Reflectivity	
Nuclear	Absorption of radiation	

In the following discussion of structure-insensitive properties, an indication will be given of the physical parameters on which the properties depend, but no attempt will be made to derive the values, a study which lies within the field of solid state physics rather than metallurgy.

3.2 Structure-Insensitive Properties

Density of the elements. The density of a crystal can be calculated from the masses of the atoms and the geometry of the crystal structure; the latter can, for this purpose, be described by the volume of the unit cell of the structure and the number of atoms per unit cell. These two parameters depend, in turn, on the type of structure (i.e., face-centered cubic, body-centered cubic, etc.) and the ionic radius appropriate to the structure. This has been discussed on p. 34. The densities of the elements are shown in Table 3.2.

Change of volume on melting. The change of volume that occurs on melting or freezing is of practical importance; it is a structure-insensitive property because it is an expression of the difference in density between the solid and the liquid at the melting point. Some values are given in Table 3.3.

Table 3.2. Densities of Some Elements

Element	Density (grams/cc)	Element	Density (grams/cc)	Element	Density (grams/cc)
Aluminum	2.699	Gallium	5.91	Radium	5.0
Antimony	6.62	Germanium	5.36	Rhenium	20.0
Arsenic	5.73	Gold	19.32	Rhodium	12.44
Barium	3.5	Indium	7.31	Rubidium	1.53
Beryllium	1.82	Iridium	22.5	Ruthenium	12.2
Bismuth	9.80	Iron	7.87	Silicon	2.33
Boron	2.3	Lead	11.34	Silver	10.49
Cadmium	8.65	Lithium	0.53	Sodium	0.97
Calcium	1.55	Magnesium	1.74	Strontium	2.6
Carbon		Manganese	7.43	Tantalum	16.6
(Graphite)	2.22	Mercury	13.55	Thorium	11.5
Cerium	6.9	Molybdenum	10.2	Tin	7.30
Cesium	1.9	Nickel	8.9	Titanium	4.54
Chromium	7.19	Osmium	22.5	Tungsten	19.3
Cobalt	8.9	Palladium	12.0	Uranium	18.7
Columbium	8.57	Platinum	21.45	Zinc	7.13
Copper	8.96	Potassium	0.86	Zirconium	6.5

Table 3.3. Change of Volume on Melting *

Element	Increase in Volume	Element	Increase in Volume
Li	1.65	Hg	3.7
Na	2.5	Al	6.0
K	2.55	Ga	−3.2
Rb	2.5	Te	3.2
Cs	2.6	Si	−12.0
		Ge	−12.0
Cu	4.15	Sn	2.8
Ag	3.8	Pb	3.5
Au	5.1	Sb	−0.95
		Bi	−3.35
Mg	4.1		
Zn	4.2		
Cd	4.7		

* From B. R. T. Frost, Chapter 3 in *Progress in Metal Physics 5*, Bruce Chalmers and R. King, Eds., Pergamon Press, London, 1954, p. 98.

Density of alloys. The density of a substitutional solid solution differs from that of the pure metal for two reasons: 1. the average mass of the atoms is changed, and 2. the lattice parameter is changed. If the lattice parameter were unchanged it would be a simple matter to calculate the density of a solid solution; however, the lattice parameter generally changes, usually increasing when an atom of larger size substitutes for one of smaller size, and vice versa. To a fair approximation, the lattice parameter of a solid solution varies linearly with the concentration (in atomic per cent). This is Vegard's law.

The effect of a solute on the lattice parameter is the sum of two components. One is a geometrical effect, by which a large atom causes expansion and a small atom causes contraction. The second effect is electronic; a solute that increases the number of valence electrons per atom causes expansion, and a solute that reduces the number of electrons per atom causes a decrease in the lattice parameter. The lattice parameter that is measured is the mean lattice parameter, since the distortion due to a solute atom is presumably quite local. A more complicated situation can arise when the size of the unit cell cannot be represented by a single lattice parameter. This is the case, for example, in the close-packed hexagonal structure; two separate lattice parameters, a and c, must be considered. In such cases the two parameters do not usually vary proportionately, and the axial ratio (c/a) therefore changes with the concentration of solute. This is attributed to the low symmetry of the Brillouin zones.

Interstitial solid solutions. The lattice parameter of a solvent crystal is always increased by the presence of an interstitial solute; the density, however, may increase or decrease, depending on the relationship between the increase of the lattice parameter and the mass of the dissolved atom. The changes in density caused by interstitial solution are small; some examples are given in Table 3.4.

Table 3.4. **Changes of Density Caused by Interstitial Solutes**

Composition	Density (grams/cc)
Fe	7.874
Fe + 0.06% C	7.871
Fe + 0.23% C	7.859
Fe + 0.43% C	7.844
Fe + 1.22% C	7.830

3.3 Elastic Properties

Elastic constants. The elastic properties are determined by the force necessary to produce a unit change in the geometry of a specimen, and are calculated as the ratio of the stress to the resulting strain.

In the simplest possible case, the elastic behavior would be represented by the slope of the "resultant" curve (Fig. 2.13) near point E. In most cases, the extent of the possible departure from point E that does not cause a permanent change in the relative positions of the atoms is so small that the force-distance curve does not depart measurably from a straight line. In such cases, Hooke's law is obeyed. Recently, however, it has been found that in certain exceptional cases the elastic strain can be so large that a marked departure from Hooke's law is observed; this occurs for strains greater than about 1%.

An important respect in which the elastic behavior of a crystal departs from the ideally simple situation represented in Fig. 2.13 is *anisotropy*. Even in crystals of the highest symmetry (i.e., those with cubic structures) the elastic moduli depend upon the direction in the crystal in which they are measured. A complete description of the elastic properties of a crystal can be approached as follows: Any applied force can be resolved into components of normal (tensile) stress, parallel to the three axes of the crystal, and three shear stresses in the three planes defined by pairs of axes. Similarly, the resulting strain can be resolved into three normal strains and three shear strains. Each component of the stress in general produces some strain in each component of the strain, and the relationship between any component of stress and any component of strain is an elastic coefficient of the crystal. If the normal stresses are represented by X_x, Y_y, and Z_z and the shear stresses by X_y, Y_z, and Z_x, and the strains by e_{xx}, e_{yy}, and e_{zz} (normal strains) and e_{xy}, e_{yz}, and e_{zx} (shear strains), then the strain produced by a given stress is

$$e_{xx} = S_{11}X_x + S_{12}Y_y + S_{13}Z_z + S_{14}Y_z + S_{15}Z_x + S_{16}X_y$$

$$e_{yy} = S_{21}X_x + S_{22}Y_y + \cdots$$

$$e_{zz} = S_{31}X_x + S_{32}Y_y + \cdots$$

$$e_{yz} = S_{41}X_x + S_{42}Y_y + \cdots$$

$$e_{zx} = S_{51}X_x + S_{52}Y_y + \cdots \qquad\qquad S_{56}X_y$$

$$e_{xy} = S_{61}X_x + S_{62}Y_y + \cdots \qquad\qquad S_{66}X_y$$

where $S_{11}, \cdots,$ are the elastic *coefficients*. A series of elastic *moduli* $C_{11}, \cdots,$ are obtained if the *stress* required to produce a given *strain* is calculated. Because of the equivalence of S_{12} and $S_{21}, \cdots,$ there are 21 independent elastic coefficients. The case discussed here is the most general one; for crystals of high symmetry the number is reduced, to 3 for cubic and to 5 for hexagonal structures.

The practical elastic constants, namely Young's modulus, the shear modulus, and the bulk modulus, can be calculated from the elastic coefficients as follows:

Young's modulus = $1/[S_{11} - 2(S_{11} - S_{12} - \frac{1}{2}S_{44})(\alpha^2\beta^2 + \beta^2\gamma^2 + \gamma^2\alpha^2)]$

Shear modulus = $1/[S_{44} + 4(S_{11} - S_{12} - \frac{1}{2}S_{44})(\alpha^2\beta^2 + \beta^2\gamma^2 + \gamma^2\alpha^2)]$

Bulk modulus = $1/[3(S_{11} + 2S_{12})]$

It follows that Young's modulus and the shear modulus depend upon the direction in which they are measured, represented here by α, β, and γ, the three direction cosines of the stress axis with respect to the major crystallographic axes.

It will be seen that the extent and sign of the anisotropy of Young's modulus and the shear modulus depend upon the value of $(S_{11} - S_{12} - \frac{1}{2}S_{44})$. When this is positive, Young's modulus is maximum in the $\langle 111 \rangle$ direction and minimum in the $\langle 100 \rangle$ direction; if it is zero these properties are isotropic.

Table 3.5 gives values for the elastic properties of some metals, and shows the extreme values of Young's modulus and the shear modulus for single crystals of these materials.

It will be evident from the foregoing discussion that there should be no difference between the Young's modulus of a particular material when it is measured in tension and when it is measured in compression. It has sometimes been reported that these values are different; careful measurements have shown that this is erroneous, and is probably due to the nonelastic behavior of the material or to the use of different instruments for making the two tests.

Elastic properties of polycrystalline aggregates

SINGLE-PHASE ALLOYS. The elastic properties of a single-phase aggregate consisting of many crystals may be calculated from the values for the single crystal if the aggregate is assumed to be isotropic. This demands a random distribution of the orientations of the crystals. The measured values for these elastic constants, given in Table 3.5, agree closely with those calculated on the basis of random orientation.

The elastic properties of solid solutions and intermediate phases differ from those of the pure metals; the variation of Young's modulus

STRUCTURE-INSENSITIVE PROPERTIES

Table 3.5. Young's Modulus and Shear Modulus for Some Metals

	Young's Modulus, 10^6 psi			Shear Modulus, 10^6 psi		
Metal	Max.	Min.	Polycrystalline	Max.	Min.	Polycrystalline
Al	11.0	9.1	10.0	4.1	3.5	3.9
Cu	27.9	9.7	16.1	13.9	4.5	6.6
Ag	16.7	6.2	10.4	6.4	2.8	4.2
Pb	5.6	1.6	2.3	2.1	0.7	0.90
αFe	41.2	19.2	30.0	16.9	8.7	12.0
W	56.5	56.5	56.5	22.0	22.0	22.0
Mg	7.4	6.3	6.3	2.6	2.4	2.5
Zn	18.0	5.0	14.5	7.1	4.0	5.6
Cd	11.8	4.1	7.2	3.6	2.6	2.8
Sn	12.4	3.8	6.6	2.6	1.5	2.4

of α brass (solid solution of zinc in copper), with composition, is given in Fig. 3.1.

POLYPHASE ALLOYS. The engineering importance of high values for Young's modulus and the shear modulus, in relation to density, has led to attempts to increase these properties. In the case of steel, the high anisotropy of iron suggests the possibility of using preferred orienta-

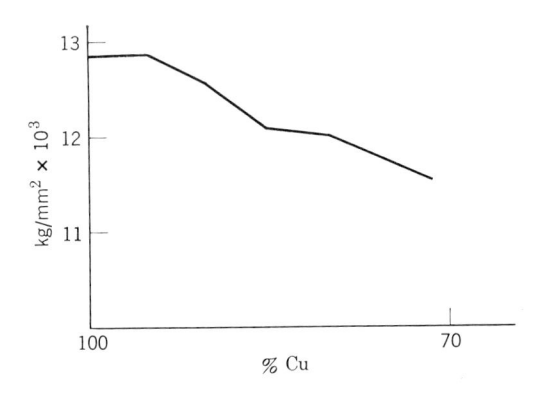

Fig. 3.1. Variation of Young's modulus with composition for solid solutions of zinc in copper. From *Metals Handbook*, American Society for Metals, Cleveland, 1948, p. 910.

tions; this would present serious technical problems. For aluminum alloys a more promising approach is to raise the elastic moduli by increasing the content of intermediate phases of high modulus. Table 3.6 shows some values that have been achieved in this way. It must be pointed out that some of the other, equally essential, properties showed marked deterioration in these alloys with high moduli.

Table 3.6. Young's Modulus of Some Aluminum Alloys *

\multicolumn{7}{c}{Composition}	Young's Modulus, 10^6 psi
Ni	Si	Cu	Mn	Cr	V	
	5.0					10.1
	10.25					10.8
	15.0					11.2
	21.3					11.4
	12.7		4.16	1.07		12.6
		4.46	3.76	0.90		11.40
		8.43			0.88	10.55
1.12			9.45			12.5
8.36					1.01	11.75
15.8	13.4					13.4

* Data from N. Dudzinski, *J. Inst. Metals*, Vol. 81, 1952, p. 50.

3.4 Melting Point

Pure metals. The melting point of a crystal is the temperature at which it is in equilibrium with the liquid form of the same material. It is the temperature T_E discussed in Section 2.3 of Chapter 2, when the two phases are the solid and liquid form of a pure metal. The condition for equilibrium as derived on p. 19 shows that $\ln (B_a/B_b) = (Q_a - Q_b)/kT_E$.

In the case now under discussion, B is defined by the accommodation coefficients, vibration frequency, and a geometrical factor for each phase. The most important variable is the accommodation coefficient, and this should be the same for all liquids, and it should have a constant value for any particular crystal structure. The ratio B_a/B_b should therefore have a constant value for each crystal structure; the value $(Q_a - Q_b)/kT_E$ should therefore also be constant for a given crystal structure. $Q_a - Q_b$ is equal to the internal energy difference per

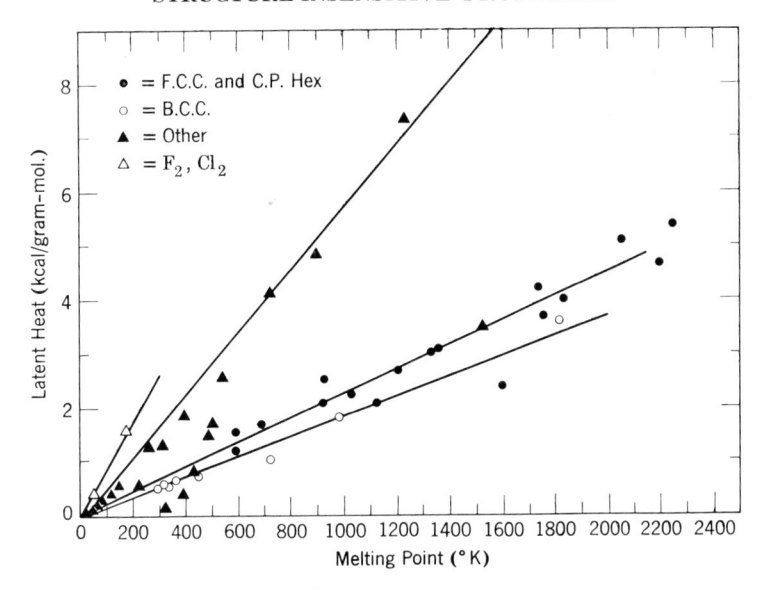

Fig. 3.2. Latent heat of fusion as a function of the absolute melting temperatures of various elements.

atom between the liquid and the crystal, and is therefore L_F, the latent heat of fusion. L_F/kT_E should therefore be a constant for a crystal structure, or $L_F = pT_E$ where p is a constant that depends only on the structure. Figure 3.2 shows the relationship between T_E and L_F for the metals.

Alloys. The liquidus and solidus lines in an equilibrium diagram define the "melting point" of a binary alloy; for a ternary alloy the liquidus and solidus are surfaces, and for systems of more than three components they cannot be represented geometrically. The data corresponding to these equilibrium conditions, as well as to other regions of equilibrium diagrams, are not structure-sensitive; this follows from the definitions implicit in the term *equilibrium* diagram. The phases that are actually present after a particular sequence of events may not be those characteristic of equilibrium.

3.5 Thermal Expansion

Variation of interatomic distance with temperature. The discussion of interatomic forces on p. 31 did not take thermal agitation into account. If we now consider the effect of temperature, we may suppose that the atoms, instead of being at rest, are vibrating about their mean positions. We will consider whether this motion contributes to the forces that determine the interatomic spacing at which the forces balance. If the resultant force-distance relationship of Fig. 2.13 were strictly linear, then the time average of the force on a vibrating atom would be equal to the value that the force would have if the atom were at rest. However, examination of Fig. 2.13 will show that the resultant curve can only be linear over a very short range (Hooke's law is obeyed for displacements of 0.1%, but not for 1%). The amplitude of thermal agitation, however, reaches a value of about 12% of the interatomic spacing at the melting point, and decreases at lower temperatures. It therefore corresponds to a range of positions in which the force-position curve is not linear, but becomes *steeper* (i.e., force increases more rapidly with displacement) as the interatomic distance decreases, and becomes *flatter* as the atoms move farther apart. The effect of vibration, therefore, is to increase the repulsive force by a greater amount than the attractive force, when averaged over a cycle of vibration. This increases the mean distance between the centers of neighboring atoms by an amount that increases with rising temperature; the result is thermal expansion.

The thermal expansion of a cubic crystal is isotropic, and a single parameter is therefore sufficient to describe it for a given substance at a given temperature. This is the linear coefficient of thermal expansion β given by the expression $l = l_0 (1 + \beta t)$, where l_0 and l are the lengths at the temperatures a and $a + t$. For many purposes a is taken as room temperature.

In crystals with noncubic structures, the thermal expansion is frequently anisotropic, and is described by two or three parameters.

Relationship between thermal expansion and melting point. The thermal expansion coefficient is related to the latent heat of fusion and the melting point in the following way: It has been shown that the equilibrium temperature (measured from absolute zero) is proportional to the latent heat of fusion per atom, fairly precisely for any crystal structure such as face-centered cubic, and approximately for all metals. The melting point, therefore, is reached when the value of kT has reached some definite fraction of the latent heat of fusion. The energy of thermal agitation is given by kT.

The latent heat of fusion is proportional to the latent heat of vaporization, and this in turn is the work done in separating the atoms from their positions in the crystal until they no longer interact. This is given schematically by the hatched area in Fig. 3.3. The amplitude of thermal agitation can be represented in this diagram by the line AB. If the force-distance curve is similar for different crystals, the

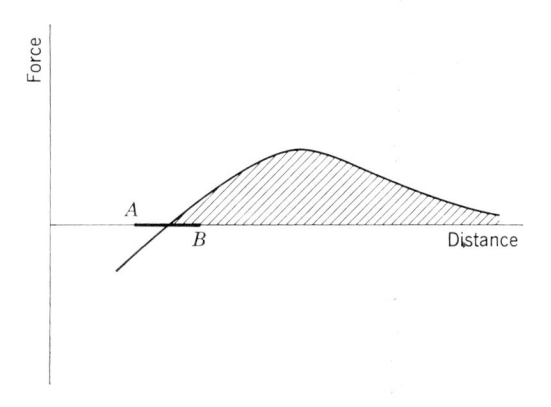

Fig. 3.3. Amplitude of thermal agitation (*AB*) in relation to the force-distance curve.

amplitude *AB* at the melting point should be a constant fraction of the interatomic distance. Measurements show that it is about 12% for many crystals. This value should correspond approximately to the same total thermal expansion from absolute zero to the melting point, irrespective of the value of the melting point. The thermal expansion per degree should therefore vary inversely as the melting point. Table 3.7 gives some values for the metals. It will be seen that, apart from a few values that fall right out of line, possibly due to incorrect data, there is reasonably good consistency among the face-centered cubic metals; calcium undergoes a phase change and therefore the phase for which the melting point is measured is not the one for which the coefficient of expansion is given. The close-packed hexagonal metals show less good agreement, probably because of their anisotropy and the resulting range of values for α. The extreme values are given for zinc. The body-centered cubic metals fall into two groups; the alkali metals, which have consistently high values for α, probably because these have the "soft" type of structure discussed on p. 32, and the transition metals, which have values comparable with the face-centered cubic metals. The metals with the less close-packed structures have low values of α in relation to the melting point, except for manganese, which undergoes phase changes before the melting point is reached.

Abnormal thermal expansion. There are several instances in which the thermal expansion does not follow the simple behavior outlined above.

Table 3.7. Relation between Melting Point and Thermal Expansion

Structure	Metal	M.P.	$\alpha \times 10^{-6}$ cm/cm/°C	$\alpha \times$ M.P. $\times 10^3$
F.C.C.	Cu	1356	16.5	22.5
	Ag	1233	19.7	18.9
	Au	1336	16.2	18.8
	Pt	2046	8.9	18.3
	Ir	2727	6.8	18.6
	Rh	2239	8.3	18.3
	Pd	1827	11.8	21.5
	Al	933	23.9	22.3
	Ni	1728	13.3	23.0
	Ca *	1123	22.0	25.0
	Pb	600	29.3	17.6
	Th	2073	11.1	21.2
C.P. Hex	Cd	594	29.8	17.8
	Zn	692	15.0	10.4
			61.8	42.5
	Mg	923	26.0	23.1
	Be	1553	12.4	15.8
	Co	1768	12.3	21.7
	Os	2973	4.6	13.7
	Te *	573	11.1	21.2
	Ti *	2193	8.5	18.6
	Zr *	2023	5.0	10.2
B.C.C.	Li	459	56	25.6
	Na	371	71	26.2
	K	336	83	27.7
	Rb	312	90	28.0
	Cs	301	97	29.2
	V	2005	7.8	15.7
	Cr	2163	6.2	13.2
	Fe *	1612	11.7	18.8
	Cb	2688	71	19.1
	Mo	2898	4.9	14.2
	Ta	3269	6.5	21.2
	W *	3783	4.3	16.3
Other	Sb	903	9.0	8.1
	Bi	793	13.3	10.5
	Ga	303	18.0	5.5
	In	429	33.0	14.1
	Sn	504	23.0	11.6
	Mn *	1518	22.0	33.5

* Denotes allotropic change between room temperature and the melting point.

THERMAL EXPANSION OF URANIUM. The crystal structure of uranium at room temperature (α uranium) is orthorhombic; it is extremely anisotropic, and although its *volume* expansion coefficient is not abnormal, one of the three linear coefficients is negative. The values of the three coefficients are $\alpha_1 = 33 \times 10^{-6}$, $\alpha_2 = -6.5 \times 10^{-6}$, $\alpha_3 = 17 \times 10^{-6}$.

INVAR. The discussion of the structure and properties of crystals has so far taken into account only the electrostatic forces between the atoms and electrons comprising the crystal. It is necessary for some purposes, however, to consider the effects that arise as a result of the magnetic moment associated with the atoms of some metals. Every electron in an orbit in an atom has a magnetic moment as a result of its spin. In most cases, each electron in a particular orbit has a partner with opposite spin, and the magnetic fields effectively cancel each other. In the cases of iron, nickel, cobalt, manganese, and gadolinium, however, there is an unpaired electron in the 3s shell, and the atom therefore has a magnetic moment.

As a result of the interaction of these magnetic moments, the spins tend to be aligned and the atoms exert forces of magnetic origin on each other unless the thermal agitation is sufficient to destroy the alignment of the spin axes. The temperature at which the alignment disappears is the Curie temperature, at which the external ferromagnetic effects become zero. One effect of the magnetic forces between the atoms is to change their equilibrium spacing in the case of iron; the spacing is *increased* when the spins are aligned, as a result of the repulsion of similar magnetic poles.

If the temperature of an alloy is raised from below to above its Curie temperature, it will undergo the normal thermal expansion, superimposed on the contraction due to the disalignment of the spin axes. The resultant thermal expansion in the temperature range around the Curie point may be either positive, zero, or negative, according to the composition of the alloy.

One of the most useful compositions for this purpose is iron 63%, nickel 36%, known as Invar, of which the thermal expansion curve is given in Fig. 3.4.

3.6 Magnetic Properties

Ferromagnetic properties are caused by the existence, in certain atoms, of an unpaired electron, as discussed in the section above.

The magnetic moment per atom is a function of the atoms that are present; the interaction between the atoms, which tends to cause align-

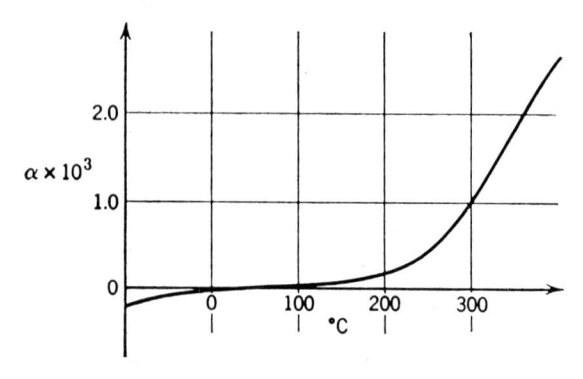

Fig. 3.4. Thermal expansion of Invar. From *The Science of Engineering Materials*, J. E. Goldman, Ed., John Wiley & Sons, New York, 1957, Fig. 5.13, p. 125.

ment, depends upon the composition of the alloy, because this affects the equilibrium separation of the ferromagnetic ions. Depending on their separation, the magnetic moments of all the atoms may be parallel, in which case the crystal is *ferromagnetic,* or they may be alternately in opposite directions, in which case there is no external magnetic effect and the material is *antiferromagnetic.*

The maximum possible amount of magnetization is achieved if all the atoms have their spins aligned. This, *the saturation intensity of magnetization,* is a structure-insensitive property. The actual intensity of magnetization that is produced by an external field, or that persists in the absence of one, is, however, sensitive to the subcrystalline structure and will be discussed in Chapter 5.

3.7 Electrical Properties

Electricity flows in metals by the motion of electrons; when the conductor is metallic, there is no limitation to the energy of an electron, and no resistance should appear in a perfectly periodic crystal. This would only be the case for zero thermal motion, since the thermal vibrations of the atoms constitute an instantaneous departure from perfect periodicity. Hence the actual resistivity has a component that increases with temperature and is a direct result of thermal agitation. This component of the resistivity is not structure-sensitive; it would account for the whole of the resistance in a perfect crystal of a pure metal. Any departure from perfection gives rise to another component of the resistance; this is the predominant part of the resistance that is observed at very low temperatures, and is known as the residual resistivity.

The electrical resistivity is, like the thermal expansion, isotropic in crystals of cubic structure but is markedly anisotropic in hexagonal and other noncubic crystals. The electrical resistivity at ordinary temperatures is considerably affected by the presence of solutes and second phases. For example, the resistivity of fairly pure aluminum is about 2.68 microhm-cm, while the addition of 0.5% of impurities raises the value to 2.80 microhm-cm; the addition of 2.25% of magnesium causes an increase to 5.3 microhm-cm.

All metals have a positive temperature coefficient of resistivity; this is because the resistivity of metals arises from the scattering of electrons from the positive ions, and increases with increasing thermal agitation. However, the temperature coefficient is strongly influenced by composition and is extremely small for some alloys; in general, the temperature coefficient of resistivity decreases as the resistivity increases, presumably because the disturbance introduced by impurities and solutes is of the same kind as that due to thermal agitation, which therefore becomes less effective.

Other electrical properties that are of importance are the thermoelectric properties, used in thermocouples, and the dependence of resistivity on elastic strain, used in resistance strain gages.

3.8 Further Reading

J. E. Goldman, Ed., *The Science of Engineering Materials,* John Wiley & Sons, New York, 1957.

W. Boas and J. K. Mackenzie, "Anisotropy in Metals," Chapter 3 in *Progress in Metal Physics 2,* Bruce Chalmers, Ed., Butterworths Scientific Publications, London, 1950.

4

Imperfections in Crystals

4.1 Types of Imperfection

This chapter is a discussion of the geometrically possible ways in which a crystal can depart from the ideal structure defined on p. 21. It might be thought that the limiting case of an imperfect crystal is the liquid; but it is found that there is a limit, far short of the liquid, to the amount of imperfection that a crystalline solid will tolerate. At this limit, the crystal structure is still recognizable, and the energy has only been increased by a few per cent of the latent heat of fusion. It is therefore valid to discuss "imperfect crystals" as aggregates of atoms that retain, with only minor changes, the crystal structure and all the structure-insensitive properties.

All possible types of imperfection can be divided, on a purely geometrical basis, into point, line, and surface imperfections. The imperfections may also be classified as thermodynamically stable and thermodynamically unstable imperfections. The former are present because the increase in entropy due to an imperfection opposes, and may outweigh, the increase in internal energy associated with it. The equilibrium content of such imperfections will increase with temperature; the actual content may not coincide with the equilibrium value.

The unstable imperfections are those which increase the free energy; they would be completely absent in the equilibrium condition. Their presence constitutes a departure from equilibrium whose existence and extent are a result of the previous history of the crystal. The *origin* of the imperfections will be discussed in the appropriate places in the following chapters; their geometrical characteristics will be described here.

Since this book is primarily concerned with metals, attention is restricted to those types of imperfections that are related to the positions of atoms, and not to the distribution of positive and negative charges.

4.2 Point Imperfections

The possible point defects can be divided into:

1. Lattice vacancies.
2. Interstitial atoms.
3. Substitutional atoms.
4. Complex point defects.

Lattice vacancies. A certain proportion of lattice sites are unoccupied in a crystal in thermal equilibrium, because the increase in energy due to the vacancies is compensated by an increase in entropy resulting from the increase in randomness of the structure.

Let E_s be the energy required to move an atom from the interior of a crystal to the surface; then the increase in the internal energy U due to the presence of n vacancies is nE_s.

The increase in entropy can be approached as follows: n atoms can be selected out of a total of N (the number in the crystal) in $N!/[(N-n)!n!]$ different ways. The entropy S due to n vacancies in a crystal containing N sites is therefore

$$S = k \log \frac{N!}{(n-n)!n!}$$

The change in free energy due to these vacancies is

$$F = U - TS = nE_s - kT \log \frac{N!}{(N-n)!n!}$$

or, using standard approximations,

$$F = nE_s - kT[N \log N - (N-n) \log (N-n) - n \log n]$$

whence
$$\left(\frac{\partial F}{\partial n}\right)_T = E_s - kT \log \frac{N-n}{n}$$

The equilibrium condition is that F is a minimum, or that

$$\left(\frac{\partial F}{\partial n}\right)_T = 0, \quad \text{when } \log \frac{n}{N-n} = -\frac{E_s}{kT}$$

When n is small compared with N, this becomes

$$n = N \exp(-E_s/kT)$$

Using the value of 1 ev for E_s, the value of n/N is about 10^{-5} at 1000°K; or one site in about 100,000 is unoccupied. The ratio n/N

varies exponentially, as shown in Fig. 4.1, which represents the variation of vacancy content with temperature for copper. The value of 1 ev for E_s is based on indirect experimental information but is unlikely to be seriously in error.

The actual vacancy content may be very different from the equilibrium value; if the temperature is changed, causing a change in equilibrium concentration, a substantial time may elapse before the number of vacancies again reaches equilibrium. An excess of vacancies may also be introduced by a variety of other processes that will be discussed at the appropriate places.

The foregoing remarks contain the implicit assumption that the vacancy content of a crystal can change. This requires the movement of vacancies, into and out of the crystal. The unit process by which this happens is the movement of an atom into an adjoining vacancy, which results in the exchange of positions between the atom and the vacancy. A sequence of such "jumps" allows a vacancy to move about within the crystal. A detailed discussion of the frequency with which such jumps occur, and of the influence of temperature and other conditions, is given in Chapter 6.

Interstitial atoms. It is possible for an atom to be moved from its lattice site into an interstitial position. It is thought that the energy of an atom of, for example, copper in an interstitial position in a copper crystal is so high that it occurs spontaneously only extremely rarely. This is because the interstitial spaces in the face-centered cubic crystal of copper are very small compared with the copper atoms and very severe local distortion would be necessary to permit the insertion of an extra atom at such a site. The energy necessary to produce this distortion is, however, available during plastic deformation

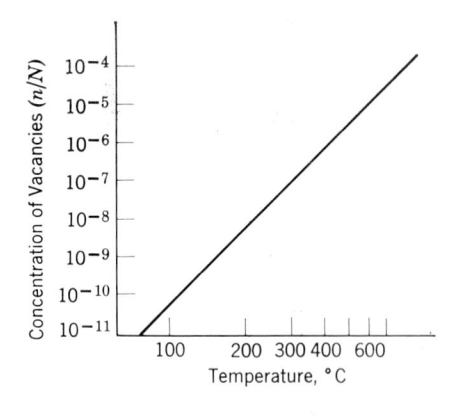

Fig. 4.1. Equilibrium vacancy content for copper as a function of temperature.

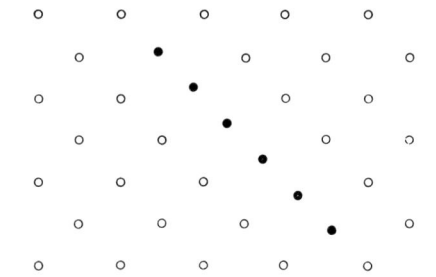

Fig. 4.2. Crowdion.

and as a result of the collision of high energy particles ("radiation damage").

It has already been pointed out (p. 53) that interstitial solid solutions only form when the solute atom is small compared with the solvent, and that the equilibrium content is small. There are some important cases to be discussed in Chapter 8 in which a very much greater interstitial solute content is obtained. These considerations apply to the close-packed structures.

In the particular case of the alkali metals, it may be possible for an extra atom to be inserted in a row of atoms whose centers lie in a close-packed $\langle 110 \rangle$ direction. This would be equivalent to an interstitial atom that is accommodated by the displacement of atoms in a single line only. This type of interstitial atom is called a "crowdion." It is shown schematically in Fig. 4.2. In the more "open" types of crystal, such as those with the diamond-cubic structure, the interstices are relatively much larger, and therefore much larger interstitial atoms are to be expected. It is believed, for example, that copper dissolves interstitially in silicon.

Substitutional atoms. The substitution of a different atom for an atom of a crystal constitutes a type of imperfection; the conditions for the existence of substitutional solid solutions were discussed in Chapter 2.

Complex point defects. There is some evidence that groups of two or more vacancies may be stable in metals, and that a vacancy and a solute atom (either substitutional or interstitial) may form a stable pair. The existence of such "complex point defects" is not firmly established but is regarded as extremely probable.

4.3 Line Imperfections

Dislocations. Although a number of point defects may exist in a line or in a plane, there is nothing in their geometry that makes it necessary that they should do so; they can exist and move independently, and they remain point defects even when arrayed in a line or in a plane. On the other hand, the line defects can exist only when they have extension in one dimension.

A line defect can be defined as follows: In a perfect crystal, a closed circuit can be described by following a path such as $ABCD$ (Fig. 4.3a), in which AB consists of n steps each of which is a vector characteristic of the crystal, in this case a nearest neighbor interatomic distance along the x axis. The distance BC is m steps in the y direction; CD and DA are $-n$ and $-m$ steps in the x and y directions respectively.

If such a loop fails to close exactly, as shown in Fig. 4.3b, it encloses a region where a line imperfection intersects the plane of the diagram. The failure to close, AA', is characteristic of the line imperfection and is known as the *Burgers vector* **b**. The case shown in Fig. 4.3 is one in which the Burgers vector is in the plane of the diagram, but the failure to close the circuit can involve a departure from this plane.

Figure 4.3 represents an oversimplified type of crystal structure, the simple cubic structure, which probably only exists for the metal polonium. However, the major concepts of the line imperfection or dislocation are more readily described in this simple case. The extension of these ideas to the more common structures will be discussed later.

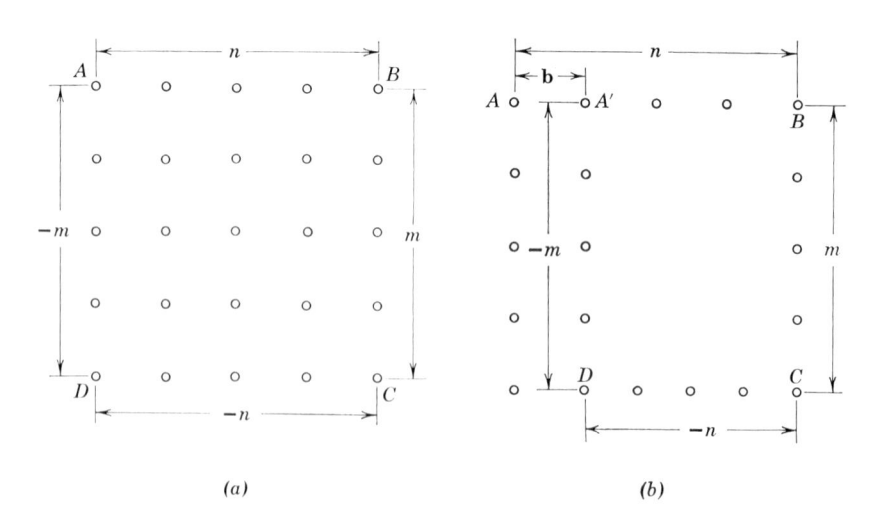

Fig. 4.3. (a) Closed circuit corresponding to a perfect crystal. (b) Unclosed circuit corresponding to a dislocation of Burgers vector **b**.

IMPERFECTIONS IN CRYSTALS 97

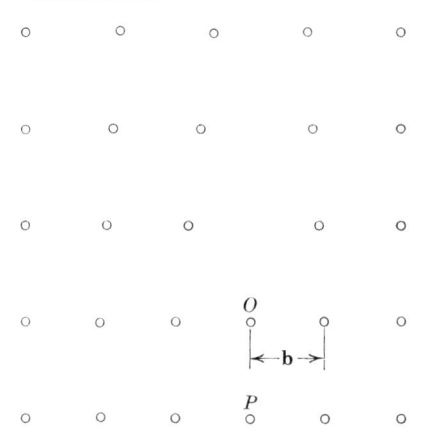

Fig. 4.4. Edge dislocation.

The area within the circuit of Fig. 4.3b can be filled in as shown in Fig. 4.4. The significant feature of this arrangement is the termination of a row of atoms at O. If a crystal is built up of a series of planes identical with that of Fig. 4.4, it contains a Taylor or *edge dislocation*, the line of which is the normal at O, and the Burgers vector is **b**. If a "perfect" layer (Fig. 4.3a) is superimposed on a series of planes like those of Fig. 4.4, there must be a misalignment between the imperfect and the perfect layers. There must therefore be a dislocation running parallel to the plane of the diagram, and forming an extension of the original edge dislocation. The component of the dislocation that is parallel to the plane of the diagram may run in any direction, but it can be regarded as being resolved into components perpendicular (OY direction) and parallel (OX direction) to the Burgers vector of the original dislocation. If the direction is OY, it terminates the "half plane" OP in the OY direction; this is then a "quarter plane." If the dislocation runs in the X direction, i.e., parallel to its Burgers vector, the dislocation changes its character and is illustrated in Fig. 4.5, in which planes close to, but *in front of*, the plane of the diagram are shown by broken lines, and planes close to, but *behind*, the plane of the diagram are represented by full lines. In the left-hand part of the diagram, the lines coincide, and the planes of atoms are continuous. In the right-hand part of the diagram, the planes are continuous both well above and well below the line OX, but each is continuous with the adjacent plane and not the same plane. OX is the axis of a *screw dislocation*, or Burgers dislocation. The name "screw dislocation" results from the fact that a continuous spiral path can be described around the screw dislocation so that the path does not at any point move from one atom plane to another. This

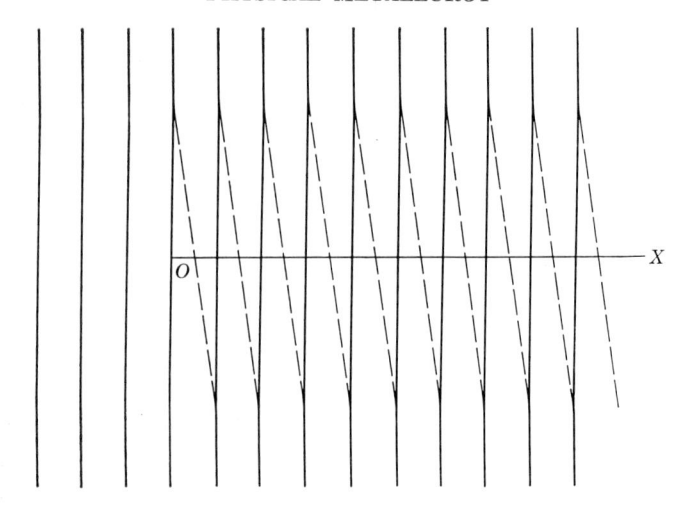

Fig. 4.5. Screw dislocation.

is illustrated in Fig. 4.6. The essential characteristic of the screw dislocation is that its Burgers vector is parallel to its axis. The "failure to close" of the circuit would be normal to the plane of Fig. 4.3b.

An alternative description of a dislocation is as follows:

A dislocation may be defined as the disturbed region between two substantially perfect parts of a crystal that are incompatible with each other. A dislocation may be described as follows: a crystal $ABCDEFGH$ (Fig. 4.7) is cut in the plane $PQRS$ so that the cut ter-

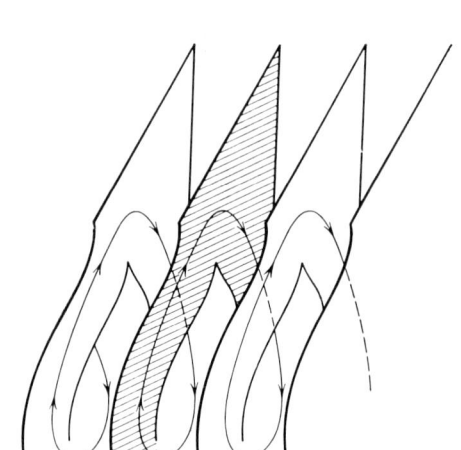

Fig. 4.6. Continuity of planes in a screw dislocation.

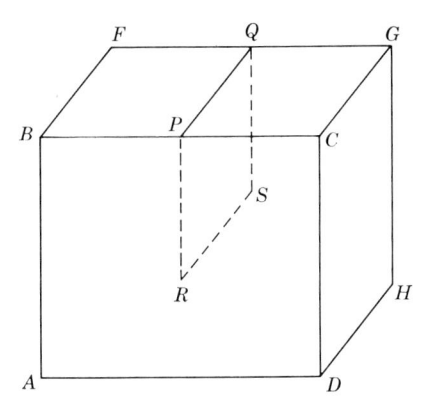

Fig. 4.7. Formation of a dislocation by cutting in the plane $PQRS$ and shearing in the direction RP or RS.

minates along the line RS. The portion of the crystal $BFQP$ is now uniformly sheared relatively to $PQGC$, by an amount, and in a direction, denoted by the Burgers vector **b**, which is in the plane $PQRS$. If the Burgers vector is in the plane $PQRS$ and normal to RS, the dislocation RS will be an edge dislocation; if it is parallel to RS, the result will be a screw dislocation.

To have any significance in relation to crystals, **b** must be a vector representing the translation from one atom site in the original crystal to another. The two parts of the crystal, which have common planes and rows of atoms, are joined together to form a crystal in which there is no longer a cut. Near the line RS, however, it is not possible to reform a perfect crystal because the part below RS still has its original alignment, while the part above has the equally perfect, but different, new alignment. The region surrounding the line PQ is a dislocation.

A dislocation can terminate only at the surface of a crystal, but it can form a closed loop within a crystal. This would be the case if the "cut" in Fig. 4.7 were entirely within the crystal; its periphery would then be a line, which would become a closed dislocation loop when the material on one side of the cut was translated with respect to the material on the other side.

Another important theorem in connection with dislocations is that the Burgers vector must be the same at all points on a dislocation line; it also follows that if three or more dislocations meet at a point the Burgers vector for any one must equal the vector sum of the Burgers vectors of the other dislocations that meet it. A dislocation need not be either perpendicular or parallel to its Burgers vector, but it is always possible to resolve it into components that are either of a pure edge or pure screw character.

It follows from these geometrical characteristics of dislocations that the dislocations in a crystal can form a three-dimensional network, consisting of dislocation *segments* which meet at dislocation *nodes*. This pattern has been described as a "three-dimensional chicken wire." There is experimental evidence, of the type to be discussed in Section 4.3 under "Decoration," for the existence of this kind of dislocation network.

Energy of a dislocation. It has already been pointed out that a crystal structure exists because it corresponds to a minimum free energy arrangement of the atoms. It follows that the energy of a crystal containing a dislocation is higher than that of a perfect crystal. It is therefore reasonable to speak of the energy of a dislocation. This is believed to be of the order of 1 ev per atom plane. It can be shown that the energy of a dislocation in a given crystal is proportional to the square of its Burgers vector, and, approximately, inversely proportional to the interplanar spacing normal to the Burgers vector. The energy may be associated with the fact that, in order to generate a dislocation, the crystal must be distorted elastically, and therefore work must be done on the crystal. This work is not recovered when the crystal is rejoined; it is only the work associated with cutting the crystal (i.e., producing two new surfaces within it) that is recovered when the two surfaces are annihilated by rejoining after the displacement.

The energy corresponds to the strain of the crystal. In the case of an edge dislocation, there are regions of compression at C in Fig. 4.8, where each atom occupies less than its normal volume, and regions of tension at T, where each atom occupies more than its normal space. There are also regions of shear strain, as at S, where the angles are

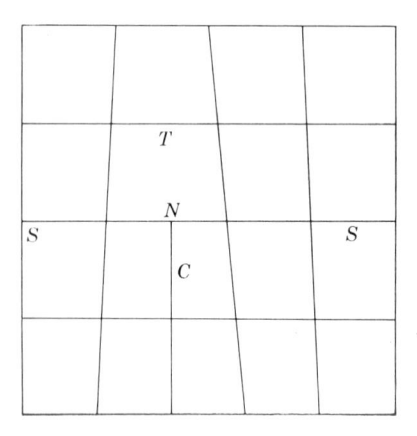

Fig. 4.8. Types of elastic strain near an edge dislocation.

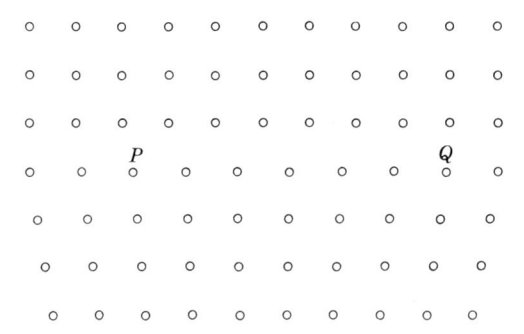

```
   o  o  o  o  o  o  o  o  o  o  o

   o  o  o  o  o  o  o  o  o  o  o

   o  o  o  o  o  o  o  o  o  o  o
            P                    Q
   o  o  o  o  o  o  o  o  o  o  o

     o  o  o  o  o  o  o  o  o  o

      o  o  o  o  o  o  o  o  o  o

       o  o  o  o  o  o  o  o  o  o
```

Fig. 4.9. Width PQ of a dislocation.

changed. There is also a region at N where the strain cannot be described as elastic strain of the perfect crystal, because the number of nearest neighbors is modified, as well as their distances.

It is usual to represent an edge dislocation by means of the symbol $\perp$, which represents the position at which a plane of atoms in the compression half of the crystal terminates. This symbol is somewhat misleading, because the actual dislocation is spread over a number of atom rows, as shown in Fig. 4.9. The distance, in the plane PQ, in which there is appreciable disturbance is the width of the dislocation.

It has been pointed out that the presence of dislocations increases the energy of a crystal. It is also evident that a crystal containing a dislocation is less ordered, and therefore has greater entropy, than a perfect crystal. The increase of entropy, however, is so small that the free energy is raised, by the increase of energy, more than it is lowered by the increase of entropy. It must therefore be concluded that dislocations, while geometrically possible, are not thermodynamically stable.

The energy associated with a dislocation is proportional to its length. It may therefore be considered to exert a tension force along its length; this line tension T is given by $T \approx G\mathbf{b}^2$, where G is the appropriate shear modulus and $\mathbf{b}$ the Burgers vector.

Motion of dislocations. One of the most important properties of a dislocation is its ability to move. There are, in general, two ways in which a dislocation can change its position: it may move in response to a shear stress (slip), or it may move by acting as a source or a sink for vacancies (climb).

Motion by slip takes place as follows: Consider the schematic edge dislocation of Fig. 4.10a. If it is acted upon by a shear stress in the direction indicated in the diagram, it will distort to the shape shown

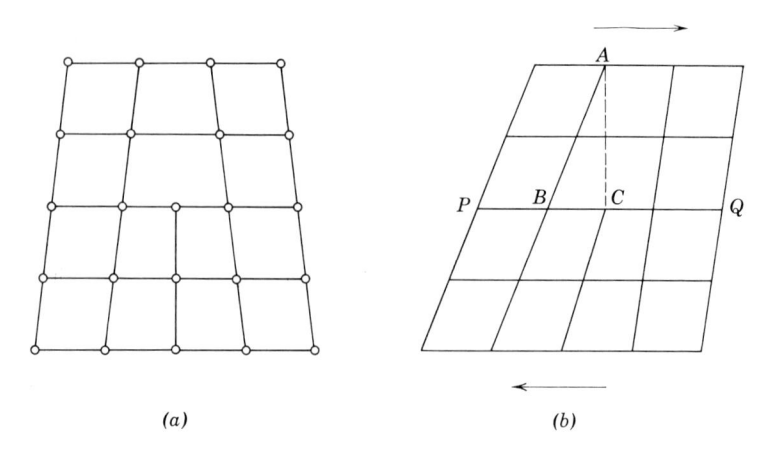

Fig. 4.10. Motion of a dislocation by slip.

(exaggerated) in Fig. 4.10b. Elastic strain is relieved if the line AB moves to the position AC; this is equivalent to moving the dislocation *from C to B*, and it corresponds to some nonelastic shear strain of the crystal. This type of motion can only take place in the plane that contains the Burgers vector and the line of the dislocation. It is the plane PQ in the diagram. In a similar way, a screw dislocation can move by slip; however, in this case the Burgers vector is parallel to the line of the dislocation and it can therefore slip in any plane in which it lies.

An edge dislocation can move normally to its slip plane only by the process known as "climb." If a vacancy arrives at point A in Fig. 4.11a, the dislocation moves to B; this must, of course, occur in all

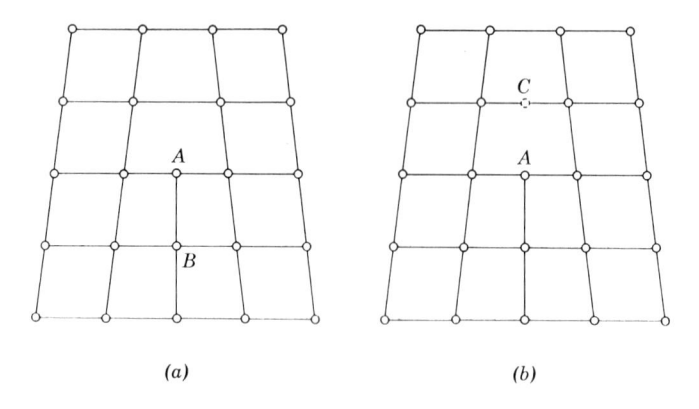

Fig. 4.11. Motion of a dislocation by climb.

parallel planes for the dislocation to climb as a whole. The converse process is that in which an atom arrives at (or a vacancy leaves) point C in Fig. 4.11b; the dislocation then climbs (in this plane) to C.

Complete and partial dislocations. In the discussion so far, the Burgers vector has been implicitly limited to values which correspond to the translation of an atom to a point normally occupied by another atom. There are an infinite number of possible vectors in a crystal, each of which represents the motion of an atom to another atom site; it is not necessary to consider Burgers vectors that are integral multiples of smaller ones because the energy of n dislocations of minimum Burgers vector is less than that of one dislocation with a Burgers vector n times as large. Hence we need only consider dislocations of which the Burgers vector corresponds to adjacent or nearly adjacent atom sites.

An assumption implicit in the foregoing discussion is that the translation of part of the crystal by the Burgers vector brings it to a position where it matches exactly with the part assumed to be at rest, so that a perfect crystal exists on each side of the dislocation. This assumption is valid for, and in fact defines, a *complete* or *perfect* dislocation. This would be true for *any* dislocation in a simple cubic crystal. However, in crystal structures that are more complicated than the simple cube, there are possible atom sites that would give Burgers vectors that do not restore a perfect crystal, and yet represent a stable end point for the translation that produces a dislocation. The most

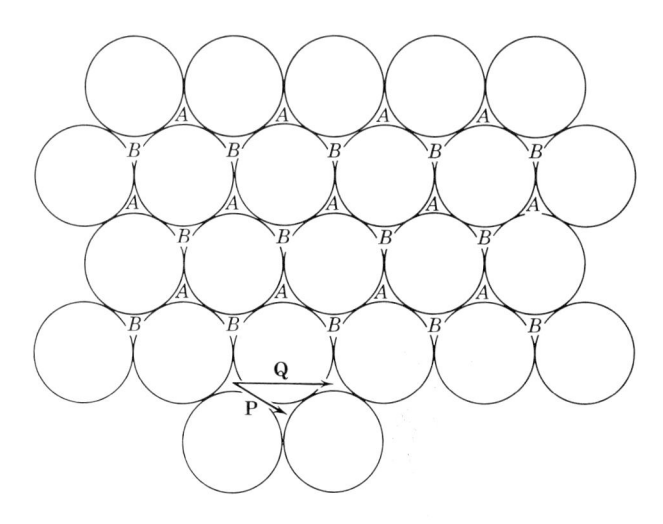

Fig. 4.12. Burgers vectors for partial dislocations.

important example of this is in the face-centered cubic structure, discussed in detail in Chapter 2, p. 26. The crystal is built up of a stack of similar layers, Fig. 4.12, the positions of the layers being such that if the atoms of the first layer *below* the plane of the diagram lie at points $B, B, \cdots$, then the atoms of the first layer *above* would be at points $A, A, \cdots$.

The Burgers vector for a complete dislocation in this structure would be represented by **Q** (Fig. 4.12). It represents a translation from an A position to another adjoining A position, and is in a $\langle 110 \rangle$ direction.

However, the translation of an atom from an A position to another A position is achieved with the least potential energy if it moves through a B position. The B position is one of relative stability, the energy of the atom at points along its path from A to A' being shown schematically in Fig. 4.13. The actual energy at the B position varies from one metal to another.

The B position is energetically less favorable than the A position, but it corresponds to a much shorter Burgers vector. A dislocation with such a Burgers vector (e.g., **P** in Fig. 4.12) is a *partial dislocation*. The Burgers vector is in a $\langle 112 \rangle$ direction. The crystal on one side of a partial dislocation is perfect, but on the other side there is an incorrect sequence of planes, described as a "stacking fault," the sequence of planes being △△△▽△△ $\cdots$. It will be seen that a complete dislocation in a face-centered cubic crystal may decrease in energy by dissociating into two partial dislocations, separated by a stacking fault. The separation of the partials depends upon the energy of the stacking fault.

The composite dislocation is represented in Fig. 4.14, in which the original crystal is represented by the sites X on the left-hand part of the diagram, and the new (perfect) part is shown by the sites X at the

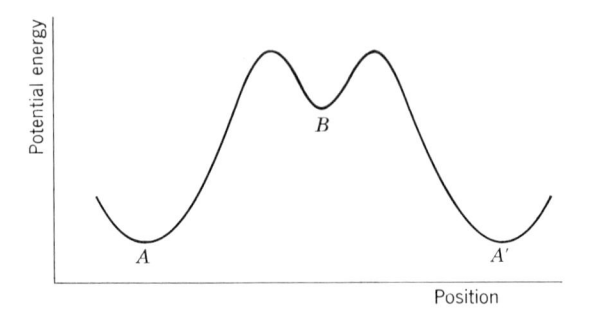

Fig. 4.13. Schematic representation of the energy of an atom during slip in face-centered cubic crystal.

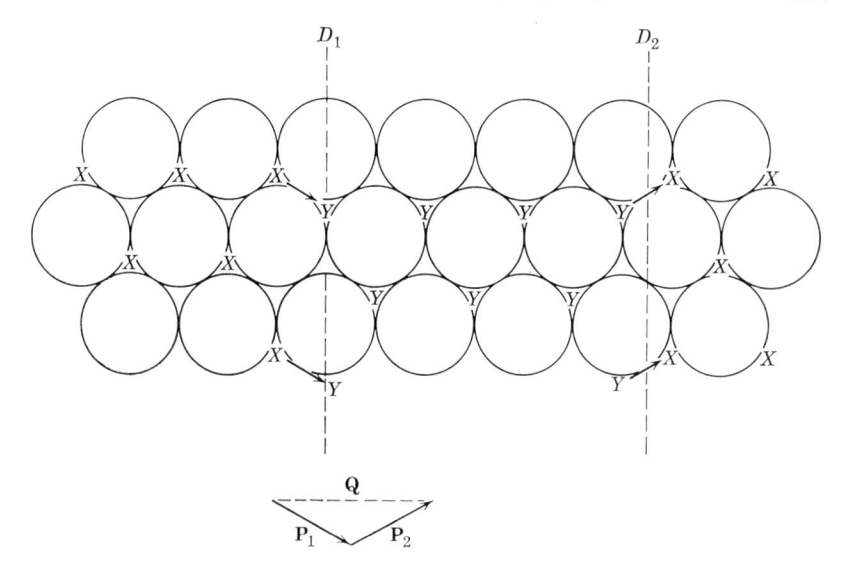

Fig. 4.14. Projection of atom sites in stacking faults.

right. The sites Y represent the region where translation by an amount $\mathbf{P}_1$ (at the partial dislocation D_1) causes a stacking fault, which is terminated by the partial dislocation of Burgers vector $\mathbf{P}_2$ at D_2. The vector sum of $\mathbf{P}_1$ and $\mathbf{P}_2$ is the Burgers vector $\mathbf{Q}$ for a complete dislocation.

It is convenient to represent the Burgers vector of a dislocation in a crystal in terms of the co-ordinates of one end if the other is at the origin. Thus the Burgers vector represented in Fig. 4.12 would be $[\frac{1}{2}\frac{1}{2}0]$, which is more conveniently expressed as $\frac{1}{2}[110]$. Similarly, the partials into which it splits are $\frac{1}{6}[121]$ and $\frac{1}{6}[21\bar{1}]$. It can be seen that the sum of the two vectors for the partials is equal to the vector for the complete dislocation.

It should be pointed out that there are two distinct types of partial dislocation in the face-centered cubic crystal. The "Shockley partial dislocation" can be obtained by the dissociation of a complete dislocation into two partials by moving them apart in the plane containing both Burgers vectors (the slip plane). A single Shockley partial can be formed by making a cut in a {111} plane, shearing the material on one side of the cut by an amount equal to the partial Burgers vector **b** which lies in the plane of the cut, and rejoining to form a stacking fault, which terminates at the partial dislocation. This type of partial dislocation can move in the slip plane, and is therefore sometimes called "glissile."

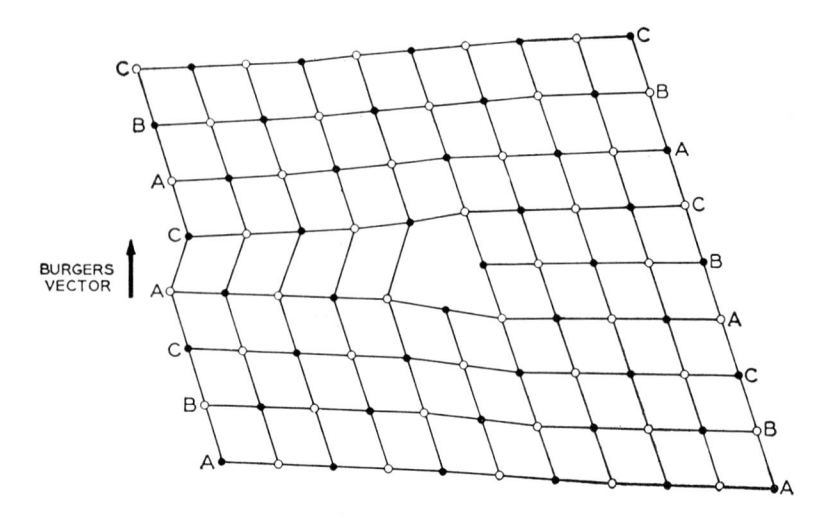

Fig. 4.15. Frank or sessile dislocation. By permission from *Dislocations in Crystals*, by W. T. Read, Jr., Fig. 7.4, p. 100. Copyright 1953. McGraw-Hill Book Co.

The "Frank partial dislocation" may be formed by removing one half-plane of atoms, and rejoining by the movement required to bring the neighboring planes together in their proper relationship. A type of stacking fault is formed where the rejoining takes place, because the proper sequence of planes cannot be maintained on both sides of the dislocation. The Burgers vector for the Frank partial is not in the same plane as the stacking fault; this type of dislocation cannot move by glide and it is therefore described as "sessile." A Frank or sessile dislocation is represented in Fig. 4.15.

A composite type of dislocation known as the "Cottrell-Lomer extended dislocation" may be formed when two dissociated dislocations in different $\{111\}$ planes combine at the line of intersection of the two $\{111\}$ planes. The two partials that coincide at the line of intersection form a new partial whose Burgers vector is $\frac{1}{6}[101]$. The result is that there are stacking faults on two intersecting planes, each terminated by a partial dislocation of the $\frac{1}{6}\langle 121 \rangle$ type at one side and by their common partial $\frac{1}{6}[101]$ where they meet. Such a dislocation cannot move either by slip or by climb; it is sometimes called a supersessile dislocation.

Segregation at dislocations. As shown in Fig. 4.8, there are regions of compression and of tension and of shear strain around an edge dislocation; similarly, there are regions of shear strain around a screw dislocation. It is therefore possible for a solute atom to occupy a site near a dislocation where it causes less elastic strain than it would elsewhere in the crystal. An interstitial atom, for example, would have less strain energy associated with its presence if it were in the dilated region near an edge dislocation or in certain interstitial sites near a screw dislocation. Similarly, a substitutional solute atom would cause the minimum strain energy by occupying sites in the compression or dilation regions according to whether it was smaller or larger than the atoms of the matrix.

It follows that the lowest energy configuration of a solid solution occurs when solute atoms are segregated at the dislocations. However, the entropy of the system is decreased by such an arrangement of the solute atoms, and it is concluded that at any temperature there is an equilibrium amount of segregation of this type; it has a maximum at low temperatures and is effectively zero at high temperatures. The existence of segregation at dislocations has experimental support from the accuracy of the predictions of the Cottrell theory of the yield point (see p. 402) which is based on the postulated existence of this kind of segregation.

Evidence for the existence of dislocations. Very persuasive evidence for the existence of dislocations has been produced recently by means of four techniques which have allowed dislocations to be "seen."

ETCHING. It has been shown that suitable etching procedures produce pits which reveal the points of emergence of line defects in crystals of various substances. The characteristics of these line defects make their identification as dislocations as conclusive as it could possibly be. Figure 4.16 shows dislocation etch figures on germanium, lithium fluoride, and silicon iron. The evidence that identifies these etch pits as dislocations is: 1. their line characteristics as shown by etching at a series of levels; 2. the correct prediction of the angular misorientation of two crystal regions separated by an array of etch pits; 3. the movement of these points under stress and at elevated temperatures, in accordance with the predicted behavior of dislocations.

DECORATION. In alkali halides and in silicon, precipitation of visible particles can be caused within the crystal. These particles form on lines that show typical predicted characteristics of dislocations. Their geometry is studied optically (by infrared microscopy in the case of silicon). Examples are shown in Fig. 4.17a and b.

ELECTRON MICROSCOPY. Direct transmission electron microscopy of thin films of aluminum and stainless steel has revealed individual dislocations in motion under the stress produced by the heating due

(a)

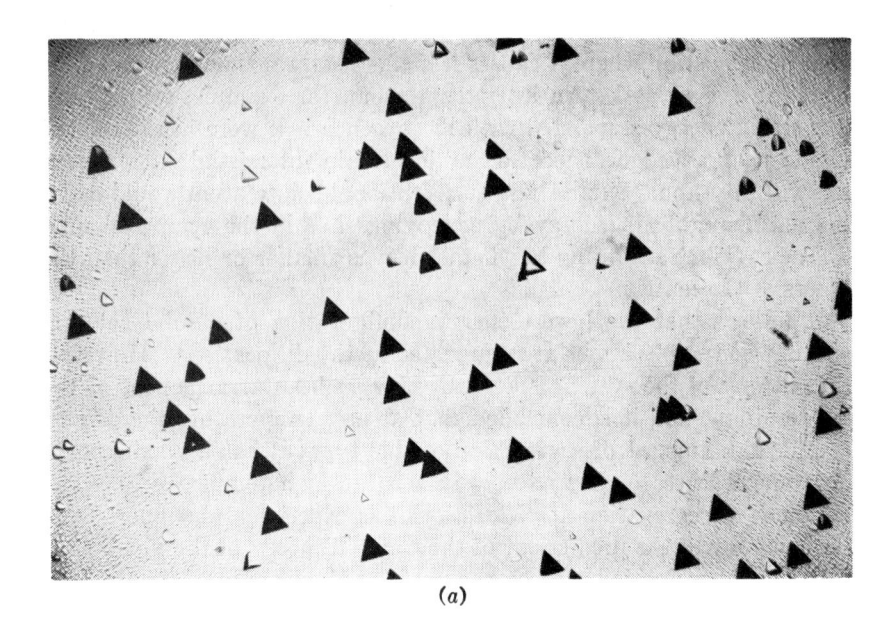

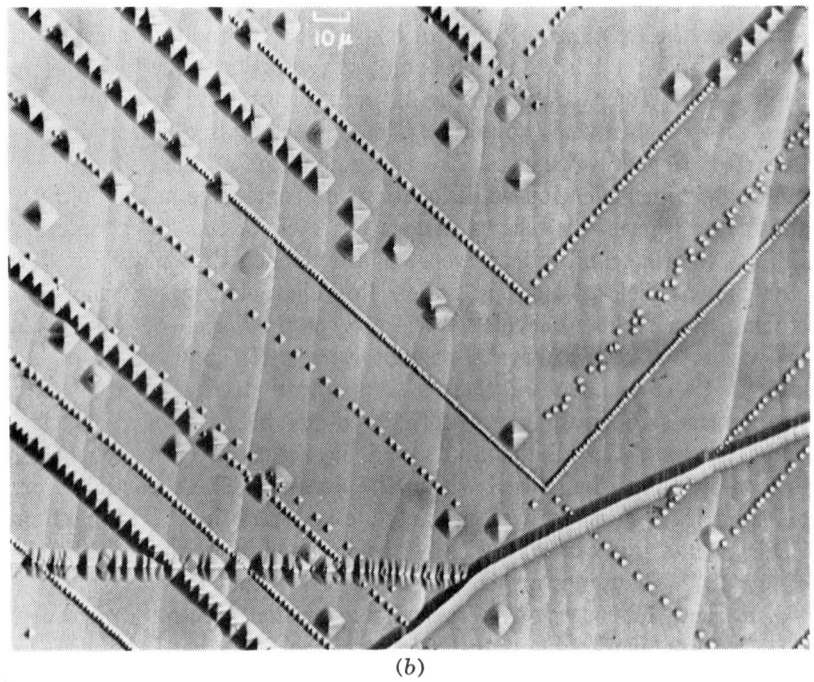

(b)

(c)

Fig. 4.16. Dislocation etch pits in (a) germanium (×190), (b) lithium fluoride (×1500), and (c) silicon iron (×900). [(a) Courtesy R. S. Wagner, Harvard University; (b) courtesy J. J. Gilman and W. G. Johnston, General Electric Company; (c) courtesy J. C. Suits, Harvard University.]

to the electron beam. Figure 4.18 shows examples in which dislocations in stainless steel can be seen.

X-Ray Diffraction. It has recently been demonstrated that individual dislocations produce X-ray diffraction effects which allow their positions within a crystal to be photographed, without cutting the crystal or "decorating" the dislocations. Figure 4.19 shows such X-ray photographs of dislocations in crystals of silicon and of aluminum.

Dislocation content of crystals. The dislocation content of a crystal is described by the number of dislocations per square centimeter of the crystal; it is assumed for this purpose that the dislocations are uniformly distributed, which is not always true. In carefully grown metal crystals the dislocation density is found to be of the order of 10^4 to 10^6 dislocations/cm^2. The density is much higher (up to 10^{12}/cm^2) in crystals that have been deformed (see Chapter 5); crystals of silicon have been grown in which there are no dislocations in quite large regions.

110   PHYSICAL METALLURGY

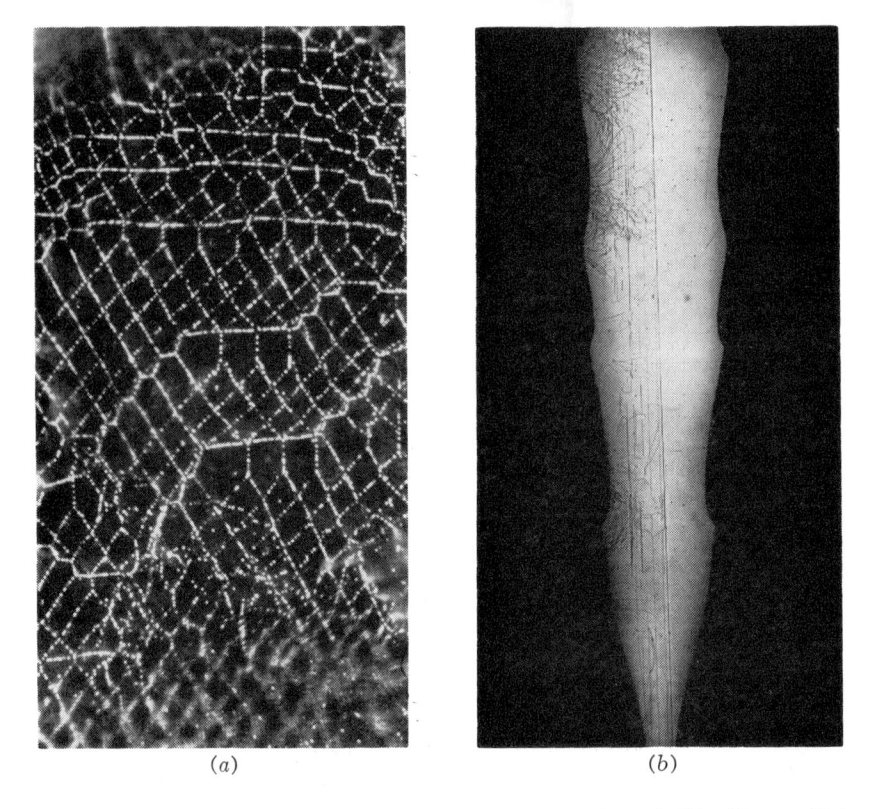

(a)   (b)

Fig. 4.17. Decorated dislocations in (a) sodium chloride and (b) silicon. [(a) Courtesy S. Amelinckx, University of Ghent; (b) courtesy W. Dash, General Electric Research Laboratory.]

4.4 Surface Imperfections

Imperfections that have extension in two dimensions are surfaces which separate a crystal or part of a crystal from material that differs from it in some way. The possible kinds of adjacent material and the resulting surface of discontinuity are listed in Table 4.1.

Before dealing with each of these types of surface separately, it is necessary to discuss certain general features of interfaces. In the first place, the energy of a crystal with any kind of interface is greater than that of the same amount of material in the interior of a crystal (i.e., without any surfaces). This may be demonstrated as follows: Let us consider the work that must be done to separate the two halves of a crystal at the plane AB of Fig. 4.20. The atoms on one side of this

IMPERFECTIONS IN CRYSTALS

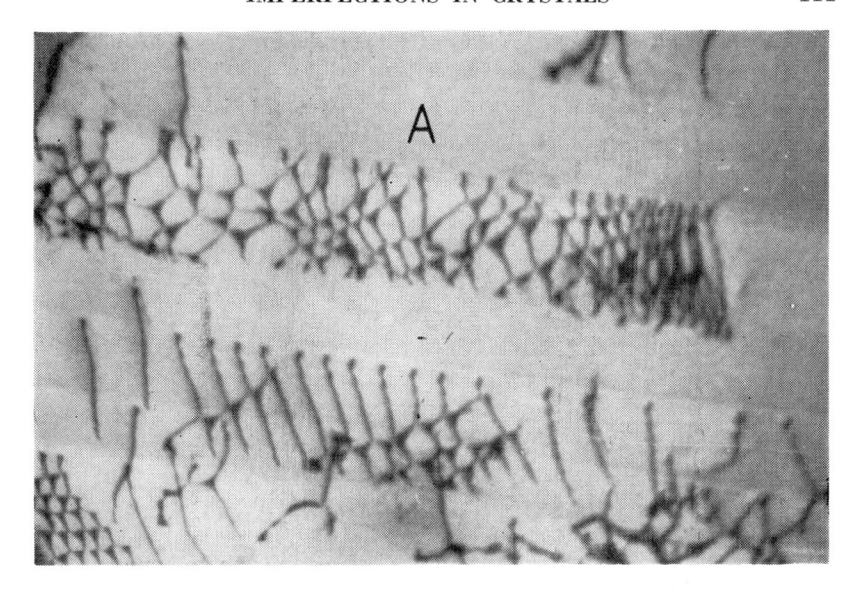

Fig. 4.18. Dislocations in stainless steel (×44,000). (Courtesy P. B. Hirsch, Cavendish Laboratory, Cambridge University.)

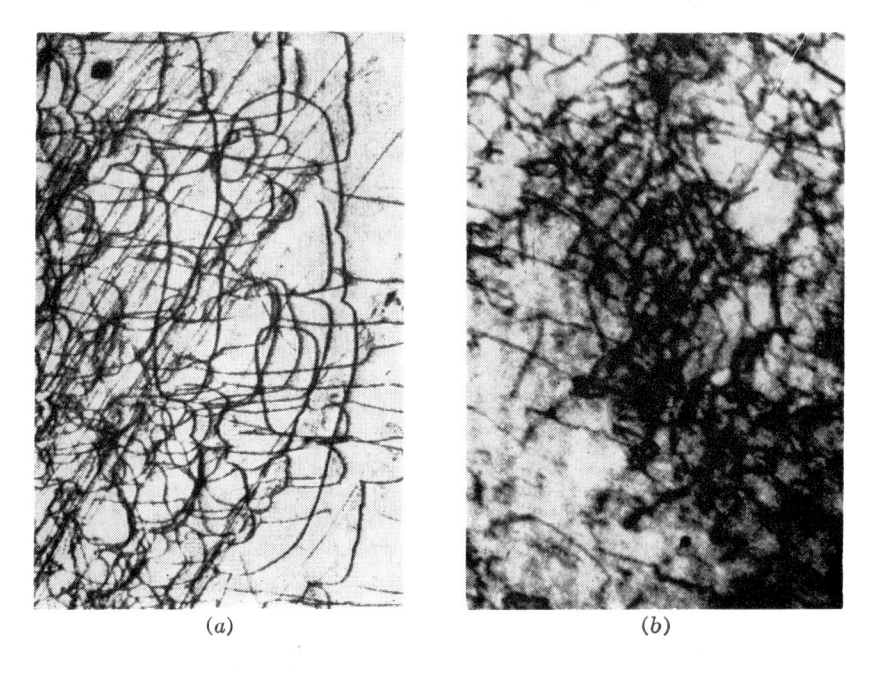

(a)   (b)

Fig. 4.19. X-ray diffraction photographs of dislocations in (a) silicon and (b) aluminum. (Courtesy A. R. Lang, Harvard University.)

Table 4.1. Nature of Adjacent Phases and Types of Surface

Nature of Adjacent Phase	Relationship	Type of Surface
Identical	Difference of orientation	Crystal boundary
Identical	Very small orientation difference	Subboundary
Identical	Difference of orientation that satisfies specific symmetry conditions	Twin boundary
Identical	Identical orientation but out of registry	Stacking fault
Differing only in direction of magnetization	Identical in structure and crystal orientation	Ferromagnetic domain wall
Crystalline, but different in structure	Related in orientation	Coherent interphase interface
	Not related in orientation	Incoherent interphase interface
Molten metal	—	Solid-liquid interface
Gas or vapor	—	Solid-vapor interface

plane are originally at their equilibrium distance d_0 from the atoms on the other side of the plane. If a force is applied, their separation increases, and work is done. The work that must be done to separate the two halves completely is given by the hatched region of Fig. 4.21. The result of performing this amount of work is to create two new surfaces, the energy of each of which is therefore one half of the work done. If the surfaces are subsequently allowed to come into contact,

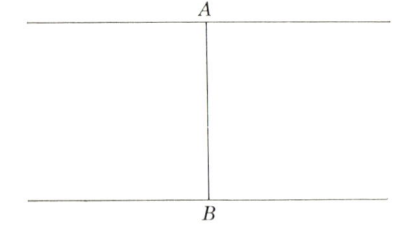

Fig. 4.20. Creation of interface AB in a crystal.

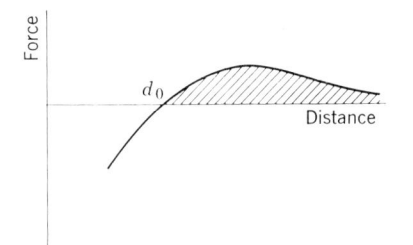

Fig. 4.21. Work done in separating two parts of a crystal.

after a change of relative orientation, an intercrystal boundary is produced, and some of the work that was done is restored. Similarly, if some other phase is allowed to come into contact with the two newly formed surfaces, some, but not all, of the work of separation is returned. It is evident that this is not how surfaces are actually formed, but it indicates in general terms why a surface must be considered to have an energy associated with it.

Crystal boundaries. We will now consider each type of surface from the point of view of structure and energy. It is evident that, in general, two crystals of the same structure cannot be joined to each other so that each atom has the environment characteristic of the perfect crystal. The junction between the two crystals could therefore be achieved only if either: (a) each crystal is perfect up to a terminating surface that coincides with that of the other crystal, or (b) if the atoms take up positions that form a transition from the structure of one crystal to that of the other, or (c) there is a layer of atoms in positions that are not related, crystallographically, to either crystal. These three distinct ways of forming a junction are shown in Fig. 4.22.

The structure represented by Fig. 4.22a can only occur when a special relationship exists between the orientations of the two crystals and their common plane. This case will be discussed in the section on "Twin Boundaries"; there is no doubt that this type of boundary has the structure represented in Fig. 4.22.

In the more general case, in which the orientations of the crystals are not simply related, the structure is not so obvious, and must be examined by indirect methods, as the resolution of available techniques does not permit the direct determination of the positions of the individual atoms in the boundary.

The rigorous approach to the problem of the structure of an intercrystal boundary would be to calculate the minimum free energy arrangement of the atoms between two crystals of any orientations. If it were found that a large number of different arrangements were

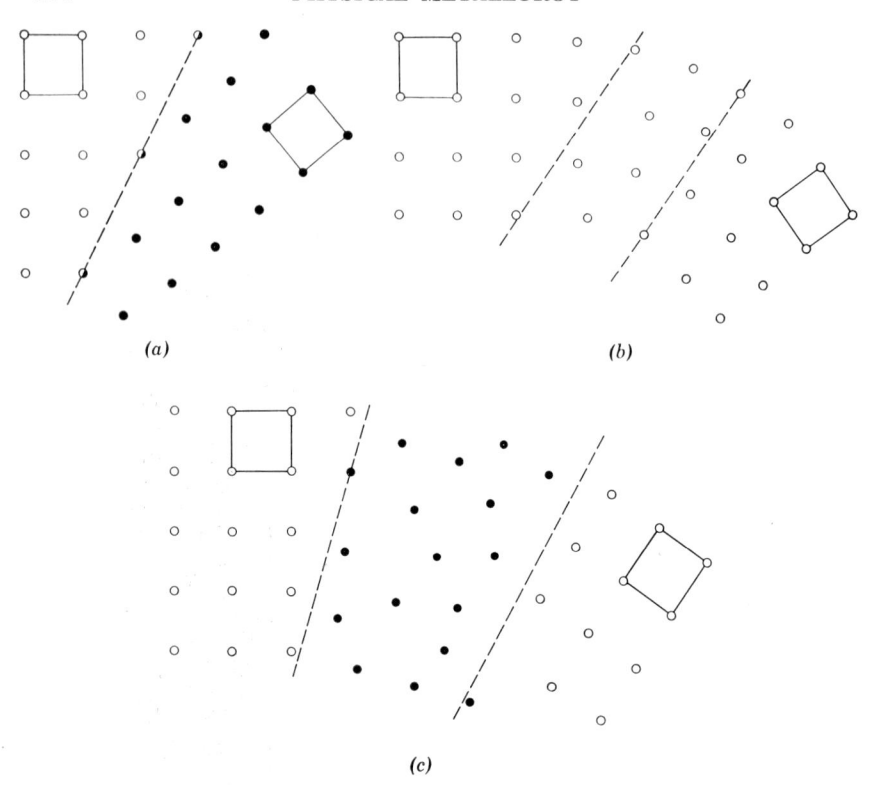

Fig. 4.22. Proposed structures for grain boundary. (a) Both crystals perfect up to plane of contact. (b) Transition structure. (c) Amorphous layer.

equally probable and were similar for all orientation relationships, it would be reasonable to describe the intercrystal material as amorphous. If, on the other hand, a specific arrangement was found for each orientation relationship, it would be possible to describe the boundary structure in more precise terms. Such calculations have not been made, and it is therefore not possible to predict by this method what the structure should be. These two possibilities are described as the "amorphous layer" and the "transition lattice" theories respectively. It is possible to distinguish between them on the following basis.

If two crystals are separated by a layer of amorphous material (i.e., in which the atoms take up positions dictated by chance, rather than uniquely defined positions), then the properties of the boundary should be independent of the difference of orientation. If, on the other hand, the properties of the boundary vary with the difference of orientation,

IMPERFECTIONS IN CRYSTALS

then it follows that the structure must also do so, and it cannot be random. A number of properties, that will be discussed at appropriate points in the text, show marked variations with the angle, and so it must be concluded that the boundary has a structure that is related to the structures of the adjoining crystals.

The alternative approach is to devise a model, calculate the implications, and test them experimentally. The only model so far proposed that has been computed in detail is that in which the boundary is considered as an array of dislocations. A very simple case is represented in Fig. 4.23, in which the difference of orientation is such that it can be represented by a simple array of edge dislocations. It is possible, however, to devise an array of dislocations to correspond to any difference of orientation and any orientation of the boundary.

It is useful for some purposes to distinguish between "tilt boundaries,"

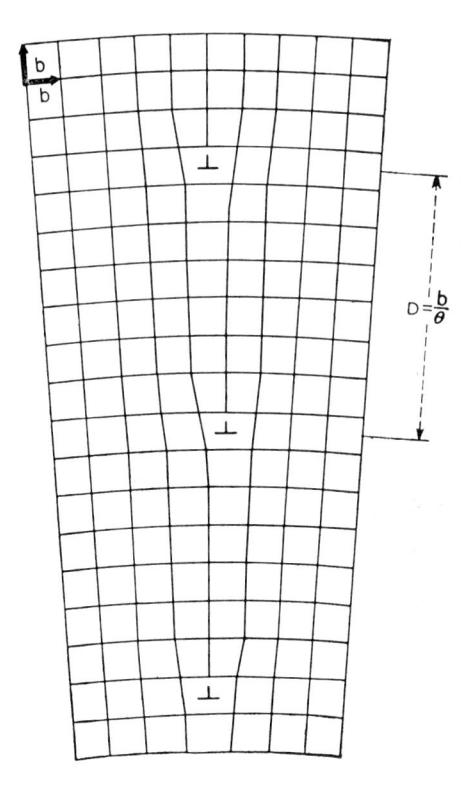

Fig. 4.23. Representation of tilt boundary as an array of edge dislocations. From *Dislocations*, Read, Fig. 11.1.(b), p. 157.

in which the axes of one crystal can be brought into coincidence with those of the other by rotation about an axis *in* the boundary, and "twist boundaries," in which coincidence can be achieved by rotation about an axis *normal to* the boundary. Boundaries will in general have components of both tilt and twist. A tilt boundary can be represented by an array of edge dislocations, the Burgers vectors of which are normal to the plane of the boundary, and the separation of which depends on the angle of tilt θ, the angle θ being approximately equal to b/h where b is the length of the Burgers vector and h is spacing of the dislocations. The shortest possible Burgers vector normal to the boundary may, however, be of much greater length than the minimum possible for the crystal structure, and the boundary would then consist of a more complicated array of several different dislocations, the vector sum of the Burgers vectors of which must be normal to the boundary.

A twist boundary cannot be composed of a single array of screw dislocations, which would be unstable; it can, however, consist of two crossed arrays as shown in Fig. 4.24, where the top plane below the boundary is shown by full lines, and the lowest plane above the boundary by broken lines. The Burgers vectors of both sets of dislocations are in the plane of the boundary; here again the dislocations corresponding to this simple arrangement might have a prohibitively high energy, in which case a more complex array containing both edge and screw components, but having the same resultant, is to be expected.

The energy of a boundary that consists of an array of dislocations can be considered as follows: The energy of a dislocation can be regarded as residing in two regions, the "core" in which the displacements of the atoms from perfect crystal positions are so drastic that elastic theory cannot be used for calculating the corresponding energy, and the outer region in which the displacements of the atoms are elastic. If the dislocations in an array are sufficiently widely spaced so that the cores do not overlap, then the variation of the energy of the array with angle can be calculated in the following manner.

The elastic energy associated with an edge dislocation is reduced if its "compression" region overlaps the "tension" region of another dislocation; this is what occurs when dislocations are arranged as indicated in Fig. 4.23. The extent to which the energy is reduced increases as the dislocations approach each other. Calculations show that the region of elastic strain adjacent to a boundary only extends outwards to a distance about equal to the spacing of the dislocations. A boundary of this type should therefore be more stable than a random arrangement of the same dislocations, and we should expect the energy of a boundary to be less than the sum of the separate energies of the dislocations by

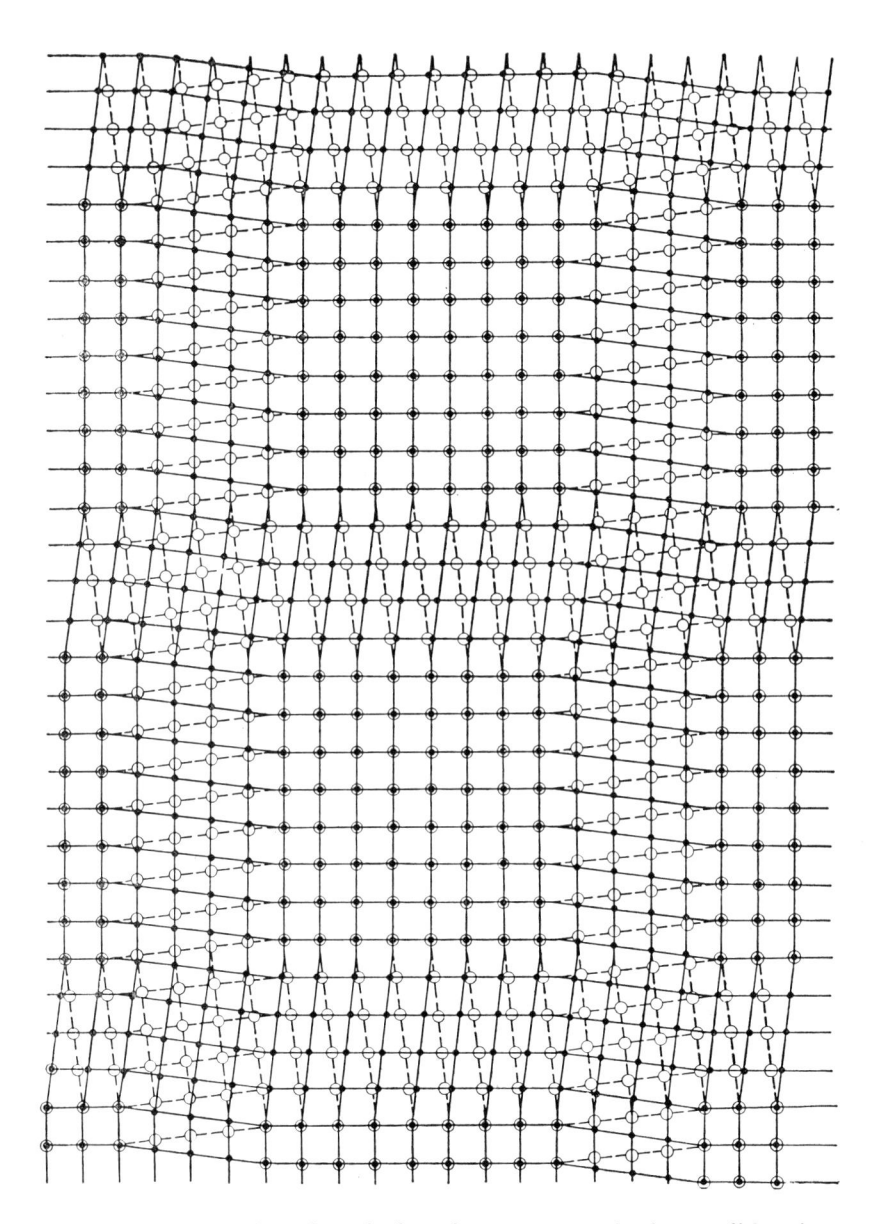

Fig. 4.24. Representation of a twist boundary as a network of screw dislocations. From *Dislocations,* Read, Fig. 12.2, p. 179.

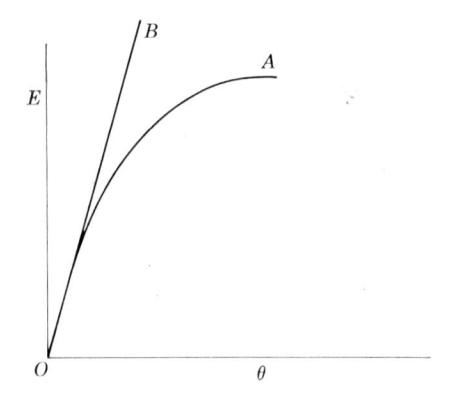

Fig. 4.25. Relationship between misorientation (θ) and grain boundary energy (E).

an amount that increases as the dislocations get closer together. The relationship between energy E and boundary angle θ therefore takes the form OA, Fig. 4.25, instead of the linear relationship OB obtained by summation of the individual energies of the dislocations. The predicted form of the curve OA for small angles agrees very closely with experimental observations on several metals and it is therefore not unreasonable to conclude that the dislocation array is a satisfactory model for the small-angle boundary.

The dislocation model of the crystal boundary cannot be applied to boundaries for which the orientation difference exceeds about 15°, because the dislocations would be so close together that their cores would overlap. They cannot then be regarded as existing as separate dislocations. The measurements of boundary energies show that the energy is nearly independent of angle in the "large-angle" range (above about 15°) except for the particular cases of twin boundaries (see p. 125). There is, however, some evidence that the atoms are arranged in a way that is dependent on the orientations of the neighboring crystals even up to quite large angles. Some of this evidence is discussed later in connection with diffusion (see p. 383). It is not yet clear what the arrangement of the atoms is; one of the possibilities is that there is an alternation of coherent regions and noncoherent regions, as illustrated in Fig. 4.26a. The shape and size of the "incoherent" regions would depend on the difference of orientation; the actual disposition of the atoms within these regions might be random or "liquid-like," or it might be "organized."

Another possibility is that the boundary has a structure of the kind that is shown schematically in two dimensions in Fig. 4.26b. According to this theory, the boundary is no longer smooth, but adjusts its position to minimize the misfit between the two crystals. It is possible, for any orientation relationship, to devise a boundary of the kind shown in Fig. 4.26b in which the interatomic distances do not depart from their normal values by a greater amount than occurs in dislocations. The misfit between adjacent crystals resembles that between adjacent regions of short range order in the liquid, and this boundary structure can therefore be regarded as a "liquid monolayer boundary." The energy of the large-angle crystal boundary is equal to the latent heat of fusion

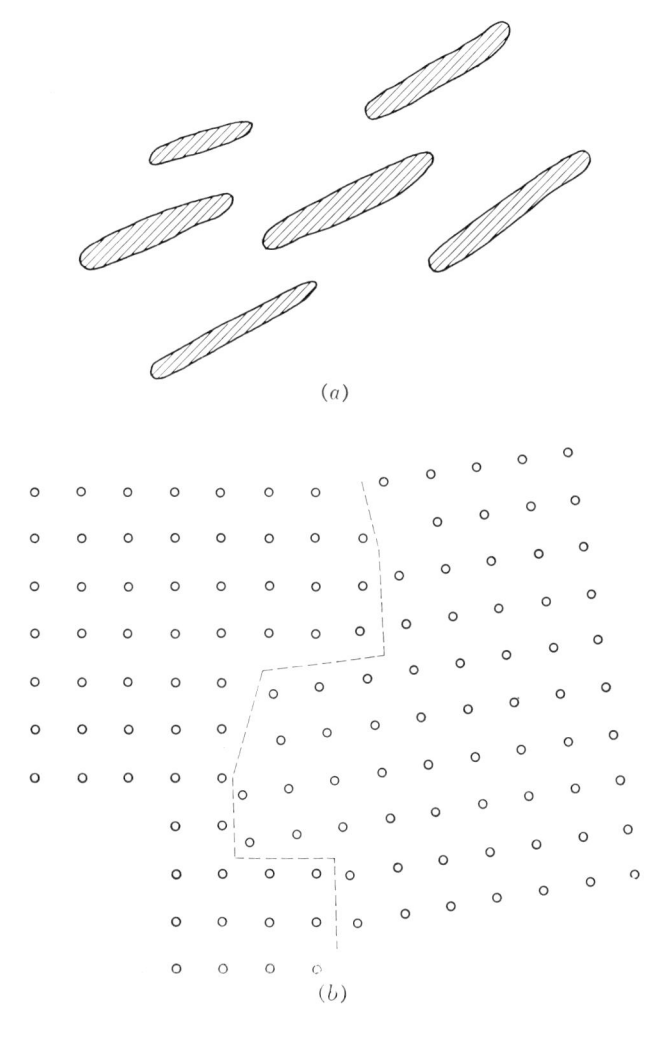

Fig. 4.26. (a) Schematic representation of possible grain boundary structure (large-angle boundary). (b) Liquid monolayer structure of crystal boundary.

of a layer whose thickness is a few atoms. This structure would be expected to display some anisotropy, and it would merge without any discontinuity into a dislocation structure at low angles.

One of the important properties of the crystal boundary is its ability to move if a suitable driving force is applied and if the temperature is high enough. The mobility of the boundary is reduced enormously by the presence of quite small amounts of impurities, as is shown in Fig. 4.27. This is evidence for the view that impurities tend to segregate at the boundaries, for the same reasons as at dislocations. The "driving force" is a reduction in free energy resulting from the motion of the

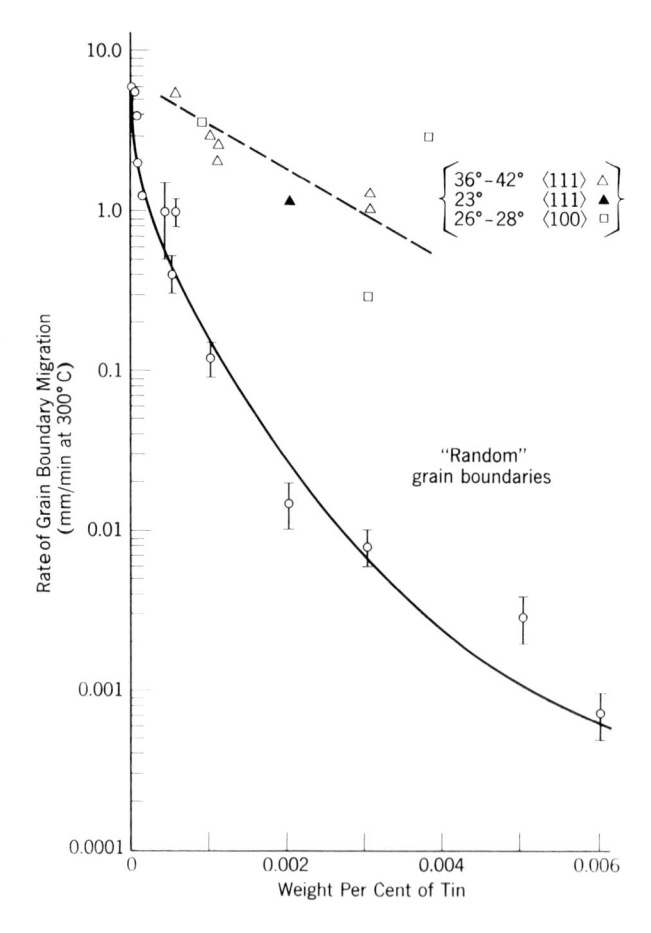

Fig. 4.27. Effect of impurities on mobility of grain boundaries in lead, for boundaries with "random" and "special" orientation differences. (Courtesy K. T. Aust and John Rutter, General Electric Research Laboratory.)

boundary; it may be a reduction in free energy of the boundary itself, resulting from a decrease in its area, or a reduction in the dislocation content of a crystal with a large content as the boundary moves across is, or it may be a decrease in the surface free energy of the specimen if the crystal that "grows" is so oriented with respect to the surface of the specimen that it has a low surface free energy. In the experiments of Rutter and Aust, on which Fig. 4.27 is based, the driving force is the energy of a lineage structure which is present in the original crystal but not in the region through which the boundary has moved. An interesting feature of these results is that certain boundaries, in which the orientation relationship of the crystals is such that some atoms at the boundary can be common to both crystals, are far less sensitive to impurities than those which do not have this characteristic. This probably means that there is less segregation at the "special" boundaries than in the more general case; this interpretation is supported by the experimental fact that the special boundaries have a rather lower energy than the "random" ones.

"Equilibrium" Disposition of Crystal Boundaries. The existence of crystal boundaries within a metal specimen increases its free energy over the value it would have if it were a single crystal. Crystal boundaries are, therefore, never present as part of the stable equilibrium structure of a metal. However, if, as is usually the case, crystal boundaries are present as a result of the previous history of the specimen, they may take up a configuration that is in metastable equilibrium, i.e., in which the system would need to pass through an intermediate stage of higher free energy before it could reach the absolute minimum that corresponds to the single crystal. The purpose of this section is to discuss the metastable equilibrium of the boundaries in a polycrystalline specimen.

If the crystals in a polycrystalline sample are randomly oriented, it can be shown by statistical analysis that it is quite a rare occurrence for two neighboring crystals to have orientations that match so closely that a low-angle boundary is formed. (It should be realized that a two-dimensional representation such as Fig. 4.43 only shows one of the three angles that may contribute to the difference of orientation.) It is therefore reasonable, for most purposes, to assume that all the boundaries are large-angle boundaries and that they have equal free energies per unit area.

If three such boundaries meet in a line, the total area of the boundaries is minimized when the dihedral angles between them are 120°, and the boundaries are planar. A network of such boundaries is shown in two dimensions in Fig. 4.28. It consists of an array of regular hexa-

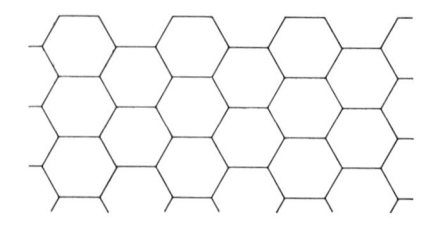

Fig. 4.28. Two-dimensional representation of metastable equilibrium of grain boundaries in a polycrystalline metal.

gons. The corresponding three-dimensional network is much more complicated, because there is no regular polyhedron with plane faces that can be packed to fill space (an obvious requirement for a crystal boundary network) which also fulfills the surface free energy requirements that the faces should meet in threes at lines with dihedral angles of 120°, and that the edges should meet at points with angles of 109°28″. (This is the angle between any two of four lines that meet symmetrically at a point.) The tetrakaidecahedron, shown in Fig. 4.29, most closely approaches those conditions; if suitable slight curvatures are introduced into the faces, the conditions for metastable equilibrium are satisfied. If the crystals do not have the shape specified, which also demands that they all be of equal size, then it is not possible to reach metastable equilibrium. The consequences of this are discussed in Chapter 7.

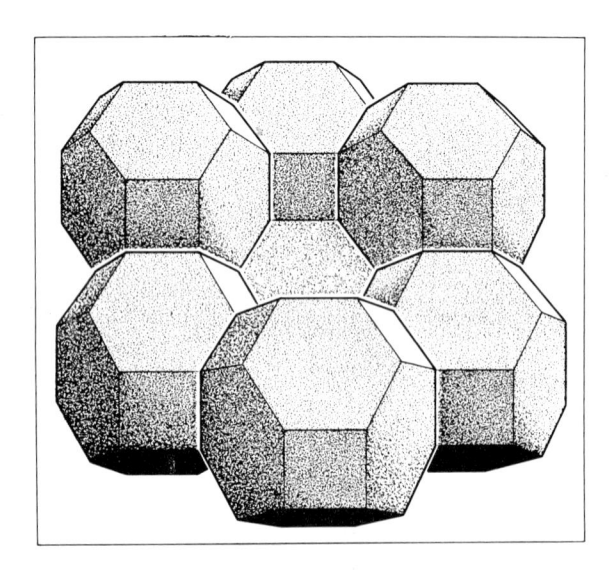

Fig. 4.29. Stack of tetrakaidecahedrons. From *Metal Interfaces,* American Society for Metals, Cleveland, 1952, Fig. 19, p. 90.

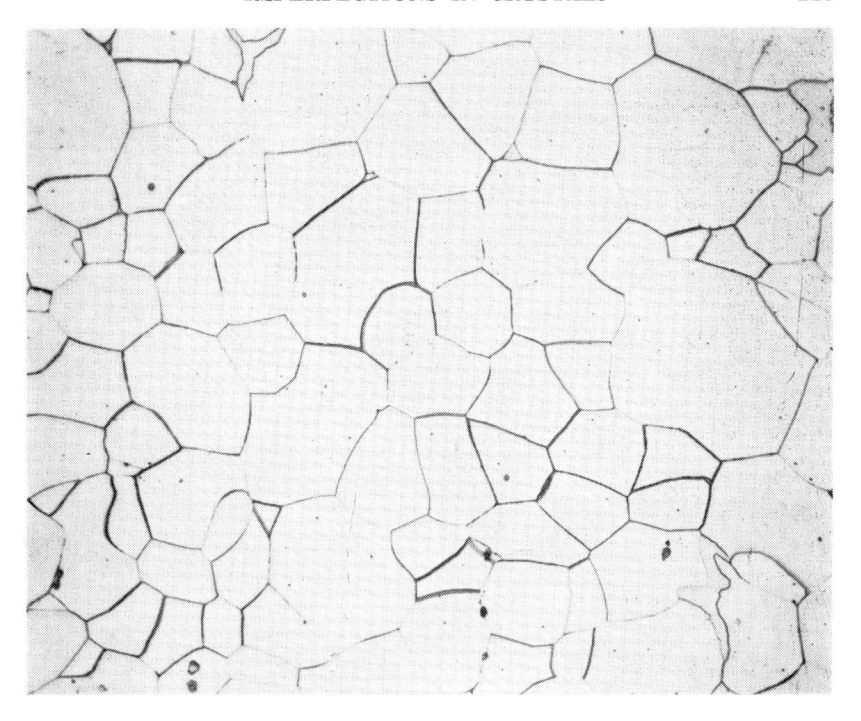

Fig. 4.30. Nearly equilibrated structure of polycrystalline specimen. Magnification (×780). (Courtesy U. S. Steel Corporation.)

When a polycrystalline specimen of a metal is examined metallographically, the intersections of the crystals with an arbitrarily chosen plane (that in which the specimen was polished) are observed. Even if the specimen has the ideal structure represented in Fig. 4.29, it would not usually have the appearance of Fig. 4.28 and the different crystals would not all appear to be of the same size. An example of a metal sample that is close to equilibrium is shown in Fig. 4.30.

The equilibrium conditions can be equally well satisfied at any crystal size; the actual size of the crystals in a specimen depends upon its history in ways that will be discussed later; the shape depends upon how close to equilibrium the specimen has been allowed to come.

For technical purposes, it is useful to define the grain size of a sample in terms of the number of grains per unit area in a photomicrograph. For example, the ASTM grain size is given by N in the formula: Number of grains per square inch at a magnification of 100 (linear) = 2^{N-1}. It is usual to determine the grain size by comparison with standard charts consisting of hexagonal networks of various sizes. It is pos-

Table 4.2. ASTM Grain Size *

ASTM Number	Grains/mm²	Grains/mm³
−3	1	0.7
−2	2	2
−1	4	5.6
0	8	16
1	16	45
2	32	128
3	64	360
4	128	1,020
5	256	2,900
6	512	8,200
7	1,024	23,000
8	2,048	65,000
9	4,096	185,000
10	8,200	520,000
11	16,400	1,500,000
12	32,800	4,200,000

* From *Metals Handbook*, American Society for Metals, Cleveland, 1948.

sible, in most cases, to allot an unambiguous grain size number to a given specimen. Table 4.2 shows how the number of grains per unit area and per unit volume vary with the ASTM grain size number.

Subboundaries. It is convenient to describe crystal boundaries corresponding to misorientations of less than about a degree as subboundaries. These differences of orientation usually arise as a result of processes (to be described later) that occur during the growth of a crystal; it is, in fact, rare for a crystal that is not very small to be entirely without a substructure of this kind. It is usual to consider a crystal as retaining its identity if its orientation does not vary between different points by more than a degree or so.

While subboundaries differ quantitatively, and in their origin, from the ordinary intercrystal boundaries discussed in the preceding section, it appears that their structure is similar; in fact, the evidence for the dislocation model of the crystal boundary is conclusive for low-angle boundaries, as a result of the techniques for revealing the positions of individual dislocations by means of etch pits. Figure 4.31 shows the etch pits formed at a subboundary in germanium. The differences of

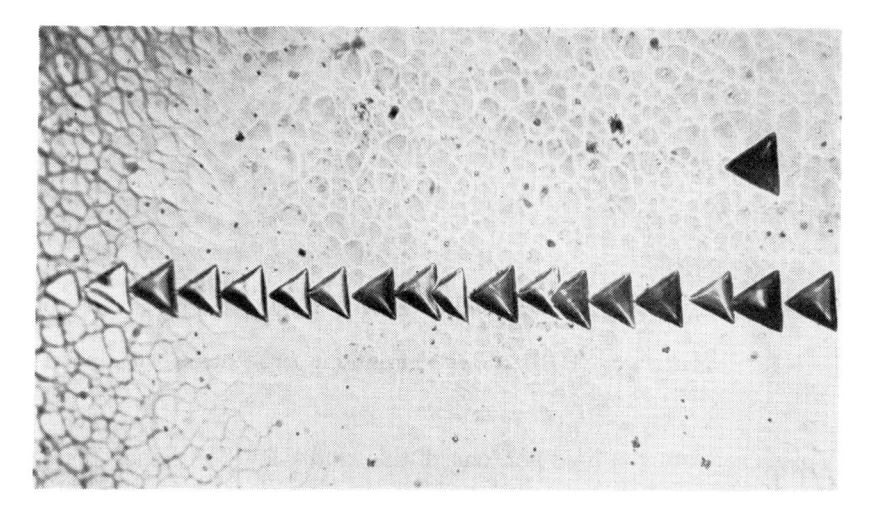

Fig. 4.31. Subboundary etch pits in germanium ($\times 750$). (Courtesy R. S. Wagner, Harvard University.)

orientation across such boundaries, calculated from the spacing of the etch pits and the Burgers vector of the dislocations, agree closely with the values obtained by X-ray diffraction methods.

It is possible that a subboundary always consists of a single array of dislocations; in this case they must always be tilt boundaries, as a single array of screw dislocations is unstable. The subboundaries that have been reported in the literature all appear to be of the tilt type, and it has been shown that, under suitable conditions, a subboundary can be made to move by the application of a relatively small stress, exactly as would be predicted if each dislocation of the array were considered as moving independently in parallel slip planes.

Twin boundaries. There are some special cases of crystal boundaries, referred to above, in which the atoms at the interface are in positions that constitute lattice sites of both crystals. The two crystals "match," plane for plane, at such an interface, and are said to be coherent with each other. The particular form in which a coherent boundary is commonly found is between crystals that have a twin, or mirror image, relationship. An important example is that of the face-centered cubic structure, in which the normal structure consists of a stack of close-packed planes, in the sequence of positions represented in Fig. 4.32 as $abcabc, \cdots$, or according to the other notation as $\triangle \triangle \triangle \triangle \triangle$. A line of atoms such as $A_1B_1C_1A_2B_2C_2$ in Fig. 4.32, in which each successive atom is in the next layer upwards, lies in a $\langle 110 \rangle$ direction in the crystal.

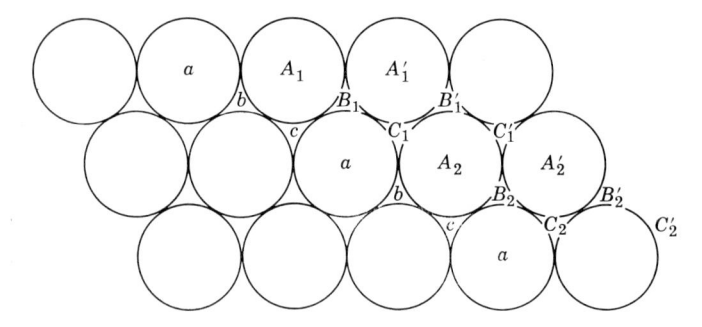

Fig. 4.32. Projection on (111) plane of atom sites in face-centered cubic crystal.

A plane consisting of lines of atoms of this kind is a {111} plane. These lines and planes are shown in Fig. 4.33. Part of the close-packed plane a of Fig. 4.32 is shown in Fig. 4.33 as the triangle containing $A_1 A_1''$.

The continuity of the lines $A_1 B_1 C_1 A_2$ depends upon the preservation of the stacking sequence $abcabc$. If at some level in the stack the order is reversed ($abcabacbacb$ or △△△△▽▽▽▽), then points corresponding to $A_1 B_1 C_1 \cdots$ of Figs. 4.32 and 4.33 will no longer be in a straight line; the line and all parallel lines will change direction to the mirror image of the former direction in the stacking plane. The cross section in the {110} plane is shown in Fig. 4.34. It will be seen that each atom, including those in the twinning plane, has the same number of nearest neighbors at the same distance as it has within the crystal. The second nearest neighbor relationships are, however, modified at the twin interface, where they are of the aba type characteristic of hexagonal close

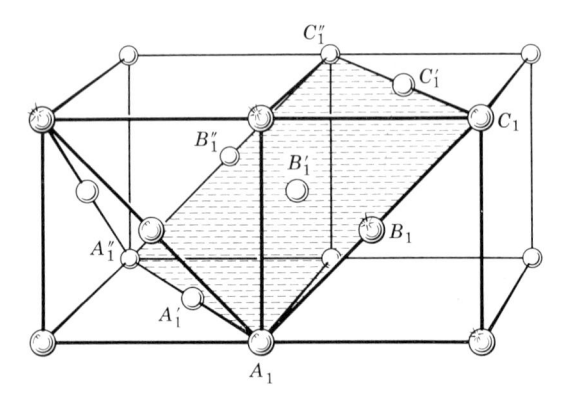

Fig. 4.33. Close-packed planes in face-centered cubic structure.

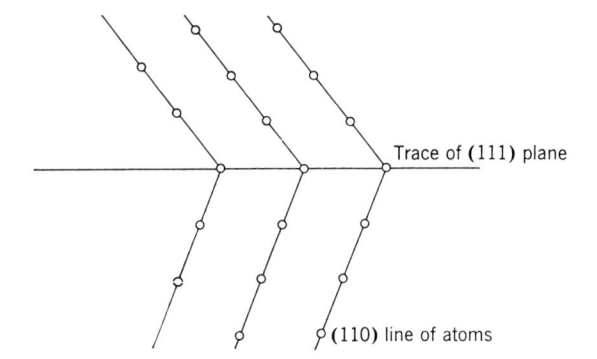

Fig. 4.34. Twinned crystal in {110} plane.

packing rather than the *abc* type of cubic close packing or face-centered cubic. The energy associated with the coherent twin boundary is much lower than for the ordinary large-angle boundary and depends upon the importance of second nearest neighbor interactions; some implications of this will be discussed in Chapter 7.

The twinning planes for various crystal structures are shown in Table 4.3. If two crystals have the twin relationship between their orientations, but are in contact in some plane that is not the twinning plane, the boundary is no longer coherent, and it no longer has the low energy and other characteristics of the coherent twin boundary.

It will be recognized that the stacking fault that separates the two parts of a dissociated dislocation is very closely related to a twin boundary; the twin boundary corresponds to a sequence *abcbacb*, while the stacking fault is of the type *abcababc;* thus the stacking fault contains two layers with abnormal second nearest neighbors, while the twin boundary contains only one. The stacking fault has a crystal of

Table 4.3. Twinning Planes

Structure	Twinning Plane
Face-centered cubic	{111}
Close-packed hexagonal	{10$\bar{1}$2}
Body-centered cubic	{112}
Rhombohedral	{001}
Tetragonal (tin)	{331}

identical orientation on each side of it, while the twin boundary does not.

Interphase interfaces. The interface between two different phases is less understood than the intercrystal boundary; however, the following generalizations may be made. An interface may be coherent or noncoherent. A coherent interface is one in which there is a relationship between the orientations of the two crystals resulting from the "matching" of a particular aspect of the structure, such as an interplanar distance, with a geometrically similar but crystallographically different aspect of the structure of the other. A very simple example, in two dimensions, is shown in Fig. 4.35, in which the side of the rectangle of one structure coincides with the diagonal of the rectangle of the other. If the matching is exact, one can expect the boundary to have quite low energy, particularly if the atoms of the two crystals are not very different in electronic configuration. It is more probable that such a coincidence of dimensions can only be achieved by introducing some elastic distortion, such as stretching one crystal near the boundary, and compressing the other, or by shear distortions. In this case, the energy associated with the boundary is increased by the elastic strain energy that is introduced. The boundary is still coherent so long as matching is achieved, even if there is strain energy. However, if excessive strain energy is required to achieve coherency, then a noncoherent boundary is formed. There will not be any coherency strain energy, but the boundary will be expected to have energy, because the atoms at the interface will not be in their lowest energy positions in relation to both crystals.

A lower energy boundary may result from the presence of an array of dislocations. An example of this is shown in Fig. 4.36, in which the uniformly distributed strain energy is replaced by the energy of the dislocations.

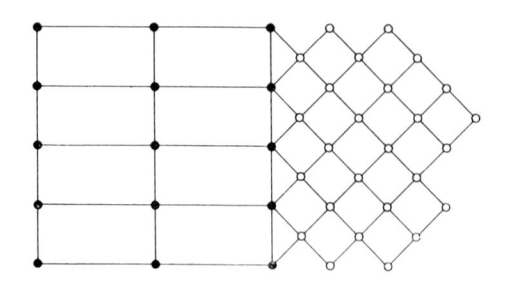

Fig. 4.35. Coherent interface between phases of different structures.

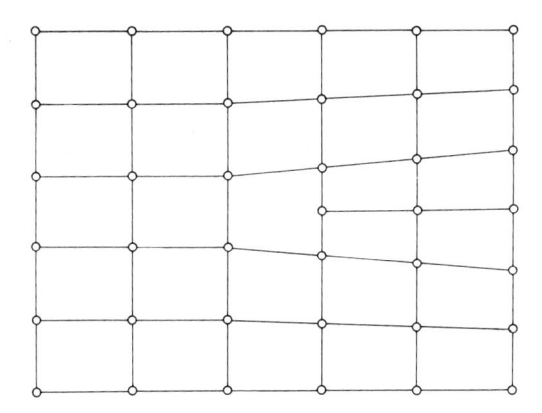

Fig. 4.36. Semicoherent interface between phases of different structures.

The above considerations are based on the assumption that the two crystals have related orientations. This is often the case when one phase is formed by transformation from the other (see Chapter 8) or when both are formed simultaneously (see Chapter 6), but there are many instances in which coherency in an arbitrary boundary plane is very improbable because of unrelated orientations. However, when the structures are different, it is always possible in principle to find a boundary surface that permits coherency. Such a surface would usually have a very large area, and a noncoherent interface of minimum area is much more probable. Whether such an interface can in general be usefully represented by an array of dislocations is open to question.

Experimentally, a comparison can be made between the energy of an

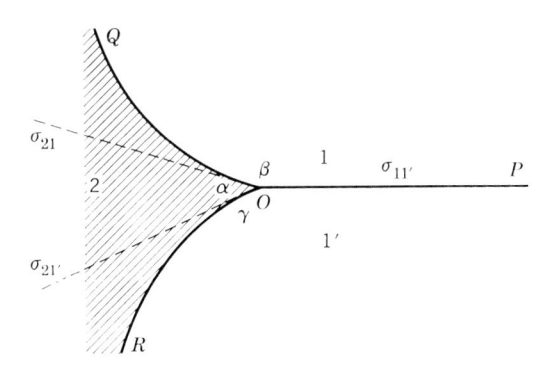

Fig. 4.37. Equilibrium between interphase interfaces (OQ and OR) and a crystal boundary (OP).

interphase boundary and that of a crystal boundary in one of the phases, by measuring the dihedral angles where the boundaries meet, as shown schematically in Fig. 4.37, in which OP is a boundary between two crystals 1 and 1' of the phase 1, and OQ and OR are the boundaries between 1 and 2 and between 1' and 2. The angles α, β, and γ are related to the energies $\sigma_{11'}$, σ_{21}, and $\sigma_{21'}$ by the expression $\sigma_{11'}/\sin \alpha = \sigma_{21'}/\sin \beta = \sigma_{21}/\sin \gamma$, if equilibrium has been established near the point O, and if the energies of the boundaries are independent of their orientations. It will be recognized that the expression given above is identical with the expression for the equilibrium of three forces at a point. Figure 4.38 is example of a photomicrograph from which measurements of this kind can be made. Table 4.4 shows some results for interphase boundary energies, compared with intraphase crystal boundaries.

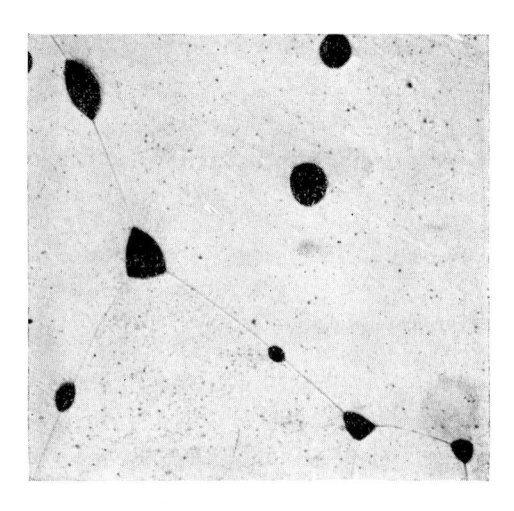

Fig. 4.38. Photomicrograph showing equilibrium shape of lead crystals in leaded brass. From C. S. Smith, Chapter 14 in *Imperfections in Nearly Perfect Crystals*, W. Shockley *et al.*, Eds., John Wiley & Sons, New York, 1952, Fig. 2, p. 379.

Solid-liquid interfaces. We will first consider the interface between a pure metal, in crystalline form, and the same metal in liquid form. It is difficult to study such an interface in its equilibrium condition because any addition of heat to, or subtraction of heat from, the system causes the interface to move so that the temperature remains the same and the heat is absorbed or supplied as latent heat of fusion. Each atom that leaves the crystal requires, on the average, the latent heat of fusion (per atom) and for a close-packed crystal it is detached, on the average, from six nearest neighbors whose relationship is characteristic of the crystal. The latent heat of fusion may therefore be re-

Table 4.4. Comparison of Interphase and Intraphase Boundary Energies *

System	Phase A	Phase B	Comparison Grain Boundary (c)	Energies $\dfrac{\gamma AB}{\gamma c}$
CuZn	α (F.C.C.)	β (B.C.C.)	α/α	0.78
CuZn	α (F.C.C.)	β (B.C.C.)	β/β	1.00
CuAl	α (F.C.C.)	β (B.C.C.)	α/α	0.71
CuAl	β (B.C.C.)	γ (Complex cubic γ brass)	γ/γ	0.78
CuSn	α (F.C.C.)	β (B.C.C.)	α/α	0.76
CuSn	α (F.C.C.)	β (B.C.C.)	β/β	0.93
FeC	α (B.C.C.)	Fe$_3$C (Orthorhombic)	α/α	0.93
FeC	α (B.C.C.)	γ (F.C.C.)	α/α	0.71
FeC	α (B.C.C.)	γ (F.C.C.)	γ/γ	0.74

* From C. S. Smith, Chapter 14 in *Imperfections*, Shockley et al., Table 2, p. 384.

garded as equivalent to the breaking of six "crystal" bonds and their replacement by the equivalent number of "liquid" bonds; or to the creation of 12 "half bonds." The existence of a half bond therefore corresponds to one-twelfth of the atomic latent heat.

On this basis we may estimate the energy of the interface between a crystal and the melt. This will depend on the crystallographic characteristics of the face, and it will be assumed at this point to be an atomically smooth low index plane. An atom in the {111} plane that forms a surface has nine nearest neighbors in the solid, six in its own plane, and three in the next plane below; there are therefore three half bonds, and the energy of each atom in such a surface should exceed that of atoms in the interior of the crystal by three-twelfths of the latent heat. Similarly, atoms in a surface plane of {001} type have eight nearest neighbors and an excess energy of four-twelfths of the latent heat. Atoms at edges and corners have higher energies that are readily calculated on the assumptions used above. The energies of other crystallographic faces would be higher, and it is therefore to be expected that the equilibrium faces would be predominantly of the {111} type. There is some experimental evidence that supports this; it will be discussed in Chapter 6.

The only experimental evidence of the energy of solid-liquid interfaces comes from nucleation data (see Chapter 6) and this is indirect; but it does indicate that for a crystal of a specific, very small, size the surface energy per surface atom is equal to about one-half of the latent heat of fusion. (See Table 4.5.) This is compatible with the point of view expressed above because of the large proportion of atoms that occupy edge and corner sites in a crystal of the size under discussion. It will be shown later (see p. 245) that the surface energy per bond is numerically consistent with the values derived from nucleation theory and experiment.

The assumption that the surface of the crystal is atomically smooth is almost certainly wrong, at least in the case of metals. It is much more likely that the extreme close-packed layer that is complete has superimposed on it at any instant a considerable proportion, perhaps one-half, of the atoms that would be required to form the next layer.

Table 4.5. **Solid-Liquid Interfacial Energy and Comparison with Atomic Latent Heat** *

Metal	Solid-Liquid Interfacial Energy (ergs/cm^2)	Interfacial Energy/Atom / Latent Heat/Atom
Mercury	24.4	0.53
Gallium	56	0.44
Tin	54.5	0.42
Bismuth	54	0.33
Lead	33	0.39
Antimony	101	0.30
Germanium	181	0.35
Silver	126	0.46
Gold	132	0.44
Copper	177	0.44
Manganese	206	0.48
Nickel	255	0.44
Cobalt	234	0.40
Iron	204	0.45
Palladium	209	0.45
Platinum	240	0.46

* From J. H. Hollomon and D. Turnbull, Chapter 7 in *Progress in Metal Physics 4*, Bruce Chalmers, Ed., Pergamon Press, London, 1953.

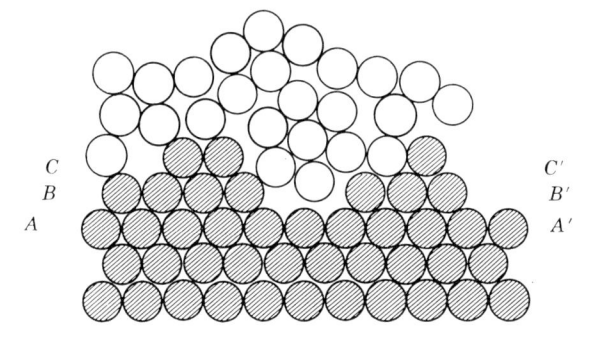

Fig. 4.39. Interface between crystal and melt.

Figure 4.39 represents such an interface; the atoms that occupy crystal sites are hatched; those which do not are open circles. AA' is the extreme complete layer, and BB' is the next layer. This layer would be complete in some regions, and these in turn would be partially covered by atoms in the positions characteristic of the layer CC'. It is likely that in the real case some of the "complete" regions would be much larger than shown in the diagrams, and that the number of partial layers would be considerably greater than the two shown in Fig. 4.39. The adoption of this more realistic model of the interface does not invalidate the estimate made above of the energy of the interface, because the average number of nearest neighbors is unaffected by this kind of surface roughness.

There is some information on the energy of the solid-liquid interface for solid solution alloys in relation to the grain boundary energy in the solid; the experimental approach is similar to that for interphase boundaries discussed in the preceding section. The equilibrium liquid phase has a different composition from the crystalline phase, and equilibrium can be achieved by prolonged exposure to a constant temperature. Some results are quoted in Table 4.6.

Solid-vapor interfaces. Relatively direct measurements of the energies of solid-vapor interfaces have been made for several metals. The method is to subject thin wires to tension stresses, which oppose the tendency of the surface energy to decrease by reducing the length and increasing the thickness. The applied tension that just balances the surface energy is used for calculating the surface energy. Some values are given in Table 4.7; the values for the heat of vaporization are also given in this table. It will be seen that the surface energy is a smaller fraction of the latent heat in this case than it is for the solid-liquid

Table 4.6. Relative Energies of Solid-Liquid Interfaces and Grain Boundary Energies *

System	Solid Grain Boundary	$\dfrac{\gamma SL}{\gamma SS}$
CuPb	α/α (F.C.C.)	0.58
CuZnPb	α/α (30% Zn, F.C.C.)	0.65
CuZnPb	β/β (49% Zn, B.C.C.)	0.87
FeCu	γ/γ (F.C.C.)	0.51
AlSn	Al/Al (F.C.C.)	0.56
FeAg	γ/γ (F.C.C.)	>4
FeFeS	γ/γ (F.C.C.)	0.52

* From C. S. Smith, Chapter 14 in *Imperfections*, Shockley et al., Table 3, p. 385.

interface. This difference probably arises because the crystal is practically undistorted both atomically and electronically right up to a solid-liquid interface, while the electron distribution probably undergoes a marked change near a solid-vapor interface because of the entirely different electron population of the vapor phase.

When some metals are held for long times at high temperatures, the surface changes its shape in two ways. The shape changes to form a "groove" whenever a crystal boundary meets the surface (Fig. 4.40a, b, c, d). When the orientation of the crystal is favorable, a series of flat surfaces, or terraces joined by steps, replaces the previous smooth surface (Fig. 4.40e and f).

These two effects are called thermal etching. They each correspond

Table 4.7. Relationship between Solid-Vapor Surface Energy and Latent Heat of Vaporization

Metal	Solid-Vapor Surface Energy, (ergs/cm^2)	Latent Heat Vaporization (kcal/mole)	Surface Energy/Atom Latent Heat/Atom
Copper	1700	73.3	0.14
Silver	1200	82	0.26
Gold	1400	60	0.18

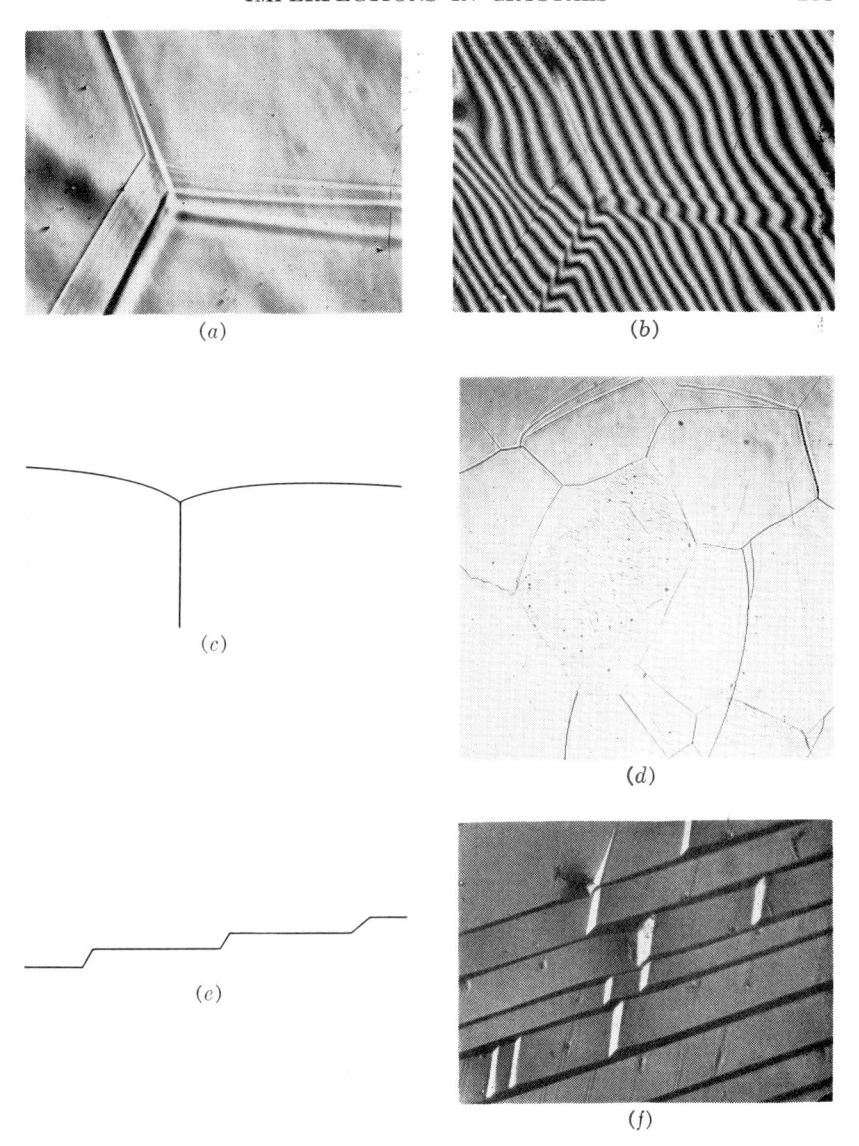

Fig. 4.40. Structures produced by thermal etching. (a) Photomicrograph of grooves at the grain boundaries in silicon iron. (b) Interference micrograph of same field as (a). (c) Transverse section of a boundary groove (schematic). (d) Surface of lead showing grooves and steps ($\times$112). (e) Transverse section of surface steps (schematic). (f) Photomicrograph of thermally etched silver ($\times$507) showing surface steps. [(a), (b) Courtesy C. S. Dunn, General Electric Research Laboratory; (d) courtesy J. Walter, General Electric Research Laboratory; (f) courtesy A. J. W. Moore, Division of Tribophysics, CSIRO, Australia.]

to an increase in surface area, but, because they occur spontaneously, the free energy must decrease. When a "boundary groove" is formed, the area, and therefore the free energy, of the boundary is decreased. Measurement of the groove angle gives data from which the ratio of surface energy to boundary energy can be calculated. It is found that the high angle boundary has about one-third the specific free energy of the external surface.

The terrace structure is formed when the surface parallel to a particular plane (in the case of silver the {111} plane) has a substantially lower specific free energy than a surface which is not a low-index crystallographic plane. It follows that, at least in some cases, the free energy of a metal surface depends on its relationship to the crystallographic planes.

It should be pointed out that it is unusual for a metal surface to be in contact with the vapor of that metal; most metals form an oxide film very rapidly when in contact with air or even a low partial pressure of oxygen (see Chapter 8), and so the surface of the metal is really an interphase interface between metal and oxide crystals. Even when a compound is not formed as a result of chemical reaction between the metal and its environment, some gas is likely to be absorbed on the surface, a process that necessarily reduces the surface free energy.

Ferromagnetic domain walls. It has been pointed out that certain types of metallic ions have an inherent magnetic moment of considerable magnitude; these are iron, nickel, and cobalt; other atoms possess comparable magnetic moments when their electronic structures are suitably modified by alloying. The energy of a crystal consisting of atoms with magnetic moments is a minimum if the magnetic fields of the atoms are either aligned parallel to each other, resulting in a ferromagnetic material, or antiparallel, resulting in antiferromagnetic properties. Only the former is of interest for the present purpose. The energy is reduced further when the magnetic fields are aligned in particular directions in the crystal. This property is called magnetocrystalline anisotropy. The extent of the anisotropy varies from a very large value for cobalt, an intermediate value for iron, to a small value for nickel; the directions of easiest and hardest magnetization are as follows:

	Easy Direction	Hard Direction
Fe	$\langle 100 \rangle$	$\langle 111 \rangle$
Ni	$\langle 111 \rangle$	$\langle 100 \rangle$
Co	$\langle 0001 \rangle$	$\langle 10\bar{1}0 \rangle$

IMPERFECTIONS IN CRYSTALS

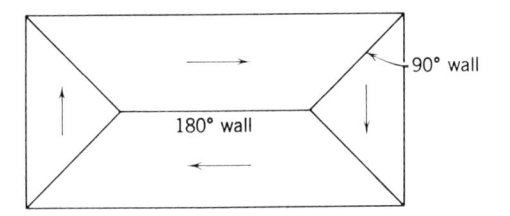

Fig. 4.41. Simple ferromagnetic domain pattern.

An ideal crystal of a ferromagnetic metal or alloy would have the atoms in all parts of the crystal so oriented that their magnetic fields would be parallel to an axis of easy magnetization. However, if the crystal is finite in extent, this would give rise to magnetic poles at some surfaces of the crystal and there would be an external magnetic field, with which some energy would be associated. This energy is reduced by dividing up the crystal into *domains* which are magnetized in different directions so as to eliminate magnetic poles and external magnetic fields. A simple case of a domain structure that satisfies these requirements is shown in Fig. 4.41. The crystal is divided into four domains, the directions of magnetization of which are shown by arrows. Each *domain wall* is so oriented that the components of magnetization normal to it are equal on both sides of the wall; there is then no unbalanced magnetic pole at any point of the crystal, either at its surface, which is everywhere parallel to the direction of magnetization, or at the domain walls. If the surfaces of the crystal are not exactly parallel to the appropriate crystal planes, this simple arrangement of domains

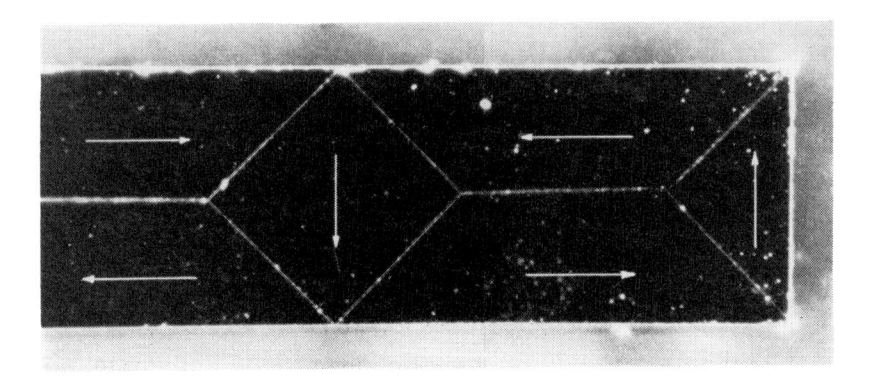

Fig. 4.42. Domain walls in an unmagnetized iron whisker. (Courtesy R. W. DeBlois and C. D. Graham, Jr., General Electric Research Laboratory.)

is not possible, and a more complicated pattern, such as that shown in Fig. 4.42, is formed. Figure 4.42 is a photomicrograph, at a magnification of $\times 250$, of the pattern formed by a colloidal suspension of a magnetic powder on the polished surface of the crystal. The domain walls, however, have some energy, since they are regions of disturbance of the uniformity of magnetization. They consist of regions of the order of 50 atoms in thickness in which there is a gradual transition from one magnetic orientation to the other. The only ferromagnetic crystals in which domain walls do not occur, in the absence of an applied magnetic field, are those that are so small that their size is comparable with the thickness of the domain wall. The properties and applications of small magnetic particles are discussed in Chapter 5. It may be pointed out that the magnetic domain wall is a thermodynamically stable type of imperfection.

4.5 Further Reading

W. T. Read, Jr., *Dislocations in Crystals,* McGraw-Hill Book Co., New York, 1953.

Metal Interfaces, American Society for Metals, Cleveland, 1952.

W. Shockley *et al.,* Eds., *Imperfections in Nearly Perfect Crystals,* John Wiley & Sons, New York, 1952.

A. H. Cottrell, *Dislocations and Plastic Flow of Metals,* Clarendon Press, Oxford, 1953.

J. G. Fisher *et al.,* Eds., *Dislocations and Mechanical Properties of Crystals,* John Wiley & Sons, New York, 1957.

5

Structure-Sensitive Properties

5.1 Stress-Strain-Time Relationships

The application of a stress to a metal may cause any of the following changes: *Elastic deformation,* in which the strain appears and disappears simultaneously with the application and removal of the stress.

Anelastic deformation, in which the strain reaches its maximum value *after* the stress has reached its maximum value, and in which the strain disappears *after* the removal of the stress. *Plastic deformation,* in which the strain occurs simultaneously with the application of the stress, but does not vanish if the stress is removed. *Creep,* in which nonrecoverable strain occurs while the stress is held at a fixed value. *Fracture,* in which physical separation takes place. These processes do not necessarily occur separately, but they may be considered individually in order to investigate the physical processes of which they are the result.

Of these processes, only the first (elastic behavior) is structure-insensitive; anelasticity and creep would not be expected to occur in a perfect crystal, and plastic deformation and fracture would occur in a perfect crystal at much higher stresses and strains than are normally found experimentally. These four processes are therefore structure-sensitive, and must be considered in relation to crystals that are imperfect.

5.2 Plastic Deformation

The very complicated plastic behavior of a polycrystalline, polyphase alloy may be considered to correspond to the properties of a single crystal of a pure metal, modified by the effects of (*a*) solutes, (*b*) crystal boundaries, and (*c*) other phases. We will therefore first consider the deformation of a single crystal.

Yield stress for the perfect single crystal. In order to limit the discussion to real cases, two experimental observations must be recognized: 1. plastic deformation may be very considerable in extent, and 2. the density and structure of the crystal are *substantially* the same after deformation as before. The only type of deformation process that can account for unchanged density and unchanged structure is shear movement of parts of the crystal with respect to other parts so that each atom finally occupies a position previously occupied by another atom. This is only possible when shear takes place in well-defined crystallographic directions. If the deformation of the crystal is macroscopically homogeneous, it must be distributed over many parallel crystallographic planes. The amount of deformation can therefore be characterized by the amount of shear (expressed as an angle), the family of parallel planes in which it occurs, and its direction. Several such processes can take place simultaneously. It would be expected that shear would occur on planes and in directions in which the resolved shear stress is a maximum; however, the structure of the crystal must be considered, as follows: Figure 5.1 represents a simple crystal structure. Assume that a stress is applied so that the maximum shear stress is parallel to the plane ABC. Shear on this plane to satisfy the condition defined above would require atom A to move to positions $B, C. \ldots$ This is a much more complicated process than if shear takes place on a plane such as $APQ \cdots$ in which atom A passes through successive similar positions $P, Q, \cdots$.

It has been shown experimentally that the deformation of single crystals can be completely described in terms of shear in specific crystallographic directions, which are always the directions of closest packing of atoms, and usually on well-defined crystallographic planes which are planes of close packing and therefore of relatively large separation.

In the perfect crystal, therefore, we must envisage the process of plastic deformation as the simultaneous movement of atoms into new but equivalent positions. The stress necessary to cause this to occur

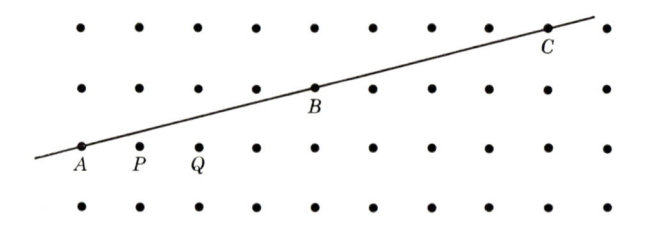

Fig. 5.1. Shear in high index plane in a crystal.

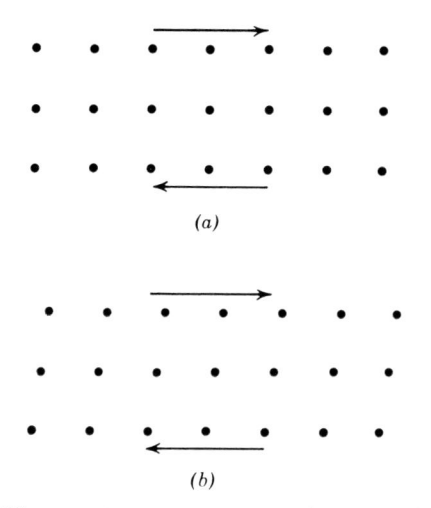

Fig. 5.2. Homogeneous shear in a crystal.

can be estimated as follows: Consider a crystal, Fig. 5.2a, to which a shear stress is applied in the direction indicated by the arrows. This will cause elastic shear as shown in Fig. 5.2b. At some value of the shear strain, each atom will move to the next equivalent position to the right of its previous position; if we consider the energy of an atom, as affected by its position relative to the underlying row, it is apparent that a minimum occurs at each point that constitutes a lattice site; at intermediate points, the energy is higher. The energy is therefore given by a curve of the kind represented in Fig. 5.3a. The force required to maintain an atom in any position is given by the slope of this curve, shown in Fig. 5.3b; the slope reaches a maximum at the points of inflection P of the curve of Fig. 5.3a. It follows that if the stress is sufficient to displace each atom beyond the points corresponding to P, the atoms will move on to the next stable sites; this will continue as long as the shear stress is maintained. The critical shear strain corresponding to the displacement of atoms from A to P depends upon the crystal structure and on the slope of the energy-displacement curve. The critical strain is that at which the applied stress is a maximum, and if it is assumed that the curve of Fig. 5.3b is sinusoidal, this occurs when the stress is $G/2\pi$ or at about one-sixth of the appropriate shear modulus. Reasonable alternative assumptions would reduce the critical shear stress to about $G/15$, corresponding to a shear strain of about $4°$.

Experimentally, the results of measurements of the shear stress required to cause plastic deformation of crystals may be divided into two groups. Most of the results give stresses smaller than the theoretical

142   PHYSICAL METALLURGY

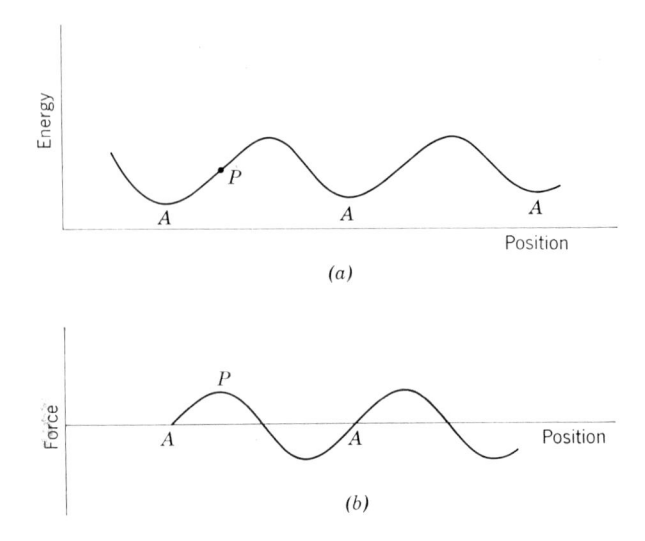

Fig. 5.3. Variation of (a) energy and (b) force with shear displacement in a crystal.

value by two or three orders of magnitude. Table 5.1 gives some examples; the criterion is the ratio of the critical shear stress to the shear modulus. These results are all for single crystals with cross sections of the order of a square millimeter, except for the "whisker." The recent discovery that some of the "whiskers" that grow spontaneously at the surface of certain metals have strengths close to the theoretical values gives very strong support to the theory, and to the conclusion that, except in the case of the whiskers, which are extremely thin, the assumed process of simultaneous shear deformation cannot be the real one. Another reason for reaching this conclusion is that the process of plastic de-

Table 5.1. Comparison of Shear Modulus and Elastic Limits

	Shear Modulus G (dynes/cm^2)	Elastic Limit B (dynes/cm^2)	G/B
Sn, single crystal	1.9×10^{11}	1.3×10^7	1.5×10^4
Ag, single crystal	2.8×10^{11}	6×10^6	4.5×10^4
Al, single crystal	2.5×10^{11}	4×10^6	6×10^4
Fe, single crystal	7.7×10^{11}	5×10^6	1.5×10^5
Fe, whisker	7.7×10^{11}	2×10^{11}	3

(a)

(b)

Fig. 5.4. Slip lines in (a) polycrystalline aluminum (×115) and (b) single crystal of α brass (×9800). [(a) Courtesy U. F. Kocks, Harvard University; (b) H. G. F. Willsdorf, Franklin Institute.]

formation considered above should occur simultaneously and uniformly throughout the crystal, because the stress should be the same everywhere, and each atom should simultaneously reach the critical position, when the applied stress reaches the critical value. It is observed, however, the plastic deformation does not occur uniformly; it occurs on relatively few individual planes of the family of planes subjected to the same shear stress. Many of these active planes are grouped into localized regions known as slip bands. The appearance of the surface of a crystal that has undergone plastic deformation is shown, at low magnification in Fig. 5.4a, and at high magnification in Fig. 5.4b.

Dislocation theory of plastic deformation. The only general explanation that has been advanced for the low values of these stresses is that, owing to crystal imperfections, the process does not take place simultaneously throughout the whole crystal, but occurs sequentially, by the motion of dislocations. It has been shown on p. 102 that a dislocation can move in response to the application of a shear stress, and that shear strain results. The stress required to cause a dislocation to move is much less than that required to produce simultaneous shear, because only one atom at a time moves through the critical position (in each plane). It will be seen from Fig. 5.5 that the movement of a dislocation right across a crystal corresponds to the relative movement of the two parts of the crystal by one interatomic spacing. It is seen that a single dislocation has moved through the crystal from right to left. It is both valid and convenient to consider a dislocation as the transition structure between the part of a crystal that has "slipped" by a unit amount and the part that has not. An edge dislocation can move by slip only in its slip plane, which is the plane containing the line of the dislocation and the Burgers vector. It can move by slip under quite small shear stresses in an otherwise perfect crystal.

A screw dislocation also moves under an applied shear stress; but in this case the effective component of the stress is parallel to the dislocation (i.e., parallel to the Burgers vector) and the dislocation moves in

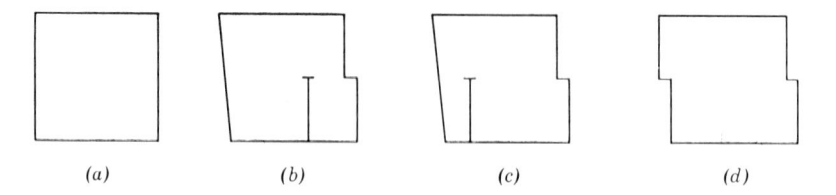

Fig. 5.5. Movement of a dislocation through a crystal.

a direction that is perpendicular to the applied stress. However, it is still correct to regard the dislocation as the transition between "slipped" and "unslipped" parts of the crystal; it moves so that the "slipped" part increases in response to the applied stress. The slip plane, however, is no longer uniquely defined by the dislocation, although it is still a plane containing the Burgers vector and the dislocation line. Since these coincide, they do not define a plane but a family of planes containing this line. The slip plane is in fact any crystallographic plane of high atomic density containing the dislocation line and also supporting a large component of the applied shear stress. The movement by slip of a screw dislocation is illustrated in Fig. 5.6.

It will be seen that a "step" of one Burgers vector height is formed whenever an edge dislocation reaches the surface of a crystal which is parallel to the line of the dislocation. No step is formed at a surface through which a screw dislocation moves as it passes out of a crystal, because the Burgers vector and consequently the slip movement are parallel to the surface. The concept that the movement of a dislocation can occur under small stresses, and can cause plastic deformation, only explains one of the experimentally observed features of plastic deformation. The other salient features of plastic deformation of single crystals of pure metal that require explanation are as follows.

The amount of deformation is far greater than can be accounted for by the movement of dislocations that were originally in the crystal. For example, it is possible to deform a single crystal of cadmium by 300%. This would mean that each atom plane must, on the average, have moved about three atom spacings over the next one, or that three dislocations moved across each atom plane and out of the crystal. If

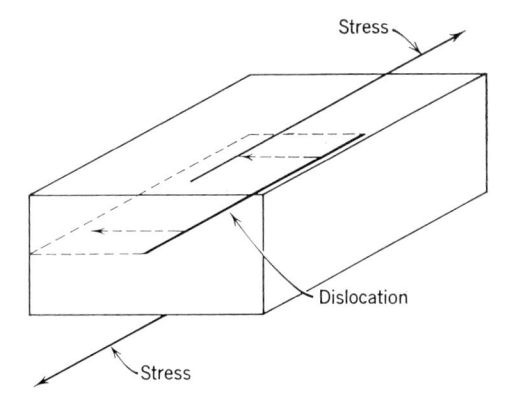

Fig. 5.6. Illustrating transverse motion of a screw dislocation.

the cross section of the crystal were 1 mm², this would correspond to about 10⁹ dislocations/cm² leaving the crystal. However, X-ray evidence shows that "good" crystals contain only $10^4 - 10^6$ dislocations/cm².

The stress required to produce further plastic deformation increases continuously as deformation proceeds; this is called *work hardening*. This would perhaps be compatible with a decrease in the number of dislocations except that (a) the crystals become increasingly distorted, as shown by X-ray diffraction, as deformation proceeds; (b) the energy of the crystal increases as work hardening occurs, as shown by calorimetric studies of work-hardened metals, which indicate that a dislocation content of 10^{12}/cm² may be attained; (c) the electrical resistance, and particularly the residual resistance, increase as work hardening proceeds; and (d) etch pit studies show an increase in dislocation content with increasing deformation. Consequently, it has to be concluded that dislocations are generated during plastic deformation, and that hardening is one of the consequences of the increase of dislocation content.

The Frank-Read source. It has been shown theoretically that an applied stress of the appropriate magnitude cannot generate new dislocations within a perfect crystal by the formation of a dislocation loop where none existed before, even with the assistance of thermal agitation. The energy per unit area of slip plane that would be required to form a very small dislocation loop, by shear of a very small amount of crystal, would be very large compared with the strain energy and the thermal energy present in the crystal under stresses that cause plastic deformation, and the same argument is believed to apply to the generation of new dislocations at crystal surfaces. However, a satisfactory mechanism, based on the expansion and subdivision of existing dislocations, has been proposed. It is known as the Frank-Read source. It can be described as follows:

Consider a dislocation segment PQ in a crystal (Fig. 5.7). It is considered to be "anchored" at points P and Q, either by meeting other dislocation components at "nodes" or by second-phase particles or impurities. The dislocation cannot terminate at these points, but the part joining them is the only part that needs to be considered. We will consider initially the case in which the Burgers vector is perpendicular to the line PQ. A shear stress parallel to the plane of the diagram, and perpendicular to PQ, will exert a force on the dislocation, causing it

Fig. 5.7. Dislocation segment.

STRUCTURE-SENSITIVE PROPERTIES 147

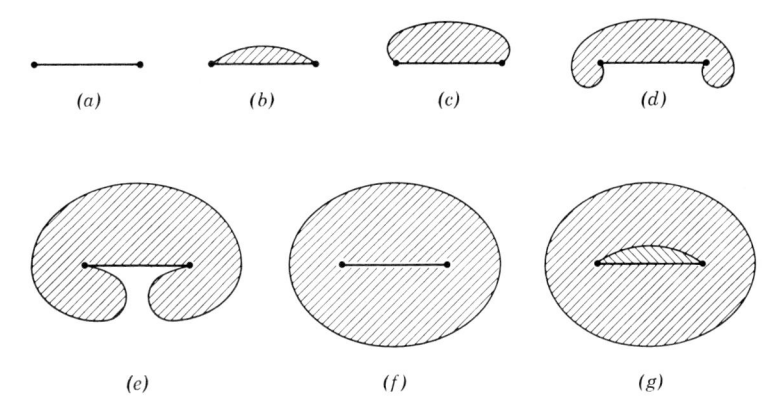

Fig. 5.8. Successive stages in the operation of a Frank-Read source.

to move so as to allow some shear of the part of the crystal above the plane of the diagram with respect to the part below. This force is opposed by the line tension (see p. 101) of the dislocation, i.e., by the fact that the energy of the dislocation is proportional to the length, and therefore work must be done to extend it. If sufficient force is applied, the "slipped" area will increase through the successive stages shown in Fig. 5.8, where the hatched region is the region through which the dislocation has moved as a result of the applied stress. It will be seen that when the original dislocation forms a closed loop, Fig. 5.8f, there is a component in the original starting position of Fig. 5.8a, and the process can therefore be repeated indefinitely, as far as the geometrical aspects of the process are concerned. It should be noted that the process depends upon movement of the dislocation in all possible directions in relation to the applied stress; but the movements are all in such directions as to increase the "slipped" area, and can therefore all result from the application of a stress in a single fixed direction. Continued application of a sufficient stress will cause the generation of a series of loops, all in the same slip plane.

Figure 5.9 shows an infrared photomicrograph of "decorated" dislocations in a silicon crystal which had been plastically deformed before decoration. It demonstrates all the features that had been predicted for the Frank-Read source.

If the dislocation loops reach a free surface of a crystal, they may break through, leaving a "step" which is visible if of sufficient size. If loops can emerge all round a crystal, as for example in a thin cylindrical crystal, the process of generating loops can continue almost

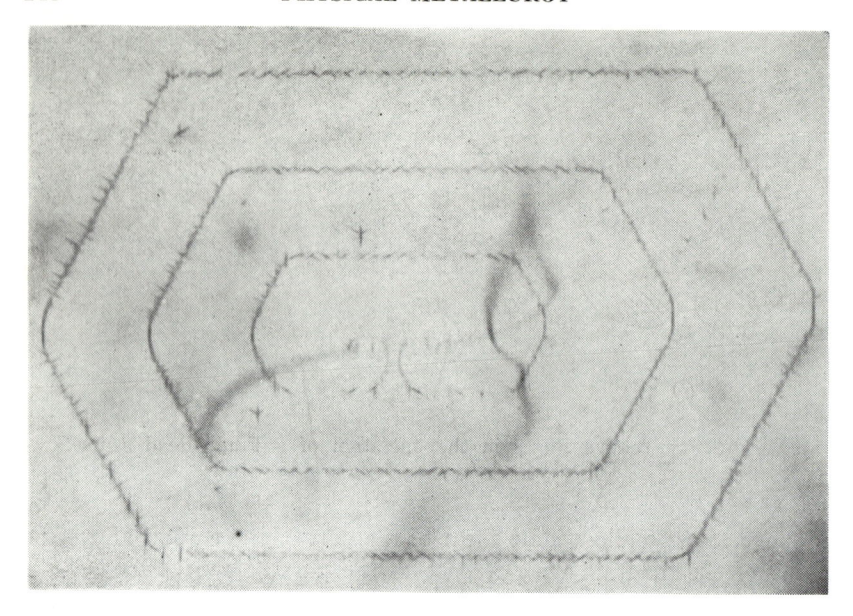

Fig. 5.9. Infrared photomicrograph of Frank-Read source in silicon. (Courtesy W. Dash, General Electric Research Laboratory.)

indefinitely. If, on the other hand, the loops cannot emerge, but are forced to remain within the crystal, they exert a "back pressure" on the source, because dislocations in the same slip plane repel each other as a result of the overlap of similar regions of neighboring dislocations. It follows that, when the dislocations cannot reach the surface at all points, the number of dislocations that can emerge from a given source is limited, or else it becomes necessary to apply a progressively increasing stress to compress more loops into the available space.

It is necessary now to consider the stress required to operate a source. Let the length of the source be L (Fig. 5.10), and assume that the dislocation always has uniform curvature, the radius of which is R. As the dislocation is "bowed-out" from its original position, its radius of curvature decreases until $R = L/2$, after which it increases

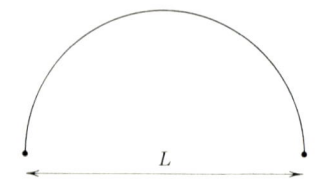

Fig. 5.10. Limit of stability of dislocation loop.

again. The application of a shear stress σ applies a force $\mathbf{b}\sigma$ to the dislocation, and this is opposed by the restoring force T/R due to the line tension of the dislocation. The dislocation takes up a curvature such that these forces balance, unless this demands a radius of curvature less than $L/2$, in which case the loop goes on expanding and the source operates. The critical stress σ_c for the operation of a source is, therefore, given by $\mathbf{b}\sigma_c = 2T/L$. But $T \approx G\mathbf{b}^2$ (see p. 101) and therefore $\sigma_c \approx 2G\mathbf{b}/L$.

The length L that corresponds to the observed values of the critical shear stress for good crystals is of the order of 10^{-4} cm.

The case that has been discussed is for a source that is completely within the crystal; if, however, a dislocation has one end that is free to move at the surface of a crystal, and the other end is "pinned" in the interior, it may act as a source; the stress required to operate such a source is one-half of the stress for an internal source of the same length. There is some evidence that surface sources may be important in plastic deformation.

Slip systems. The plane or planes in which slip takes place in a crystal can be determined by a study of the surface markings which correspond to the intersection of the slip planes with the surface. Unambiguous identification requires the examination of at least two surfaces of a crystal. The identification of the slip direction can only be made with certainty by studying the change of crystal orientation with respect to the external geometry of the specimen. Table 5.2 shows the slip directions and slip planes for the structures for which they have been determined. It can be seen that the slip direction is always that which corresponds to the shortest Burgers vector, i.e., the closest linear packing of the atoms.

The stress-strain curve. We will now consider the various parts of the stress-strain curve in tension for a single crystal of a pure metal, and its interpretation in terms of dislocations.

Table 5.2. **Slip Directions and Planes**

Structure	Slip Direction	Slip Plane
F.C.C.	$\langle 110 \rangle$	$\{111\}$
B.C.C.	$\langle 111 \rangle$	$\{100\}, \{112\}, \{123\}$
C.P. Hex	$\langle 11\bar{2}0 \rangle$	(0001)
Tetragonal (tin)	$\langle 001 \rangle$	$\{100\}, \{110\}$

Fig. 5.11. Stress-strain curve for zinc single crystal; basal plane at 45° to direction of tension stress.

The simplest case is that of the close-packed hexagonal crystals in which the axial ratio is large (zinc and cadmium) with the result that slip normally only takes place on a single family of planes [the basal or (0001) planes]. The typical shear-stress strain curve for zinc or cadmium, originally oriented with the slip direction and slip plane at 45° to the axis of tension, is shown in Fig. 5.11. The corresponding load-elongation curve differs from this because: 1. the cross section decreases as shear takes place, and 2. the angles between the slip plane and direction, and the axis of tension, change as shear takes place. These changes are illustrated in Fig. 5.12. If the tension stress is S, and the slip plane and slip direction make angles θ and λ respectively with the direction of tension, then the resolved shear stress σ is given by

$$\sigma = S \cos \lambda \sin \theta$$

The factor $\cos \lambda \sin \theta$ which is used for converting tensile to resolved shear stress is called the Schmid factor. The Schmid factor has a maximum value of 0.5 when θ and λ are both 45°; they both increase during slip in tension. It follows that if angles θ and λ are greater than 45° initially, the Schmid factor increases as slip takes place, and on this basis it should be easier for slip to continue than for it to start.

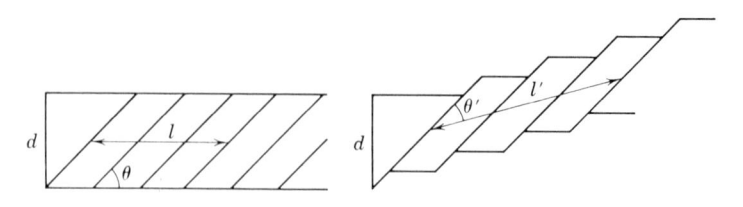

Fig. 5.12. Change of orientation during slip.

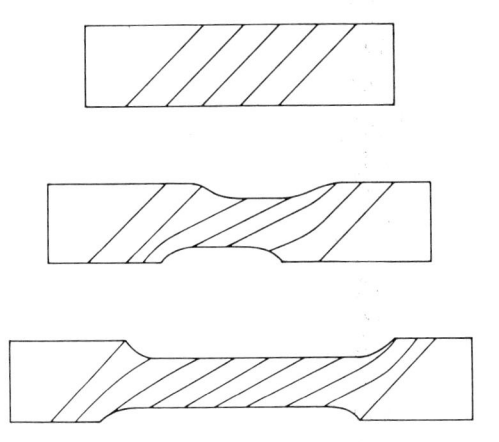

Fig. 5.13. Inhomogeneous slip accompanied by bending.

This "geometrical softening" effect has been observed in crystals of cadmium with large initial values of θ and λ; the resulting deformation is not uniform, but takes place by a type of yield phenomenon in which the first part of the crystal to yield undergoes considerable extension; the "stretched" region then gradually spreads along the crystal under substantially constant stress if the stress is maintained. Several successive stages of the process are shown diagrammatically in Fig. 5.13. It is evident that this process cannot happen without the slip plane at the ends of the slipped region being considerably bent, and microscopical examination shows this to be so. The bend can be accounted for by a "pile-up" of dislocations, which subsequently leave the crystal as the slip planes again become straight in the stretched part of the crystal.

If the values of θ and λ are initially less than 45°, so that the Schmid factor decreases as the crystal is stretched, then a progressively higher stress is required to cause increased deformation. This is called "geometrical hardening." When the initial values of θ and λ for a good cadmium crystal are about 45°, it is possible to stretch it by amounts of the order of 300%. The stress increases during the process by an amount rather greater than that required by geometrical hardening. This additional hardening is an example of work hardening. The process of extension by slip is limited in the case under discussion either by the incidence of twinning or by fracture; these processes will be discussed on pp. 158 and 194.

It has been demonstrated that plastic deformation by slip begins when the resolved shear stress reaches a value that is characteristic of

152 PHYSICAL METALLURGY

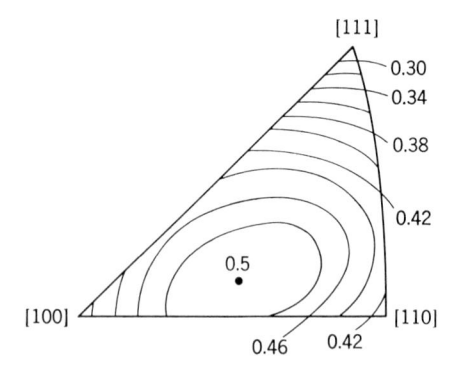

Fig. 5.14. Contours for equal Schmid factor.

the material and independent of the orientation and, therefore, of the stress normal to the slip plane. This is the critical resolved shear stress.

In a crystal of face-centered cubic structure, there are 12 possible slip systems, each of which is a {111} plane combined with a ⟨110⟩ direction that lies in it. When a stress is applied to such a crystal, the Schmid factor will in general be different for the various systems. The values depend on the orientation of the crystal in relation to the axis of tension. An indication of the variation of the highest Schmid factor with orientation is given in Fig. 5.14. Each of these Schmid factors is associated with 11 other ones, in general of smaller value. However, there are special orientations, corresponding to the sides and corners of the typical stereographic "triangle," for which the Schmid factor is the same for two or more systems. Figure 5.15 indicates the multiplicity of systems with equal Schmid factors. In the most common case, therefore, a single crystal of face-centered cubic structure subjected to a tensile stress will have a greater resolved shear stress on one system than on any other.

When such a crystal is subjected to a tension stress, its stress-strain

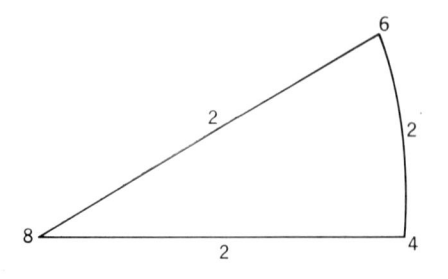

Fig. 5.15. Multiplicity of systems with equal Schmid factors.

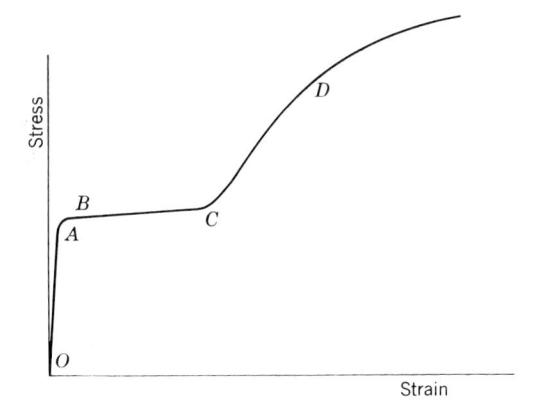

Fig. 5.16. Stress-strain curve for face-centered cubic crystal in tension (schematic).

curve is as shown in Fig. 5.16, in which the part OA corresponds to elastic behavior, and AB is a region of transition from the elastic region to the region of "easy glide" BC. At C, the slope increases to a much greater value, and thereafter the curve is either linear, followed by a parabolic region, or else the parabolic region begins immediately.

The interpretation of this curve is as follows: Region AB is one in which the stress reaches a level high enough to activate some Frank-Read sources in the system with the highest Schmid factor, so that they give out some dislocation loops. Near point B the stress is sufficient to cause the dislocations to break through the surface or other barriers (such as subboundaries), and from B to C the deformation is a result of the operation of sources in the system that has the greatest Schmid factor. The slight increase of stress that occurs between B and C is of unknown origin, but it may be due to the interaction of the moving dislocations with other existing dislocations that play no part directly in the deformation process, as follows. We will first examine the conditions that exist when a moving dislocation encounters a dislocation which is stationary. There are several quite distinct cases, depending on the relationship between the Burgers vectors of the two dislocations. When a moving dislocation intersects a stationary one, each acquires a "jog" AB (see Fig. 5.17) equal in length and direction to the Burgers

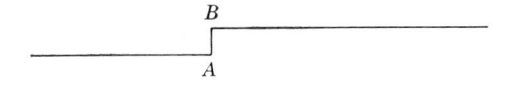

Fig. 5.17. "Jog" in a dislocation.

vector of the other dislocation, since each is subsequently in a region consisting of two parts displaced, by the Burgers vector, from the corresponding parts of the original crystal.

A jog is a step in the *direction* of the dislocation line; the Burgers vector of the dislocation is the same at every point on it, including the jog. There are three types of jog that can be produced by the intersection of two dislocations: 1. A jog whose slip plane contains the direction of motion of the moving dislocation; this does not hinder the motion of the dislocation. 2. A jog which cannot move by slip in the direction of motion of the dislocation, but can move by slip along the line of the dislocation. There is some doubt whether such jogs must move with the dislocation, generating vacancies or interstitials as they move (forced climb) or whether they move along the dislocation by slip and rapidly cease to have any influence on the process. If the former possibility is valid, then an additional stress is necessary to move the dislocation. 3. A jog may be formed in a plane which is not a slip plane of the crystal [e.g., a (100) plane in a face-centered cubic crystal]. This is a sessile jog and should make the dislocation more difficult to move.

Thus it is to be expected that a dislocation moving through a "forest" of stationary dislocations should become progressively more and more jogged, which would account for the slight work hardening that occurs during the operation of a single slip system in easy glide. There is experimental evidence, based on etch pit studies of dislocations, that loops from the sources expand much faster in the "edge" than in "screw" directions: the loops are elongated in the direction perpendicular to the Burgers vector.

The termination of easy glide corresponds to the initiation of slip on a second system. This must operate on a different plane from the primary system, and it must have the largest Schmid factor that is compatible with this condition. The criterion for the activation of the second slip system is not consistent; in some cases, such as aluminum, it occurs well before the change of orientation due to slip has reached the point at which the shear stress on the second system is equal to that on the first; on the other hand, some crystals, such as α brass, continue to slip on the primary system well beyond the point at which a second system is equally stressed. This difference may depend on the extent to which the secondary system is "hardened" by the action of the primary system. The stress on the secondary system should be considered as including any stresses that occur as a result of the constraints applied to the specimen by the machine in which it is

STRUCTURE-SENSITIVE PROPERTIES

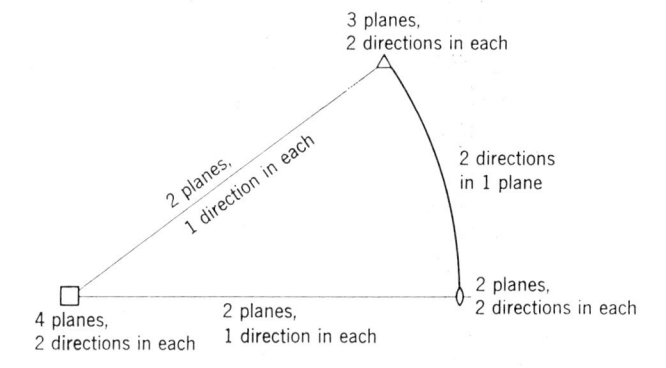

[Figure: stereographic triangle with labels at vertices and edges:
- Top vertex: 3 planes, 2 directions in each
- Left vertex: 4 planes, 2 directions in each
- Right vertex: 2 planes, 2 directions in each
- Left edge: 2 planes, 1 direction in each
- Bottom edge: 2 planes, 1 direction in each
- Right edge: 2 directions in 1 plane]

Fig. 5.18. Slip systems at special orientations.

being tested; these "parasitic" stresses depend on the geometry of the specimen and may be quite significant for relatively short, thick specimens.

The part of the curve beyond the end of easy glide is a region of rapid but decreasing work hardening. The structure becomes progressively more imperfect, as shown by X-ray diffraction. (In the region of easy glide, there is no increase of asterism.) The work hardening that occurs when a second slip system becomes active (the region CD of Fig. 5.16) is probably due to the following sequence of events. Dislocations on the two active systems interact to form Cottrell-Lomer supersessile dislocations, and each of these acts as an obstruction to the dislocations which follow in the same planes. This causes dislocation "pile-ups," each of which is surrounded by a stress field which opposes the motion of other dislocations in adjacent planes. The sources that gave rise to the dislocations that have piled up cease to operate, and previously inactive sources, which require higher stresses, come into play. Eventually the stresses become high enough for dislocations to bypass the obstructions by cross slip,* and this leads to a decrease in the rate of work hardening, as observed experimentally.

The foregoing discussion relates to a crystal that is so oriented that it has one preferred slip system, i.e., in which one combination of slip direction and slip plane has a larger Schmid factor than any other. There are, however, special orientations in which this is not the case. These are represented by points on the perimeter of the stereographic triangle. Figure 5.18 shows the multiplicity of slip systems for the

---

* Cross slip is the motion of a dislocation on a plane inclined to that on which it was moving previously. It can only do so if it is a screw dislocation and if it is not dissociated.

sides and corners of the triangle. As would be expected from any theory that relates work hardening to the intersection of moving dislocations, the easy-glide portion of the stress-strain curve is not found in any case in which there are two or more active planes. It would also be expected that the rate of work hardening should increase with increasing multiplicity of equally stressed slip systems. Both of these expectations are confirmed, as is shown in Fig. 5.19.

It should be noted that each slip system corresponds to the movement, in a particular plane, of a procession of dislocations whose

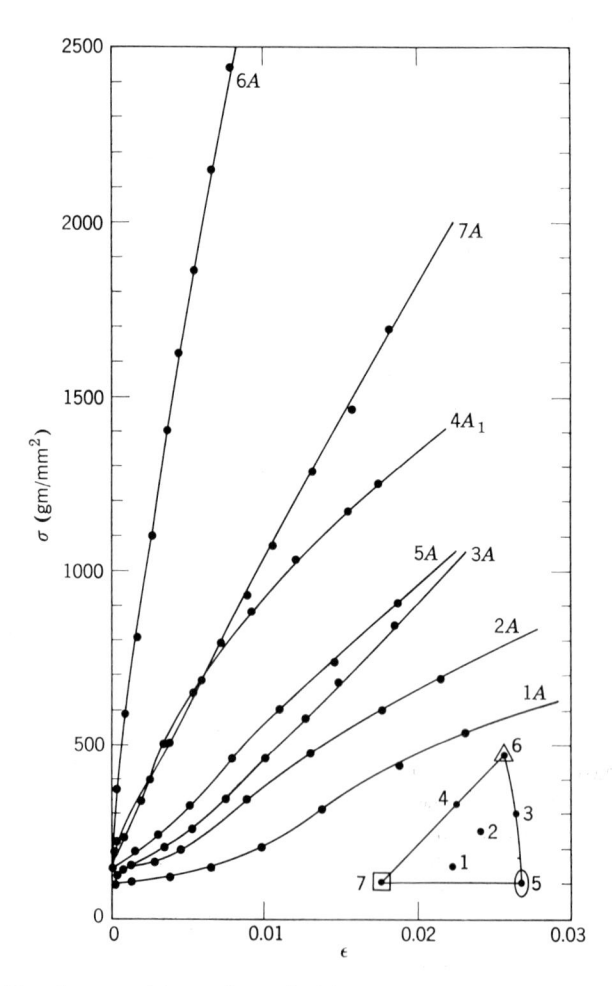

Fig. 5.19. Tensile stress (σ)–tensile strain (ϵ) curves and axial orientations of single crystals of aluminum.

STRUCTURE-SENSITIVE PROPERTIES

Burgers vector is the slip direction. The dislocations will, in general, be partly of edge and partly of screw character.

In the face-centered cubic crystals, the slip system is, with possible minor exceptions, well defined as regards both direction and plane. However, in the body-centered cubic system, the situation is more complicated. Slip is apparently always in the ⟨111⟩ direction, which is the direction of closest packing of the atoms, but the slip plane is by no means so well defined. It has been identified as (110), (112), and (123). The surface markings that are caused by slip (slip bands) are not as straight or as regular in body-centered cubic crystals as they are for face-centered cubic structures.

Nonhomogeneous deformation processes. Reference has already been made to the heterogeneous deformation of cadmium single crystals (see p. 151). A somewhat similar process of heterogeneous deformation, which also occurs at the beginning of plastic deformation, is that associated with the formation of Lüders bands in iron. These are the regions in which substantial plastic deformation has already taken place. Figure 5.20 shows a single crystal of iron in which deformation by Lüders bands has occurred. Deformation begins by the formation of a Lüders band, and continues by the propagation laterally of the band, and by the formation and growth of additional bands. As in the case of cadmium, this type of heterogeneous deformation is due to the fact that less stress is required to continue deformation than to start it. The reason for this is different in the case of iron from that of cadmium; it will be discussed on p. 402.

The late stages of plastic deformation are often notable for marked heterogeneity in the deformation process. The processes that lead to the subdivision of the volume into microscopically or macroscopically

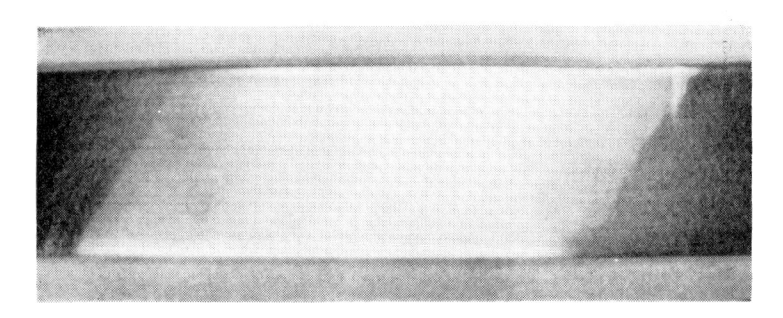

Fig. 5.20. Lüders band in steel (×1). From R. B. Liss, *Acta Metallurgica*, Vol. 5, No. 6, June 1957, Fig. 3, p. 342.

observable regions that differ significantly in orientation are (a) *twinning* and (b) *the formation of deformation bands.*

TWINNING. The formation of twinned regions as a result of stress appears to depend on the achievement of a critical shear stress for twinning in a suitable direction in a crystal. The details vary from one structure to another. In the face-centered cubic structure, it was thought until recently that twinning did not take place as a result of stress. It has been proved recently, however, that copper single crystals twin under tensile stress at extremely low temperatures (4.3°K).

In the close-packed hexagonal metals with a large c/a ratio (zinc, cadmium, magnesium) twinning occurs under tension when the Schmid factor for the slip plane (0001) is low, and the Schmid factor for the twinning plane is high. This condition is satisfied when the slip plane is nearly parallel to the axis of tension.

In the case of body-centered cubic structures, twinning occurs only after some plastic deformation or when the stress is applied rapidly. This is apparently due to the increase in the stress required to cause deformation by slip when strain is rapid; this allows the stress to reach the value required for twinning. Twin bands in iron are sometimes called Neumann bands. The same conditions are necessary for the twinning of tin, which has a tetragonal structure.

Twinning appears to take place very rapidly, and only after considerable elastic energy has been stored in the crystal. The process of twinning is characteristically accompanied by a sound of high pitch and short duration, an example of which is the well-known "cry of tin" which is heard when a bar of tin is bent. A dislocation mechanism for twinning has been proposed, as follows. Twinning requires a shear displacement that moves each atom to a new position that is not equivalent to its original position; it is therefore equivalent to the passage of a suitable partial dislocation. A twin of observable size can only be formed if the necessary shear movement takes place in a large number of consecutive lattice planes. The dislocation that causes twinning must therefore sweep the crystal a large number of times, its plane being displaced by one atomic plane each time. This can be achieved by a partial dislocation line OA (Fig. 5.21), whose Burgers vector is the twinning shear. It is "anchored" to the dislocations PO and PQ, which have a Burgers vector with a screw component equal to the interplanar spacing of the twinning plane. The dislocation OA rotates around point O, which, however, moves up or down by one plane each time the twinning dislocation goes around it.

It is probable that a "double-ended" source analogous to that shown

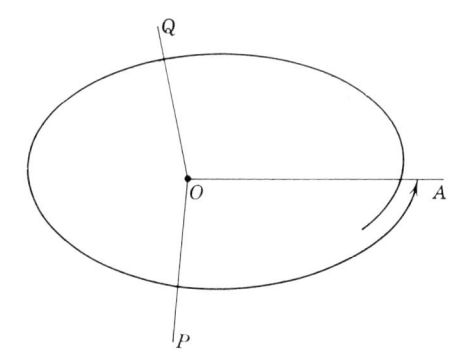

Fig. 5.21. Twinning dislocation.

in Fig. 5.8 could be constructed to account for the end A of the rotating dislocation OA.

The twinning process causes a change in the geometry as well as in the crystallography of the specimen, and may therefore be a component part of the process of plastic deformation. Twinning cannot occur as a result of tension unless the twinned crystal is longer, in the stress direction, than the untwinned, and vice versa for compression.

The boundary between a twinned and an untwinned part of a crystal is usually a plane surface, and its section in a photomicrograph is therefore a straight line; the boundaries at the two sides of a twin appear as parallel lines; this is because the energy of a twin boundary is greatly increased if it loses coherence by departing from the position of symmetry between the two lattices. If, however, the change in shape that accompanies twinning is opposed, for example by another, previously formed, twin, then the boundaries may be somewhat curved, and no longer parallel to each other. An example is shown in Fig. 5.22.

DEFORMATION BANDS. The remaining type of inhomogeneous deformation is that in which definite bands of the crystal change in orientation differently from the remainder of the crystal. An example is shown in Fig. 5.23.

Deformation bands have been studied in most detail in aluminum, in which metal they have the following characteristics: The bands are thin sheets, of the order of 0.05 mm thick, about 1 mm apart; they are roughly perpendicular to the active slip direction. The crystal axes in the deformation band are inclined to those in the remainder of the crystal by rotation about an axis in the slip plane perpendicular

160   PHYSICAL METALLURGY

Fig. 5.22. Shape of twin crystals. From A. J. W. Moore, *Acta Metallurgica*, Vol. 3, No. 2, Mar. 1955, Fig. 6b, p. 166.

to the slip direction. The sense of this inclination is that the crystal axes in the band are *less* rotated from the original orientation of the crystal than in the remainder of the crystal. It appears that the bands begin to form at the end of easy glide, and they may be the regions in which a second slip system first becomes operative. It has also been suggested that the deformation bands are regions that have undergone less slip than the bulk of the crystal owing to some defect that prevents dislocations from passing through such regions. Dislocations of oppo-

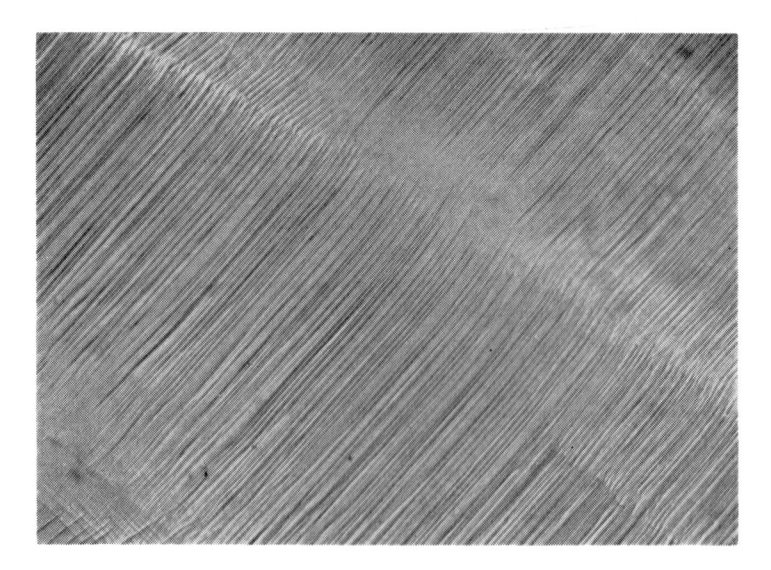

Fig. 5.23. Photomicrograph of deformation bands in aluminum ($\times 100$) produced by 10% tensile deformation. (Courtesy U. F. Kocks, Harvard University.)

site sign would pile up at opposite sides of each band; this would cause lattice curvature, which is consistent with the observations.

Effect of size. Many experiments have shown that the plastic properties of single crystals depend on their size; the general conclusion is that the yield stress decreases as the size increases, and there is some indication that the rate of work hardening in the easy-glide region increases with increasing size of the crystal. It also appears to have been established that the appropriate parameter for "size" is the length of the slip direction in the crystal.

It is probable that the effect of size is due at least in part to the oxide film that is normally present on a metal surface. The curves shown in Fig. 5.24 demonstrate that the thickness of the oxide film has a significant effect on the yield stress. This may be because the oxide film prevents the escape of dislocations at the surface until the stress is sufficient to cause local fracture of the oxide film. The stress that must be considered is that in the neighborhood of the active slip plane due to a pile-up of dislocations.

The shear stress exerted by the leading dislocation of a pile-up of n dislocations in a slip plane is n times the shear stress σ that is applied to the slip plane. This local stress, which is exerted on the oxide film or other obstacle, may be much higher than the applied stress, and its value should increase as the area of the slip plane increases. Thus the strengthening effect of the surface film decreases as the size of the specimen increases. The importance of the *length in the slip direction* suggests that it is the emergence of edge components at surfaces normal to the slip direction that constitutes the limiting factor.

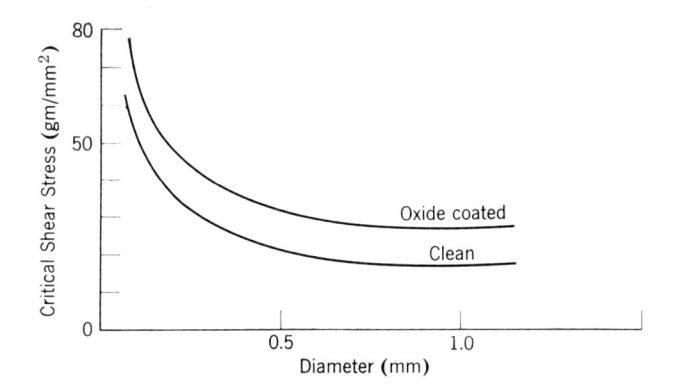

Fig. 5.24. Effect of size and oxide film on yield stress for cadmium. From E. N. da C. Andrade, The Effect of Surface Condition on the Mechanical Properties of Metals, in *Properties of Metallic Surfaces,* Institute of Metals, London, 1953, p. 139.

The effect of size on the work-hardening rate does not appear to be a result of the oxide film, and is probably associated with the fundamental mechanism of work hardening; for example, the probability of a dislocation from an active source getting close enough to a second dislocation to form a "Cottrell-Lomer immobile dislocation" increases as the distance of travel increases; each time this happens, the source becomes inoperative and a higher stress is needed to start another source. This would constitute work hardening. Other speculative explanations have been offered.

Effect of environment. It has been reported that the electrical and mechanical properties of single crystals are modified by contact with a surface-active solute in a nonpolar solvent. This is referred to as the Rehbinder effect. The electrical effect has not been found by other investigators and it has been demonstrated that the mechanical effect is due to the disruption of the surface oxide film by the surface-active material. Thus there is apparently no confirmation of the existence of an effect attributable to the penetration of polar molecules into the crystal.

Effect of temperature. The effect of temperature on the plastic deformation of single crystals has not been investigated thoroughly enough for detailed relationships to be quoted; the following generalizations, however, appear to be justified. The effect of raising the temperature is generally to lower the flow stress at any strain; however, the extent of easy glide usually increases at lower temperatures, and this may cause the stress-strain curve to be lower for some strains

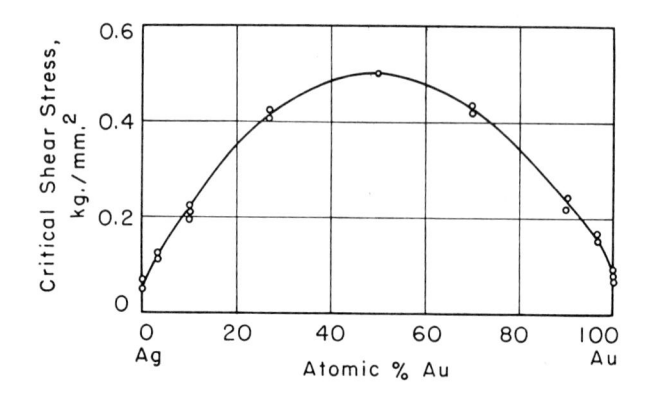

Fig. 5.25. Effect of composition on critical resolved shear stress for crystals of copper-gold alloys. From *Relation of Properties to Microstructure*, American Society for Metals, Cleveland, 1954, Fig. 1, p. 31.

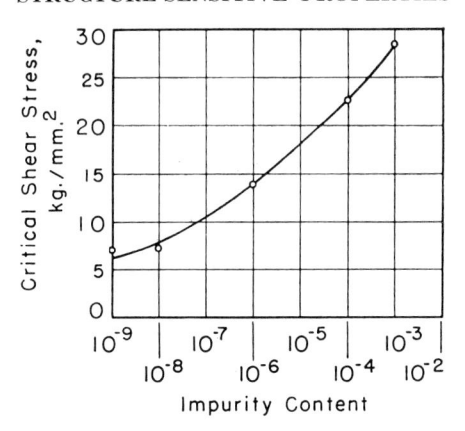

Fig. 5.26. Effect of silver on the yield stress of single crystals of mercury. From *Relation of Properties,* ASM, Fig. 2, p. 31.

at a lower than at a higher temperature. The extent of the "linear" part of the stress-strain curve increases at lower temperatures, and the rate of hardening in the parabolic region also increases at lower temperatures.

A specific effect of temperature is as follows: Aluminum single crystals tested at room temperature show no linear hardening region, whereas copper crystals give the typical curve consisting of easy glide followed by linear and parabolic regions. Aluminum tested at liquid nitrogen temperature behaves similarly to copper at room temperature.

Plastic properties of solid solutions. It is always found that a solid solution offers more resistance to plastic deformation than the pure solvent metal; for example, the critical shear stress for single crystals of copper-gold alloys is shown in Fig. 5.25. Even very small amounts of impurities have quite large effects, as shown in Fig. 5.26. It is not only the critical resolved shear stress that is raised by the presence of solutes; Fig. 5.27, for example, shows stress-strain curves for a series of alloys of aluminum and magnesium; these tests were made on polycrystalline samples but they serve to illustrate the increase of both the work hardening and the critical stress with increasing solute content. The explanation of these effects is still in some doubt; the current ideas are outlined below.

The presence of a substitutional solute atom in a crystal causes a stress field because of the difference in size and, in suitable cases, electron contribution between solvent and solute. The stress field interacts with the stress field around a dislocation, the result being a force either

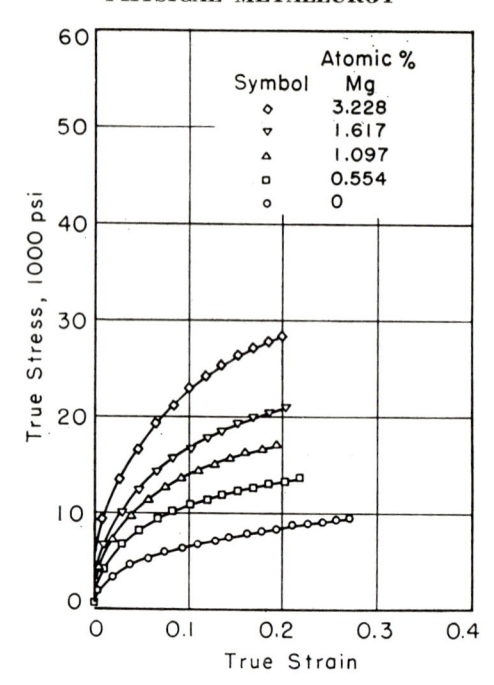

Fig. 5.27. Effect of composition on the stress-strain curve for solid solutions of magnesium in aluminum. From *Relation of Properties,* ASM, Fig. 9, p. 35.

of attraction or of repulsion. In either case, a dislocation moving past a solute atom will experience in turn a force that opposes its motion and a force that assists its motion.

If the dislocation were completely "flexible," its motion would be opposed by all the solute atoms it encounters, as shown in Fig. 5.28a. In this case the force required to move the dislocation would be greater than if the solute atoms were not present. If, however, the dislocation were regarded as "rigid," as in Fig. 5.28b, its motion would be assisted and opposed by equal numbers of solute atoms, and would require no more force than in the crystal of pure solvent.

The flexibility of a dislocation is limited by the fact that it has energy associated with it, and therefore work must be done in increasing its length by bending. If the "wavelength" of the bend is too small it becomes energetically very expensive, and the dislocation is effectively rigid for stress disturbances that are only a few atoms apart, as is the case with randomly distributed atoms in solid solution. This may be expressed more precisely in the form that the limiting radius of curvature R to which a dislocation can be bent by an applied shear stress σ is $R = G\mathbf{b}/\sigma$, where G is the shear modulus and $\mathbf{b}$ the Burgers vector of the dislocation. For the radius of curvature to be small enough for the dislocation to behave as a flexible line on the *atomic* scale, the internal stress (i.e., stress applied to the dislocation by the solute atom) would have to be comparable with the shear modulus G. It does not appear that this stress can normally exceed $G/100$. It is therefore concluded that randomly distributed atoms in substitutional solid solution do not increase the yield or flow stress. Two distinct cases arise: 1. in which the solute atoms are, apart from their interaction with dislocations, randomly distributed; and 2. in which the solute atoms are segregated into regions of high concentration, separated by regions of low concentration. This is not an equilibrium distribution but it can be produced in some alloys by a suitable treatment, as discussed in Chapter 8. The essential difference between these two cases is the distance between the stress centers; in case 1, the distance is much less than the radius to which the dislocation can be bent by the applied stress, while in case 2 it is comparable to or greater than this radius. Case 2 will be discussed below under the heading "Hardening by a Second Phase."

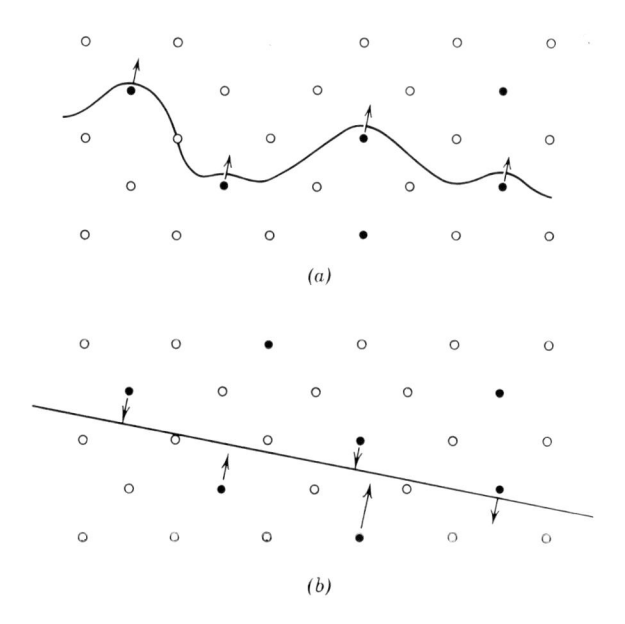

Fig. 5.28. Effect of solute atoms on (a) "flexible" and (b) "rigid" dislocation.

Returning to case 1, there should be no strengthening due to randomly disposed solute atoms. It is, however, found experimentally (see p. 164) that the presence of solute atoms does raise the whole stress-strain curve. This must be due to nonrandom distribution of the solute atoms. Three mechanisms have been proposed to account for the effects: 1. Locking of dislocations by an "atmosphere" of solute atoms (Cottrell). 2. The existence of short range order (Fisher). 3. The adsorption of solute atoms on the stacking fault parts of extended dislocations (Suzuki). Each of these will be discussed in detail.

THE COTTRELL MECHANISM. It has been shown by Cottrell that the energy of a crystal containing dislocations and solute atoms is reduced if the solute atoms segregate at the dislocations. There are two kinds of segregation that can occur. One, described as a dilute atmosphere, has a concentration C, given by the expression $C = C_0 \exp(-V/kT)$, where V is the binding energy between the dislocation and a solute atom, and C_0 is the concentration of solute in the crystal as a whole. The second type of segregation, which occurs below a critical temperature T_0, is that in which the solute atoms condense into a "mono-row" along the dislocation line (in the case of an edge dislocation). It is evident that it is necessary to perform work in order to pull a dislocation away from its segregated atoms; the force required to overcome the attraction between the dislocation and a *condensed* atmosphere is quite large, while the value is much smaller when the atmosphere is dilute.

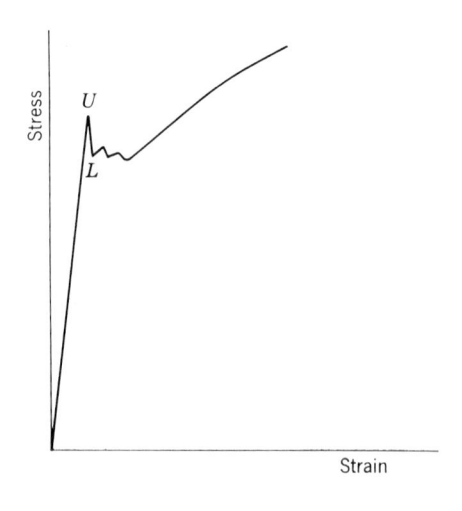

Fig. 5.29. Stress-strain curve showing yield point.

This extra force, which corresponds to a higher applied stress, will cause the "softest" source to operate, and this causes slip to occur in a single plane. If all other sources operate entirely independently, each of them will require the same enhanced stress and the stress-strain curve will be higher than for the pure solvent. It is clear that the condition of independent operation is not always satisfied; in the case of body-centered cubic iron containing a small amount of carbon in interstitial solid solution, for example, it is easier to propagate a Lüders band than to initiate one, and the resulting stress-strain curve is shown in Fig. 5.29, in which the upper yield point U corresponds to the stress for the formation of the first Lüders band and the lower yield region L corresponds to the propagation of Lüders bands. It has been shown that the yield point is only evident if some carbon is present, and that its existence depends on the segregation of carbon. The operation of one source must in some way "trigger" the operation of sources in the other planes. This behavior is typical of iron containing small amounts of carbon in solution, zinc and cadmium containing nitrogen (these are interstitial solutes), and for certain dilute substitutional solutions in aluminum and copper. The observation that the yield point is absent if a specimen is immediately retested, and that it returns after the lapse of sufficient time to allow the atmosphere to reform by diffusion (see Chapter 8), provides strong support for the validity of this explanation of the yield point.

The yield point is not found at temperatures above T_0; however, there can still be some solution strengthening above this temperature, because the "dilute" atmosphere will exert some force on each dislocation that comes from a source; this force is considerably less than that exerted by a condensed atmosphere, but it may still be significant.

THE FISHER MECHANISM. It has been shown, both theoretically and experimentally, that the atoms in a substitutional solid solution are not always disposed randomly, but that there is often a tendency for each atom to surround itself by unlike neighbors (see p. 45). This may be explained on the basis of the better geometrical packing that can be achieved by alternating atoms of two different sizes than by mixing them randomly. The lattice strains are reduced by this type of arrangement, which, however, in dilute solutions can only occur locally. Fisher has pointed out that work must be done to propagate slip through such regions of "short range order," because order, which is presumably the lower energy configuration, is destroyed by the process of slip. Figure 5.30 shows how the order across the plane PQ is progressively destroyed by shear in this plane.

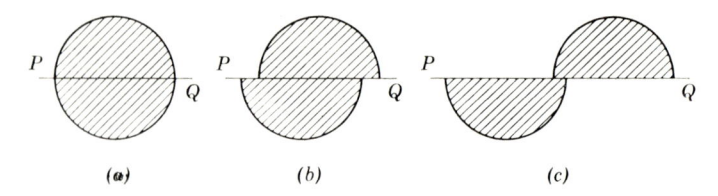

Fig. 5.30. Decrease in short-range order as a result of slip.

THE SUZUKI MECHANISM. Suzuki has shown that segregation of solute atoms should take place on the stacking fault plane that separates the two partial dislocations of an extended dislocation. Figure 5.31 illustrates this type of segregation. It should be noted that, while the Cottrell type of segregation is energetically favorable because of geometrical effects (i.e., lattice strain is reduced by putting a "misfit" atom in a position where the lattice is distorted by the presence of a dislocation), the Suzuki type must depend on a chemical type of interaction, since the sites on the "hexagonal" or stacking fault plane do not differ in size from other sites; their special feature is that they separate two layers whose second nearest neighbor interaction is abnormal.

As in the case of the Cottrell mechanism, it is necessary to do work in order to move the dislocation away from its segregate and it should therefore require more force to move such a dislocation than when there is no solute present. It is not clear how this applies to all the dislocations from a source, as it would appear as if only the first one would have a condensed segregate of this kind. This would lead to a yield point effect. The Suzuki type of hardening, however, is invoked to account for solution strengthening in the absence of a yield point. It is predicted that the Suzuki type of interaction between dislocations and solute atoms should be almost independent of temperature.

EXPERIMENTAL EVIDENCE. The experimental evidence on solution strengthening is not sufficiently consistent to provide a definite criterion

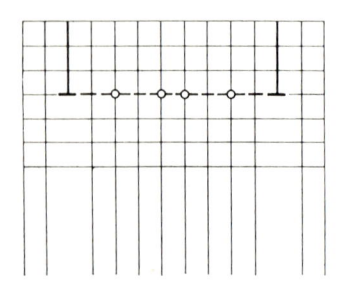

Fig. 5.31. Segregation on stacking fault.

STRUCTURE-SENSITIVE PROPERTIES

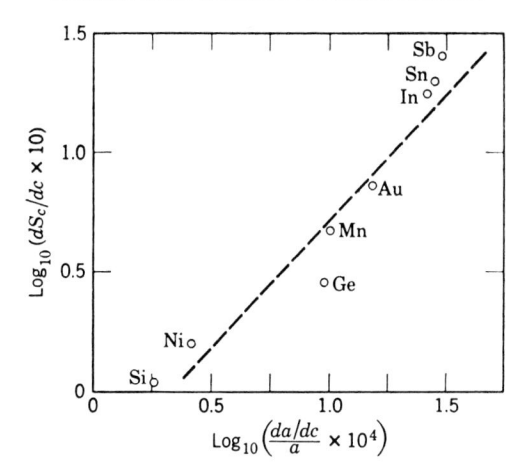

Fig. 5.32. Relationship between lattice parameter and yield stress for solid solutions. From *Dislocations and Mechanical Properties of Crystals*, J. C. Fisher et al., Eds., John Wiley & Sons, New York, 1957, Fig. 1, p. 393.

for the prediction of the effect of a given solute. This is partly because most of the work has been performed on polycrystalline samples, in which there may be other effects as well as the effect of solutes on the plastic properties of the single crystal; however, it is clear that there are two important considerations that probably apply to single crystals as well as to the polycrystalline alloys in which they were investigated. The two effects are: change of lattice parameter and change of electron-atom ratio.

There is evidence of fairly good, but not perfect, correlation between the amount by which the lattice parameter is changed by a solute and the extent to which the yield stress is raised (Fig. 5.32). Figure 5.33 shows the superposition of the stress-strain curves for alloys of the same electron-atom ratio obtained by alloying copper with appropriate amounts of zinc, gallium, germanium, and arsenic. Figure 5.34 shows that the yield stress is well correlated with the electron-atom ratio in the same series of alloys. The size factor is ignored in these correlations, and this is a strong indication that chemical interaction (Suzuki) may be a much more potent source of hardening than lattice strain, although the latter is undoubtedly of some importance.

Further support for the significance of the electron density as a parameter for the prediction of solution hardening has been given by Hibbard, who showed that the yield stresses of a series of copper alloys,

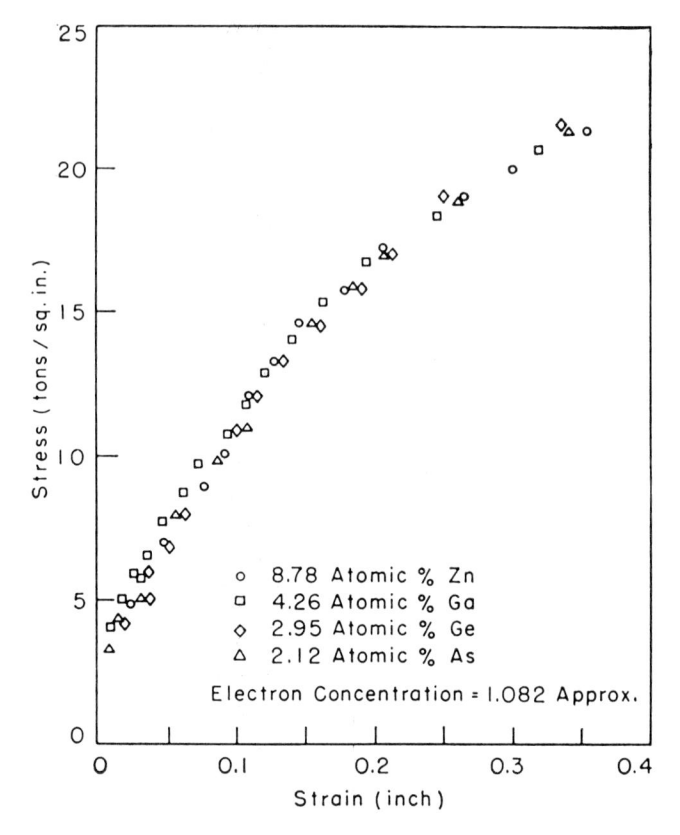

Fig. 5.33. Stress-strain curves for alloys with the same electron-atom ratio. From *Relation of Properties*, ASM, Fig. 36, p. 55.

with the same lattice parameter, varied linearly with the electron-atom ratio (Fig. 5.35).

The hardening effects of solutes that have been discussed so far relate to solid solutions in which the distribution of solute is that of equilibrium, except insofar as it is affected by the presence of dislocations; the only departures from randomness are those corresponding to short range order. Under some conditions, to be discussed in Chapter 8, it is possible for the solute atoms to depart much farther from randomness by the formation of zones containing a very high concentration of the solute. Although a crystal containing these zones may still be regarded as single-phase, the discussion of the resulting plastic properties is more properly considered under the effects of a second phase, because the effect of a "zone" on the movement of dislocations is similar to that of a second-phase particle.

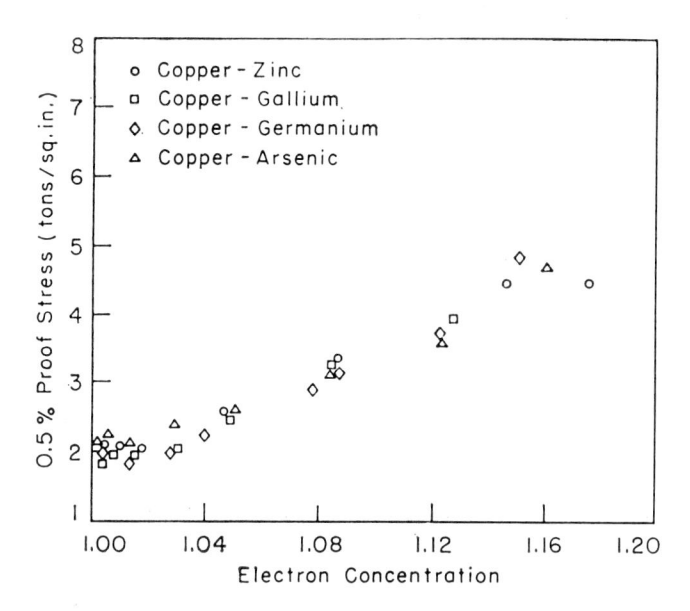

Fig. 5.34. Relationship between yield stress and electron-atom ratio. From *Relation of Properties*, ASM, Fig. 37, p. 56.

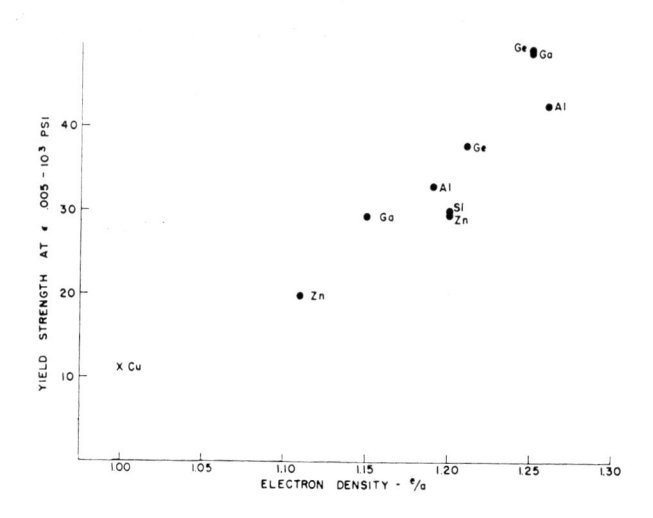

Fig. 5.35. Yield strength as a function of electrons per atom in copper and copper base alloys (299°K). From W. R. Hibbard, Jr., *Transactions of the Metallurgical Society of AIME*, Vol. 212, No. 1, Feb. 1958, Fig. 10, p. 4.

Hardening by a second phase. Although some alloys contain two phases in comparable amounts, the majority of technically interesting alloys contain a second phase that is relatively small in amount dispersed in a matrix. In many cases the dispersed phase is an intermediate phase and it may be much harder than the matrix. It will be shown in Chapter 8 that the size and distribution of second-phase particles can in some important cases be controlled by means of heat treatment. It is therefore of interest to consider the influence of such particles on the plastic properties of the alloy.

The theory of the effects of dispersions on the properties can be outlined as follows: If a dislocation line is subjected to a small shear stress, it will move until it reaches "obstacles." If the stress increases, it will begin to "bow out" between the obstacles, and a sufficient increase of stress will cause it either to cut through the obstacles or to "break through" the spaces between them, as illustrated in Fig. 5.36. The stress required for the latter is given by $\sigma = G\mathbf{b}/d$, where d is the distance between neighboring obstacles (see p. 101).

There is a maximum value of σ when d has the value that makes it equally likely that the dislocation will cut through the obstacles or that it will break through between them. If the separation is increased, the critical radius for breaking through is increased, and a smaller stress is then required; if it is decreased, the dislocation line becomes more rigid, i.e., it is supported by fewer obstacles per unit length and so can break through them more easily.

When yielding takes place by the process illustrated in Fig. 5.36, a dislocation loop is left around each obstacle. This increases the effective size of the particle and consequently decreases the distance d between them. This causes an increase in flow stress, and consequently it would be expected that work hardening should be more rapid in a dispersion-hardened system than in one that does not depend on this mechanism.

If the second-phase particles in a crystal are of constant total volume,

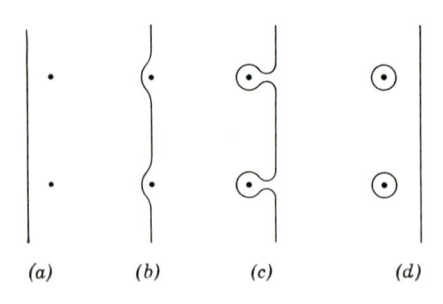

Fig. 5.36. Motion of dislocation past second-phase particles.

(a) (b) (c) (d)

STRUCTURE-SENSITIVE PROPERTIES

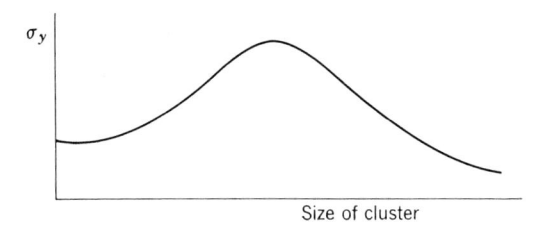

Fig. 5.37. Effect of size of particle on yield strength.

the size of particles, the number of particles, and their distances apart can vary. The variation of yield stress with the average size of particle should be of the kind shown in Fig. 5.37. Figure 5.38 shows an experimental curve for the variation of yield stress with particle size for cobalt particles in a copper matrix. The size of the cobalt particles was determined from measurements of their magnetic properties. The effect of the particles is small when they are very small because the conditions approach those of a random solid solution, the dislocations being effectively rigid; when the particles are large and, therefore, far apart, the value of d is large and so $G\mathbf{b}/d$ is small. Hence the dislocations can move under relatively small stresses by the mechanism represented in Fig. 5.36.

It should be appreciated that the maximum of the curve of Fig. 5.37

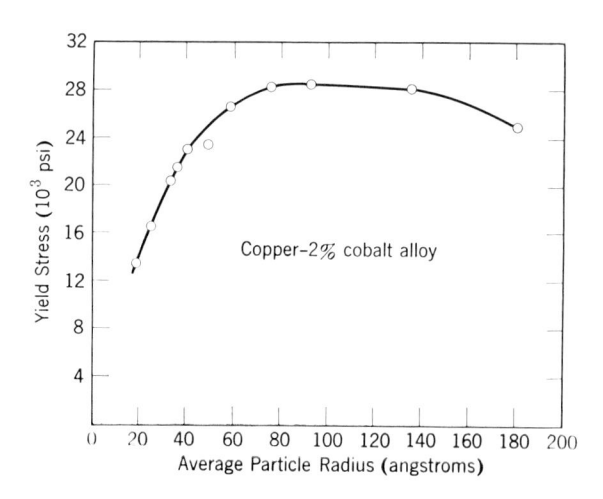

Fig. 5.38. Relationship between yield stress and particle size for copper containing a constant amount of cobalt in the form of nearly spherical particles. (Courtesy J. D. Livingston, General Electric Research Laboratory.)

occurs when the particles are about 1 μ apart. The particles are necessarily extremely small, and can only be resolved by means of the electron microscope.

When a study is made of the effects of a dispersion of cementite (Fe_3C) in ferrite (solid solution of carbon in α iron), the cementite particles are separated by distances that are large compared with the distance corresponding to the maximum of Fig. 5.37. In such cases the plastic properties correlate better with the average distance between cementite particles, also referred to as the mean free path in ferrite. A relationship obtained from experiment is shown in Fig. 5.39. The decrease of yield stress with increase in the distance between second-phase particles corresponds to the right-hand part of the curve of Fig. 5.37. The shape of the cementite particle appears to have no influence on this relationship, since several quite different shapes were included in this study. It should be remarked that the effective size of a second-phase particle may be considerably larger than the actual size of the particle if the neighboring region of the matrix is in a state of elastic strain as a result of coherency between itself and the particle.

Effects of crystal boundaries

THE SINGLE BOUNDARY. The crystallographic nature of the plastic deformation of single crystals has two geometrical consequences. 1. Deformation is not isotropic; a sphere is deformed into an ellipsoid.

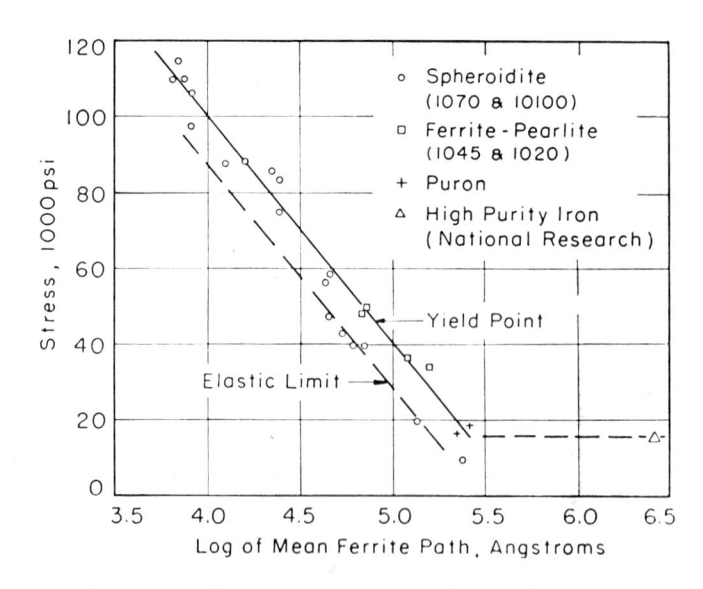

Fig. 5.39. Effect of the distance between cementite particles on the yield stress of the ferrite matrix. From *Relation of Properties,* ASM, Fig. 7, p. 83.

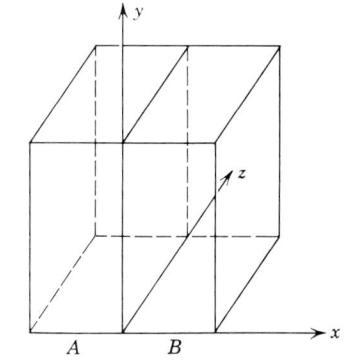

Fig. 5.40. Simultaneous distortion of two crystals with a common boundary.

2. Deformation is not homogeneous; slip is largely confined to a very small proportion of the material, the remainder (the parts between the active slip planes) undergoing little or no deformation.

It follows that, if two crystals that adjoin at a boundary are deformed simultaneously, either they do not remain in contact all over their boundary or the deformation of one or both is different from that of the crystals when deformed separately. Experiments show that contact is maintained at the boundary in practically all cases. It must be concluded that each crystal deforms in a way which is influenced by the other crystal. Two distinct conditions must be satisfied; the *macroscopic continuity* condition is that the surface of each crystal at the boundary must deform similarly; this requires that, if the two crystals A and B of Fig. 5.40 are extended equally in the y direction, they must contract equally in the z direction and they must shear through equal angles in the yz plane. They can shear differently in the xy and xz planes, but if their volumes are to be unchanged they must contract equally in the x direction.

It is in general impossible for these conditions to be satisfied by slip on one system in each crystal; they can, however, be satisfied by a total of four slip systems, either two in each crystal, or one in A and three in B or vice versa.

The second condition for continuity, that of *microscopic continuity*, is that a dislocation cannot, except at high temperatures, leave a crystal at a boundary by the formation of a "slip step" since this would require the local separation of the two crystals; when a dislocation approaches a boundary, therefore, it experiences a force which opposes the motion of the dislocation. The material just beyond the boundary experiences a corresponding shear stress. If a number of dislocations on the same slip plane are piled up at a boundary, a large stress can be generated in the other crystal; this can cause slip to occur in the other crystal on a

⊥   ⊥   ⊥   ⊥   ⊥ ⊥ | C
                        T

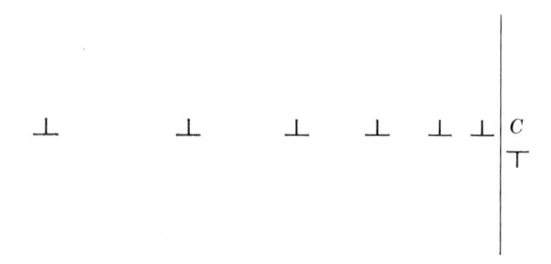

Fig. 5.41. Pile-up of dislocations at a grain boundary.

system that is not necessarily highly stressed by the external stress that is applied. The transmission of a shear stress across the boundary is a result of the strength of the boundary in relation to normal tension stresses, which can be very large in the region T of Fig. 5.41.

Experiments show that a bicrystal such as the one illustrated in Fig. 5.40 stressed in the y direction has a different stress-strain curve from a similarly oriented single crystal *only if each of them would deform by slip on a single plane,* if isolated from each other. If the single crystals have "symmetrical" orientations such that each would deform on more than one plane, then the bicrystal has a stress-strain curve similar to that of the single crystals. Typical curves are shown in Fig. 5.42.

The results of such investigations to date can be summarized as follows: Macroscopic continuity demands that each crystal shall (on the average) deform on at least two systems, and microscopic continuity determines which slip systems shall operate near the boundary. At regions remote from the boundary, neither condition need apply, but

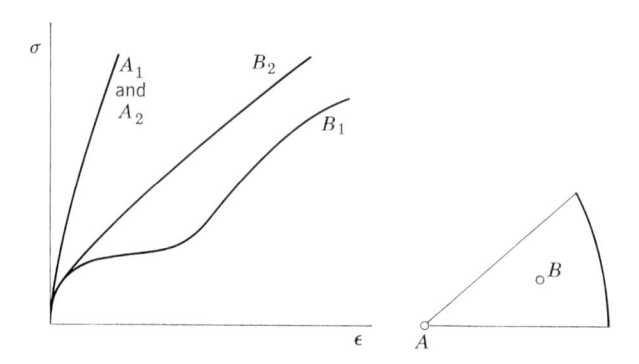

Fig. 5.42. Effect of interaction between neighboring crystals on their stress-strain curve. A_1 and B_1, single crystals, A_2 and B_2, bicrystals, with orientations represented by A and B in the stereographic triangle.

if it does not, there must be regions of crystal curvature corresponding to the transition region where double slip changes to single slip. This requires a pile-up of dislocations that corresponds to local curvature.

In close-packed hexagonal crystals such as zinc, which normally has only one slip plane at room temperature, a boundary has been shown in some cases to cause slip on an $(10\bar{1}1)$ plane, which does not normally operate.

POLYCRYSTALLINE AGGREGATES. In a polycrystalline aggregate, most of the crystals are completely surrounded by other crystals. If macroscopic continuity is assumed to apply when an aggregate is deformed, each crystal is constrained by several neighbors; it was shown by Taylor that, if it is assumed that the aggregate deforms homogeneously, each crystal must slip on five systems (three planes, with two directions in two of them) in order to deform in the competely prescribed manner. (This differs from the conditions of the bicrystal (Fig. 5.40) in that shear in both the xy and yz planes is now specified.) Taylor showed that the appropriate planes and directions could be selected for any given orientation by means of the principle of least work, still on the assumption of homogeneity of deformation. The directions and amounts of the resulting changes of orientation were predicted; they were partly but not completely confirmed by subsequent experiments. The reason for the discrepancies probably lies in the invalidity of the assumption of homogeneous deformation. There is considerable evidence that, as in the bicrystal, the deformation near a boundary differs from that in the more remote parts of the crystal.

An important result of the much more complex nature of slip in the crystals in a polycrystalline aggregate than in the single crystal is that the stress-strain curve is higher than for any except the hardest single crystals. Figure 5.43 shows the relationship for aluminum. It is found

Fig. 5.43. Relationship between stress-strain curves for single crystal and polycrystal.

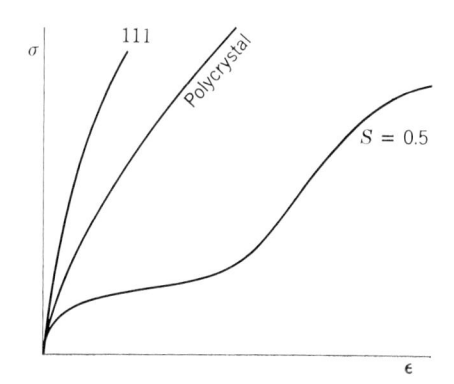

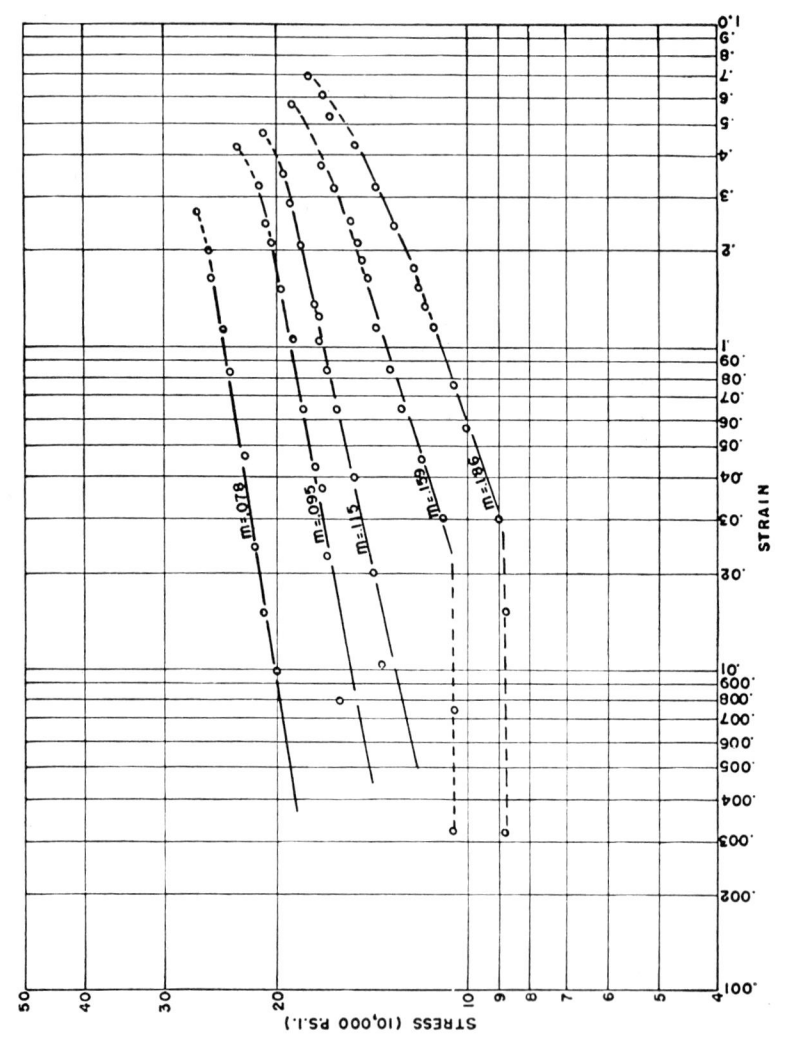

Fig. 5.44. Logarithmic plot of stress-strain relationship. From *Ferrous Metallurgical Design*, J. H. Hollomon and L. D. Jaffee, John Wiley & Sons, New York, 1947, Fig. 27, p. 84

that the plastic deformation of a polycrystalline aggregate can often be represented by a straight line when the true stress σ is plotted against the true strain ϵ on a logarithmic plot. The true strain at length l' is $\int_{l_0}^{l} \frac{dl}{l}$, which is $\ln \frac{l}{l_0}$; this is preferable to the strain $\frac{\Delta l}{l}$ because the gage length changes throughout the test, and the instantaneous strain should be related to the gage length at that time. An example of the log ϵ–log σ relationship is given in Fig. 5.44. The expression that represents this curve is $\sigma = \sigma_0 \epsilon^m$. m is the *strain-hardening exponent* of the material.

EFFECT OF CRYSTAL SIZE ON PLASTIC PROPERTIES. There is some evidence that the number of crystals in the cross section of a specimen, rather than the absolute crystal size, is a significant parameter; this would follow because a crystal that is constrained on all sides by other crystals behaves differently from one which has a free surface.

Most of the experiments on the influence of grain size have been made by means of hardness testing, a method which gives no information on the details of the stress-strain curve, and which may be influenced by the relationship between the grain size and the size of the indenta-

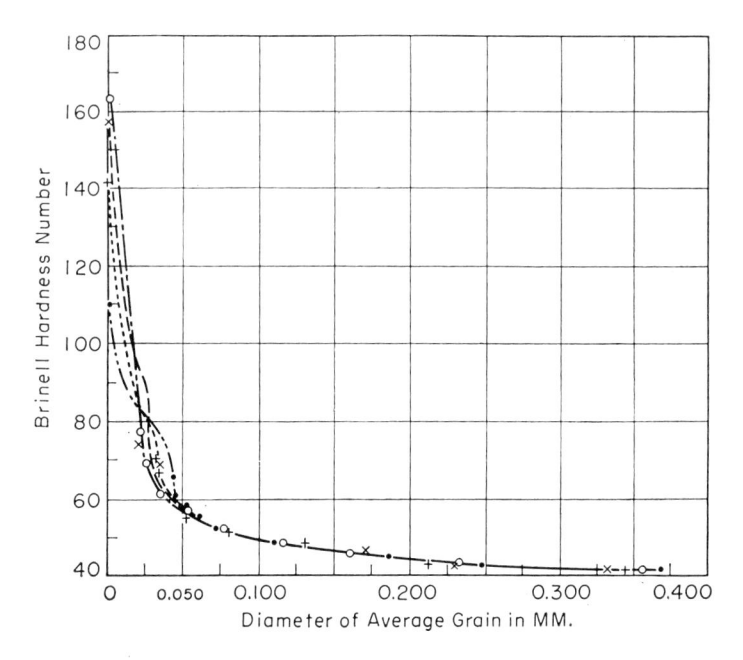

Fig. 5.45. Effect of grain size on hardness. From *Relation of Properties*, ASM, Fig. 2, p. 18.

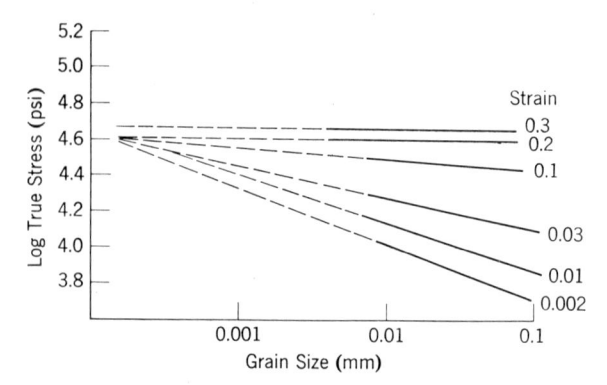

Fig. 5.46. Effect of grain size on flow stress for copper. (After Carreker and Hibbard, *Acta Metallurgica*, Vol. 1, 1953, p. 661.)

tion. (See p. 220 for a discussion of the measurement of hardness.) An example of the effect of grain size on hardness is given in Fig. 5.45.

There is, however, some evidence of the effect of grain size on the tensile stress-strain curve. The effect is largest at small strains, and vanishes at large strains. Figure 5.46 shows the relationship between the stress required to produce a given strain and the grain size for copper at room temperature. The effect of decreasing the grain size may be due to the increase in the proportion of the material that is close enough to the grain boundaries to be forced to deform "turbulently."

In cases in which there is a definite yield point (see "The Cottrell Mechanism") the effect of grain size is particularly strong. The upper yield point rises as the crystal size decreases. In a polycrystalline sample, the yield stress may be related to the propagation of slip from one crystal to the next, in which case the increase of stress due to pile-up should be less for smaller crystals than for larger ones because a given source would be subject to a back pressure that would suppress further activity when the number of dislocations that have emerged is proportional to the distance through which they can travel before reaching a boundary. The occurrence of very slight, but measurable, plastic deformation before yield is evidence in favor of this explanation.

Substructure hardening. The low-angle boundaries that occur either as a result of crystal growth processes (see p. 260) or as a result of polygonization (see p. 322) have a marked effect on the plastic properties. It has been shown that the small-angle boundaries that can be introduced into, for example, nickel or zinc by straining followed

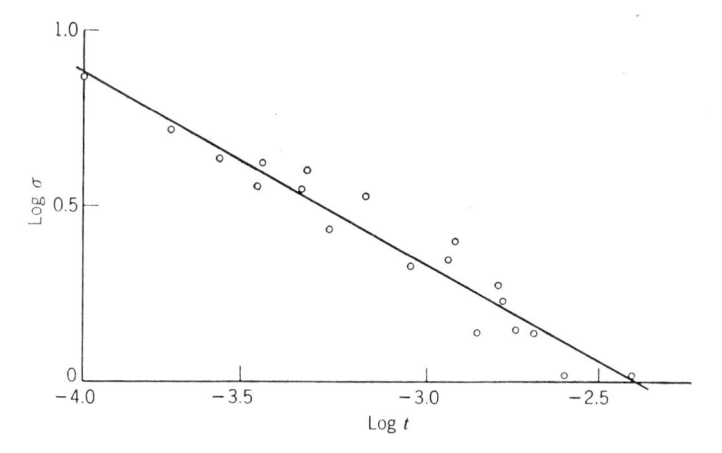

Fig. 5.47. Effect of subgrain size (t) on yield stress (σ). From *Dislocations and Mechanical Properties*, Fisher et al., Fig. 2, p. 355.

by suitable heat treatment can raise the stress-strain curve by substantial amounts.

An example of the effect of "naturally occurring" subboundaries is shown in Fig. 5.47, in which the subgrain size is plotted against the yield stress. The relationship can be accounted for if it is assumed that the stress required to cause dislocations to break through a subboundary is constant, and that the number of dislocations produced by a source is proportional to the size of the subgrain (this is analogous to the effect of grain size on the yield point, see p. 180). A further example is shown in Fig. 5.48; the substructure in this case is introduced by plastic deformation followed by annealing, and its extent increases with increasing prestrain.

5.3 Creep

The influence of time on plastic deformation. It has been assumed in the discussion up to this point that plastic deformation is independent of time; i.e., 1. that deformation occurs concurrently with the application of stress, and 2. that no further deformation occurs if the stress is maintained at a constant value.

The first of these assumptions is never absolutely true, but it only becomes important when the stress is increased extremely rapidly. For most purposes no appreciable error is caused by making this assumption.

Whether the second assumption is substantially justified or not depends on the temperature of the metal when it is being tested. For

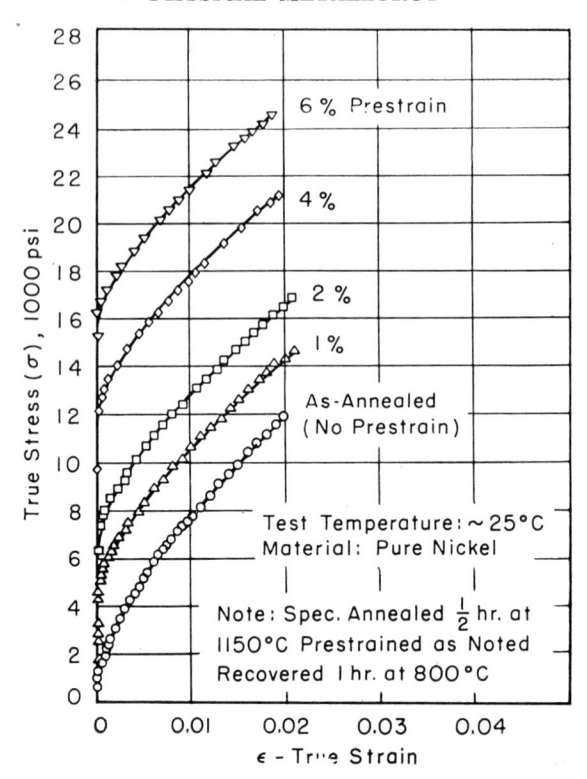

Fig. 5.48. Effect of substructure on stress-strain curve. From *Impurities and Imperfections,* American Society for Metals, Cleveland, 1955, Fig. 6, p. 149.

purposes such as this, it is necessary to consider the temperature T not in conventional terms but as a fraction of the melting point of the metal, T_E. Both temperatures are expressed on the Kelvin or absolute scale of temperature. The ratio T/T_E can be regarded as an equivalent temperature. If the equivalent temperature is less than about 0.4 for pure metals, or rather higher for alloys, then the effect of time on plastic deformation can usually be neglected. This is satisfied for the structural materials of engineering practice when they are used at or near ordinary atmospheric temperatures. For example, $T/T_E = 0.4$ for iron at about 450°C, while the corresponding temperatures for aluminum, lead, and tungsten are about 100°C, -30°C, and 1200°C respectively. However, when metals or alloys are used at higher temperatures, the time effects become increasingly important. In these circumstances the state of strain is a function not only of the stress but also of the time. Qualitatively, the behavior under stress of a metal at a high temperature can be represented in a three-dimensional diagram by means of the surface illustrated in Fig. 5.49, which shows strain increasing with time as well as with stress. The stress-strain curve at zero time is shown as OAB, and the curves PQ and $P'Q'$ show the increase of strain with time at fixed values of the stress. It is evident that the stress for a given strain is lower if the measurement is made slowly, instead of in zero time. This is shown by the curve $OA'B'$, drawn in a plane in which the time increases linearly with the strain. The stress-strain curve is no longer a unique curve for the material, but depends on the rate at which the test is performed.

The stress-strain-time surface of Fig. 5.49 cannot be a precise representation of the actual behavior of any material, because the strain at any value of stress and time is not independent of the stress-time path by which that point was reached; but the stress-time surface does give a qualitative idea of the relationship between the three kinds of measurements that are usually made. These are: 1. stress-strain curve at constant strain rate $(OA'B')$; 2. strain-time at constant load or stress (creep); and 3. stress-time at constant strain (the projection of such

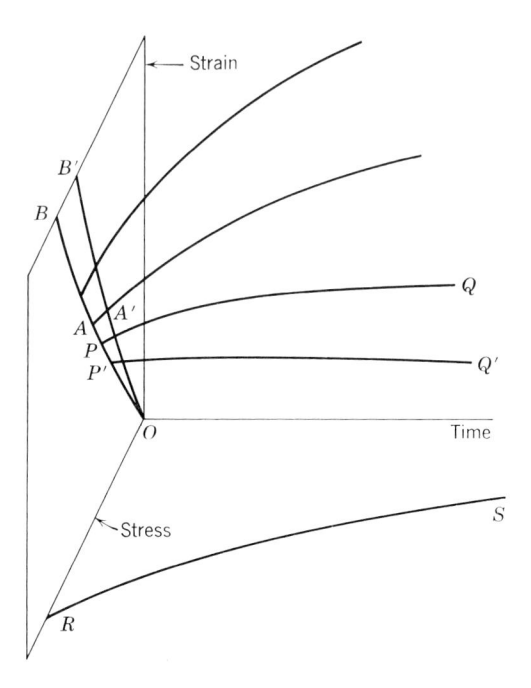

Fig. 5.49. Stress-strain-time relationships.

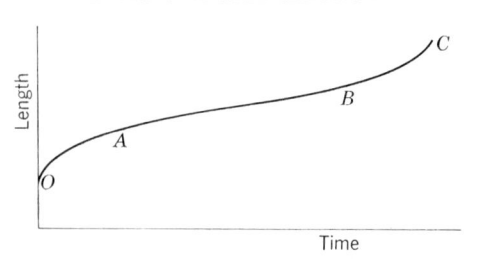

Fig. 5.50. Typical creep curve.

a curve on the stress-time plane is represented by RS), the relaxation curve.

Creep. Most of the experimental work, as well as most of the practical interest, has been concerned with the strain-time relationship. The process of creep is apparently a complicated one, and the following discussion indicates the main phenomena that are observed when a tensile load is applied to a specimen at a "high" temperature. Concurrently with the application of a force, some deformation takes place; if the stress is maintained, the specimen continues to elongate, at a rate that decreases with time, OA, Fig. 5.50. This is the so-called "primary" or "transient" stage of creep. The rate eventually becomes essentially constant, AB; this is called "secondary," "steady-state," or "quasi-viscous" creep. In many tests the rate subsequently increases again (tertiary stage, BC), and the increase continues until fracture occurs. All these stages are not always observable; at relatively low stresses and temperatures, at which creep occurs at low rates, the "steady-state" part of the process may be unobservable; fracture may terminate the test at any stage.

When a creep test is made in order to provide information of direct engineering usefulness, it is usual to subject the specimen to a constant force, often by means of dead loading. As the specimen increases in length, it decreases in cross section (its volume remains essentially constant) and the stress therefore increases as the test proceeds. For the study of the more fundamental aspects of creep, it is desirable to arrange for the *stress* to remain constant by reducing the force so that it remains proportional to the cross section throughout the test. There are a number of simple devices that can be used for this purpose.

Attempts to describe the process of creep under constant stress by functional relationships have led to two contradictory results. Andrade proposed the relationship $L = L_0(1 + \beta t^{1/3}) \exp(kt)$ to represent the length L as a function of time t. L_0 is the length immediately after the application of the stress, $\beta t^{1/3}$ represents the transient stage, and exp (kt) corresponds to steady state. This relationship gives very good agreement with many experimental results obtained at relatively fast creep rates. On the other hand, the relationship $E = At^a$ provides much better agreement with many results obtained at low creep rates. This does not contain a term corresponding to a steady state, although the exponential form of the curve represents a progressive approach to a linear rate.

Neither of the expressions quoted above includes the tertiary stage; it has been observed that tertiary creep occurs only if (a) the *stress increases* during the test, or (b) if some abnormal structural change takes place during the test.

Many observations have been made of structural changes that appear to be a normal accompaniment to the "transient and steady state," or the "exponential" stages of creep. In the first place, it has been found that there are two distinct types of deformation; the crystals themselves deform, and they slide over each other at their boundaries. It seems that intercrystalline slip is of comparatively little importance except at very high temperatures. Most attention has been concentrated on the changes that take place within the crystals of a polycrystalline aggregate. The crystals retain their identity, change somewhat in orientation, and become subdivided into subcrystals, or cells, of high perfection, separated by small-angle boundaries. The "cells" or subcrystals have a characteristic size for any given temperature and creep rate; the cells are larger for high temperatures and for low creep rate. There is evidence which shows, however, that the formation of a cell structure is not an inevitable consequence of creep, and so it may be concluded that the cells form as a result, but not a necessary result, of the creep process.

Effects of stress and temperature on creep. The most comprehensive experiments on the effects of stress and temperature on creep rates have been made on "pure" metals. The results can be summarized as follows: The effect of temperature on the creep of high purity aluminum is shown in Fig. 5.51, from which it will be seen that the effect of temperature is to displace the creep curve along the time axis. The extent of this displacement as a function of temperature can be interpreted as an activation energy for the rate-controlling aspect of the process of creep. Figure 5.52 shows this activation energy, and the activation energy for self-diffusion (see p. 373) in relation to the melting point for a number of metals. The good agreement of these two activation energies suggests that the rate of creep is controlled by the rate at which dislocations can climb out of pile-ups

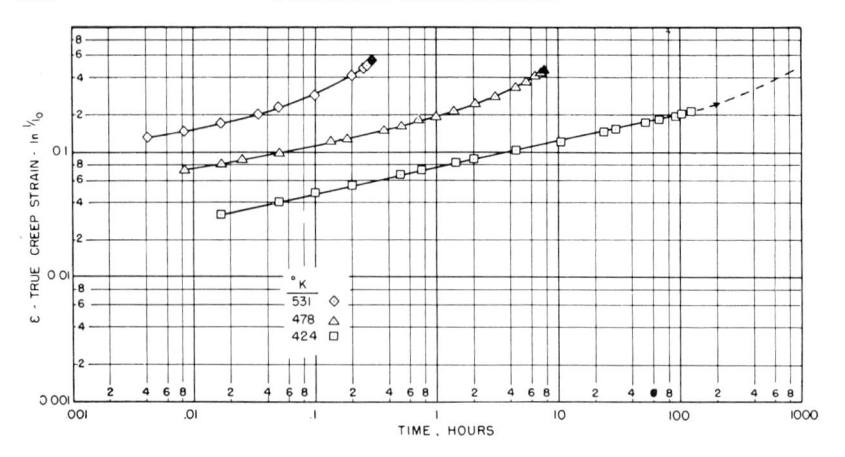

Fig. 5.51. Effect of temperature on the creep of aluminum. From *Creep and Fracture of Metals at High Temperatures,* National Physical Laboratory, Philosophical Library, New York, 1957, Fig. 2, p. 104.

against grain boundaries or other obstacles (Fig. 5.53). Creep rate is also very sensitive to stress; some data are given in Fig. 5.54, in which data for several temperatures are combined by plotting the parameter Z (the Zener-Holloman parameter), which is the product of the creep rate $\dot{\epsilon}$, and the term $\exp(-Q/kT)$, which corrects for temperature by the use of activation energy Q. The effect of stress is also well shown by Fig. 5.55, which shows creep strain as a function of time for an alloy of aluminum with 3.1 atomic per cent magnesium.

The effect of stress is presumably to increase the speed of climb by contributing some of the energy that is required for a vacancy to leave a dislocation, i.e., for climb to take place.

Effect of structure on creep. The effect of atoms in solid solution is usually to reduce the rate of creep at a given stress and temperature. A useful parameter for making these comparisons is the slope B of the log $\dot{\epsilon}$ versus stress curve; the reciprocal of this slope, $1/B$, is plotted against the concentration for various solid solutions in aluminum in Fig. 5.56. It will be seen that increasing the solute content decreases the rate of creep. The effect of dispersed second phase is usually also to decrease the creep rate; this is presumably because dispersed particles increase the stress necessary to move dislocations. It has frequently been observed that creep rates are somewhat greater for a smaller grain size in a given alloy.

It was pointed out on p. 170 that the most effective distribution of a second phase, from the point of view of raising the stress required

STRUCTURE-SENSITIVE PROPERTIES

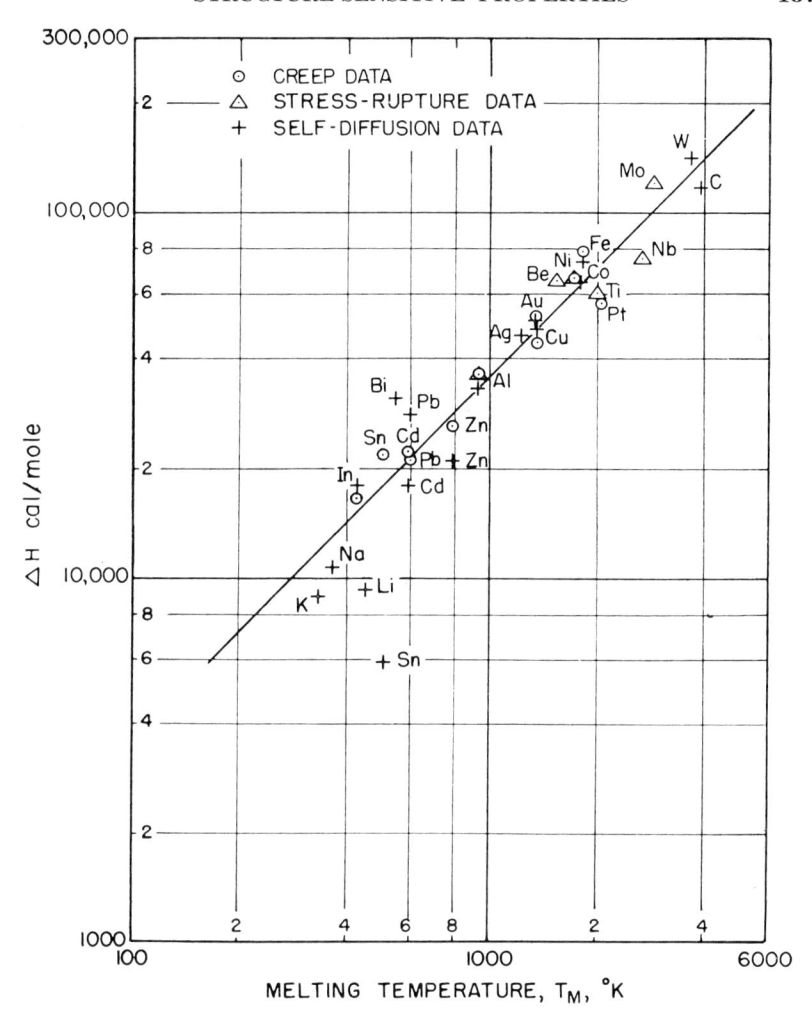

Fig. 5.52. Activation energy for creep as a function of melting point. From *Creep and Fracture*, NPL, Fig. 15, p. 117.

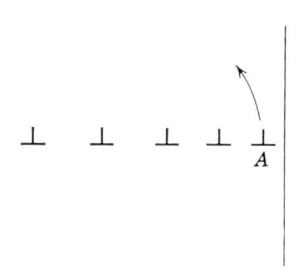

Fig. 5.53. Dislocation climb during creep.

188   PHYSICAL METALLURGY

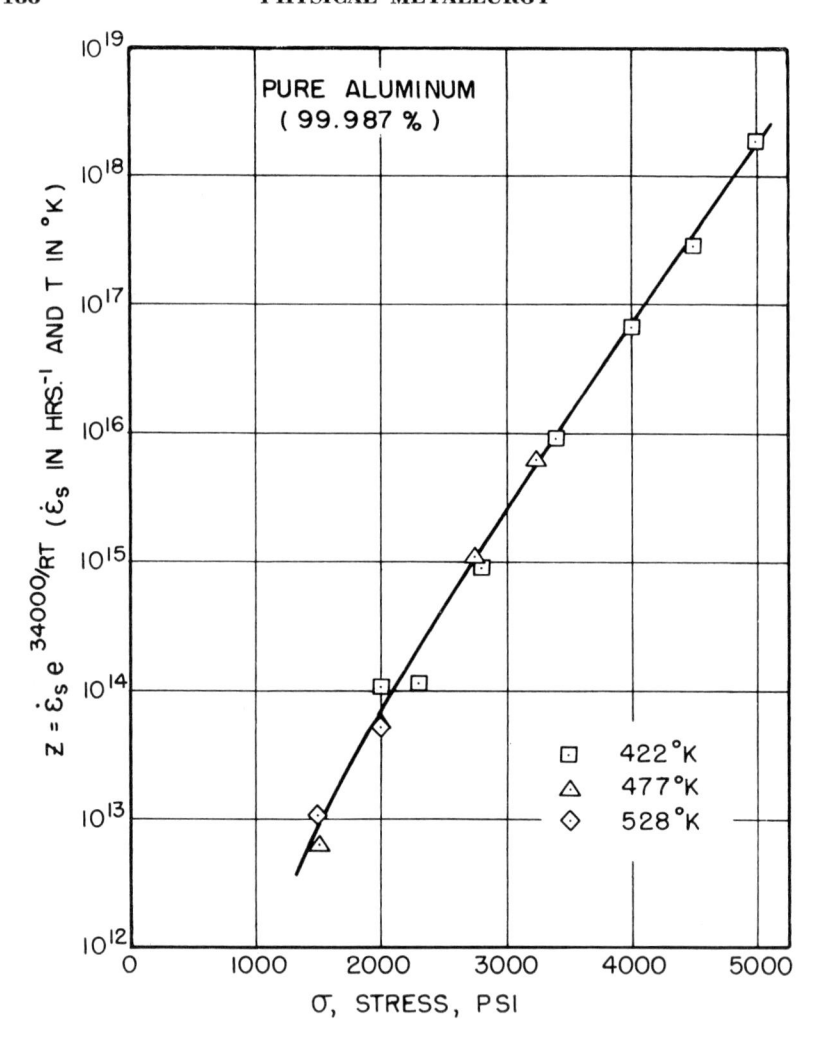

Fig. 5.54. Effect of stress on creep rate. From *Creep and Fracture*, NPL, Fig. 17, p. 119.

for dislocation movement, is usually a metastable one. Creep takes place at temperatures at which such metastable structures do not persist; the major problem in devising alloys that are highly creep resistant, therefore, is to arrive at a good compromise between optimum distribution and stability. This is achieved in practice by the use of a material of high melting point, strengthened by a substantial amount of several solutes, and further strengthened by a dispersion of particles

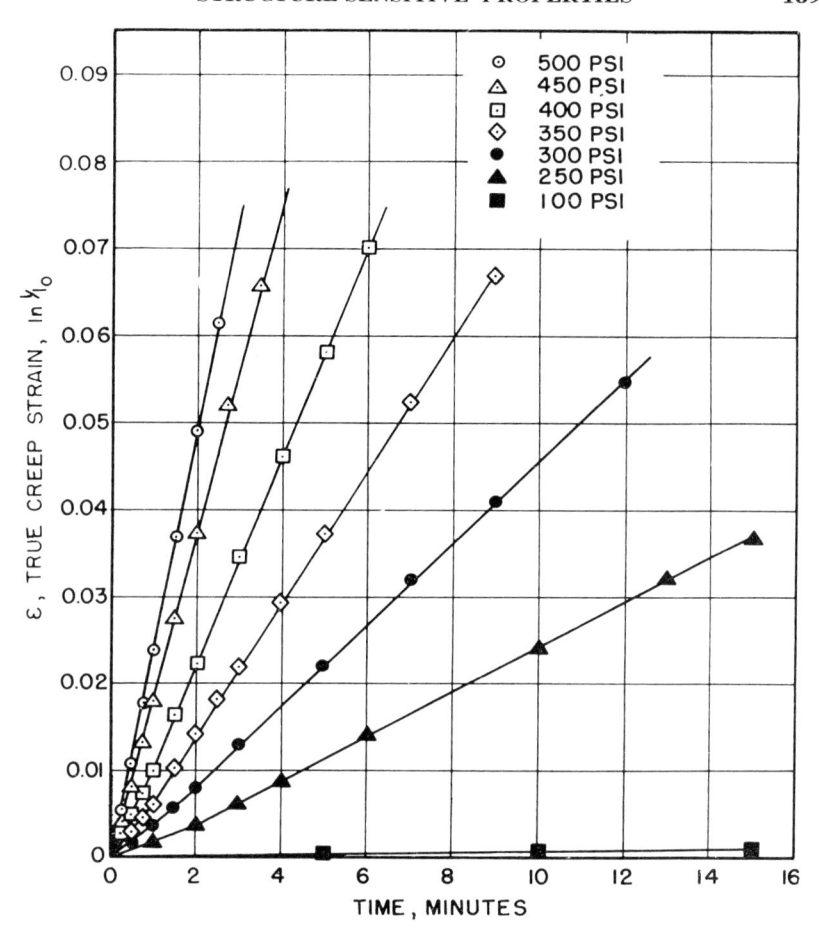

Fig. 5.55. Effect of stress on creep. From *Creep and Fracture*, NPL, Fig. 23, p. 125.

of phases of high stability produced at temperatures above the operating temperature of the material.

It should be emphasized that most of the experimental work that is aimed at elucidating the mechanism of creep has been carried out at high strain rates compared with those of practical interest. It is found that the relative importance of different aspects of creep depends on the strain rate; at very low strain rates, for example, the importance of crystal boundary sliding is apparently increased, while the importance of steady-state creep decreases. These differences may correspond to the fact that a creep rate of 10^{-6} in./in./hr with an alloy of

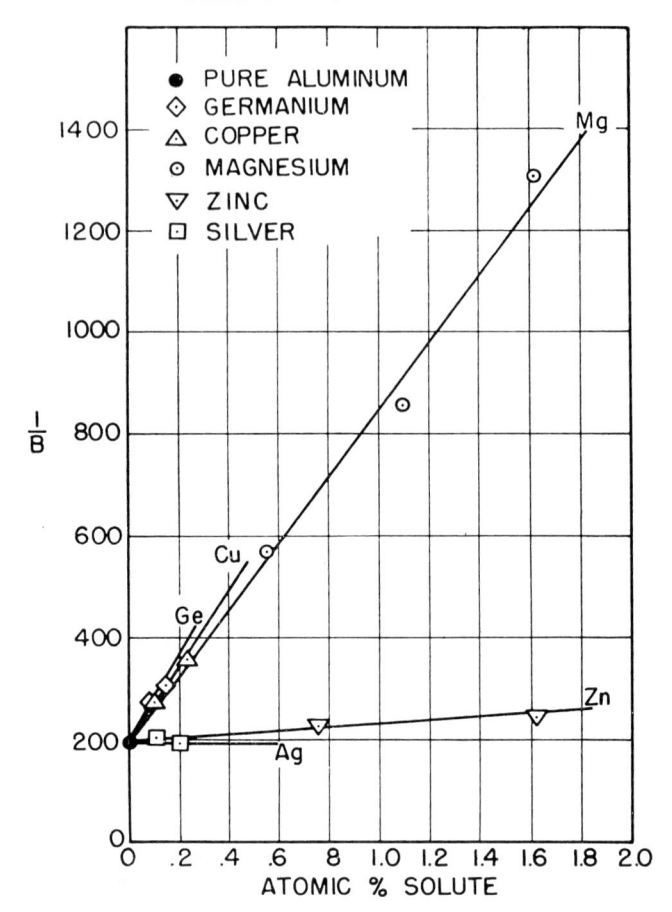

Fig. 5.56. Effect of solute content on creep rate. From *Creep and Fracture,* NPL, Fig. 22, p. 124.

mean grain size 0.01 cm corresponds to the movement through each crystal of an average of 1 dislocation/hr. This may mean that very few of the crystals deform by slip under these conditions, and that the remainder of the deformation is by the relative movement of crystals, with some adjustment of the positions of the boundaries to allow intercrystalline slipping without loss of continuity. The practical problems of creep involve both the deformation that occurs and the fact that fracture may terminate the process. Fracture as a result of creep will be discussed in Section 5.5.

It should be pointed out that the empirical relationships between creep strain and time, and between creep rate, stress, and temperature, should be used with the utmost caution for the prediction of long-term creep behavior from short-term tests. The most important consideration is that the constants in the empirical formulas give no information about the process that results in fracture, which is governed by different physical laws from creep strain; it is also to be remembered that the empirical formulas are mainly based on short-term tests.

5.4 Anelasticity

If a metal behaved perfectly elastically, Hooke's law would be obeyed and a sample would vibrate or undergo stress cycles within the elastic limit without any loss of energy except to the surrounding air. None of the energy of the vibration would be converted into heat within the specimen; there would be no *internal friction*. That metals do not behave in this ideal fashion can be shown in various ways; sound waves are attenuated when they pass through a metal; vibrations are damped out faster than can be accounted for by the external loss of energy. The extent of these effects varies greatly from one metal or alloy to another; the existence of internal friction is in itself important in some engineering applications (e.g., when forced or resonant vibrations are limited in amplitude by the internal conversion of vibrational energy into heat), and in some acoustic and musical applications (e.g., bells).

Detailed studies have shown that there are many processes which can cause internal friction. In general they all correspond to a phase lag between the applied stress and the resulting strain. This phase lag may be caused by plastic deformation of the kind discussed above; but if the stress is too low for this, it may be caused by thermal, magnetic, or atomic effects. The thermal effects are a result of the fact that if a solid is subjected to an elastic stress, such as compression, its temperature rises instantaneously. If heat then leaves the sample, by conduction to the surroundings, there is a subsequent decrease in size corresponding to the temperature change. The response to the stress is therefore in two stages, one of which is immediate and the other takes time. If the sample is taken through cycles of alternate tension and compression, the strain will always lag behind the stress, unless the cycle is so fast that there is no time for any heat conduction, or so slow that the temperature is always the same as that of the surroundings. The amount of energy that is converted into heat is the area of the stress-strain loop (Fig. 5.57). The variation of this area with frequency is shown schematically in Fig. 5.58 and this depends upon the time per cycle and on the properties and geometry of the

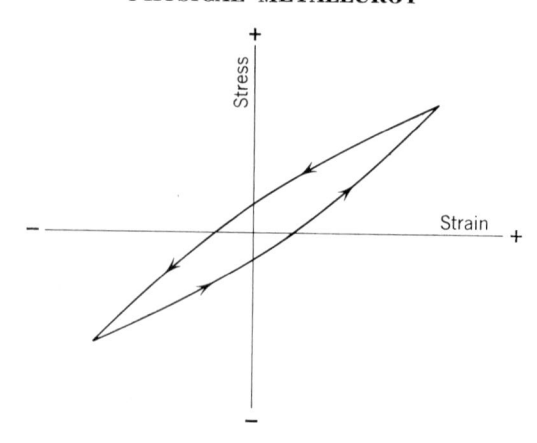

Fig. 5.57. Stress-strain loop.

specimen. Thermoelastic effects can also exist as a result of inhomogeneity of the material, for example, a polycrystalline specimen in which neighboring crystals have different elastic properties in the direction of the stress, and therefore undergo different amounts of adiabatic changes of temperature; and as a result of inhomogeneity of stress, as in a sample subjected to bending, in which one part is heated by compression while another part is cooled by tension.

Internal friction resulting from thermoelastic effects is qualitatively similar to that resulting from many other effects; a system in vibration suffers a decrease in amplitude which can best be represented by its logarithmic decrement δ, which is the natural logarithm of the ratio of the amplitudes of successive vibrations. Another measure of the internal friction is Q^{-1}, which is δ/π. The value of δ is frequency-dependent, having a maximum value when the time period of the cycle is comparable with the time constant of the relaxation process (such as temperature equalization by heat conduction) that is responsible for

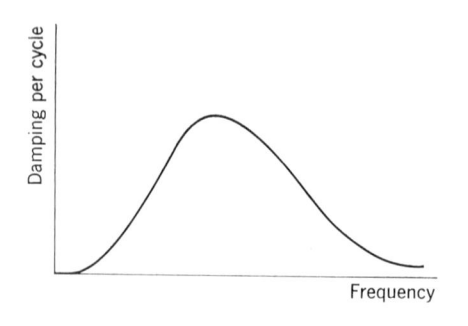

Fig. 5.58. Effect of frequency on damping.

the effect. Since some of these relaxation processes are thermally activated, the internal friction may also be temperature-dependent.

An exception is internal friction caused by magnetic effects, which is due in part to the movement of domain walls caused by the application of a stress. The movement of domain walls takes place as a result of the change of dimensions with magnetization (magnetostriction). Any "frictional" type of force that tends to prevent motion of the domain walls causes a lag between the strain and the stress. The result is frequency-independent internal friction.

The various internal friction effects associated with local structural changes have been of value in identifying the nature and measuring the characteristics of these processes. Each process has a characteristic time of relaxation, and this corresponds to a peak on the curve relating internal friction with frequency of vibration. Since the relaxation times vary with temperature, it is also possible, and much more convenient experimentally, to vary the temperature while the frequency remains constant. A single sample may show several different relaxation peaks. An example is shown in Fig. 5.59, in which two peaks and

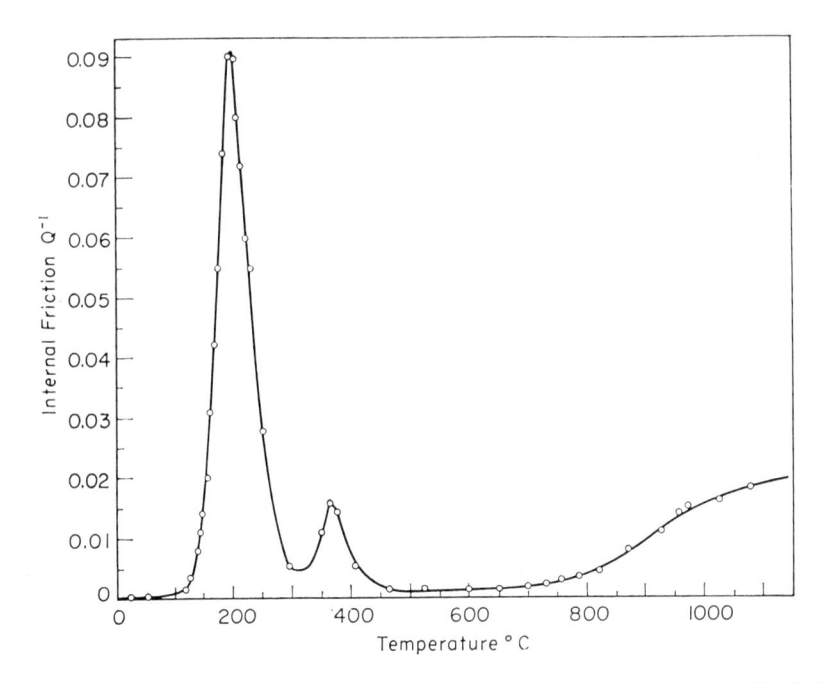

Fig. 5.59. Relaxation peaks. From *Modern Research Techniques in Physical Metallurgy,* American Society for Metals, Cleveland, 1953, Fig. 1, p. 228.

part of a third may be seen. The following effects have been studied by means of internal friction: movement of interstitial solute atoms from one interstitial site to another; rearrangement of positions of pairs of solute atoms (both these give information about diffusion, see p. 369); measurement of the concentration of interstitial solutes within the crystal; extent and rate of ordering; relative motion of grains at grain boundary; movement of dislocations.

Two other effects are closely associated with internal friction, and are also caused by relaxation effects. One is the variation of the elastic moduli due to the contribution of the relaxation, which is normally included in the strain, but is not part of the elastic strain. The measured value, therefore, is lower than the true value. The second effect is the relaxation of stress if the specimen is strained "elastically" and then held in the strained position. The stress tending to restore it to its original size or shape decreases with time.

5.5 Fracture

Theoretical strength. The theoretical strength of a solid can be defined as the maximum stress that would be required to separate two parts from each other if such separation were made to occur simultaneously over the whole of a cross section. If separation occurs sequentially, for example by the propagation of a crack, the theoretical strength only needs to be exceeded at one place at a time, and the mean applied stress can therefore be much lower than the theoretical strength. The theoretical strength can be calculated as follows from the energy of the two new surfaces that are created when a piece of solid is broken into two parts.

The cohesive force between two neighboring atoms varies as shown in Fig. 5.60, in which σ_{th} is the theoretical strength; the initial part of this curve can be approximated by

$$\sigma = \sigma_{th} \sin (2\pi x/\lambda)$$

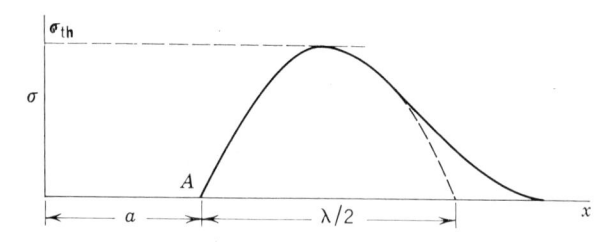

Fig. 5.60. Theoretical fracture strength.

The work done per unit area in separating two planes of atoms is

$$\int_0^{\lambda/2} \sigma_{th} (\sin 2\pi x/\lambda)\, dx = \lambda \sigma_{th}/\pi$$

This work is equal to the surface energy $2S$ of the two new surfaces that are produced. Therefore

$$\sigma_{th} = 2S/\lambda$$

but, for the part of the curve near A,

$$\sigma = Ex/a$$

where E is Young's modulus; also, since $\cos(2\pi x/\lambda) \approx 1$ for this part of the curve,

$$d\sigma/dx = 2\pi \sigma_{th}/\lambda$$

Therefore

$$2\pi \sigma_{th}/\lambda = E/a \quad \text{or} \quad \sigma_{th} = \left(\frac{ES}{a}\right)^{1/2}$$

Using $E = 10^{11}$ dynes/cm^2, $S = 10^3$ ergs/cm^2, and $a = 3 \times 10^{-8}$ cm:

$$\sigma_{th} = 10^{11} \text{ dynes/cm}^2 \text{ or about } 10^6 \text{ psi}$$

This stress is much higher than that normally observed for metals; however, recent experiments on metal "whiskers" have shown that occasional samples have a strength of 10^6 psi to 2×10^6 psi, and that such samples deviate from Hooke's law to a marked extent before they fracture. This is to be expected for strengths close to σ_{th}. It is possible that such samples fracture by the simultaneous separation of two planes of atoms; but it is clear that this is not the mode of fracture in more ordinary cases.

Types of fracture. It is necessary, therefore, to consider the geometrical aspects of fracture as it actually occurs, as distinct from the ideal case discussed above. It immediately becomes apparent that there are at least four distinct types of fracture that can be distinguished by their geometry. They are as follows:

1. Fracture by uninterrupted plastic deformation. Some metals that do not work-harden much, and many very viscous liquids, can be broken without any process other than plastic deformation. This is illustrated in Fig. 5.61.

2. Cleavage. Many crystalline substances, including metals with body-centered cubic or hexagonal structures, may fracture by separation normal to crystallographic planes of high density. In the face-centered cubic structure, fracture by normal separation does not appear

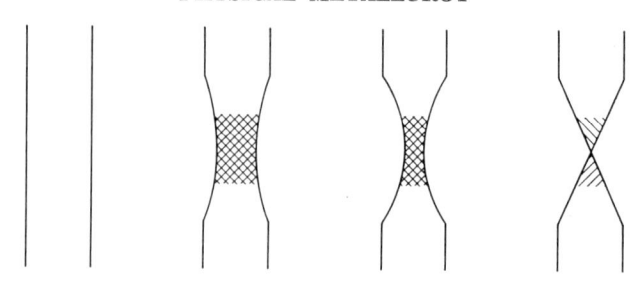

Fig. 5.61. Ductile fracture.

to be crystallographically controlled, but takes place in planes subjected to maximum tensile stress.

3. Intercrystalline separation. Separation of crystals from each other is a distinct type of fracture, as it does not involve the fracture of the actual crystals.

4. Shear fracture. It seems that fracture can occur by catastrophic local shear deformation in polycrystalline materials.

The understanding of fracture has been retarded by the fact that different criteria control the initiation of fracture and the propagation of fracture. The former of these is undoubtedly structure-sensitive, while the latter may depend on locally abnormal conditions that are caused by proximity of the existing crack, or partially developed fracture. If, however, there is a suitable imperfection, it may contribute to the propagation as well as to the inception of a crack.

Criterion for crack propagation. In order to appreciate the conditions that must be met for a crack to have passed the "embryo" stage, we must first discuss how the propagation of an existing crack takes place. This discussion is general, referring to any kind of crack that corresponds to the formation of two new surfaces. This excludes fracture by shear and fracture by plastic deformation.

The following situation will be considered: A homogeneous rectangular specimen is subjected to a tensile stress that is insufficient to cause fracture. Following the point of view developed by Griffith, we will consider the effect of introducing a discontinuity in the material. This discontinuity is assumed to be elliptical in section, as shown in Fig. 5.62. The specimen is assumed to be of unit thickness normal to the plane of the diagram. The problem is to relate the size ($2C$) of the crack with the stress σ necessary to extend it. The Griffith criterion is that the crack will spread if the decrease in elastic strain energy resulting from the increase in C is greater than the increase in surface energy due to the increase in the surface area of the crack. The presence of the crack decreases the strain energy because the crack itself does not transmit stress to the material adjacent to it. Various estimates have been made of the strain energy; they agree to within a small numerical factor which is not important in the application of the results; we will adopt the value given by Petch:

$$\text{Strain energy} = \frac{\pi C^2 \sigma^2}{E}$$

where E is Young's modulus.

The surface energy of the crack is approximately $4CS$, where S is the specific surface energy. Equilibrium between the reduction in elastic energy and the increase in surface energy is attained when

$$\frac{d}{dC}\left(4DS - \frac{\pi C^2 \sigma^2}{E}\right) = 0 \quad \text{or} \quad \sigma_c = \left(\frac{2ES}{\pi C}\right)^{1/2}$$

where σ_c is the critical stress; if the stress is greater than σ_c, the crack will spread. The resulting increase of C reduces the critical value of σ_c, and so a crack, once started, should require progressively less stress to extend it. It is not believed that this is necessarily true for metals, since under some conditions a progressively increasing stress is required to cause a crack to propagate. It has been proposed by Orowan that, in the case of a metal that has some capacity for plastic deformation, the plastic work done on the metal at the point of fracture should be added to the energy corresponding to the new surface in order to establish the criterion for crack propagation.

The problem may also be considered in terms of stresses rather than energies. The highest tensile stress, which occurs at point A in Fig. 5.62, is given by

$$\sigma_{\max} = 2\sigma_c \left(\frac{C}{\rho}\right)^{1/2}$$

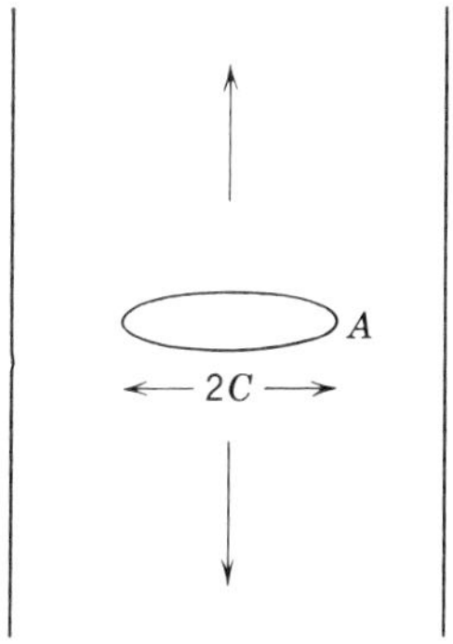

Fig. 5.62. Griffith crack.

where ρ is the radius of curvature of the end of the crack, at A. The most severe possible case will exist if ρ is comparable with a, the interatomic spacing, as atoms much closer than $2a$ will exert a significant attractive force on each other and this will reduce the value σ_{max}. The value of σ_{max} is therefore given by

$$\sigma_{max} = 2\sigma_c \left(\frac{C}{a}\right)^{1/2}$$

If σ_{max} is replaced by the theoretical strength σ_{th}, which is given by $\sigma_{th} = [(ES)/a]^{1/2}$, the result is that $\sigma_c = [(ES)/4C]^{1/2}$, which agrees very closely with the result obtained from energy considerations. In a ductile material the value of σ_{max} would be less than that given above because of plastic yielding just ahead of the fracture, and so the critical stress σ_c would be increased.

The effect of speed must now be considered. There are two distinct effects. In the first place, plastic deformation is not instantaneous, and therefore it may be expected that the amount of plastic deformation just ahead of the crack will increase as the rate of propagation is reduced. Thus a very slow crack requires the maximum amount of work. Secondly, the yield stress increases as the strain rate increases, if the latter is sufficiently high. At very high speeds, therefore, the amount of plastic deformation is further reduced because less of the material is above the enhanced yield stress. A crack which reaches a critical speed should therefore accelerate.

It may therefore be concluded that for any given size of crack there is a critical stress, in the form of a tensile stress normal to the plane of the crack, which will cause it to propagate.

In some types of material that fail by crack propagation, the cracks are initially present as a structural feature of the material.

The critical size of crack, calculated from the formula derived above for several materials, is shown in Table 5.3. It will be seen that such

Table 5.3. The Calculated Size of Griffith Cracks

Material	σ_c (dynes/cm^2)	S (ergs/cm^2)	E (dynes/cm^2)	e (cm)
Glass	1.8×10^9	210	6.2×10^{11}	2.6×10^{-5}
Fe (polycrystalline)	7×10^9	1220	20.5×10^{11}	7.8×10^{-5}
Zn (polycrystalline)	1.8×10^7	800	3.5×10^{11}	0.55
NaCl	2.2×10^7	150	4.9×10^{11}	0.10

cracks could exist in glass and in iron without being evident, but that they cannot exist before the application of a stress in ordinary specimens of zinc or of sodium chloride. An example of the propagation of pre-existing cracks is found in the fracture of gray cast iron. Figure 5.63a shows cast iron containing graphite in flake form. The black regions are graphite, which has very low strength; the cavities in which it occurs have suitable shapes to act as crack initiators; it is found that this material is relatively weak in tension and that it fractures without perceptible plastic deformation. The shape of the graphite regions is important; if they are produced in spherical (see p. 268), instead of flake, form, they no longer produce sufficient stress concentration to initiate cracks, and the resulting material can be deformed plastically without fracturing. A sample of cast iron containing graphite in nodular form is shown in Fig. 5.63b.

A somewhat similar type of fracture is that which occurs at crystal boundaries under conditions in which the crystal boundary is either (a) brittle, or (b) weak. The former appears to be the case for many alloys and impure metals in which a layer or a partial layer of a brittle compound forms at the crystal boundaries. The embrittlement of titanium by rather small amounts of oxygen or nitrogen or hydrogen may be caused in this way. It should be appreciated that the fracture originating from an easily fractured region of crystal boundary may not be obviously intercrystalline; this is because it may be able to propagate in a direction dictated by the stress pattern, which would mostly be transcrystalline, after it had started at a crystal boundary. The intercrystalline origin would not necessarily be seen in a metallographic section. Crystal boundaries may be weak as a result of preferential corrosive attack (see p. 303); they also tend to be weak at high temperatures; this may be a result of the process that often leads to fracture under prolonged application of a stress, as in a creep test. It is thought that fracture initiates at grain boundaries under such conditions as a result of the formation of voids by the aggregation of vacancies which are generated by the climb of dislocations, as discussed on p. 102. The boundaries should be the most suitable places for vacancies to aggregate, and there is experimental evidence that such voids do actually form during creep (Fig. 5.64). The practical importance of the time under creep conditions which elapses before fracture is shown by the wide use of the so-called stress-rupture type of creep test. In this test, the elongation of the sample is not measured; the time to failure is determined as a function of load and temperature. An alternative proposal for the initiation of a crack during

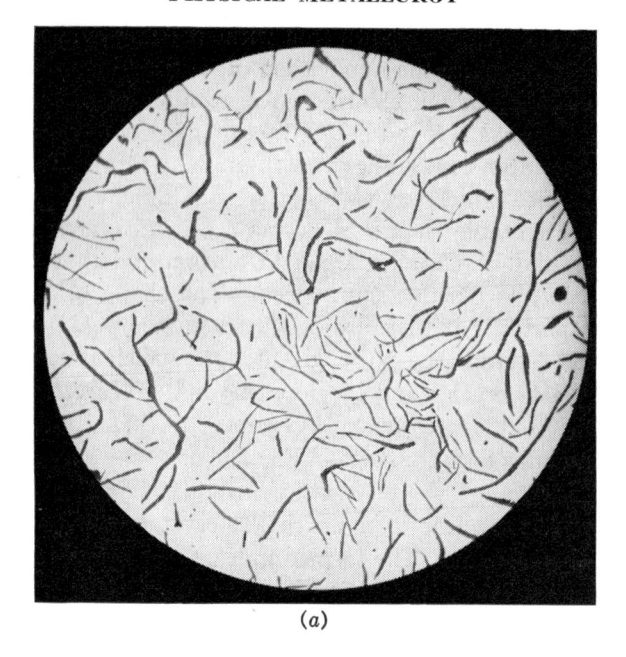

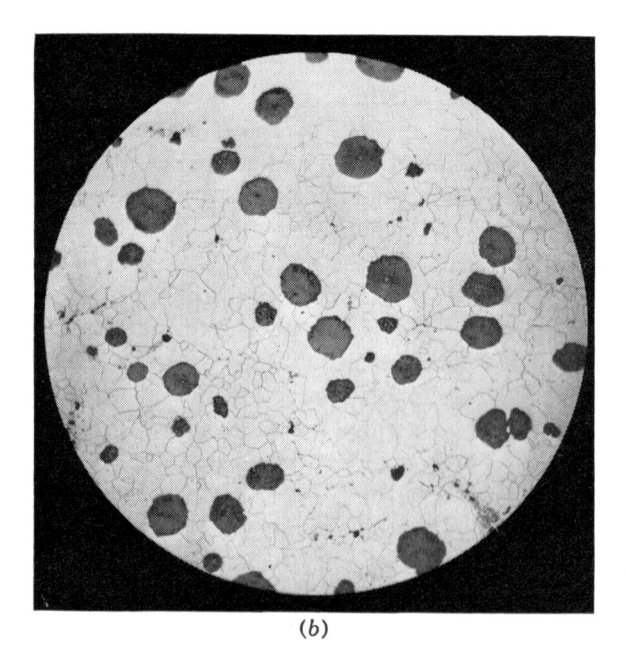

Fig. 5.63. Photomicrographs of cast iron: (a) Flake graphite (×96). (b) Nodular graphite (×385). (Courtesy International Nickel Company.)

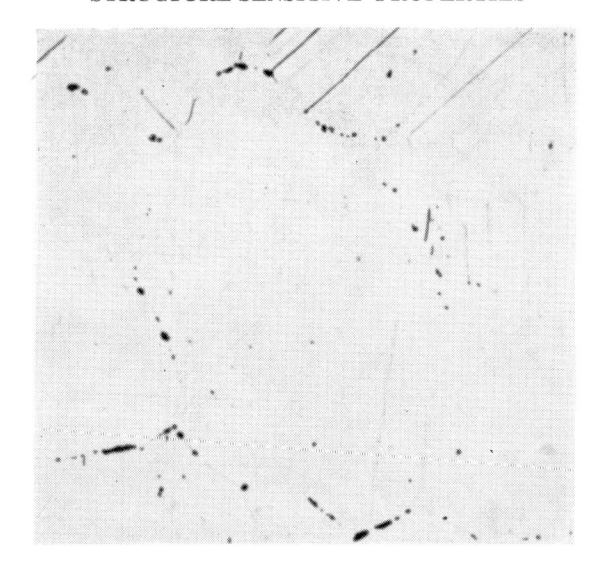

Fig. 5.64. Formation of voids at grain boundaries. From J. N. Greenwood *et al.*, *Acta Metallurgica,* Vol. 2, No. 2, Mar. 1954, Fig. 5, p. 253.

creep is that it results from the sliding of grains over each other at a discontinuity such as another boundary or a slip step. Once a void has reached a critical size it could propagate as a Griffith crack.

Origin of cracks. The least obvious aspect of fracture is the problem of how a crack originates within a crystal, as must be the case for fracture by cleavage. This may happen when the stress reaches a specific level; this is often preceded by plastic deformation, which may itself change the critical level for initiation of fracture. It is believed by some investigators that some plastic deformation is a prerequisite for cleavage fracture, and that it is inherent in the mechanism.

A type of fracture of great interest is the brittle fracture of materials such as structural steels. It is most probable that the initial stage of fracture in these materials is the initiation of slip, i.e., the operation of dislocation sources. The best evidence for this conclusion is that, under suitable conditions, fracture occurs under a tension stress which is equal in magnitude to the compressive stress that just causes yielding. The initiation of a crack probably occurs by the interaction of the dislocations on two intersecting slip planes; the following detailed mechanism, proposed by Cottrell, explains why the phenomenon under discussion occurs in body-centered cubic but not in face-centered cubic metals. Consider a (001) plane intersected by (101) and (10$\bar{1}$) slip planes (Fig.

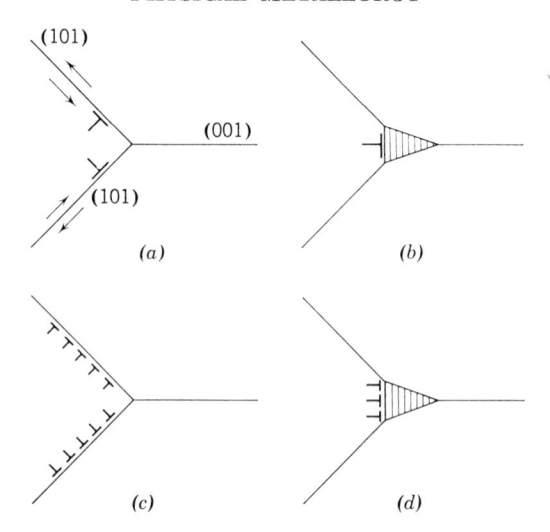

Fig. 5.65. Coalescence of two slip dislocations to form a crack dislocation (after Cottrell).

5.65a). A slip dislocation from each plane meets on the (001) plane. The slip dislocations have Burgers vectors $\frac{a}{2}[\bar{1}\bar{1}1]$ in the (101) plane, and $\frac{a}{2}[111]$ in the (10$\bar{1}$) plane. They react to form a new dislocation, $\frac{a}{2}[\bar{1}\bar{1}1] + \frac{a}{2}[111] \to a[001]$. It can be shown that this reaction is energetically favorable in the body-centered cubic structure but not in the face-centered cubic. The new dislocation is a pure edge dislocation of large Burgers vector in the (001) plane, and is equivalent to a "cleavage knife" one lattice constant thick (Fig. 5.65b). This forms the nucleus of a crack that can grow by the arrival of more dislocations on the same slip planes (Fig. 5.65c and d). This occurs as a result of shear stresses on the slip planes, while a tensile stress normal to the (001) plane is required to propagate the crack. It is probable that less stress is required to nucleate a crack than to cause it to propagate; the behavior of the material depends upon the relative values of the yield stress and the crack-propagation stress; if the former is higher than the latter, brittle fracture must occur.

The fracture curve. Whether this mechanism is valid or not, it is useful to represent the condition for fracture in a form originally proposed by Ludwik, as in Fig. 5.66. Here σ is the tensile stress that produces the shear stress required to cause plastic deformation.

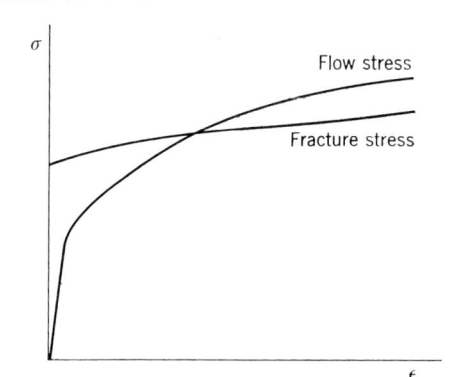

Fig. 5.66. Flow stress and fracture stress.

Because the material work hardens, σ increases as the strain ϵ increases; the resulting curve is the flow curve. σ also represents the tensile stress that would cause fracture. It is assumed that the fracture stress depends on the strain; this gives the fracture curve.

A significant feature of the diagram is the fact that neither the shear stress nor the tensile stress is uniquely related to the applied stress (assumed to be in tension) because they are modified by the geometry of the specimen. Let us consider the case of a completely ductile material; this is represented in Fig. 5.67, in which σ represents the stress (shear or normal tension) resulting from the application of a tension force, and ϵ represents the strain. The flow curve does not reach the fracture curve and the material is therefore ductile.

There are some materials which are normally ductile, but are brittle if a sharp notch is present. This is because the presence of a notch in the specimen can raise the actual tensile stress locally to a value that can approach three times the mean stress; the tension force required to produce the fracture stress is therefore reduced, in the extreme case to one-third of the nominal value. The result is shown in

Fig. 5.67. Flow stress and fracture stress for ductile fracture.

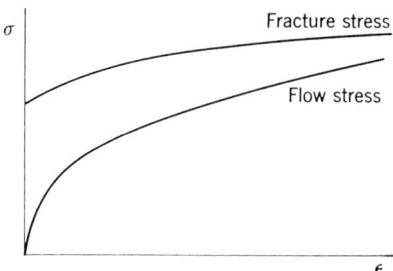

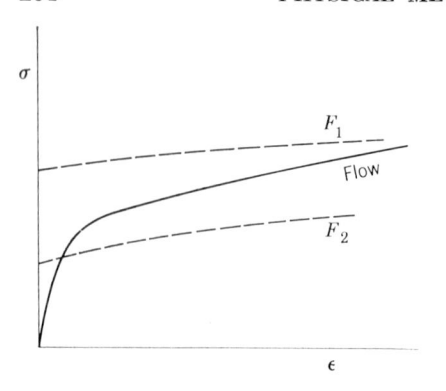

Fig. 5.68. Embrittling effect of a notch.

Fig. 5.68, in which F_1 is the original fracture stress and F_2 corresponds to the applied stress required to cause the stress to reach locally the value F_1. Such a material would be brittle if a notch of suitable severity were present but not otherwise. A second effect of geometry is that the shear stress may be reduced, for a given applied stress, as a result of a hydrostatic or triaxial stress condition. This is represented in Fig. 5.69, in which the stress distribution represented in a and b shows how part of the stress may be regarded as triaxial tension, which does not contribute to shear stresses. The resultant effective shear stresses therefore are reduced, and it is necessary to apply *more* tension to produce the shear stress corresponding to the flow curve. The normal tension is not affected by the radial forces. The result is shown in Fig. 5.70, in which the applied stress σ is shown. Again the result is to reach the fracture curve after less strain than would be required in an unnotched specimen. If the direction of the radial

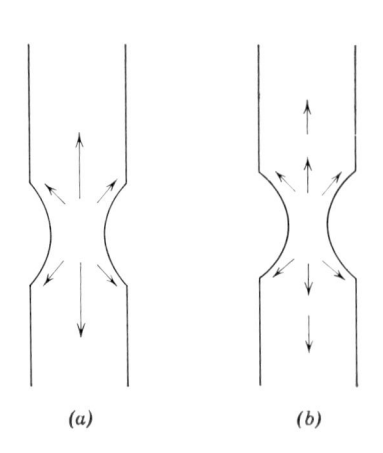

Fig. 5.69. Triaxial stresses caused by a notch.

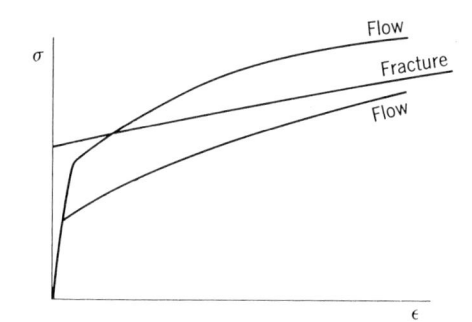

Fig. 5.70. Embrittlement due to triaxial stresses.

components were reversed, which would occur with an applied hydrostatic pressure, the result would also be reversed; ductility (plastic shear strain) can be expected when a brittle material is subjected to hydrostatic pressure and tension. This has been observed by Bridgman.

Another example of the effect of hydrostatic stress on fracture is the strength of thin layers of relatively weak metals. Consider the system represented in Fig. 5.71, where A and B are of steel and C is a thin layer of tin or of solder, bonded to each by the appropriate layer or layers of intermediate phase. It is found that the strength of the soft layer is much higher than would be calculated from the properties of the soft metal as normally measured. The reason is that a hydrostatic or triaxial tension develops when the system is subjected to tension, and this reduces the shear components of the tensile stress. The hydrostatic tension develops because the thin layer of soft metal is prevented from flowing inwards at the edges by its adhesion to the steel.

Effect of temperature on fracture. It is characteristic of the body-centered cubic and the close-packed hexagonal crystal structures that

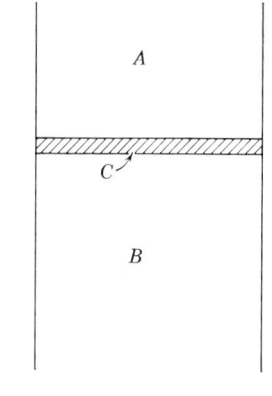

Fig. 5.71. Strength of a thin layer of a soft metal.

there is a rather abrupt transition from ductile behavior at higher temperatures to brittle behavior at lower ones. Brittleness does not occur in metals with the face-centered cubic structure at any temperature, except as a result of the presence of a second phase.

The occurrence of low temperature embrittlement in body-centered cubic metals is probably due to the effect of temperature on the flow curve, and not on the fracture curve; this is supported by the fact that grain size and composition, which, in the case of certain steels, have known effects on the flow curve, have corresponding effects on the transition temperature. The effects of temperature and of notches are additive, in the sense that notched specimens of a particular steel may be brittle at a temperature above the transition temperature for an unnotched specimen. These facts are compatible with the hypothesis that brittleness can occur as a result of the increase of the stress required to cause flow, or to cause further flow, either by lowering the temperature or by the presence of notches, or both.

These matters are of practical importance because the transition temperatures for some steels used for shipbuilding, for example, have "notch transition temperatures" that are sometimes within the temperature range encountered in service. Catastrophic failures have occurred in welded ship hulls, as a result of stress concentrations, a steel below its "brittle transition temperature," and a method of construction that allows a crack to propagate from one plate to the next. The "brittle transition temperatures" for the other body-centered cubic metals also show some variation with composition and grain size, but there is an approximate proportionality between transition temperature and melting point.

Effect of time; delayed fracture. If a stress is applied to a brittle material, it is sometimes found that fracture takes place after a delay period, which may be as long as an hour. This may either be due to the contribution of thermal activation to the propagation of the Griffith crack which is just subcritical, or to the redistribution of stress as a result of movement of dislocations; or else to the effect of adsorbed films which lower the surface energy of a Griffith crack to such an extent that it can propagate. This could only occur at the speed at which the adsorbed film could spread over the newly formed surface.

Hydrogen embrittlement. Hydrogen is frequently present in steel in interstitial solution as a result of electrolytic or other effects, and it is found to cause a complete loss of ductility and reduction in strength under some conditions. It is generally believed that hydrogen comes out of solution and forms hydrogen molecules at particularly favorable points, which may be the surfaces of second-phase particles, or they may be dislocation nodes; once a "bubble" of hydrogen has started to form, it can continue to grow by the arrival of more hydrogen by diffusion and a high enough pressure can be developed to cause a crack to initiate locally. The high pressure is due to the fact that equilibrium between molecular hydrogen and dissolved atomic (or ionic) hydrogen is only established at extremely high pressures unless the dissolved hydrogen is at a very low concentration.

Fatigue. Fatigue failure is fracture that occurs as a result of many applications of a stress which, applied once, would not cause failure. The number of times the stress must be applied to cause failure is related to the stress σ as shown in Fig. 5.72. The following features of fatigue behavior have been established:

1. The life (i.e., number of applications of stress before failure) increases as the stress decreases.

2. There is, for steel and some other metals, a stress level below which failure never occurs. This is the "fatigue limit," and is often about 0.4 times the ultimate strength. The fatigue limit does not appear to be related to the ductility.

3. There is marked "scatter" in the results of any series of tests.

4. The "time scale" is unimportant; the life (expressed as the number of applications) for a given stress is not affected by the frequency of application of the stress.

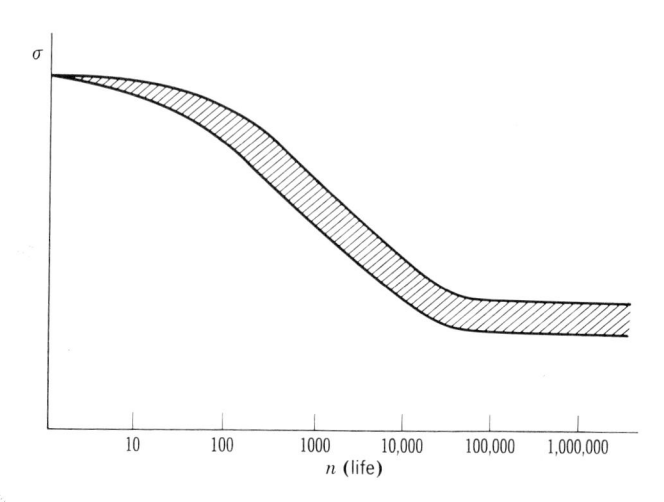

Fig. 5.72. Relationship between fatigue stress and endurance.

5. Stress concentrations have a marked effect; the maximum stress that can be applied for a given life is very much reduced if a severe "stress raiser" is present.

6. The fatigue limit depends upon the range through which the stress varies, as well as on the maximum stress. This variation is given approximately by the "Goodman diagram" shown in Fig. 5.73, in which pairs of points such as A and B give the maximum and minimum for a stress cycle that is just at the fatigue limit.

As in any other type of fracture, the criteria for initiating and for propagating a crack are quite different. The propagation of a crack depends, for example, upon whether the stress cycle or the strain cycle remains constant; if the latter, then formation of a crack may so relieve the stress that the crack fails to propagate. This applies particularly when the sample is subjected to alternating bending through a fixed amplitude. Recent work appears to show that the crack exists and can be detected after a rather small number of applications of a stress cycle, and that it originates at the surface at a slip band. This indicates that slip band formation is in some respects irreversible.

The mechanism of the nucleation of fracture may be as follows: Suppose that the stress applied during each cycle is sufficient to activate some Frank-Read sources; a few dislocations would be generated by such sources, one of which is represented in Fig. 5.74. Let us suppose that a single edge dislocation has emerged at A. When the stress is

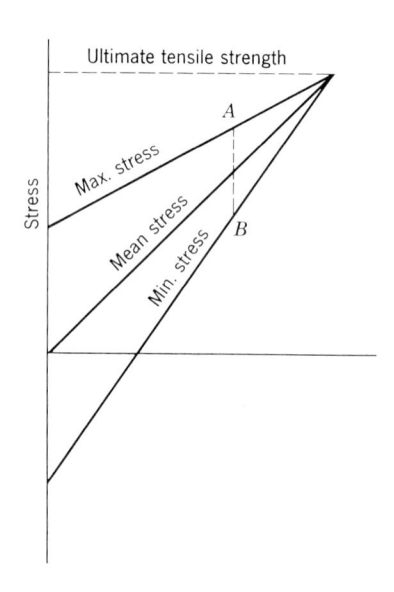

Fig. 5.73. Goodman diagram.

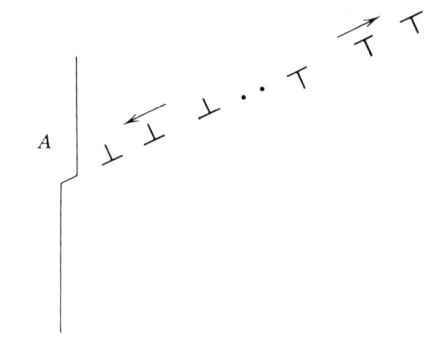

Fig. 5.74. Slip step at surface.

removed or reversed, the dislocations will either return to their source or be held up as a result of "jog formation." However, the dislocation that has emerged at A can only retreat if the new element of surface at A can recombine with the appropriate part of the slip plane. This may be opposed by (a) oxide or other compound formations, (b) adsorbed gas film, or (c) readjustment of atomic positions because the element is now a surface. If there is a normal force tending to separate the parts above and below the slip plane, the result may be a failure to re-form the slip plane; this could be the origin of a crack, particularly if the process were repeated sufficiently often. A second possibility, that is apparently consistent with experimental findings, is that the dislocation may return, but in a slightly displaced plane. This could be achieved by the addition of extra components of screw dislocation to connect the returning dislocation to the rest of the loop to which it belongs. If this process occurred as in Fig. 5.75, it could represent the origin of crack, especially if successive dislocations followed the same path as the one depicted. It is still necessary to account for the fact that this process, if it occurred once every cycle, would progress too rapidly. This process would not proceed beyond the point at which the crack reaches the Frank-Read source, which is

Fig. 5.75. Irreversible slip step.

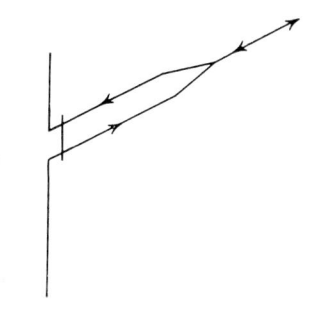

probably very close to the surface. Further propagation of the crack, which would occupy most of the life of the specimen, occurs on planes approximately normal to the applied tension stress, and must correspond to the propagation of a crack that is below the Griffith size for catastrophic propagation, but large enough to propagate by the occasional irreversibility of atom movements at the "growing point" of the crack.

A stress below the fatigue limit would either not cause any sources to be activated, or else would allow active sources to be suppressed by work hardening, which would proceed more rapidly than the increase of local stress due to the irreversible emergence of dislocations.

The important feature of fatigue is that it can cause fracture even if the applied stress has never exceeded one-half of the value that would cause failure in a "static" test; this ratio can be much lower if notches are present; however, some steps can be taken to prevent fatigue. The most obvious and useful of these is to avoid stress concentrations in regions that are to be subjected to fluctuating stresses even if these are well below the static strength of the material. For some purposes, even the irregularities left by machining can act as stress concentrators. The second approach depends upon the fact that fatigue cracks nearly always start at a surface. If the surface material is (a) strengthened and (b) put into a state of compression by adding an interstitial solute or by deformation, some improvement in fatigue properties is usually found.

Characteristic appearances of fractures. Much information can often be obtained about the cause of a fracture from its appearance.

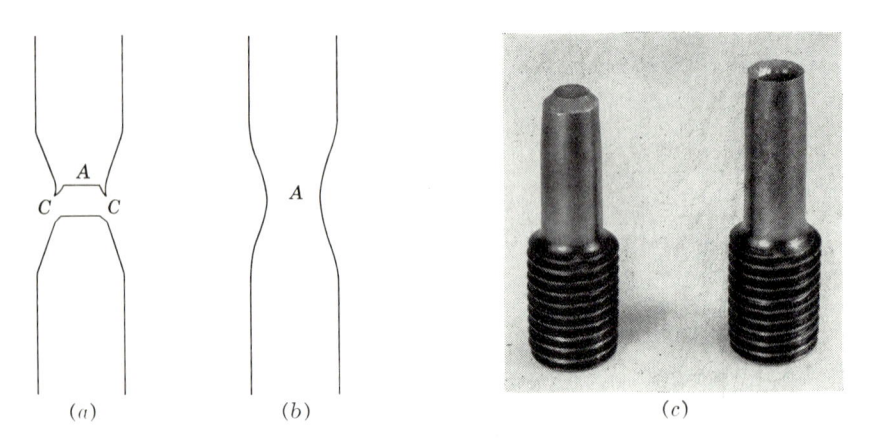

Fig. 5.76. Cup and cone fracture.

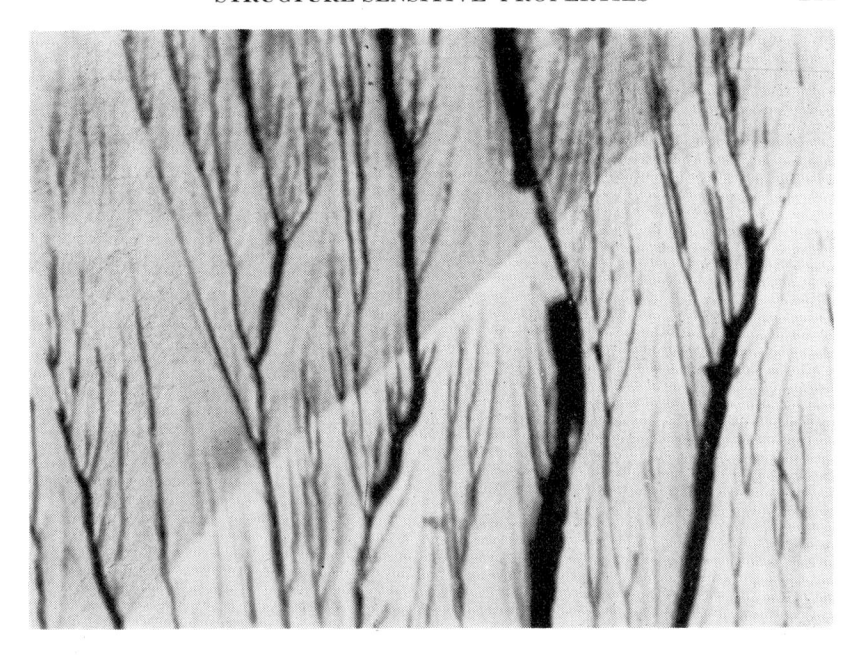

Fig. 5.77. Fracture surface of a single crystal of iron–3½% silicon, cleaved at 78°K. (Courtesy J. R. Low, Jr., General Electric Research Laboratory.)

The first question is whether the fractured material suffered any distortion in the neighborhood of the fracture; if it did, then either the fracture is all of the same appearance and is a shear fracture in a plane approximating to a plane of maximum shear stress, or else the deformation prior to fracture so changed the stress distribution that a brittle failure was initiated; such a failure may finish as a shear failure. A good example of this type of failure is the "cup and cone" fracture typical of the tensile test of a reasonably ductile metal. A cross section of the separated parts of such a failure is shown diagrammatically in Fig. 5.76a. The same specimen, just prior to failure, is shown in Fig. 5.76b. An actual fracture of this kind is shown in Fig. 5.76c. The "neck" that has formed in the specimen as a result of its ductility sets up a component of hydrostatic stress; this causes brittle failure to start at A; the fracture in this region is normal to the direction of tension. The formation of the crack at A decreases the hydrostatic tension and the crack propagates as shear failure at about 45° to the axis of tension. Fracture terminates at the periphery CC.

212   PHYSICAL METALLURGY

If fracture occurs without visible distortion, it may be (a) brittle fracture, or (b) fatigue failure. Brittle fracture may occur in a material that shows normal ductility as a result of either stress concentration, biaxial or triaxial stressing, low temperature, or impact conditions. A brittle fracture usually has a consistent appearance over the whole fracture surface; the direction of propagation of the fracture is often revealed by the characteristic "feather" markings, which converge in the direction of propagation of the crack. An example is shown in Fig. 5.77.

A fatigue failure, on the other hand, usually contains two distinct zones; one is a smooth region in which the fatigue crack propagates outwards from its point of origin, often a stress concentration, while the remainder, usually of "granular" or "rough" appearance, is the part

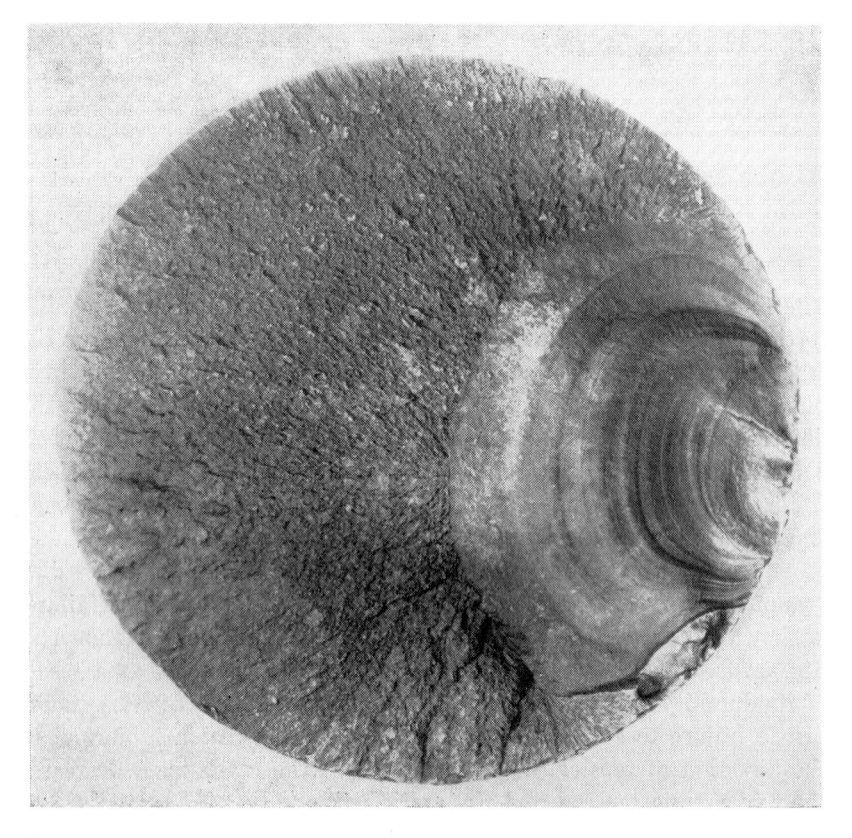

Fig. 5.78. Photograph of a fatigue failure in a 10-in.-diameter piston rod. (Courtesy John M. Lessells, Lessells Associates, Inc.)

Fig. 5.79. Intercrystalline fracture in steel. (Courtesy J. R. Low, Jr., General Electric Research Laboratory.)

of the failure that occurs when the fatigue crack has so reduced the section that the remaining material cannot sustain the load. The "fatigue" region often has concentric marks showing a series of positions at which the crack was stationary for some period. A typical fatigue failure is shown in Fig. 5.78. When embrittlement is due to intercrystalline weakness, microscopical examination usually reveals other cracks (besides the one that formed the fracture). A photomicrograph of a typical intergranular failure is shown in Fig. 5.79.

5.6 Mechanical Properties

The mechanical properties of metals and alloys are of practical interest from two distinct, and sometimes opposed, points of view. These relate to the problems of deforming a metal into the shape required for use and to its subsequent behavior in service.

Mechanical testing. The objects of mechanical testing are: 1. to provide a criterion for the comparison of experimental procedures, alloys, etc.; 2. to provide data for use in the design of engineering structures; 3. to determine whether a particular sample conforms to the properties assumed in design; or 4. to determine how a given sample will behave during fabrication.

Of these four objectives, only the first is directly achieved by the type of measurement implicitly discussed above, i.e., by the determination of stress-strain-time relationships with simple stress systems. For the other three purposes, it must be recognized that the behavior of a sample depends on its geometry as well as on the properties of the material, and that the properties of the material are by no means fully determined by the behavior under a simple stress such as tension.

For most engineering purposes, the design data that have been required have been the elastic properties, the lowest stress at which observable plastic deformation takes place (the yield stress) and the maximum tensile load that can be withstood without failure (the ultimate tensile strength). A further requirement is that the material must be able to undergo some plastic deformation before fracture. For special purposes, ductilities have been used as design data, but the fact that many metals undergo plastic deformation before fracture has been used mainly as an extra safeguard, and to permit assembly of parts which require some adjustment of shape. Some account is taken, qualitatively, of the fact that certain metals are more "notch sensitive" than others, i.e., are embrittled by the presence of a stress-raising geometrical configuration.

Many other criteria have been developed, some for predicting the response of a metal to a forming operation, and others for comparing a particular critical property of the material with the corresponding property of a satisfactory sample. Tests of these types will be discussed on p. 222; it is necessary first to discuss the tensile test, which has become the basic mechanical test. It should be noted that when mechanical testing was first introduced, engineering design was such that most critically stressed components were in tension; a direct test of the behavior of a material in tension was therefore completely relevant. More modern design frequently depends on critical stressing in compression or in shear or in a combination of stresses and the tensile test is therefore no longer so closely related to service conditions.

The tensile test. In the typical tensile test, the length of a test piece is increased at a constant rate, and the load necessary to do this is measured. The result is plotted directly as the load-elongation curve. The stress-strain curve can also be plotted from the same data if the load is divided by the cross section to give the stress, and the strain is calculated as the sum of incremental strains each of which is based on *unit* gage length, rather than on the length of the same part of the specimen, which increases during the test. It is necessary to distinguish between the "engineering stress," in which the load is divided by the original cross section, and the "true stress," in which it is divided by the actual cross section.

The following features of the load-elongation curve are recognized. The region OA, Fig. 5.80, is the elastic region and its slope corresponds to Young's modulus of the material. At A, the curve begins to deviate from a straight line. This point, the position of which depends on the sensitivity of the elongation measurement, is called the elastic limit. A point B, the yield stress, is defined either as the load at which 0.2% permanent extension has taken place, divided by the *initial* area of cross section, or as the load at which the departure from the linear continuation of the elastic line is 0.2%, also divided by the initial cross section. These two definitions are indistinguishable in practice; they are equivalent if, as is usually the case, the load-elongation line produced by unloading from B is straight and parallel to OA.

The next recognized point is C, corresponding to the maximum load. The maximum load divided by the *initial* cross section is the ultimate tensile strength (U.T.S.).

The load decreases beyond this point and eventually fracture occurs, at D. The increase in length, divided by the original length, of a defined length of the test piece ($4\pi D$, where D is the diameter) is the *elongation*, and the decrease in area of cross section divided by the original area at the point of fracture, is the *reduction of area*. Fracture can occur at any point on the curve; if the load-elongation curve does not become horizontal before fracture, then fracture occurs at the

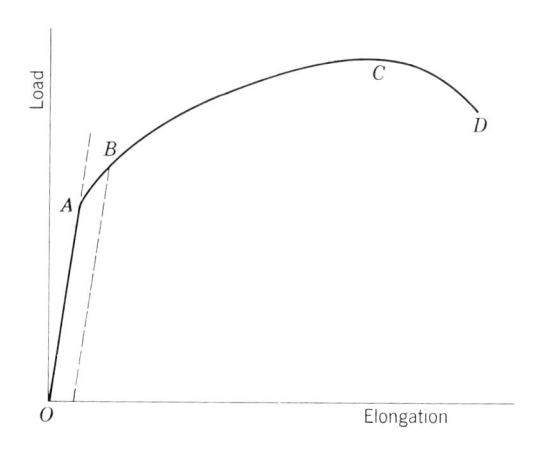

Fig. 5.80. Load-elongation curve, showing yield stress and U.T.S.

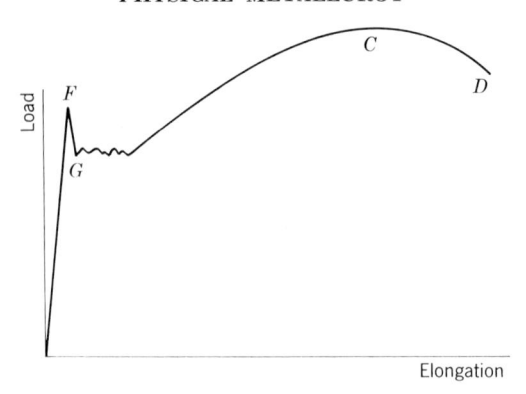

Fig. 5.81. Load-elongation curve showing yield point.

maximum stress reached, and again the U.T.S. is the maximum load divided by the original area. The corresponding curve for some materials which exhibit upper and lower yield points, such as mild steel, is shown in Fig. 5.81. In this case, points F and G are the upper and lower yield points respectively. C and D have their previous significance.

It must be pointed out that this method of test is somewhat arbitrary. The rate of elongation is controlled by the testing machine, and is the independent variable, while the load, which is the dependent variable, is measured. This type of machine is described as "hard." An equally legitimate method of testing is to increase the load at a predetermined, controlled rate, and to measure the resulting deformation. In this case, the load is the independent variable, and the elongation is dependent. A machine that conducts this type of test is described as "soft."

As might be expected, the load-elongation curves would not be the same for these two types of machine. A comparison is shown in Fig. 5.82. There are intermediate types of machines, such as the classical "beam" type, in which a decrease in load can be observed but not readily measured.

We must next discuss the significance of the U.T.S. If fracture occurs before the curve has become horizontal, the U.T.S. corresponds to the load required to cause fracture, divided by the initial cross section of the specimen; it is not the maximum stress, because the cross section no longer has its initial value.

If, however, the maximum is passed before fracture, the significance is entirely different. In the first place, the U.T.S. in this case has no connection with the stress required to cause fracture; it is the stress

STRUCTURE-SENSITIVE PROPERTIES

required to produce the strain beyond which uniform elongation no longer occurs. The maximum of the load-elongation curve is the point at which local elongation (necking) starts, usually at the center of the specimen. The condition for a cylindrical test piece to elongate as a cylinder is that, if any part of it elongates more than the rest, a higher *load* is required to cause further extension. This may be roughly expressed by saying that the material must work-harden more rapidly than it decreases in cross section. In terms of the stress-strain curve, proportional work hardening, $\delta\sigma/\sigma$, resulting from an increase in length, δl, must be greater than the proportional decrease in cross section, $\delta l/l$, for a cylindrical shape to be stable. The required stress for further elongation increases more rapidly than the stress resulting from the same load acting over a reduced area.

If $\delta l/l$ becomes greater than $\delta\sigma/\sigma$, then further extension will occur at the point at which reduction in cross section has taken place, and the required load will decrease. The point at which the two slopes are equal is the limit of stability. It is given by the construction of Fig. 5.83. At A, $\delta l/l = \delta\sigma/\sigma$. The point A on the stress-strain curve corresponds to the maximum load point on the load-elongation curve.

There is no discontinuity in the stress-strain curve at point A, although this point represents the major engineering criterion for the strength of a metal. It will be appreciated that a much more significant criterion from a fundamental point of view is a parameter that

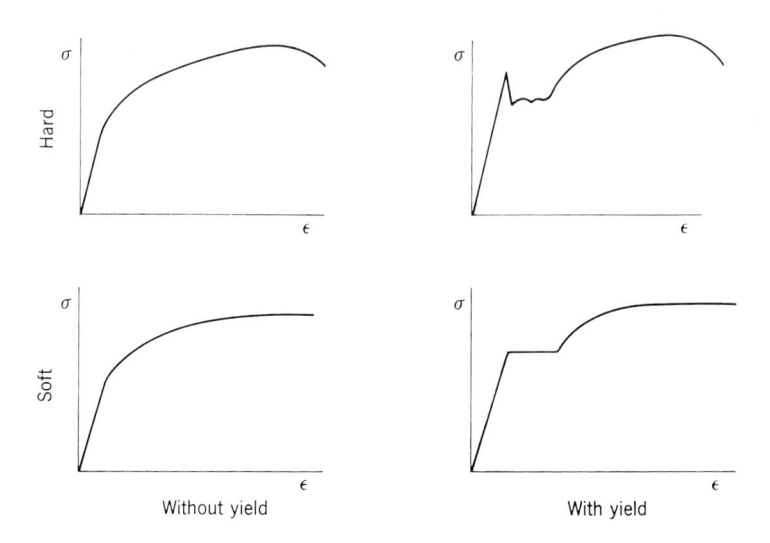

Fig. 5.82. Stress-strain curves for hard and soft testing machines

represents the rate of work hardening; the work-hardening exponent n is an example.

The elongation and the reduction of area are convenient methods for obtaining information on the total amount of deformation that occurs before fracture. These criteria include two distinct parts if nonuniform deformation (necking) has taken place, since the change of length that constitutes uniform elongation is included with the local elongation that is observed at the neck. The value obtained for the elongation is sensitive to the gage length, while the reduction in area is not.

The practical importance of elongation is twofold. In the first place, many engineering applications demand that a material should be able to undergo some deformation without fracturing, either to permit conventional fabrication and assembly practices, or to allow for some redistribution of stress in service, or to allow some overload without catastrophic failure. The amount of elongation that is required for these purposes depends upon the application, and is not normally derived from design calculations. It is usually selected on the basis of previous experience or tradition. There has been a tendency to reduce the required elongation for many purposes. The second aspect of ductility is that it is a relatively easily measured property that is sensitive to errors in composition, mechanical treatment, or heat treatment.

The discussion has been developed in terms of the load-deformation curve for tension. It will be evident that the corresponding curve for compression loading will be different, because the result of compressive strain is to *increase* the cross section and therefore reduce the stress resulting from the application of a given load. A specimen stressed

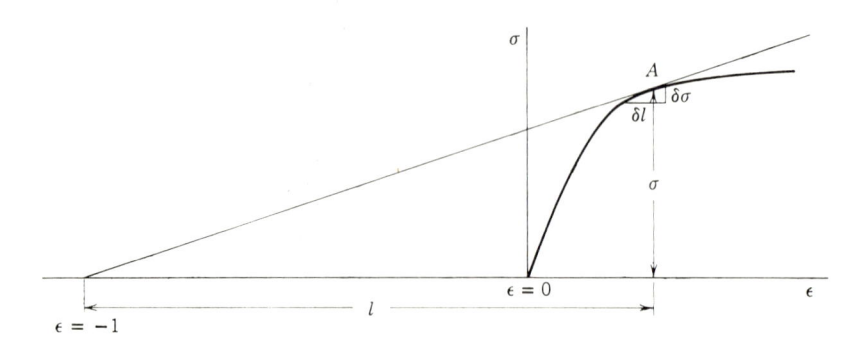

Fig. 5.83. Maximum load criterion from stress-strain curve.

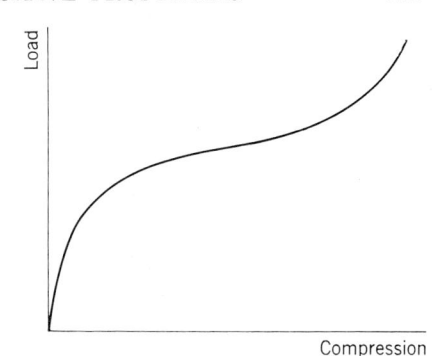

Fig. 5.84. Force-displacement curve for compression.

in compression is therefore stable as far as plastic deformation is concerned. The typical load-deformation curve in compression is shown in Fig. 5.84. Failure in such a test is never by cleavage but may be by shear at about 45° to the axis of compression.

A third simple type of test is the torsion test. The torsional stress on a cylinder varies from zero at the center to a maximum at the outside, and therefore a torsion test on such a specimen gives the integrated result of a whole series of "torque-twist" curves. In order to investigate the properties of the material, rather than of the specimen, in torsion, it is necessary to use a thin-walled tube in which the stress and the strain are nearly constant throughout the material. The cross section of the specimen does not change during the torsion test, which can therefore be continued far beyond the point of instability of the tensile test. The resulting curve closely resembles the stress-strain curve. It is, however, not always possible to obtain a thin-walled cylinder in which the material is similar in all respects to that of a sample that has been subjected to any particular sequence of processes.

In a uniform isotropic material, the stress-strain curves derived from tension, compression, and torsion tests should be the same. However, metals are not always isotropic, because the individual crystals have preferred directions of deformation and if they are not distributed randomly, as regards orientation or shape, the stress-strain curve may vary with the direction of application of the stress. This is discussed further in connection with deformation processes.

A second property of some metals that is also related to the direction of stressing is the Bauschinger effect. This effect may be described as follows: If a specimen has been strained in tension, its yield stress in compression is reduced, and vice versa. Thus the result of applying a stress first in tension and then in compression would be as shown in

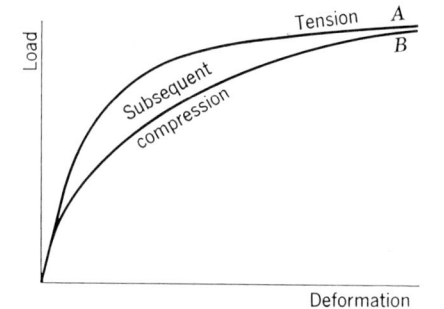

Fig. 5.85. Bauschinger effect.

Fig. 5.85. If the first load had been in compression, instead of tension, the two curves A and B would be interchanged.

Hardness testing. The advantages of a mechanical test that is both rapid and nondestructive are so great that some sacrifice in fundamental significance can often be tolerated. The widespread use of one or another form of hardness testing is in this category; hardness, defined as *resistance to deformation,* is usually measured by the size of indentation produced when an indentor of specified shape is pressed into the surface by means of a specified force. The result is seldom of direct interest from a design point of view, because indentation is not often the type of damage which a metal has to resist; from a research point of view, hardness does not provide a measure of any one of the properties, such as yield stress, that are usually of interest. However, hardness measurements are of great value for the following reasons. For a given class of materials, in which the stress-strain curve is generally similar, the hardness correlates well with the U.T.S., the correlation being linear for some methods of measuring the hardness. The correlation for steels is given in Fig. 5.86. The existence of such a relationship can be understood in terms of the geometry of the hardness test. We will consider a simple case of the spherical indentation, Fig. 5.87, in which the original shape of the surface is represented by the full line; the final shape, after the indentation has been made, is shown by the broken line. It is evident that some plastic deformation has taken place, and that there is also a region in which the strain was elastic. The force required to produce the indentation will therefore depend on the yield stress and on the rate of work hardening after the yield stress has been exceeded. Since the U.T.S. also depends on these two properties, we should expect some correlation. As an indication of whether a particular material has been correctly treated, either mechanically or thermally, the hardness test is satisfactory, except that it gives little or no indication of the presence of cracks or other potential

STRUCTURE-SENSITIVE PROPERTIES

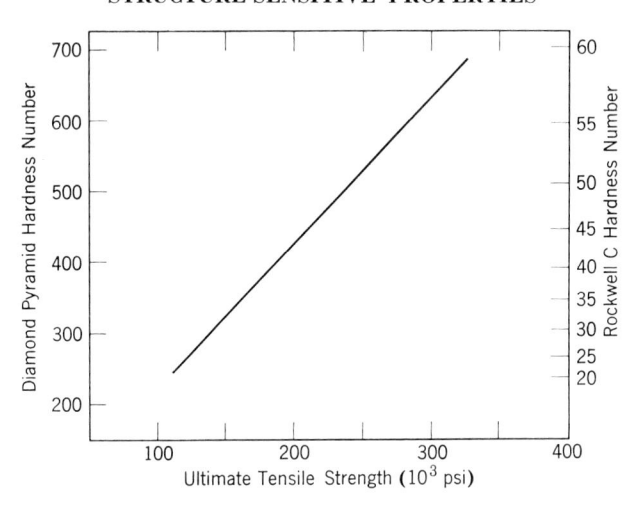

Fig. 5.86. Correlation between hardness and U.T.S. for steel.

fracture sites. The use of impact tests for detecting such defects will be discussed below.

Hardness testing has a second major value. It can be applied to quite small amounts of material (especially the micro-hardness type of test), and so can be used for detecting and measuring local differences, such as between the matrix and a particle of a second phase, in an alloy, or variations in the matrix itself.

If the hardness of a metal is measured by means of a spherical indentor at several different loads, it is found that the diameter d of the impression is related to the load L by the expression $L = ad^n$, where a represents the hardness, and n, the Meyer index, gives an indication of the rate of strain hardening. The value of n is decreased, for a given material, by prior work hardening; this corresponds to the decreased slope of the stress-strain curve with increasing deformation.

Impact testing. The impact test, used for measuring the energy required to break a notched bar of specified dimensions, has about the

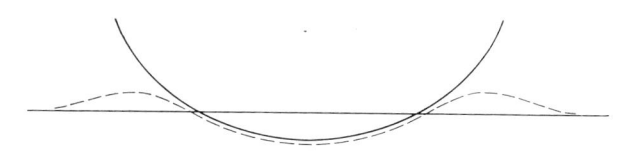

Fig. 5.87. Flow during indentation hardness testing.

same relationship to the stress-strain curve as has the hardness test. Impact testing by itself does not provide a good criterion for engineering design, except in the very unusual case in which the anticipated service closely resembles the test. It does, however, provide an excellent indication of whether a particular sample contains "stress raisers" that cause it to fracture more easily than another sample that is sound. This is because the combination of the notch and the high speed of the test provides the maximum increase in the stress level that can be applied before plastic deformation begins, and so increases the probability that fracture will occur first. Useful results in this test are obtained by comparing the impact value of a test specimen with that of a sample that is known to behave satisfactorily in service. It is also a useful method for measuring the "notch sensitivity" of a material; i.e., whether the flow curve is close enough to the fracture curve for them to intersect at low strain when the geometry is conducive to such behavior.

Special mechanical tests. Many tests have been devised for the inspection of materials for their suitability for particular fabrication techniques. For example, the Erichsen cupping test (Fig. 5.88) determines whether a sheet of metal has sufficient ductility to be satisfactorily "drawn." The ductility is measured by the depth to which the plunger can deform the sheet before it tears. The stresses imposed on the material are complicated and the result of this empirical type of test cannot, at present, be predicted from the more fundamental data. Another test that is often applied to sheet material is the bend test, which determines the smallest radius to which the sheeet can be bent without cracking. This test relates to the practical limitations of some forming procedures.

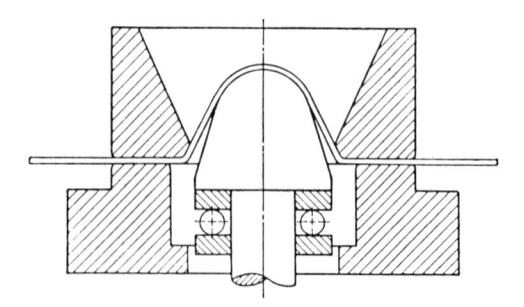

Fig. 5.88. Erichsen cupping test. From *Metals Handbook*, American Society for Metals, Cleveland, 1948, Fig. 1, p. 129.

Nondestructive testing. In order to test a component for serviceability, it is in principle necessary to stress it more highly than is expected ever to occur in practice. This procedure, the "proof test," is acceptable in some circumstances, such as for pressure vessels, where there can be a considerable difference between the actual strength and the maximum service loading. The test is carried out within this range. In some engineering applications, particularly those in which weight is an important criterion, the difference between actual and required strength is much less. In such cases proof testing may have to be at a stress level at which damage can occur, and an actual overload test therefore cannot be made. The problem then is to apply tests that do not destroy or even damage the component in question. There are three methods that may be used: 1. To employ nondestructive methods of seeking defects that would cause weakness. 2. To measure, nondestructively, some property that is known to correlate with the mechanical properties. 3. To apply destructive mechanical tests to samples and apply statistical methods to determine the probability of other individuals falling below an acceptable level of strength. These three methods are not mutually exclusive. The methods that are used for seeking defects are: visual, radiographic, sonic, ultrasonic, magnetic, and hardness; electrical and magnetic properties sometimes correlate with the critical mechanical properties.

5.7 Magnetic Properties

It has been shown on p. 136 that a crystal of a magnetic material has an equilibrium domain configuration in the absence of an external magnetic field. When an external magnetic field is applied, the energy of the system is reduced by the formation of magnetic poles at the ends of the specimen, and the specimen becomes magnetized. The relationship between the magnetization induced in the specimen (magnetic induction, **B**) and the external field **H** that is applied to it is shown by means of the **B-H** or magnetization curve (Fig. 5.89). The significant features of this curve are the permeability μ (ratio of **B** to **H**) and the saturation magnetization $\mathbf{B}_s$. The value of μ obviously varies with **H**, and for many purposes its initial value (i.e., at low field strength) is significant.

The magnetization curve consists of three regions. In region OA, magnetization takes place by the reversible movement of domain walls, in such directions that domains whose directions of magnetization have a component in the same direction as the applied field grow at the expense of those with magnetization in the opposite direction. The

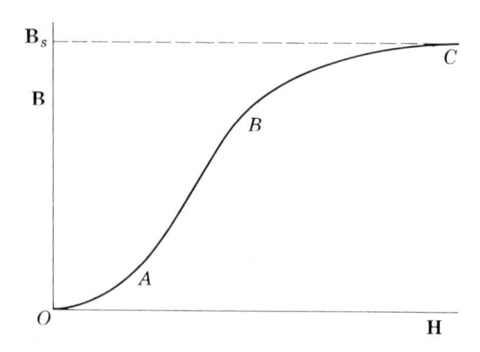

Fig. 5.89. Magnetization curve.

motion of domain walls is reversible (in the region OA) in the sense that they return to their original position of equilibrium if the field is removed. Region AB corresponds to more drastic motion of the domain walls; they move past obstacles, to be discussed below, which do not allow them to return spontaneously. Region BC, in which saturation is approached, corresponds to the rotation of the direction of magnetization of the domains, from directions of easy magnetization into the direction of the field. The detailed shape of the curve depends on the obstacles to domain movement; these also control the shape of the demagnetization curve that is obtained if the field is first reduced to zero and then increased in the opposite direction.

A complete cycle, resulting from increasing the field to the saturation value first in one direction and then in the other and then again in the first, is shown in Fig. 5.90. There are three important features of this curve. The first is the residual induction $\mathbf{B}_r$, which is the magnetization that remains, in the absence of a demagnetizing field, when the magnetizing field is removed. The second is the coercive force $\mathbf{H}_c$, which is

Fig. 5.90. Hysteresis loop.

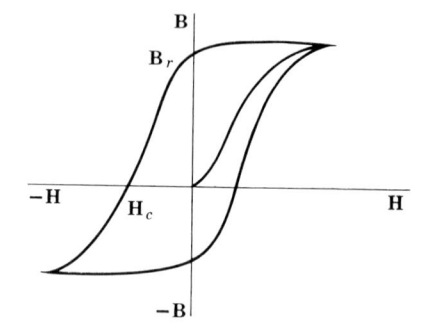

the field required for demagnetization. The product of these two quantities is a measure of the effectiveness of the material as a permanent magnet. The third important characteristic of the curve is the area which it encloses; this corresponds to the hysteresis loss, which is the amount of energy that is converted into heat within the material during a complete cycle.

There are two main ways in which magnetic materials are used. They are employed in transformer cores, for example, where the requirement is to produce the highest possible induction for a given applied field, combined with the minimum hysteresis loss; materials that satisfy these requirements are called *magnetically soft* materials. Other magnetic materials are used for permanent magnets, in which the residual induction and the coercive force must be as high as possible. These are *magnetically hard* materials.

Magnetically soft materials. Consideration so far has been confined to single crystals. It is necessary now to consider the effect of crystal size and orientation on the domain pattern. In the most general case, neighboring crystals will be oriented independently, and the grain boundary between them will be at an angle determined by conditions not related to the magnetic properties. As a result, the domain pattern in the vicinity of a boundary will be complex, and the two crystals may interact to form a domain pattern that does not produce any external field. A result of this is that a material with smaller grain size will have more domain walls per unit volume, and the coercive force should therefore increase with decreasing grain size. This has been shown experimentally to be the case in Fig. 5.91. Complicated domain patterns would also occur at the surface if it were not parallel to an easy direction of magnetization; hence, the coercive force is a minimum when the crystals are so oriented that the direction of easy magnetization that is to be used is parallel to the surface of the specimen. This is of practical importance because magnetically soft materials are usually used in sheet form in order to minimize eddy current losses.

Fabrication and heat treatment procedures (see p. 312) have long been known for producing iron-silicon alloy strip with almost all the crystals oriented so that a (110) plane is parallel to the plane of the strip, and a [100] direction parallel to the length of the strip. This texture is described as "grain-oriented." This gives one easy direction of magnetization, $\langle 100 \rangle$, parallel to the plane of the sheet, the two other $\langle 100 \rangle$ directions being inclined to it at 45°. Methods have recently been discovered for producing strip in which the crystals are oriented with a (100) plane parallel to the surface of the strip, a $\langle 100 \rangle$ direction still being parallel to its length. This texture, referred to as "cube texture," gives two ⟨100⟩ directions in the plane of the sheet, which is of great advantage for many applications, particularly those in which magnetization cannot be confined to a single direction, as in a transformer core. Figure 5.92 shows hysteresis curves for the two types of oriented strip, both in the longitudinal direction, in which there is little difference, because both have ⟨100⟩ directions parallel to the rolling direction, and transversely, where large differences are apparent.

A major source of coercive force in magnetically soft materials is the resistance to domain wall movement caused by nonmagnetic particles, such as oxides, dispersed within the crystals. A domain wall is decreased in energy when it passes through (or includes) a nonmagnetic inclusion, because the volume of wall is reduced, and therefore work must be done in moving the wall away from such inclusion. There is some analogy with the effect of dispersions on the movement of dislocations; however, in this case, the wall is rigid, or very nearly so, and is two-dimensional. The effect is largest for a critical size and

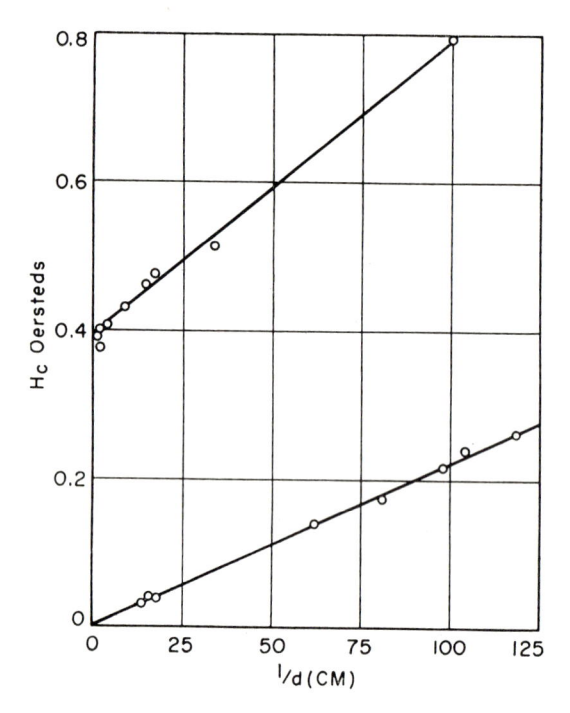

Fig. 5.91. Effect of grain size on coercive force. (Two independent investigations.) From *Relation of Properties*, ASM, Fig. 6, p. 221.

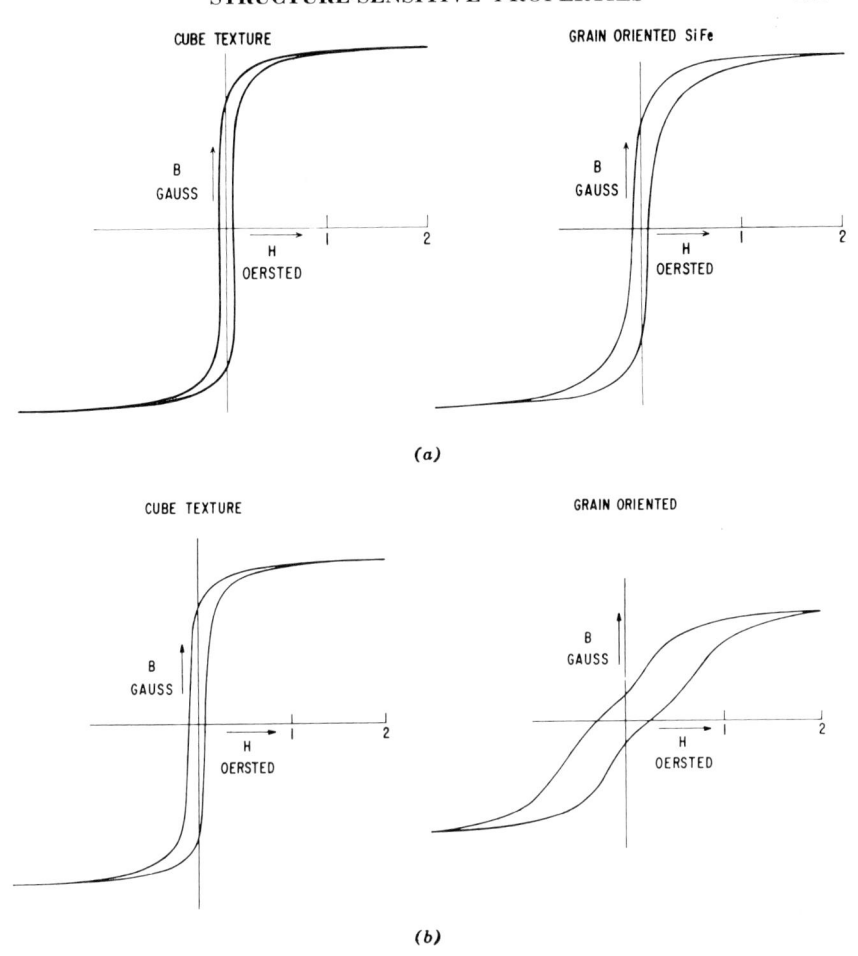

Fig. 5.92. D-c hysteresis loops for cube texture and grain-oriented silicon iron; (a) parallel to rolling direction; (b) transverse to rolling direction. (Courtesy General Electric Research Laboratory.)

spacing of particles; Fig. 5.93 shows the variation of coercive force with particle diameter for dispersions of the same total concentration of Fe_3C in iron. As in the case of dislocation movements, a random solid solution has little effect, because the domain wall would gain as many solute atoms as it loses during motion. The effect of pores on the coercive force is similar to that of inclusions.

The important property of the silicon-iron strip material is its

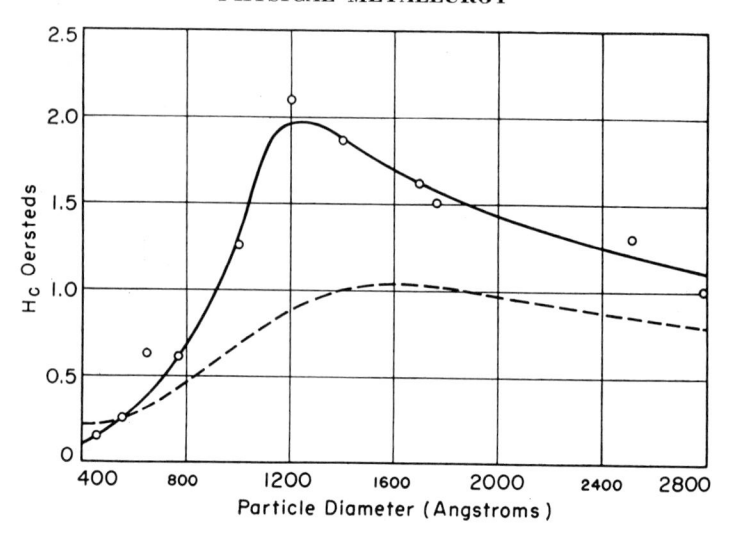

Fig. 5.93. Variation of coercive force with particle size. From *Relation of Properties*, ASM, Fig. 12, p. 230.

low hysteresis loss, and its main application is in electrical power engineering. Another important use for magnetically soft materials is in transformers which handle very small amounts of power, as in communication engineering. In such applications, the primary criterion is the initial permeability. The most interesting series of alloys from this point of view are the alloys of iron and nickel. The initial permeability depends on the composition as shown in Fig. 5.94, in which the three curves show the difference between material that has been cooled fast enough to prevent "ordering" of alloys of composition near $FeNi_3$ (upper curve) and those in which ordering has been allowed to occur. In these alloys the directions of easy magnetization are $\langle 111 \rangle$, and the anisotropy and the magnetostriction both reach a minimum for the disordered alloy at about 75% nickel. The work done in moving a domain wall is a function on the energy of the domain wall, which depends on the magnetic anisotropy and the magnetostriction of the alloy.

Magnetically hard materials. There are two distinct approaches to the problem of magnetically hard materials. One is to depend on particles that are so small that each consists of a single domain which can only change its magnetization by rotation of the direction of magnetization everywhere at once; there is a critical size of particle, about 2×10^{-6} cm for iron, below which they behave as single domains.

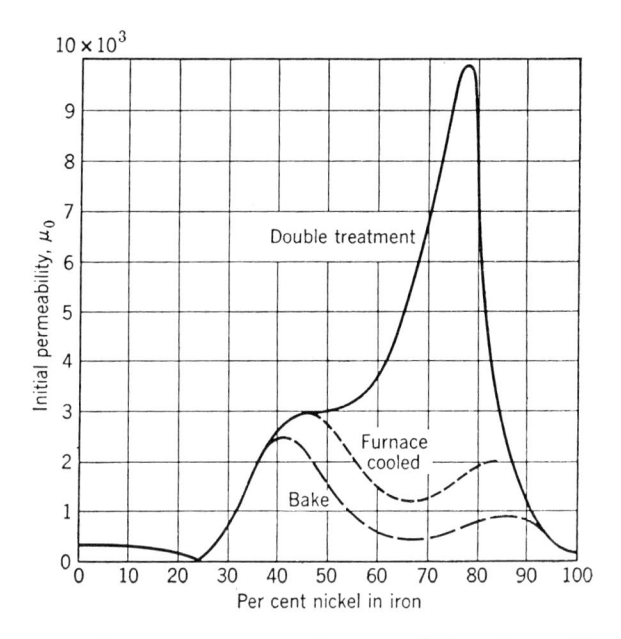

Fig. 5.94. Effect of composition on initial permeability. From *The Science of Engineering Materials*, J. E. Goldman, Ed., John Wiley & Sons, New York, 1957, Fig. 13.26, p. 331.

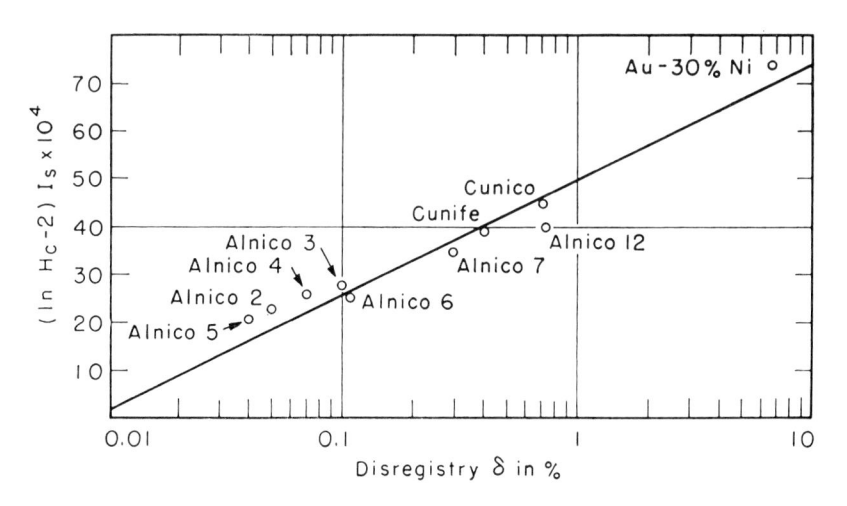

Fig. 5.95. Relationship between coercive force and lattice mismatch. From *Relation of Properties*, ASM, Fig. 11, p. 247.

The coercive force is a maximum at about the critical size; at smaller sizes the coercive force decreases for reasons that are not clear; at larger sizes the decrease is due to the presence of domain walls.

The other approach is to produce large local lattice strains, preferably by the coherency strains caused by particles of a second phase that differ in lattice parameter from the matrix. The strains caused by coherence may have their effect on domain boundary movement by means of local anisotropy of magnetic properties resulting from the strain. The effect of the shape of the second-phase particles is probably important in determining the position and extent of this effect. An empirical relationship between the coercive force and the "lattice mismatch" for a number of hard magnetic materials is given in Fig. 5.95.

5.8 Further Reading

A. H. Cottrell, *Dislocations and Plastic Flow in Crystals,* Oxford University Press, New York, 1953.

J. G. Fisher et al., *Dislocations and Mechanical Properties of Crystals,* John Wiley & Sons, New York, 1957.

R. Maddin and N. K. Chen, "Geometrical Aspects of the Plastic Deformation of Metal Single Crystals," Chapter 2 in *Progress in Metal Physics 5,* Bruce Chalmers and R. King, Eds., Pergamon Press, London, 1954.

N. J. Petch, "The Fracture of Metals," Chapter 1, *ibid.*

A. H. Sully, "Recent Advances in Knowledge Concerning the Process of Creep in Metals," Chapter 4 in *Progress in Metal Physics 6,* Bruce Chalmers and R. King, Eds., Pergamon Press, New York, 1956.

A. S. Nowick, "Internal Friction in Metals," Chapter 1 in *Progress in Metal Physics 4,* Bruce Chalmers, Ed., Pergamon Press, London, 1953.

C. Zener, *Elasticity and Anelasticity of Metals,* University of Chicago Press, Chicago, 1948.

6

Change of State

6.1 Control of Shape and Structure

The universal problem in the engineering or industrial use of metals is that of producing the required shape; for some purposes the properties are of secondary importance, but the challenging metallurgical problem is to combine the economic production of a required shape with the development of the required properties. There are in general three approaches to this problem. The first is to cause the metal to become solid (usually but not necessarily from the molten state) with the required shape and to have a structure that corresponds to the required properties. The second is to produce some other shape and subsequently to change the shape (by plastic deformation) in such a way that the shape and the structure simultaneously reach the required state. The third technique is to start with one or the other of the preceding methods and then to cause changes of the phases that are present and their distribution by controlled changes of temperature, i.e., heat treatment.

This chapter discusses the first of these approaches, the problem of the simultaneous control of shape and structure, i.e., the processes by which solid metal is produced from liquid, solution, or vapor, the resulting structure and properties, and the extent to which they can be controlled. The converse processes (melting, corrosion, etc.) will also be discussed here because the same mechanisms and ideas are applicable.

6.2 Solidification

Of these processes, the most important technically is that of solidification or freezing.

The process of freezing. The process of freezing is the growth of crystals by the addition of single atoms or, in many nonmetallic substances, molecules to the surface of an already existing crystal. The observable, or net, process is the difference between the number of atoms that join a surface in a given time and the number that leave it. At the equilibrium temperature, or melting point, these rates are equal. The two rates are given by kinetic expressions of the kind discussed in Chapter 2. It follows that, for freezing or melting to occur at a finite rate, one rate must be faster than the other and there must be a departure from the temperature of equilibrium. This departure is very small for ordinary rates of freezing or melting, and can be ignored for most purposes. It is, however, important from the following point of view:

Let us suppose that a crystal and a liquid are in contact at a temperature slightly below their equilibrium temperature. At this temperature, freezing should occur at a definite rate, which is characteristic of this temperature. However, one of the consequences of transferring an atom from liquid to solid is that the atomic latent heat of fusion is transformed into thermal energy, which raises the temperature towards the equilibrium value. The atomic latent heat of fusion is the difference in internal energy of an atom in the solid and in the liquid. This amount of heat must on the average be supplied to an atom in a crystal to detach it from its neighbors to the extent necessary to convert it to liquid. The process of freezing cannot continue unless the temperature of the interface is maintained below the equilibrium temperature. This can only be achieved by the continuous removal of heat from the interface, by conduction into the solid or the liquid; the interface otherwise rapidly reaches its equilibrium temperature and becomes stationary. The rate of freezing is therefore controlled by the rate at which latent heat is extracted, and, if one compares different metals, by the value of the latent heat.

The influence of crystallography. If it is assumed that the structure of a growing crystal is the same at the surface as in the interior, it follows that we cannot assign any arbitrary shape to the crystal sur-

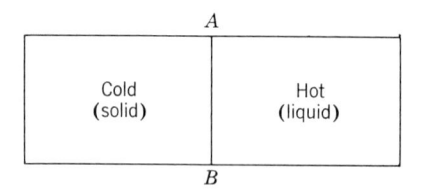

Fig. 6.1. Unidirectional freezing (schematic).

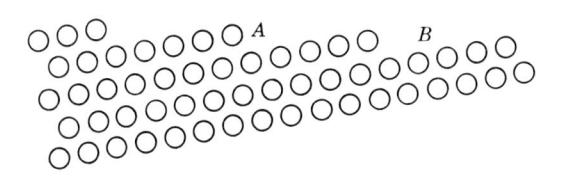

Fig. 6.2. Idealized surface of a crystal.

face, when we consider it on an atomic scale of sizes. Freezing usually takes place under conditions in which the temperature distribution in the system is imposed, as for example when a mass of molten metal is cooled by heat loss from its surface, so that part is solid and part is liquid (Fig. 6.1). In this case the isothermal surface AB corresponding to the limit of freezing cannot in general coincide with any particular crystallographic plane; the surface must then consist of "steps," an idealized case of which is shown in Fig. 6.2. The actual "steps" have a "height" of one atom layer, although they may not always be of equal lateral extent. It is emphasized that this topography represents the best possible approach to ideal smoothness of an arbitrarily selected surface; the actual surface is probably less smooth than this, as shown in Fig. 4.39 and discussed on p. 132.

It will be evident that all the possible atom sites on the surface are not equally favorable for the addition of atoms, and that atoms cannot leave all sites equally readily. It is probable that growth takes place mainly by the addition of atoms at points such as A (Fig. 6.2), because an atom arriving at this point has more nearest neighbor atoms than one arriving at a point such as B; its energy is nearer to that of the solid, and it therefore is less likely to leave because it requires more activation energy to do so. It follows that growth of a crystal in such conditions is largely by "edgewise" propagation of atomic layers, although the frequent formation of new layers is probable, and there is therefore always an adequate number of favorable growth steps. The result will be observable as growth of the crystal at a rate which depends on how many steps there are and on how fast they grow laterally.

Direct observation of the individual steps is not possible, because each is only as high as a single atom layer; but under some conditions the steps do not remain equally spaced but form microscopically visible multiple steps, separated by smooth regions that correspond in orientation to the close-packed planes; these indications of stepwise growth can be seen on the surfaces of crystals whose growth has been interrupted by very rapid removal from the melt.

"Smooth" and "dendritic" freezing. In the conditions represented in Fig. 6.1, the imposed temperature gradient is such that heat can only flow from right to left. The latent heat is therefore removed by conduction through the solid. The interface advances to the right at a speed that is determined by the rate of removal of latent heat, and this in turn depends on the temperature gradients that are maintained in the liquid and the solid. If changes of specific heat are ignored, the latent heat of fusion corresponds to the difference between the temperature gradients in the solid and the liquid, corrected for the difference in thermal conductivity.

In order that freezing may take place at a finite rate, it is necessary that the interface should be below the temperature at which the crystal would be in equilibrium with the liquid. The rate of freezing increases as this amount of supercooling is increased.

Under the conditions of Fig. 6.1, a "smooth interface" is stable; if any part of it is, for any reason, further to the right than the remainder, it would be slowed down because it would be at a higher temperature and therefore closer to its equilibrium temperature.

An entirely different condition is that in which the liquid is cooler than the solid, and the latent heat is therefore conducted from the interface into the liquid. A smooth interface is then unstable because any part of the interface that grows ahead of the remainder is in a region of lower temperature and therefore of more rapid growth. The result is a "spike" that grows out into the liquid. It remains isolated because its latent heat is conducted into the immediately surrounding liquid, which consequently becomes less favorable for crystal growth. The interface therefore develops into a series of discrete spikes, or sometimes plates, that grow into the liquid at a speed that depends on the amount of supercooling, but can be very rapid compared with the rate of growth attainable by conduction through the solid. The high speed is due to the fact that the growing tip is always very close to a heat sink. The geometry of the "dendritic" spikes is very closely related to the crystallography of the metal; the axis of the dendrite is very accurately parallel to the crystallographic direction given in Table 6.1.

The condition that gives rise to dendrite growth, i.e., a positive gradient of supercooling from the interface into the melt, also causes branches to grow from the primary dendrites; the axes of these branches are crystallographically similar to those of the primary dendrite spikes. Under favorable conditions, tertiary and even quaternary branching may occur.

Table 6.1. Directions of Dendritic Growth

Structure	Direction
Face-centered cubic	$\langle 001 \rangle$
Body-centered cubic	$\langle 001 \rangle$
Close-packed hexagonal	[0001]
Tetragonal (tin)	$\langle 110 \rangle$
Diamond cubic	$\langle 112 \rangle$

Figure 6.3 shows some examples of dendritic freezing.

There is no satisfactory general explanation for the crystallography of the dendrites; however, the very precise adherence of the growth directions to the crystallographic axes indicates that the growing surface is not, as has sometimes been suggested, a random one, but that it is somewhat crystallographic in character, as is implicit in the stepwise growth mechanism discussed above.

6.3 Nucleation

It is necessary at this point to distinguish between the conditions that are necessary to cause freezing to begin in a liquid that is not in contact with any crystals and the conditions that allow the process to continue. It is generally true that more severe conditions must be met to intiate a process than to continue it. We will first discuss the problem of nucleation in general terms, and then apply the principles to the particular case of freezing.

Qualitative discussion. Once a transformation has started, there is usually an interface between the material that has transformed and that which has not. The subsequent development of the process can be considered in terms of the movement of atoms into positions in which their free energy is lower than it was in their previous positions in the untransformed material. The new sites have the low free energy characteristic of the growing phase of which they form a part. If this were not so, the new phase would not be stable.

The establishment of the interface, however, is not achieved so easily, because there are no low free energy sites available until the interface has been established. It is therefore necessary to discuss the conditions that govern the formation of the first stable particle of the

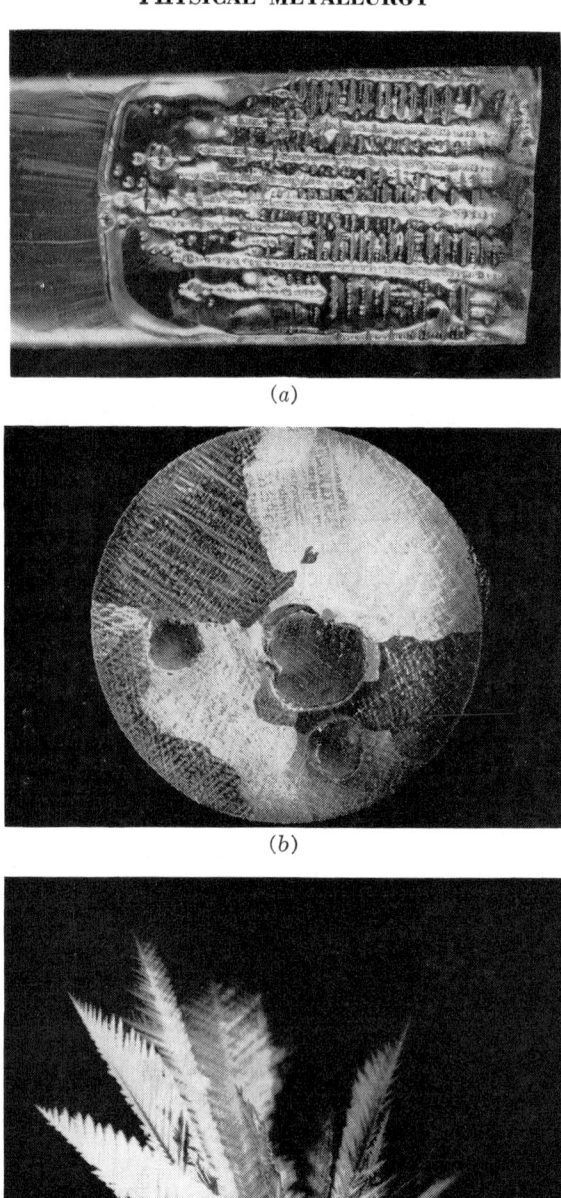

Fig. 6.3. Examples of dendritic freezing. (a) Lead, decanted ($\times 1.1$). (b) Stainless steel, etched ($\times \frac{3}{4}$). (c) Ice ($\times 1.4$). [(b) Courtesy S. L. Walker, General Electric Research Laboratory.]

new phase, in general terms at this point, and later in more detail in relation to specific types of transformations.

The following qualitative considerations define the nucleation problem. Consider two phases A and B, which are in equilibrium at temperature T_E. This is equivalent to stating that the free energies of the same amount of the two phases are equal at that temperature. At temperatures above T_E, A has the lower free energy, while at temperatures below T_E, B is the stable phase. Nucleation is the process by which, for example, phase B begins to form if phase A is cooled below T_E. B does not in general form spontaneously as soon as the temperature falls below T_E because it must pass through the stage at which it consists of very small particles, and a very small particle of B is not necessarily a stable phase, compared with A, at temperatures below T_E. This is because the change in free energy corresponding to the transformation of A to B includes the free energy of the interface as well as the specific free energies of the two phases. If the particle of B is large, the free energy of the interface can be ignored in comparison with the volume free energy, but this is not the case when the particle is very small. From a kinetic point of view, the same result may be arrived at on the basis that the atoms at the surface of a very small crystal have a higher energy, because many of them are at edges or corners, than the surface atoms of a larger crystal. This decreases the average activation energy for their escape from this phase, and therefore increases the average rate of the reverse transformation $(B \to A)$. Therefore the temperature of equilibrium, at which atoms arrive and leave at the same rate, is lower for a very small crystal than for a large one.

At any given temperature below T_E, a particle of B will be in equilibrium with A when its radius of curvature has a particular value, the *critical radius*. This equilibrium is, however, unstable, since the particle will grow if it exceeds the critical radius and will decrease in size if it is below that size. The critical radius decreases with increasing supercooling because there is more volume free energy difference to compensate for the surface free energy. The critical radius is infinite at T_E.

The second component of the nucleation concept is that at any temperature there will be within phase A some groups of atoms or molecules that are spatially related to each other in the way that characterizes phase B. These groups or clusters are called "embryos" and at any given temperature there is a statistical distribution of the sizes of embryos. At any temperature there is a maximum size of embryo that is likely to exist. This size increases as the temperature is lowered.

Nucleation occurs when the supercooling is such that there are embryos that exceed the critical radius and are therefore more likely to grow than to revert to the parent phase. The supercooling required to provide an embryo of critical size is reduced if a solid surface is present on which embryos can form easily. This is described as heterogeneous nucleation, as contrasted to homogeneous nucleation when the critical nucleus forms without the help of a substrate.

Calculation of critical size. The formal treatment of nucleation theory is as follows:

Let the difference in volume free energy of the two phases A and B be F_v ergs/cm³, and let the free energy of the interface between A and B be σ ergs/cm². The change in free energy F corresponding to the formation of a sphere of B of radius r cm is given by

$$\Delta F = -\tfrac{4}{3}\pi r^3 \, \Delta F_v + 4\pi r^2$$

The form of this variation is shown in Fig. 6.4, from which it is seen that, as r increases, ΔF passes through a maximum and then decreases and becomes negative. At the temperature to which this diagram relates, the critical radius r^* is the value of r corresponding to the maximum of ΔF, because the free energy of the system will decrease at the same rate whether the embryo grows or shrinks from this size. The value of r^* is therefore obtained by equating to zero the derivative of ΔF with respect to radius: thus

$$\frac{d(\Delta F)}{dr} = -4\pi r^2 \, \Delta F_v + 8\pi r\sigma = 0$$

or

$$r^* = \frac{2\sigma}{\Delta F_v}$$

It is usually assumed that σ is independent of temperature; however, ΔF_v is temperature-dependent, being zero at the equilibrium temperature, at which r^* is infinite.

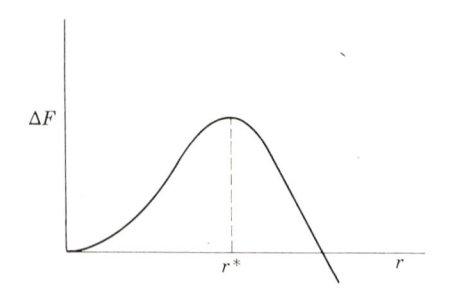

Fig. 6.4. Variation of ΔF with r showing critical radius r^*.

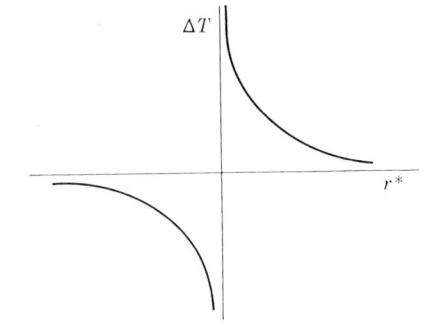

Fig. 6.5. Variation of critical radius with temperature.

The variation of ΔF_v with temperature can be derived as follows: ΔF_v can be written as

$$\Delta F_v = (U_A - TS_A) - (U_B - TS_B)$$
$$= U_A - U_B - T(S_A - S_B)$$

At T_E $\quad\Delta F_v = 0$

or $\quad U_A - U_B = T_E(S_A - S_B)$

Hence $\quad S_A - S_B = \dfrac{U_A - U_B}{T_E}$

if $S_A - S_B$ and $U_A - U_B$ do not vary with temperature. $U_A - U_B$ is the heat of transformation or latent heat, ΔH. Then

$$\Delta F_v = -\Delta H + T\frac{\Delta H}{T_E} = \frac{\Delta H \cdot \Delta T}{T_E}$$

where ΔT is the amount of supercooling. Thus

$$r^* = \frac{2\sigma T_E}{\Delta H \cdot \Delta T}$$

The variation of r^* with ΔT is shown in Fig. 6.5. Both branches of the curve are physically significant; the upper branch corresponds to nucleation of B in A at temperatures below T_E, the radius of curvature of the B phase being taken as positive to correspond to a convex surface of B; the lower branch corresponds to nucleation of A in B above T_E, the interface now being curved so that the B phase has a *concave* surface, and therefore a negative radius of curvature. An example in which both branches of the curve exist in the same system is found in the condensation of water drops in supersaturated (or supercooled) water vapor, and the formation of bubbles of water vapor in superheated water (boiling).

The derivation of the relationship between r^* and ΔT, given above, involves the condition that there is no free energy change other than those specified. If the formation of the new phase involves a change of composition, or the appearance of strain energy resulting from the new phase having a different shape or size in a crystalline matrix, then the derivation is more complicated. The result is different in detail but remains essentially the same in form.

Distribution of embryo sizes. The calculation of the number of embryos as a function of their size is more complicated; it may be approached from the point of view that an embryo of phase B of a given size will simultaneously lose atoms to phase A, and gain atoms from A. If it is assumed that these processes occur by the addition or subtraction of single atoms, then an equilibrium distribution can be calculated, either thermodynamically or kinetically.

The equilibrium distribution of sizes is that at which the number of embryos of each size remains constant, just as many embryos reaching a particular size as leaving it. The kinetic condition for the equilibrium is that if there are n embryos containing i atoms, then the number of embryos containing $i - 1$, $i + 1$, $\cdots$, must be such that the number of embryos changing *from i* is equal to the number changing *to i*. Because a given embryo of less than critical size is more likely to lose atoms than to gain them, there must be fewer embryos of larger size than of smaller size. The important criterion is the maximum size of embryo that has a reasonable probability of existing; this depends on (a) the temperature, (b) the size of sample, and (c) the criterion for "reasonable probability." If arbitrary values are taken for the second and third of these conditions, then the maximum embryo size

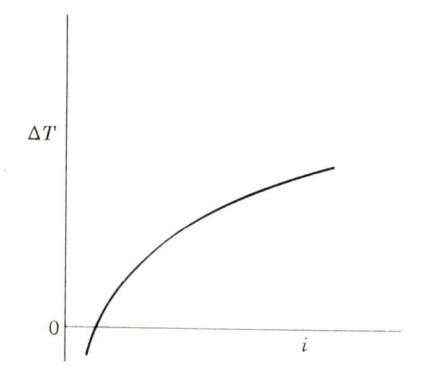

Fig. 6.6. Maximum embryo size as a function of temperature.

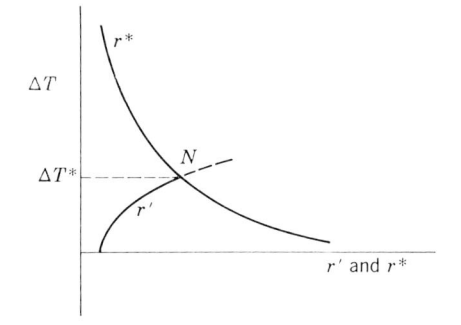

Fig. 6.7. Condition for homogeneous nucleation.

can be computed as a function of temperature. The result of such a calculation is shown in Fig. 6.6, in which ΔT is the amount of supercooling and i the number of atoms in the largest cluster.

In Fig. 6.7 both the critical radius r^* and the radius r' of the largest probable cluster are plotted against ΔT. The point N at which these two curves intersect represents the value of ΔT at which homogeneous nucleation will occur in a sample of the size assumed, in the time that corresponds to the assumed probability. A very small increase in the amount of supercooling causes a very large increase in the rate of formation of critical nuclei, and it is therefore permissible to speak of T^*, the temperature of homogeneous nucleation, or ΔT^*, the supercooling for homogeneous nucleation.

Heterogeneous nucleation. It has been assumed in the two preceding sections that the nucleus is a complete sphere; in fact, the temperature at which a given radius is critical depends only on the radius of curvature of the surface, and not on what proportion of a sphere is present. The thermodynamic derivation outlined above does not make this clear, but consideration of the problem in terms of the equality of rates of atoms crossing the interface in both directions shows that it is only the curvature of the interface, and not its extent or that of the particle, that is relevant. On the other hand, the probability of an aggregate of atoms reaching a certain size depends almost entirely on the *number* of atoms, and much less on the *shape* of the cluster. Consequently, we may consider a nonspherical cluster part of whose surface has spherical curvature. Such an embryo is represented in Fig. 6.8. If such an embryo were at the critical temperature for r^*, the curved part of its surface would be as likely to grow as to shrink. The regions of very small radius of curvature at the periphery PQ, however, would be below critical radius and would shrink very quickly, *unless they were stabilized.* This may occur by contact with a suitable sub-

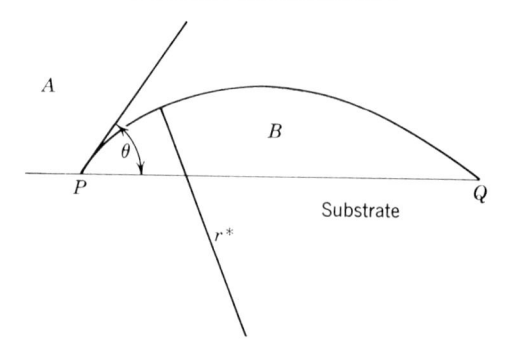

Fig. 6.8. Heterogeneous nucleation.

strate, which provides relatively low energy sites for atoms in phase B. The efficacy of the substrate determines the minimum angle θ at which atoms are stabilized in the B sites. This angle depends on the relative values of the free energies of the interfaces A—substrate (σ_{AS}); B—substrate (σ_{BS}); and A—B (σ_{AB}).

The equilibrium angle θ is determined by the condition that

$$\sigma_{AS} = \sigma_{BS} + \sigma_{AB} \cos \theta$$

A small angle of θ corresponds to the case in which the interface between B and the substrate has a low free energy compared with that between A and the substrate. This can occur when coherence is possible between B and the substrate. Close matching of characteristic interatomic distances in the two structures may allow a low energy interface to be formed (see p. 128). However, heterogeneous nucleation can occur when θ has any value less than 180°; this corresponds to the condition that

$$\sigma_{AS} < \sigma_{BS} + \sigma_{AB}$$

when σ_{AS} and σ_{BS} are the interfacial energies between the substrate and phases A and B respectively, and σ_{AB} is the interfacial energy between phases A and B.

The angle θ determines the volume of the spherical cap of radius r^*, and therefore the number of atoms i^* in the critical nucleus. For any value of θ less than 180°, i^* for heterogeneous nucleation is less than i^* for homogeneous nucleation. Approximately, therefore, the nucleation condition on a substrate of sufficient size can be represented as in Fig. 6.9, in which r' is the radius of curvature of the largest embryo, if it is assumed to be spherical, and r'' is the radius of curvature of an embryo of the same size in the form of a spherical cap bounded by the angle θ as in Fig. 6.8. Curves for r'' are shown for values of $\theta = 45°$ and $5°$. It follows that nucleation takes place at much smaller amounts of supercooling when a suitable substrate is present. This is *heterogeneous nucleation*. The potency of the nucleation catalyst increases as θ decreases.

Nucleation of freezing. Extensive studies of the nucleation of crystals in supercooled melts have shown that it is possible to induce homogeneous nucleation under special conditions, but that heterogeneous nucleation, under far less supercooling, usually intervenes.

HOMOGENEOUS NUCLEATION. The prediction of homogeneous nucleation conditions requires a knowledge of the interfacial energy σ; this can be estimated as a fraction of the latent heat of fusion by a consideration of the deficiency of nearest neighbor atoms at the surface as discussed on p. 131. A flat, close-packed surface should have a surface energy equal to one-quarter of the latent heat per atom, because each atom in a close-packed surface has 9 nearest neighbors instead of the 12 that are characteristic of an atom in the interior. A convex interface has a greater deficit of nearest neighbors; the extent depends on the radius of curvature. Assuming the energy of the surface to vary linearly with the nearest neighbor deficit, the surface energy per surface atom varies with the number of atoms (in a spherical crystal) as shown in Fig. 6.10. This allows us to insert actual numbers in the relationship between ΔT and n. This is given, in Fig. 6.11, for a close-packed metal

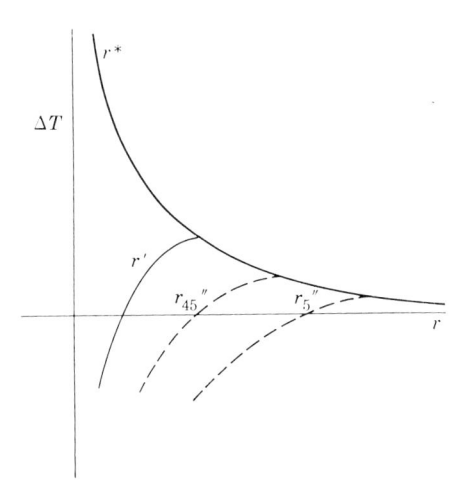

Fig. 6.9. Influence of angle of contact on heterogeneous nucleation.

244 **PHYSICAL METALLURGY**

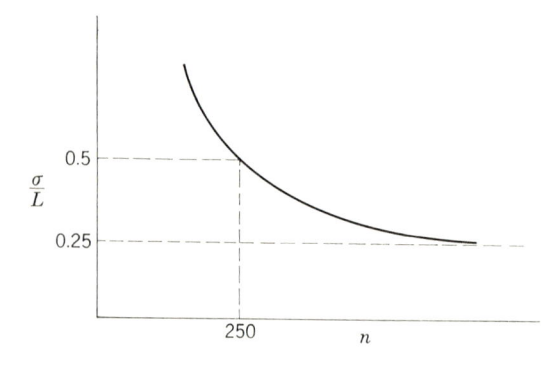

Fig. 6.10. Relationship between size of crystal and its specific surface energy.

using the assumption of linear dependence of the surface energy on the nearest neighbor deficit.

The other datum required for the prediction of the homogeneous nucleation temperature is the relationship between the "largest probable embryo" n_E and the amount of supercooling. The value of n_E can be calculated from the equilibrium of embryos of different sizes, if the size of sample is assumed; a curve for n_E as a function of $\Delta T/T_E$ is shown in Fig. 6.11. The point H corresponds to the conditions for homogeneous nucleation, for at this temperature the largest probable nucleus is of critical size. The experimental study of homogeneous nucleation in metals has shown that when a metal sample is sufficiently subdivided into isolated drops, some supercool to the extent of about $0.2T_E$ before nucleation occurs; more precise values are given in Table 6.2. Using an argument that is essentially the converse of that

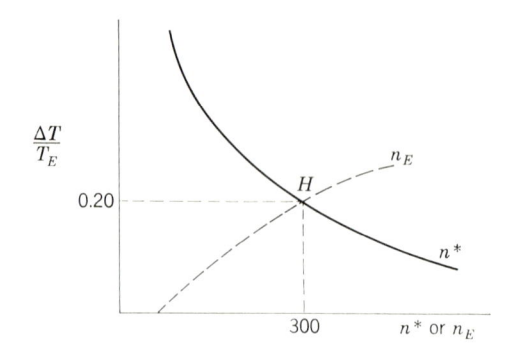

Fig. 6.11. Homogeneous nucleation of freezing.

Table 6.2. Supercooling for Homogeneous Nucleation *

Metal	Supercooling (ΔH)	$\dfrac{\Delta H}{T_E}$
Mercury	77	0.33
Gallium	76	0.25
Tin	118	0.23
Bismuth	90	0.17
Lead	80	0.13
Germanium	227	0.18
Aluminum	195	0.21
Antimony	135	0.15
Silver	227	0.18
Gold	230	0.17
Copper	236	0.17
Nickel	319	0.18
Cobalt	330	0.19
Iron	295	0.16
Palladium	332	0.18
Platinum	370	0.18

* Data from J. H. Hollomon and D. Turnbull, Chapter 7 in *Progress in Metal Physics 4*, Bruce Chalmers, Ed., Pergamon Press, London, 1953, p. 356.

presented above, it may be seen that the nucleation data predict values of σ/L equal to about 0.5.

It follows from the foregoing discussion that the critical nucleus contains about 300 atoms; there is, however, one interesting inconsistency: it is impossible to arrange 300 atoms to form a sphere and at the same time to preserve the crystal structure. The nearest neighbor count assumed a crystalline material, and the value of the latent heat also assumes unmodified crystal properties. Let us therefore consider a *crystal* containing about 300 atoms. The discussion will be limited to close-packed crystals. A crystal of the required size, with a face-centered cubic structure, may have the shape shown in Fig. 6.12. This is a truncated cube (kaidodekahedron) bounded by six square {001} faces and eight hexagonal {111} faces. The former contain 16 and the latter 27 atoms each. The total number of atoms is 305. An important feature of this nucleus is that it can only grow by the addition of complete layers superimposed on its faces; it can only melt by the removal of such layers. At $T/T_E = 0.2$, it is found from nucleation

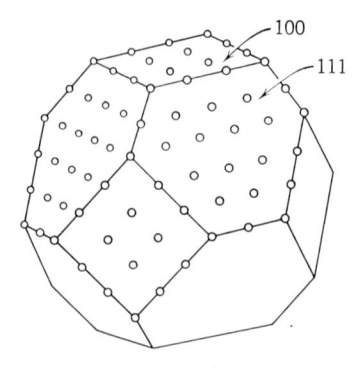

Fig. 6.12. Homogeneous nucleus for face-centered cubic crystal.

theory that the {001} faces are supercritical, but that a superimposed layer, which would only contain 9 atoms, would be subcritical; the {111} layers are supercritical and so is the superimposed layer, which would contain 18 atoms. The addition of such a layer would increase the size of the adjacent {001} faces which could then maintain a superimposed layer which would in turn increase the {111} faces to such a size that they could grow. It must be concluded that a critical crystal nucleus can be described as either having a critical volume (or radius) or as being bounded by faces, some of which can support critical-sized additional layers. These are alternative descriptions of the same condition.

HETEROGENEOUS NUCLEATION. Heterogeneous nucleation occurs when part of a critical nucleus is stabilized by contact with a suitable substrate. The critical radius is not changed, but the required number of atoms in it is reduced and so the probability of achieving criticality is increased.

The criterion for a heterogeneous nucleus is, from the thermodynamic point of view, that the interfacial energy of the crystal-substrate interface should be lower than that for the sum of the liquid-substrate interface and the liquid-crystal interface. The greater the difference in interfacial energies, the more effective will be the substrate. It may be noted that *any* suitable solid in contact with the liquid may act as a heterogeneous nucleation catalyst; this includes the container and any oxide or other film at the surface.

SURFACE NUCLEATION. The problem of nucleation has been looked at so far in terms of a three-dimensional embryo which constitutes a nucleus when its radius of curvature reaches the critical value. It is necessary also to consider the nucleation of a new two-dimensional layer on an existing surface. This is important because it relates to

CHANGE OF STATE 247

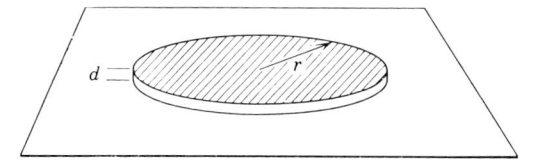

Fig. 6.13. Surface nucleation.

the question of the amount of supercooling that is necessary to cause a crystal to grow if it is bounded by atomically perfect crystal planes. The formal statement of the nucleation condition is analogous to that for the three-dimensional case: Consider a circular layer (Fig. 6.13) of radius r, and thickness d, situated on a substrate. The difference in volume free energies per unit volume is ΔF_v, which gives a "volume" term $\pi r^2 d\, \Delta F_v$, while the "surface" term is $2\pi r d \sigma_E$, where σ_E is the surface energy of the "edge" of the disk. Using the same procedure as for volume nucleation, we find that $r^* = \sigma_E/\Delta F_v$. If σ_E is calculated on the basis of the nearest neighbor count, it is found to vary with r in much the same way as σ; the resulting relationship between r^* and ΔT is shown in Fig. 6.14. The negative branch of the curve corresponds to the process of nucleating the removal of part of a complete layer. The problem of the maximum probable size of cluster is more complicated; the probability of finding a cluster on the surface that has critical size depends on the proportion of surface sites that are filled at any time. This is estimated to be about 0.5 for a metal in contact with its melt, and 10^{-5} for a crystal in contact with its vapor. It is concluded that, even for very small supercooling, a critical-sized cluster is likely to be found on a metal surface, while, as will be seen on p. 293, the conditions are entirely different for a vapor or a solution.

Fig. 6.14. Negative branch of ΔT versus r^* curve.

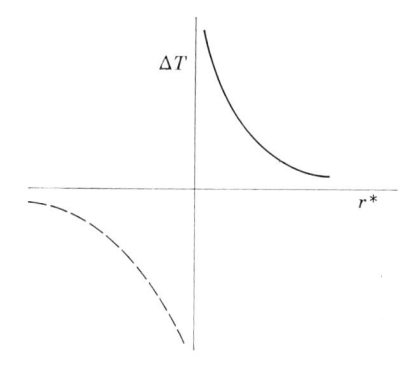

6.4 Imperfections Resulting from Freezing

Dislocations. When a mass of liquid metal is allowed to freeze, it may do so as one or more crystals, depending on how frequently the conditions for nucleation are satisfied. The effect of imposed conditions on the shape, size, and orientation of the resulting crystals will be discussed on pp. 232–234; the individual crystals, however, are not perfect; they almost always contain dislocations and often low-angle boundaries are present also. The existence of these imperfections appears to be closely related to the purity of the metal, the low-angle boundaries being more pronounced for less pure metals. It is not known whether they would be eliminated in a completely pure metal, but the evidence derived from work on silicon suggests that this might also be so for metals. High purity is not the only requirement for high perfection; it is also necessary to avoid the introduction of stresses by nonparallel heat flow. The dislocation content of metallic crystals grown from the melt depends on the growth conditions, and has been reported to be as low as $10^4/cm^2$ and as high as $10^8/cm^2$.

The origin of the dislocations has not been established; it is possible that the mechanism may be the following: Consider the process of growth by the addition of single atoms at step edges (Fig. 6.15). The next atom to be added will occupy the site X. This atom normally comes from the liquid; however, it may instead come from the adjacent solid (atoms P or Q), in which case a vacancy is formed at that point. It is probable that a proportion that may reach 1% of the sites will be filled in this way. This will produce a concentration of vacancies far in excess of the equilibrium value at the melting point. The excess vacancies can leave the crystal by several processes: (a) they can diffuse to the interface or any other available surface, (b) they can cause climb of edge dislocations, or (c) they can condense to form clusters of vacancies that can be either roughly spherical or disk-shaped. A disk-shaped aggregate of vacancies can collapse to form a dislocation loop. If a collapsed disk is at, or is able to expand to reach, the interface, but is not parallel to it, a half-ring of dislocation (or of partial dislocation) will result, the two ends of which are at the interface. The interface is then as shown in Fig. 6.16, as seen from the liquid. Further growth would perpetuate the two dislocation lines.

Fig. 6.15. Formation of vacancy at interface.

Fig. 6.16. Collapse of vacancy disk to form dislocation loop.

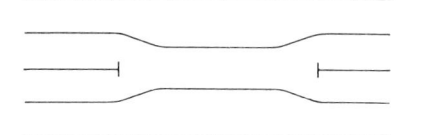

The dislocation content would be expected to increase until the excess vacancies have a sufficiently high probability of condensing on dislocations that the vacancies never reach the concentration required for disk formation. It is possible that the formation of vacancy disks may be nucleated by the aggregation of impurity atoms.

6.5 Freezing of Liquids Containing More than One Component

The discussion of freezing up to this point has implicitly assumed that all the atoms are identical, and are interchangeable. If more than one species of atom is present, however, this no longer applies, and we must consider the consequences of the fact that solute atoms are in general either more or less soluble in the liquid than in the solid. This has been discussed in terms of the kinetics of the interface in Chapter 2. The discussion will be confined to binary systems. We may generalize the result as follows: When the solute has a higher concentration in the liquid than in the crystal, then the liquidus and solidus temperatures decrease with increasing concentration (Fig. 6.17a); conversely, greater solubility in the solid causes a rise in liquidus and solidus (Fig. 6.17b). The reason for this may be restated qualitatively as follows:

1. If the concentrations of B atoms in the solid and liquid are equal, at the temperature of equilibrium, more atoms go from solid to liquid than from liquid to solid, until the concentrations are such that the

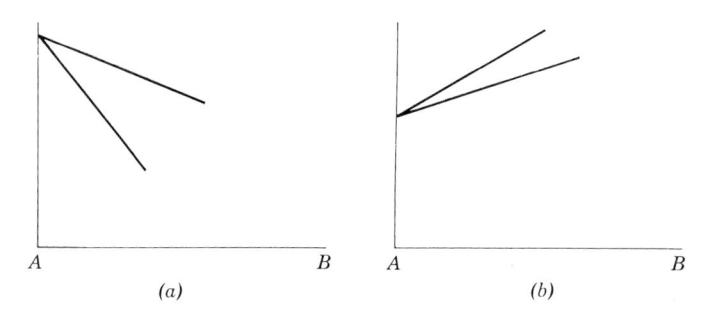

Fig. 6.17. Typical parts of binary phase diagrams.

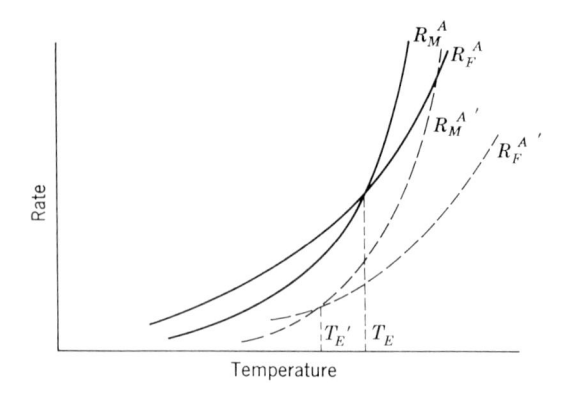

Fig. 6.18. Effect of solute on freezing and on melting.

rates are equal. This corresponds to a lower activation energy for melting for B than for A, and is a consequence of the energy associated with the presence of A in the lattice of B.

2. The equilibrium temperature for pure A is the temperature at which the melting and freezing rates are equal, when both solid and liquid are 100% A.

3. When solid and liquid are "diluted" by the addition of B the rates of melting and freezing are decreased in proportion to the respective amounts of dilution. Thus at the equilibrium temperature for pure A, the freezing process, for liquids A and B, is decreased more than the melting process, because there is more "dilution" of A in the liquid than in the solid.

4. The equilibrium temperature is therefore decreased. The appropriate rate curves for the component A are shown schematically in Fig. 6.18, in which $R_M{}^A$ and $R_F{}^A$ are for pure A, and $R_M{}^{A\prime}$ and $R_F{}^{A\prime}$ are for the AB alloy. T_E and $T_E{}'$ are the equilibrium temperatures for pure A and the alloy.

In many cases, the ratio of the concentration of solute in the solid to its concentration in the liquid is constant over a wide range of temperature. This ratio, C_S/C_L, is the *distribution coefficient k* (Fig. 6.19). In most alloy systems, k is less than unity; there are a few examples, however, in which k is greater than unity. The discussion that follows refers to the case of $k < 1$.

Redistribution of solute. It will be evident that, if a small fraction of a liquid containing a concentration C_L of the component B is frozen under equilibrium conditions, the solid will contain C_S of B,

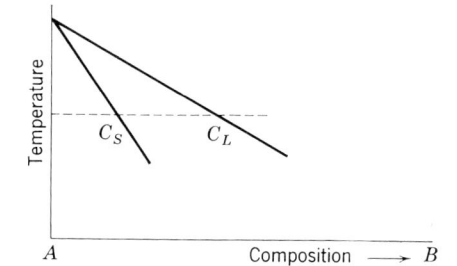

Fig. 6.19. Equilibrium compositions of solid and liquid solutions.

where $C_S/C_L = k$. There is therefore an amount of the solute B given by $C_L - C_S$, or $C_L(1 - k)$, that is unaccounted for. This amount of solute is rejected by the solid and remains in the liquid. This increases the concentration of the liquid above the value C_L. The result of this, in terms of the composition of the subsequently forming solid, depends upon the way in which the rejected solute is distributed in the available liquid.

Before considering the probable results of real conditions, it is desirable to consider three simple cases, as follows:

1. Equilibrium maintained at all times. Consider a liquid of composition C_0, Fig. 6.20; this starts to freeze at temperature T_1, forming an infinitesimal quantity of solid of composition C_S. The composition of the liquid changes slightly, to C_0', as a result and further freezing takes place at a slightly lower temperature (T_1'); the solid that is now formed has a composition C_S'. Complete equilibrium demands that all of the solid has this composition, which can only be achieved by diffusion in the solid. In the same way, the composition of the liquid moves along the liquidus line, and the composition of the whole

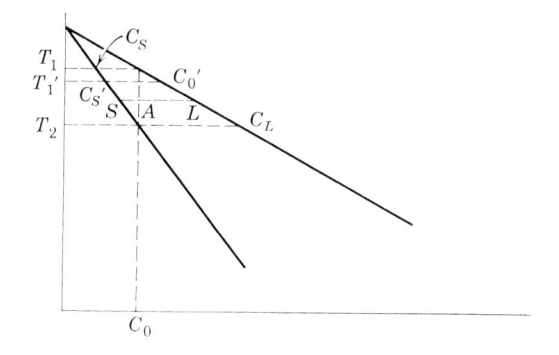

Fig. 6.20. Change of composition during freezing of alloy. Equilibrium case.

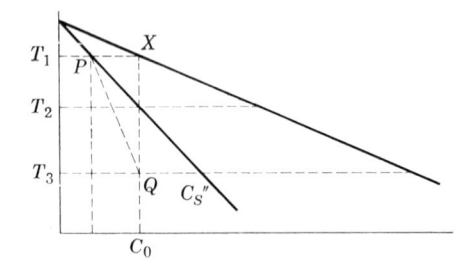

Fig. 6.21. Change of composition during freezing of alloy. No solid diffusion.

of the solid moves along the solidus line. The relative amounts of solid and liquid are given by the ratio of the lengths AL to AS. It follows that when the temperature of the liquid has fallen to T_2, and the compositions of solid and liquid are C_0 and C_L respectively, the amount of liquid has just become zero, and the process is therefore complete. The whole of the material is solid with composition C_0.

2. Liquid uniform in composition at all times, no solid diffusion. As in case 1, freezing starts at temperature T_1, and a solid of composition C_S is formed. Subsequently, the liquid changes composition along the liquidus curve, but the first formed solid remains at the composition C_S, the subsequently formed solid having a progressively higher concentration of solute. The mean composition of the solid no longer corresponds to the liquid (i.e., S in Fig. 6.20) but to a point on a line such as PQ of Fig. 6.21. Consequently, the liquid does not disappear when T_2 is reached: the material is all solid only when the mean composition of the solid is equal to C_0, i.e., at T_3. The last solid to be formed has a composition C_S'', the value of which can be found if it is assumed that PQ is the locus of the arithmetic mean of C_S and the

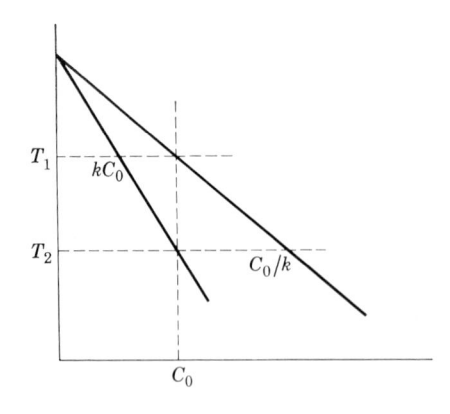

Fig. 6.22. Steady-state condition for freezing of alloy.

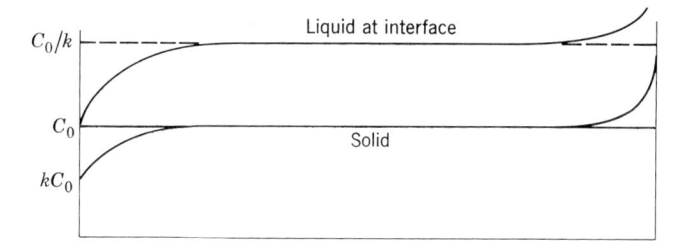

Fig. 6.23. Change of composition during freezing. Solute redistributed by liquid diffusion only.

solidus composition. Thus $QC_s'' = XP = C_0(1 - k)$ whence $C_s'' = C_0 + C_0(1 - k) = C_0(2 - k)$.

3. No solid diffusion, no mixing except by diffusion. In this case the distribution of solute in the liquid depends upon the rate of freezing R and the diffusion coefficient in the liquid D as well as on the distribution coefficient k. If freezing is assumed to take place linearly, with a plane interface, the sequence of events is as follows. Freezing starts at temperature T_1 (Fig. 6.22) and the solute begins to accumulate in front of the interface; it also begins to diffuse forward into the liquid. The concentration of the liquid at the interface increases in an almost exponential manner until it reaches C_0/k, at which point a balance is achieved between the rejection of the solute and its forward diffusion. The solid formed from this liquid has a concentration C_0, Fig. 6.23. A steady state has now been reached, and the conditions at and near the interface remain the same until the end of the liquid is approached. When forward diffusion is restricted by the end of the liquid, the composition of the liquid at the interface again starts to rise; this continues until all the liquid is frozen, and the concentration of the last part to freeze may be very much higher than that of the initial liquid.

It is clear that the final distribution of solute in the solid will differ considerably according to which of the above assumptions is made. A comparison of the three results is shown in Fig. 6.24. In practice, the result is always somewhere between case 2 and case 3, as diffusion in the liquid always occurs, and it is likely that there will be some mixing by convection, etc., which would lead to greater uniformity of the liquid than would result from diffusion alone. Effectively complete mixing is not likely to occur, except under specially designed conditions, for the following reason: When a liquid flows past a solid sur-

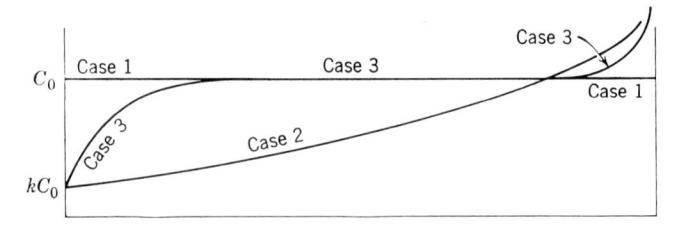

Fig. 6.24. Limiting cases for distribution of solute.

face, the velocity of the liquid is always zero at the surface, and increases with increasing distance from the surface. There is, therefore, always, in contact with the solid, a layer of liquid that may be assumed to be stationary. This layer may be thin, but the only way in which solute can move through it is by diffusion. The relative importance of diffusion and convection depends upon the extent to which the stagnant layer is a barrier to diffusion, and therefore on the concentration difference across it.

It is of interest to consider the thickness of the diffusion zone if there is no bulk movement of the liquid. The concentration in the diffusion zone decreases exponentially from the value C_0/k at the interface (if a steady state has been reached) to the value C_0 for the liquid that has received no excess solute (Fig. 6.25). The important parameter is the characteristic distance X' from the interface at which the concentration has changed by some proportion (e.g., $1/e$) of the total amount. This distance depends only on the speed of freezing R and the diffusion constant D. All diffusion constants for liquid metals are thought to be roughly equal, so a single curve can be used to relate

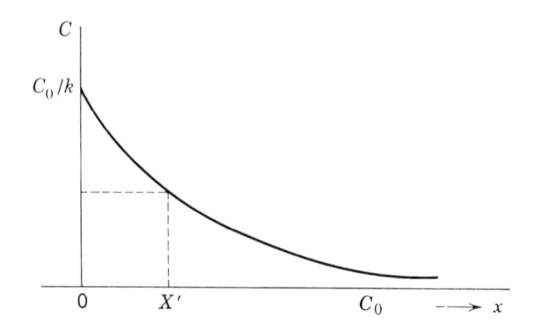

Fig. 6.25. Distribution of solute ahead of the interface.

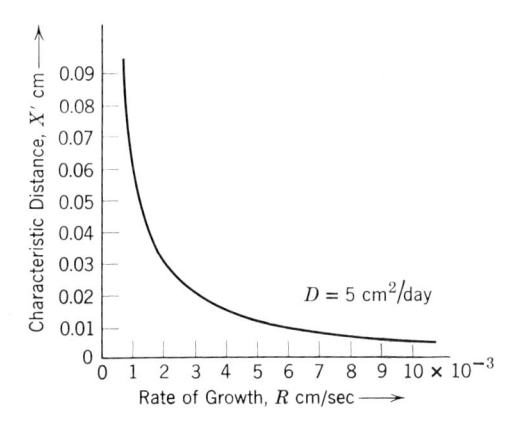

Fig. 6.26. Thickness of diffusion layer as a function of speed of freezing. From *Acta Metallurgica,* Vol. 1, 1953, p. 429.

X' to the rate of advance R. Such a curve is given in Fig. 6.26. The actual distribution of solute depends on the relative values of X', as defined above, and X'', the thickness of the stagnant layer. The value of X'' depends on the velocity of the liquid parallel to the surface. The relationship between X'' and v is such that unless the value of X' is very large (as a result of extremely slow growth) any probable convection or stirring will fail to remove the major part of the diffusion layer. It follows that the liquid at the interface, from which freezing actually takes place, will usually have a composition that is different from that of the bulk liquid.

The significant qualitative result is that the liquid at the interface is more concentrated than the bulk liquid; its liquidus temperature is therefore lower than that of the undisturbed liquid. The liquidus temperature is important as the temperature above which the liquid is stable and below which the solid can nucleate and the liquid is, therefore, supercooled. The amount by which the temperature of the liquid is below the liquidus temperature will be defined as the amount of supercooling ΔT. A typical curve for the variation of liquidus temperature with distance from the freezing interface is shown in Fig. 6.27. If a steady state is assumed, the interface temperature is that corresponding to a solidus composition of C_0, i.e., T_2 in Fig. 6.22, and is independent of R; the liquidus temperature-distance relationship, while always exponential, does depend on R.

Zone refining. It is evident that the first part of the metal to freeze is of higher purity (or contains less solute) than the liquid. This purification only continues until the steady state is reached; this is

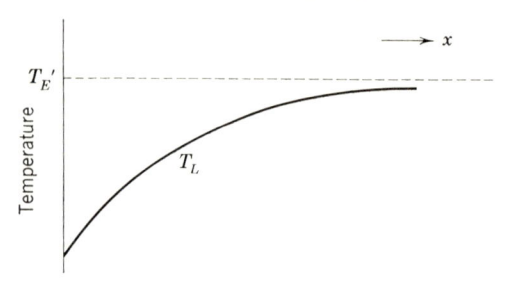

Fig. 6.27. Variation of liquidus temperature with distance x from interface.

postponed or prevented by mechanical mixing of the liquid ahead of the interface. There are advantages in mixing the rejected solute with a relatively small, but constant, part of the melt, because this allows a number of successive stages of purification without recontaminating the purer material near the beginning of the sample by mixing with it the less pure material at the end of the sample.

These principles are applied in *zone refining*, a technique devised by Pfann for use with semiconductors, but also used for producing extremely pure metals. Zone refining is a multistage process for removing solute by controlled freezing. A sample of metal, assumed cylindrical (Fig. 6.28) is subjected to a series of identical processes, of which the following is typical. A short length of the cylinder is melted, and the molten zone is moved along from one end of the bar to the other. The molten zone is subjected to considerable stirring (usually electromagnetic) and it may be assumed to a fair approximation that the composition of the molten zone is uniform at all times. If this is so, the concentration of solutes increases by the rejection into the molten zone, and the solid that is formed is purer, by a factor equal to k, than the liquid then in the zone. The result is a transfer of solute along the bar in the direction in which the zone moves. The distribution after one to ten zones have passed is shown in Fig. 6.29. Successive zones cause progressive movement of the solute from left to right, resulting in a very high degree of purification of part of the material. Any solute that has a distribution coefficient greater than 1 will be moved to the left.

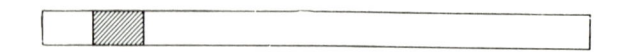

Fig. 6.28. Zone refining, schematic diagram for single zone.

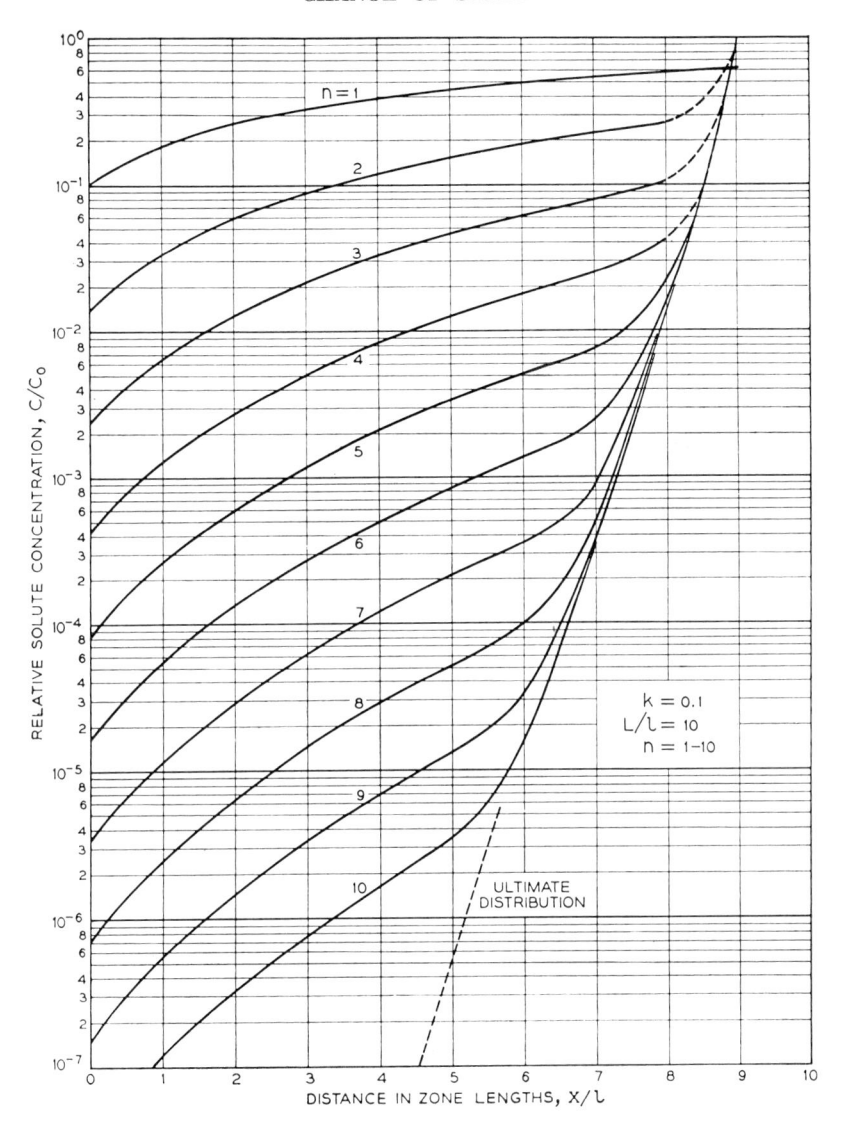

Fig. 6.29. Variation of solute concentration after various numbers of zone refining passes. From *Zone Melting*, W. G. Pfann, John Wiley & Sons, New York, 1958, Fig. A.1, p. 211.

It should be recognized that the solutes that are removed most effectively by this method are those with very small values of k. The process is not efficient for solutes for which k is close to unity.

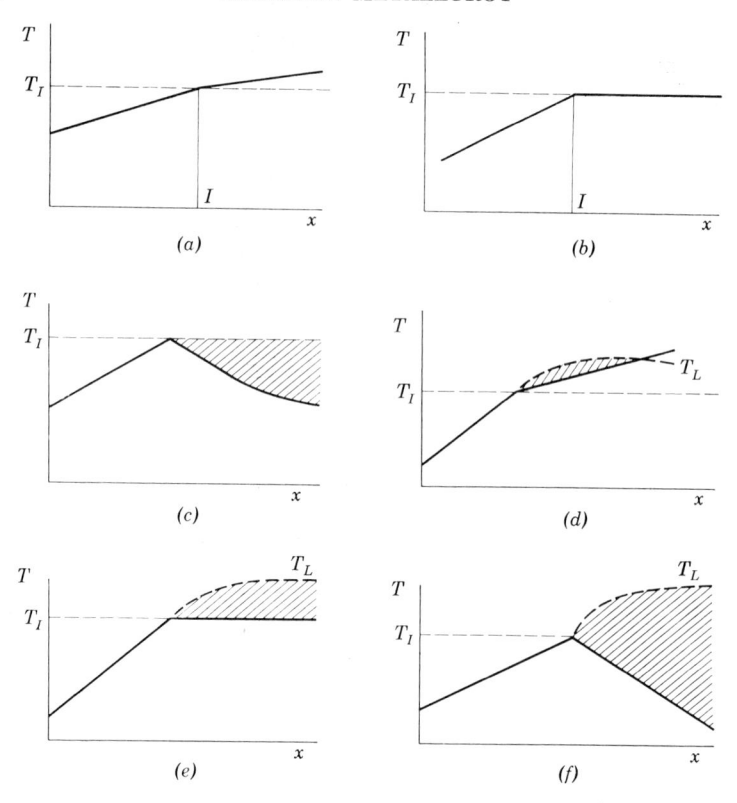

Fig. 6.30. Relationship between temperature and liquidus temperature for a pure metal (a, b, and c) and for an alloy (d, e, and f).

Constitutional supercooling. According to the imposed conditions, the temperature of the liquid may be higher than, equal to, or lower than the temperature at the interface. This will depend upon whether heat is stored in, supplied to, or extracted from the liquid. These three conditions are illustrated in Fig. 6.30 for a pure metal and for an alloy of the type discussed in the section on "Redistribution of Solute." The liquidus temperature is given in each case by a broken line; regions of supercooling are hatched. It will be seen that, for a pure metal, there is supercooling only when the liquid is below the interface temperature. This condition is only attained in special cases. In the alloy, however, there may be a region of supercooling even if the liquid is at a higher temperature than the interface, a condition frequently encountered in the freezing of a mass of metal when heat is extracted through the solid that has already formed.

Supercooling that occurs as a result of the depression of the interface temperature is referred to as *"constitutional supercooling."* It is convenient to discuss the influence of the temperature gradient in the liquid in terms of the supercooling, rather than the temperature, because this takes into account the effect of changes of composition.

Two important conditions may be distinguished. In one, the temperature of the liquid is everywhere above the liquidus except for the very small effect due to the finite rate of freezing. In the other, the amount of supercooling increases progressively from the interface into the liquid, at least for a limited distance (temperature inversion). The mode of growth of the crystal is different in the two cases. In the first, the liquid ahead of the interface is at a temperature corresponding to *slower* crystal growth than the liquid at the interface. This stabilizes an interface that is as smooth as crystallographic considerations permit. In the second case, if supercooling increases with the distance from the interface, any part of the surface that advances ahead of the remainder reaches a region corresponding to *faster* crystal growth and dendritic freezing occurs.

The smooth interface is therefore unstable when a "temperature inversion" is present. The result of the breakdown of the smooth interface depends upon the gradient of supercooling and the extent of the supercooled region. If the supercooled region is small in extent, and if the gradient is also small, the interface breaks up into dome-shaped growth cells, the profile of which is shown in Fig. 6.31a. The cells are usually hexagonal in shape, as illustrated in Fig. 6.31b. The center of each cell is the "growing point"; rejection of solute at such points (G) sets up lateral concentration gradients that cause diffusion of solute towards the cell walls. This lowers the liquidus temperature of the liquid ahead of the cell walls, and so stabilizes the shape of the surface. The cell walls contain a higher concentration of solute than the cell centers.

Under conditions in which the cellular structure is formed, a second substructure, the "lineage" or "macromosaic" structure, is often also present. In this type of structure, different parts of a crystal that originated from a single nucleus differ in orientation by amounts of the order of a degree. The boundaries between these component parts are roughly parallel to the direction of growth, usually coinciding in position and direction with cell walls. The lineage boundaries form gradually, the difference of orientation increasing from zero up to the maximum over a distance that is of the order of a centimeter. The relative orientations are such that each boundary is equivalent to an array of edge dislocations running parallel to the growth direction. It is possible

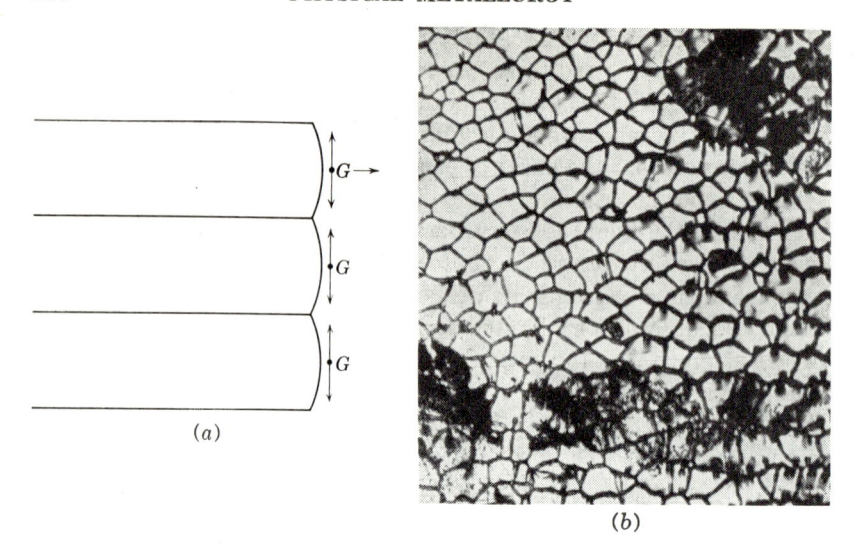

Fig. 6.31. (a) Transverse diffusion of solute during formation of cellular structure. (b) Photomicrograph of cellular structure (×50). (Courtesy D. Walton, Harvard University.)

that the low-angle boundaries form by the "condensation" of dislocations into low-energy arrays. The lineage boundaries are 10 to 100 times as far apart as the cell walls.

When the supercooling extends a considerable distance into the liquid, the cellular structure is not stable, and a few cells grow rapidly as dendrites. These are several differences in detail between dendrites that grow as a result of constitutional supercooling, in an alloy, and those that grow in a pure metal. In an alloy, the changes of composition that occur in the liquid near the growing point of a dendrite tend to slow down its growth, as solute must diffuse away for the liquid in contact with the dendrite to remain supercooled; secondly, there is no "forward" temperature gradient in the case of a constitutionally supercooled alloy; for these reasons, the growth of dendrites in an alloy is not as fast as in a pure metal. Further, dendritic growth does not extend beyond the supercooled zone, which may be limited in extent.

The amount of liquid that can freeze dendritically is limited, and depends on the extent of supercooling. When sufficient latent heat has been liberated to raise the temperature of the whole of the liquid to the liquidus temperature, then no more freezing can occur except as a result of the extraction of heat through the solid. Thus the final stage of freezing is usually the advance of the smooth interface through the spaces between the dendrites.

It should be recognized that zone refining is reduced in efficiency if cellular growth takes place, because of the solute that diffuses laterally and is subsequently found at the cell boundaries. Zone refining is almost completely ineffective when dendritic growth occurs, because nearly all the rejection of solute takes place laterally into the interdendritic spaces, in which it is virtually "trapped."

Nucleation in supercooled melt. If the amount of constitutional supercooling is sufficient, nucleation of new crystals may occur within the melt. These nuclei presumably form, in most cases, on heterogeneous catalysts, and are randomly oriented. They grow dendritically because they occur in a supercooled melt. Their growth is limited by the fact that the latent heat that is evolved must eventually raise the temperature so that the supercooling vanishes; this is accelerated by the enrichment of the remaining liquid, which lowers its liquidus temperature. A result is that the new crystals grow to a limited extent, and then stop growing, forming a fairly stable suspension or skeleton of dendritic crystals in the liquid. This is the explanation of the so-called pasty condition that some alloys pass through while they are freezing. The liquid in the interdendritic spaces eventually freezes as heat is conducted outwards.

The possible extent of constitutional supercooling can be derived from the phase diagram, if it is assumed that some of the liquid maintains its original composition, and that all the liquid takes up the inter-

Fig. 6.32. Maximum possible extent of constitutional supercooling. From *Metals Handbook*, American Society for Metals, Cleveland, 1948, p. 1204.

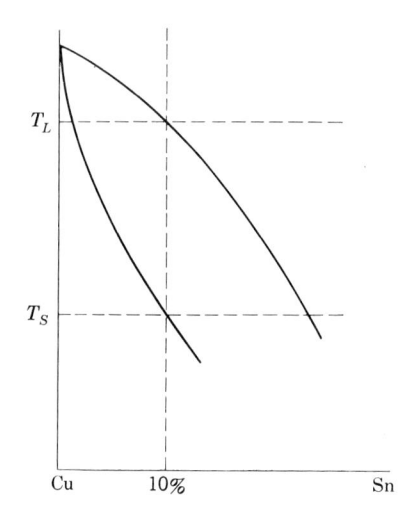

face temperature. The maximum amount of supercooling on this basis is the vertical separation of the liquidus and the solidus at the composition of the original liquid, since the liquidus temperature of the original liquid is T_L, and, at the steady state, the interface temperature is T_S. In many cases nucleation would occur before this amount of supercooling has occurred. An example is shown in Fig. 6.32; an alloy of 10% tin in copper could supercool by 190°C unless nucleation occurred first.

It will be seen that constitutional supercooling cannot occur when the liquidus and the solidus coincide. This applies to (a) a pure metal, (b) a eutectic alloy, (c) a minimum or a maximum M in solidus and liquidus (Fig. 6.33). In these cases neither "pasty" freezing nor dendritic freezing occurs, unless the liquid is cooled to a temperature below that of the interface, which cannot occur by conduction of heat through the interface.

6.6 Two-Phase Alloys

The discussion of freezing has been limited so far to cases in which the solid consists of a single phase. It is necessary to discuss now the processes that occur when more than one solid phase is formed. We may distinguish between several cases as follows:

1. The liquid is of such a composition that the two solid phases are simultaneously in equilibrium with it at the appropriate temperature. This is a liquid of eutectic composition.

2. The liquid is of such a composition that it reaches the eutectic composition by enrichment in one component as a result of rejection of solute at the interface. This is a eutectic system at a noneutectic composition; or a liquid of such a composition that part of the first-formed solid phase reacts with the liquid to form a second solid phase.

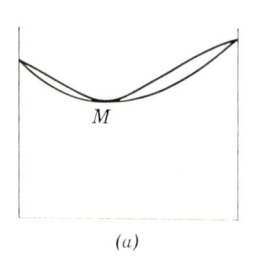

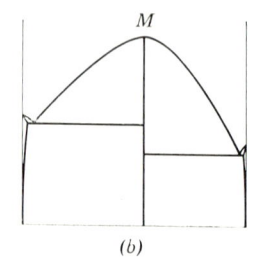

Fig. 6.33. Special cases in which no constitutional supercooling can occur.

CHANGE OF STATE 263

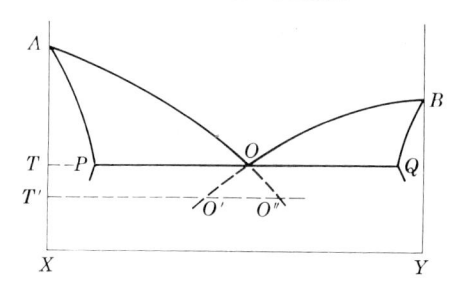

Fig. 6.34. Eutectic freezing.

This is typical of a peritectic system. In each of these cases, the main interest attaches to nonequilibrium freezing, and it will be shown how the resulting structures depend upon the extent of the departure from equilibrium. Each of the cases defined above will be discussed.

Liquid of eutectic composition. A eutectic system occurs as a result of the intersection of two liquidus curves, such as AO and BO in Fig. 6.34. At point O, the two solid phases of composition P and Q respectively are in equilibrium with the liquid. The solid that is in equilibrium with the liquid is a two-phase solid that has the same mean composition as the liquid. It follows that freezing of this liquid can proceed without the liquid as a whole changing in composition. There is, therefore, no necessity for redistribution of solute by forward diffusion. However, two solid phases of widely differing composition are formed, and experiments show that in the typical case they are formed simultaneously. The necessary processes can be considered in terms of the elementary case of Fig. 6.35. The formation of a crystal of

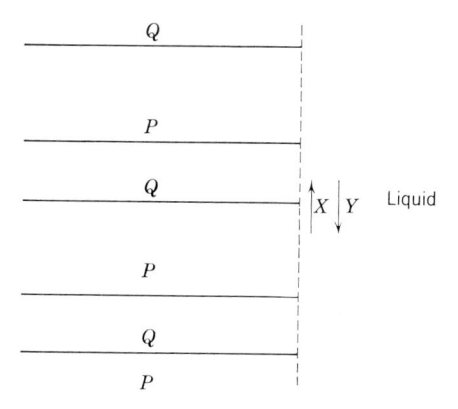

Fig. 6.35. Transverse diffusion during growth of lamellar eutectic.

phase P from the liquid causes the adjacent liquid to become enriched in the component Y, and similarly the liquid ahead of phase Q is enriched in X. A *transverse* concentration gradient is therefore established in the liquid, in addition to the longitudinal concentration gradients that would exist if a single-phase solid were forming. Each phase is therefore in contact with a liquid that differs from O in composition, e.g., O' and O'' in Fig. 6.34. Freezing of these liquids takes place at a temperature T', and the solid phases that are formed differ slightly in composition from the equilibrium phases P and Q. A steady state is established when the diffusion of the two components in the liquid just balances the rejection of solutes by the solids.

In detail, the composition of the liquid at a point such as M, Fig. 6.36, is determined by the rate of rejection of Y by the growing crystal of P, and the diffusion away of Y and its replacement by X. The actual local composition of the liquid must vary from point to point, being essentially equal to O at the junction J of the two crystals, and departing progressively as the lateral diffusion path increases. The temperature of the interface varies similarly, and consequently its geometry is as indicated in Fig. 6.36. There is a limit to the extent to which the liquid ahead of P can be enriched in Y, because any such enrichment corresponds to supercooling with respect to the equilibrium of Q. If the supercooling becomes excessive, therefore, Q will nucleate and a new crystal of the phase Q will form. The lateral extent of the crystals of P and Q is therefore limited; the limiting width is less for higher speeds because less diffusion occurs and this corresponds to more supercooling at a given distance from J.

The geometry of the crystals of the two phases depends on several factors, of which the most important are the ratio of the volumes of the two crystalline phases and the speed of growth. A third factor is the extent to which the crystals tend to grow with crystallographic

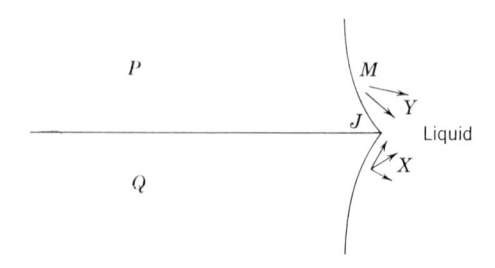

Fig. 6.36. Possible detailed shape of junction between lamellae.

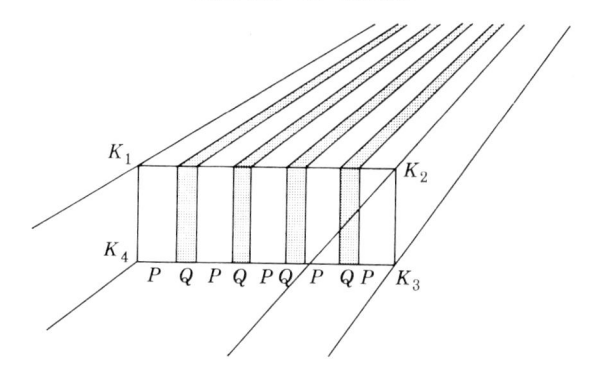

Fig. 6.37. Growth of lamellar eutectic.

bounding surfaces of high atomic density. A fourth factor is the presence or absence of small amounts of specific third components.

The most usual case is that in which the two phases are roughly equal in volume. In such systems, the eutectic is usually lamellar in structure; this structure is shown schematically in Fig. 6.37, in which alternate layers are of phase P and phase Q. The surface $K_1K_2K_3K_4$ represents the interface between the two-phase, or eutectic, solid and the liquid. It is often found that each phase in such a structure has a crystallographic orientation that is common to all the lamellae; this indicates that they all originated from the same nucleus, and therefore are continuous. The origin of such a eutectic colony is probably as follows. The liquid of composition O is supercooled until a nucleus of one of the phases forms. Suppose that this is the phase P. The crystal of P grows, enriching the surrounding liquid with Y until the supersaturation is sufficient to cause Q to nucleate. It will normally do so heterogeneously on the crystal of P. The two crystals grow outwards side by side. Diffusion distances are limited by the "spreading" of each phase over part of the other, instead of by nucleation. The geometry of this stage is complicated, but it is feasible in view of the fact that it is geometrically possible to have two interpenetrating continuous phases. A solid formed by eutectic freezing usually consists of a number of eutectic colonies, each of which has grown from a single nucleus. An example of the structure of a lamellar eutectic is shown in Fig. 6.38.

The spacing of the lamellae depends on the speed of freezing, decreasing with increasing speed.

When one of the phases occupies a much smaller volume than the other, it may occur as isolated plates, rods, or spheres. It is useful

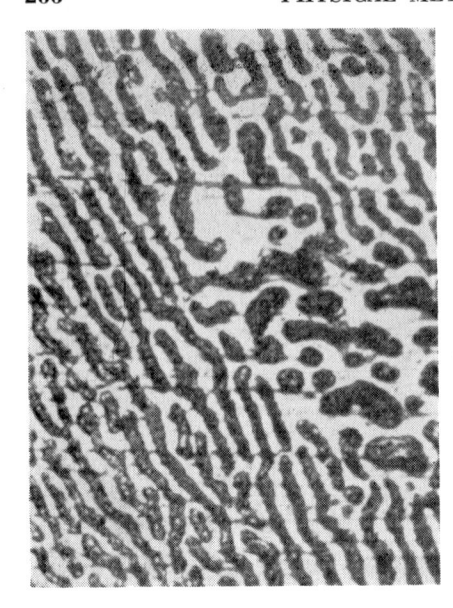

Fig. 6.38. Structure of lamellar eutectic. Section normal to growth direction.

to distinguish between those eutectic structures that are continuous in the direction of freezing, and which correspond to steady-state conditions, and those which are not continuous, and which require frequent nucleation of crystals of one of the phases. The latter necessarily consist of a continuous phase and a discontinuous phase, and usually occur when one of the phases occupies a much smaller volume than the other. There appear to be two distinct kinds of discontinuous eutectic structure; those in which the discontinuous phase consists of "needles" or "plates," and those in which it consists of spheres. The former kind, of which the aluminum-silicon eutectic is a good example, is apparently discontinuous because the needles of the silicon-rich phase grow in directions that are dictated by their crystallography and the orientation in which they happen to nucleate. The needles therefore do not maintain the uniformity of spacing and direction that would be necessary for continuous growth. A second example of a discontinuous eutectic is the iron-carbon eutectic found in cast iron in which the carbon-rich phase is graphite. The graphite normally occurs as flakes (see Fig. 5.63a) the geometry of which again bears little or no relationship to the direction of freezing.

Both the aluminum-silicon and the iron-carbon eutectic have mechanical properties that are affected adversely by the presence of a brittle (silicon) or weak (graphite) constituent in a shape that leads to serious stress concentration. In both cases, the properties, in particular the ductility, are improved by causing the discontinuous phase to occur as spheres rather than as needles or plates. In each case this is achieved by the addition of a small amount of a third component. This does not cause a third phase to appear, but apparently it causes the discontinuous phase to stop growing much sooner than in the normal eutectic process. A possible explanation is that the third component is rejected much more by the discontinuous phase than by the continuous one, and that, consequently, the interface temperature of the discontinuous phase is lowered more than that of the continuous phase. This would produce an interface shaped as in Fig. 6.39. The continuous phase C would then grow across in front of the discontinuous phase D, which would then consist of very small particles instead of the much larger ones that normally occur. The addition of sodium to the aluminum-silicon alloy, or magnesium or cerium to the iron-carbon alloy, in quite small amounts produces these effects.

In the latter case the graphite has a special crystallographic structure (spherulitic) that is not accounted for by the explanation given above. The structure of graphite is extremely anisotropic, consisting of layers in which the atoms are covalently bonded; the layers are bonded to each other by van der Waals forces. It follows that a surface that is parallel to the basal plane, i.e., to the layers, should have a much lower surface energy than one which passes through the layers. The spherulitic structure is one in which the interface between the graphite and its environment is everywhere parallel to the basal plane. This cannot occur without either strain energy resulting from a curvature of the basal layers, as in Fig. 6.40a, or by the existence of boundaries between differently oriented parts of the spherulet, as in Fig. 6.40b. It appears that the decrease in interfacial energy ex-

Fig. 6.39. Suppression of second-phase crystals by third component.

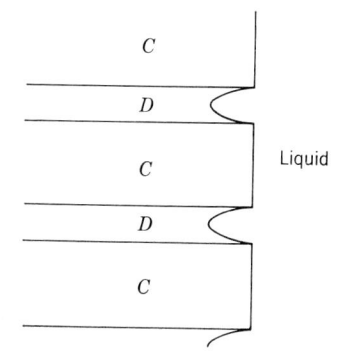

268   PHYSICAL METALLURGY

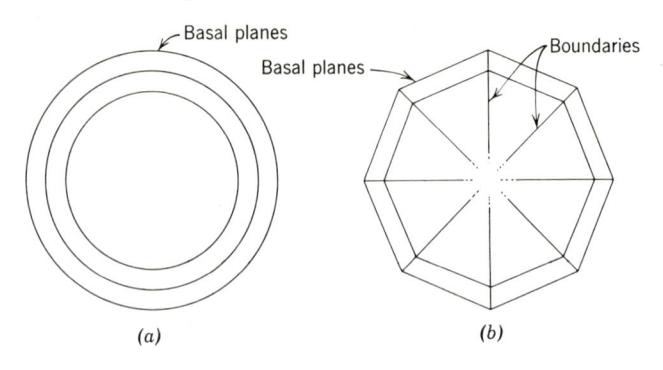

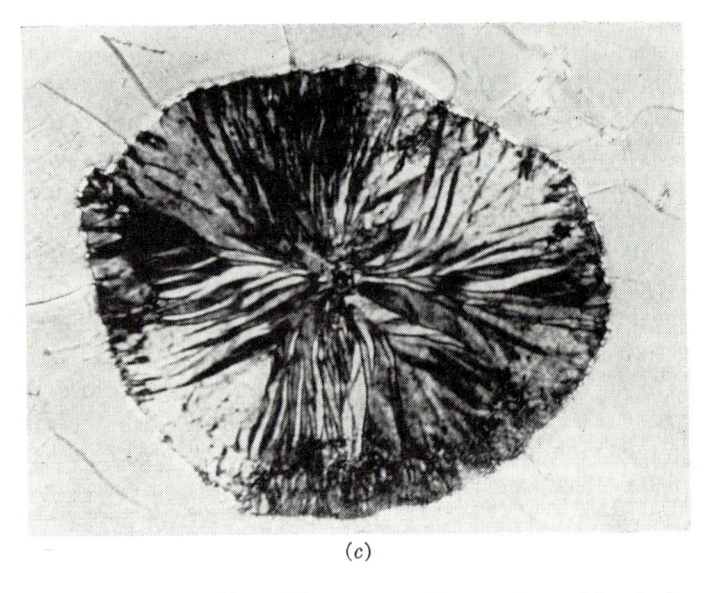

Fig. 6.40. Spheroidal graphite. (a) represents the curvature of basal planes, (b) the presence of boundaries and flat basal planes. (c) is a photomicrograph ($\times 250$) of a spherulet. (Courtesy International Nickel Company.)

ceeds the energy corresponding to the curvature or boundaries, and that the spherulet may be the stable form.

The mechanical properties of lamellar eutectic alloys are in general better than those of the individual phases, unless one of them is brittle, when the alloy may also be brittle; however, alloys of eutectic composition are seldom used where mechanical properties are important.

As an example, the properties of the tin-lead eutectic are compared with those of the two equilibrium phases in Table 6.3.

CHANGE OF STATE 269

Table 6.3. Mechanical Properties of Some Tin-Lead Alloys

Alloy	Yield Stress (psi)	Ultimate Tensile Strength (psi)	Elongation (%)
Sn + 2.5% Pb	—	3000	20
Pb + 20% Sn	3650	5800	16
Eutectic	4800	7500	32

Eutectic system, noneutectic composition. The process to be considered is the freezing of an alloy of composition C_0 (Fig. 6.41). Freezing begins with the formation of phase A, of composition S, at temperature T_1. If it is assumed that there is no mixing except by diffusion, it follows that the primary crystals of phase A will grow until the liquid ahead of them reaches the composition and temperature O' at which phase B nucleates. The primary crystals will normally be dendritic, because considerable constitutional supercooling is likely to occur. It follows that when nucleation of phase B occurs, and if the whole liquid is at the same temperature, the composition of the liquid varies from C', at the interface of the dendrites of phase A and the liquid, to a value C'', the limiting supercooling below the liquidus of phase A. C'' may be very close to C', in which case the mean composition of this liquid will not be exactly that of the eutectic. The remaining liquid nevertheless freezes as a eutectic, usually with a slightly abnormal ratio of the amounts of the two phases. It was assumed in the foregoing that there were no temperature gradients; in the presence of a gradient, the same sequence of events will occur

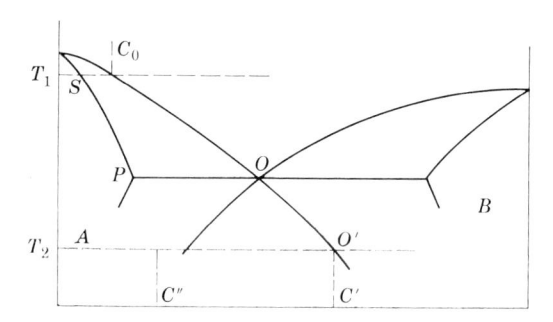

Fig. 6.41. Freezing of eutectic system at noneutectic composition.

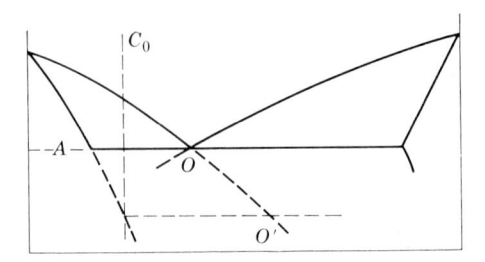

Fig. 6.42. Phase diagram for divorced eutectic.

in each portion of the liquid as it reaches the appropriate temperature.
If phase B has not nucleated by the time the solidus of the A phase has reached the initial composition C_0 (Fig. 6.42), then the condition for steady-state freezing of A is achieved; in this case phase B is not formed until the terminal region of limited diffusion is reached; phase B then forms in the final region, but the typical eutectic structure may not appear. This case is known as a "divorced eutectic."

It should be remarked that the very rapid rise in the solute content of the liquid as freezing approaches completion (see p. 254) may cause the eutectic composition to be reached in a small proportion of the liquid even if the initial composition corresponds to a single-phase solid, i.e., to the left of point P in Fig. 6.41. Thus an aluminum-magnesium alloy containing only 5% of magnesium may form some solid of eutectic composition and structure. The equilibrium diagram is shown in Fig. 6.43, from which it is seen that under equilibrium conditions the

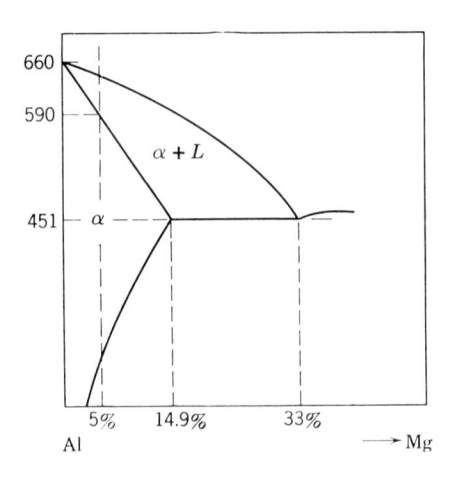

Fig. 6.43. Aluminum-magnesium phase diagram. From *Metals Handbook*, ASM, p. 1163.

CHANGE OF STATE 271

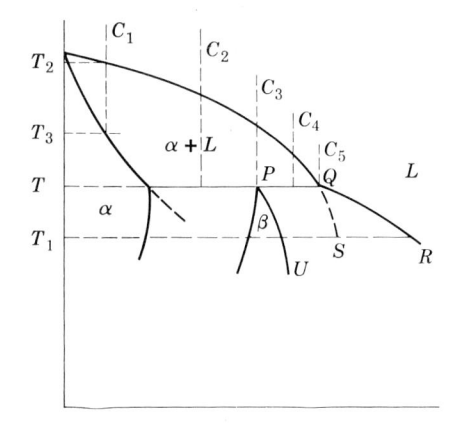

Fig. 6.44. Peritectic reaction; copper-tin phase diagram.

alloy should be solid long before the temperature has fallen to 451°C, the eutectic temperature.

Peritectic system. A typical peritectic system is that of copper-tin, the relevant part of which is shown in Fig. 6.44. Above the peritectic temperature T, the liquid is in equilibrium with the α phase, which is a solid solution of tin in copper; below T, the β phase is in equilibrium, at different compositions, with the α phase and with the liquid. *Under equilibrium conditions,* the result of cooling liquids of compositions C_1, C_2, C_3, and C_4 to the temperature T_1 would be to produce a solid consisting of an α phase only, a solid consisting of α and β phases, β phase only, and a mixture of β phase and the liquid, respectively. Under nonequilibrium conditions, it is necessary to consider the sequence of events in detail in order to determine the result. It will be assumed that there is no diffusion in the solid, and that the liquid mixes only by diffusion. The modifications required if the latter assumption is not true will be evident. The four cases will be considered separately.

1. When an alloy of composition C_1 is cooled to T_2, nucleation of crystals of the α phase may occur. Their growth causes a rejection of tin into the melt; the interface temperature falls, and the composition of the α phase that is being formed moves to the right. Constitutional supercooling will occur unless a very steep temperature gradient is maintained, and further crystals of α phase are nucleated. The liquid between these crystals becomes progressively enriched in tin, and the composition of the liquid may reach C_5. If it does so, β crystals may nucleate. The result is a two-phase solid, consisting of α and β.

If there is insufficient constitutional supercooling to cause nucleation ahead of the interface, a steady state is reached in which the α phase of composition C_1 is formed at a temperature T_3. This is maintained until the terminal transient is reached, when the compositions of the liquid at the interface and of the solid being formed become progressively richer in solute. The β phase forms if the composition of the liquid reaches the value corresponding to the point Q.

2. For an initial composition C_2, the steady state corresponds to the formation of both α and β phases; it is to be expected that the simultaneous growth of these two phases should produce structures resembling those of eutectics.

3. If the initial composition is C_3, a steady state can be reached in which the β phase forms from the liquid at the interface, of which the composition is Q. The terminal transient causes the composition to move to the right, along the liquidus line QR, with the result that the final solid is formed at a temperature below T_1, and has a composition on the line PU.

4. At C_4, α is again the first phase to be formed and β is formed in any liquid that exceeds Q in solute content. The resulting structure consists of primary crystals of α surrounded by the β phase.

It is usually stated that in a peritectic system the liquid reacts with the primary phase to form the secondary phase. This can in fact only occur to an extremely limited extent without diffusion in the β phase, which immediately forms at the interface of the liquid and the primary phase. It follows that unless very long times are available during freezing the primary (α) phase will not be converted to the secondary (β) phase by reaction with the liquid. Thus in most peritectic systems, a primary phase persists although it is not an equilibrium phase.

It will be shown in Chapter 8 that diffusion in crystals is an extremely slow process, and it follows that processes that depend on *solid* diffusion cannot have much influence on the structures formed in ordinary freezing processes.

6.7 Structure Resulting from Solidification

Although metals and alloys are caused to solidify in many different shapes and in many different environments, the structure resulting from such procedures can in general be understood or predicted in terms of the principles discussed above. These considerations will be described first in terms of a geometrically very simple case; then they will be extended to the conditions encountered in the various kinds of practical processes that depend on freezing a molten metal.

Size, shape, and orientation of crystals. It is assumed that a molten metal or alloy is poured into a container or mold, a square

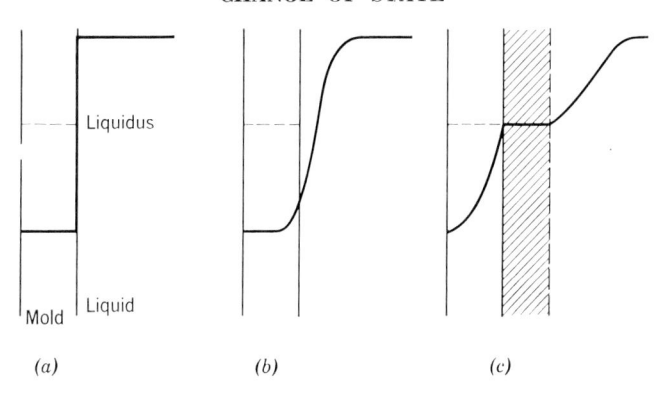

Fig. 6.45. Temperature distribution during freezing in a mold.

prism in shape, with which it does not react in any way. The mold is assumed to be at a temperature well below the melting point of the metal. The following events occur:

The metal adjacent to the mold is cooled very rapidly to a temperature well below its liquidus. The supercooling is sufficient to cause the nucleation of many crystals, which grow dendritically into the surrounding liquid. The distribution of temperature is shown in Fig. 6.45a, which represents the temperature immediately after filling the mold with liquid metal, before there has been time for the temperature to change; part b shows the temperatures a short time later, when the outer layer of liquid has cooled and the mold has started to heat up; part c shows the temperatures after the nucleation and growth of dendrites in the part of the liquid that was below the liquidus temperature. The actual nucleation process is likely to occur on the mold wall. The hatched region of Fig. 6.45c contains a skeleton structure of dendritic crystals, and the temperature in this region has risen nearly to the liquidus temperature as a result of the evolution of latent heat by the growing dendritic crystals. As more heat is extracted, the dendritic skeleton fills in by the growth of a continuous solid mass of crystals, which everywhere have the same orientation as the nearest dendrite. These crystals are usually small and randomly oriented. They form the so-called "chill zone" of the structure. When the extraction of heat from the surface is very rapid, and the extent of the liquid metal relatively small, the chill zone may extend throughout the entire volume of the liquid.

The filling in of the interdendritic spaces by the growth of a continuous polycrystalline mass sets up constitutional supercooling if the

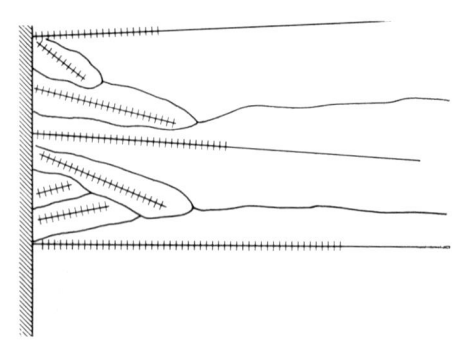

Fig. 6.46. Transition from chill zone to columnar zone, showing directions of dendritic growth in each crystal.

liquid is a suitable alloy. In this case, the crystals continue to grow dendritically; those with their axes of dendritic growth more nearly perpendicular to the mold wall (or, more strictly, to the isotherms) grow with slightly less supercooling than those that are less nearly perpendicular: the boundaries formed between the growing crystals are inclined *away* from the crystals that have their dendrite axis normal to the isotherms, as shown in Fig. 6.46. It follows that all other crystals will gradually be suppressed. The result is that the crystals become progressively fewer, larger, and with a more precise preferred orientation. Typical curves for crystal cross section and spread of orientation as a function of distance from the mold wall are given in Fig. 6.47. As would be concluded from this mechanism, the preferred orientation is such that the crystals have the dendrite axis (see p. 235) normal to the mold wall.

The crystals that survive this selection process may continue to grow with essentially constant cross section for a considerable distance. They constitute the "columnar zone." This stage of the process is limited by the nucleation of new crystals in the remaining liquid, which takes place when its constitutional supercooling reaches some value that depends on the potency of the nucleation catalysts that are present. The crystals that are nucleated at this stage grow at first as dendritic skeletons and finally by the advance through them of a continuous interface. These crystals are equiaxed in shape and random in orientation; the region occupied by these crystals is called the "equiaxed zone." Figure 6.48a shows a cross section of a structure containing all three zones. One of the interesting features of this structure is that the columnar crystal boundaries clearly are not normal to the isotherms near the corners. This is because the boundaries

CHANGE OF STATE 275

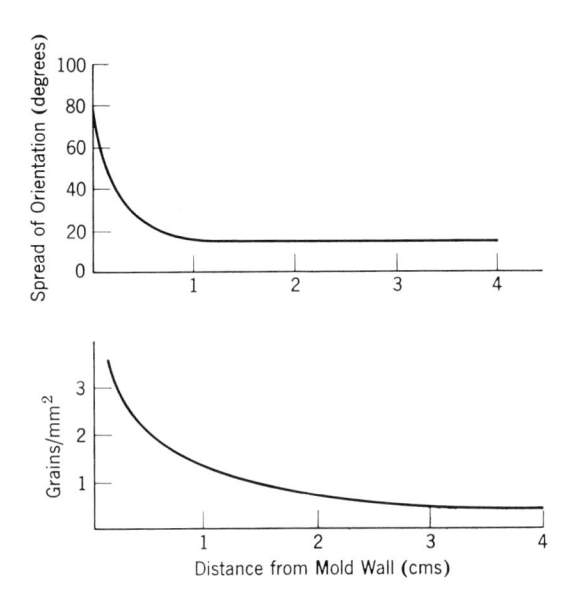

Fig. 6.47. Development of preferred orientation and columnar grain size. (Courtesy D. Walton, Harvard University.)

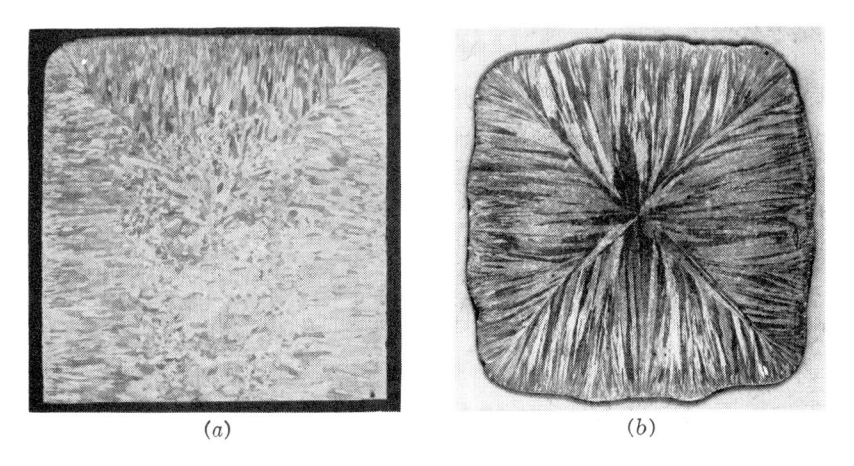

(a) (b)

Fig. 6.48. Cross sections of ingots (a) with equiaxed zone and (b) without equiaxed zone. (Courtesy J. L. Walker, General Electric Research Laboratory.)

between parallel dendritically growing crystals form in the direction of the dendrite axes and not of the heat flow.

In the case of a "pure" metal, there is not sufficient constitutional supercooling for equiaxed nucleation to occur, and the columnar structure persists to the center. The preferred orientation is unchanged, however, except in metals of the very highest purity (zone-refined), because dendritic growth is not suppressed completely unless very high purity is reached. The structure of an ingot of a "pure" metal is shown in Fig. 6.48b. When a second phase is produced, the structure may be much more complicated; the simplest case is when only a small amount is formed. It is found at the intergrain corners of the equiaxed zone.

It is often desirable to produce a fine grain size: this may be done by (a) reducing the length of the columnar zone and (b) producing a fine grain size in the equiaxed zone. The former is achieved by using an alloy with a large difference between liquidus and solidus, and by decreasing the temperature of the melt so as to produce the required amount of constitutional supercooling as soon as possible. The grain size depends upon the number of nucleation catalyst particles that become effective. This depends upon the number of particles that are present, their potency, and the rate of cooling.

The effect of rate of cooling can be explained on the assumption that all particles do not nucleate at precisely the same time, either because they nucleate at different temperatures, or because all of the liquid does not reach the same temperature at the same time. The first few crystals to nucleate grow until the conditions have been reached for the nucleation of some more crystals; if the time is relatively long, they will have grown to a considerable extent and fewer catalyst particles remain for further nucleation. With faster cooling, a larger number of particles are put into operation before they are suppressed by the arrival of the growth interface of other crystals.

The extreme case of slow cooling is that employed in several techniques that are used for growing single crystals, in which a temperature gradient is used to ensure that none of the liquid is supercooled except that which is in contact with the growing crystal. Thus every potential crystallization catalyst, except the first one, is absorbed into the growing crystal before its nucleating temperature is reached.

Several addition agents have been discovered empirically for the production of fine grain sizes in suitable alloys; for example, carbon in certain magnesium alloys, and titanium, which apparently forms titanium nitride, for some aluminum alloys. The improvement in properties that is obtained in this way is sometimes quite important:

Table 6.4. Variation of Properties with Grain Size in a Cast Magnesium Alloy

Grain Size (Average Grain Diameter, cm)	Ultimate Tensile Strength (psi)	Elongation (%)
0.011	40,700	10.6
0.094	31,200	4.5

an example is the case of a magnesium, alloy A292, for which the properties at two grain sizes are compared in Table 6.4.

There is also adequate evidence that suitable vibration, applied during freezing, sometimes produces a finer grain size than would otherwise be obtained. It is not clear whether the effect is due to an increase in the potency of catalysts by the vibration, or whether the number of catalyst particles or crystals is increased, perhaps by fragmentation of the growing dendrites.

6.8 Porosity

One of the most important aspects of the "structure" of a sample of a metal or alloy is the presence or absence of discontinuities or holes. Such defects may occur during freezing as a result of the decrease in volume that accompanies freezing in most metals, or of the rejection at the interface of gas dissolved in the liquid, or a combination of the two.

Effects of shrinkage. All the metals that have close-packed or body-centered cubic structures contract when they freeze; many covalently or partly covalently bonded crystals, with co-ordination numbers less than 8, expand on freezing. The extent of the expansion or contraction is shown in Table 3.3. Comparison of these amounts of expansion or contraction with the bulk modulus of elasticity shows that it could only be prevented by the application of extremely large triaxial forces of compression or tension. For example, in the case of copper, which contracts by about 5% when it freezes, over 70,000 atmospheres *negative* pressure would be required to cause the solid to expand elastically by this amount. It follows that contraction on freezing cannot be prevented, except by the choice of an alloy that inherently does not contract. Some of the so-called "fusible alloys" show some expansion on freezing, a property that occasionally is useful when the alloy is used for holding an object for machining or other mechanical operations. The fusible alloys that contain more than about 55% bismuth expand on freezing; an example is one which melts at 174°C, and consists of 57% Bi, 17.3% Sn, and 25.2% In.

The simplest case of shrinkage is that in which the metal freezes in a mold of the shape represented in Fig. 6.49, in which successive stages of freezing are shown schematically. It will be seen that a "pipe" is formed in the upper part of the ingot. It is evident that the existence or extent of piping depends upon the extent to which the isothermal surface corresponding to the interface becomes concave. The concavity of the interface in turn depends upon the relative amounts of heat conducted in the vertical and the horizontal directions. If no heat is lost laterally, the isothermal surface is a horizontal plane and there is no pipe, the shrinkage being accounted for by a difference between the original level of the surface of the liquid and the final level of the surface of the solid. It is not usually practicable to prevent lateral heat flow, but its effect can be minimized by supplying heat at the top of the ingot, thereby maintaining a vertical temperature gradient and increasing the amount of heat that is conducted downwards.

If freezing takes place from the top as well as from the sides and bottom, a totally enclosed pipe is formed. Its shape is shown in Fig.

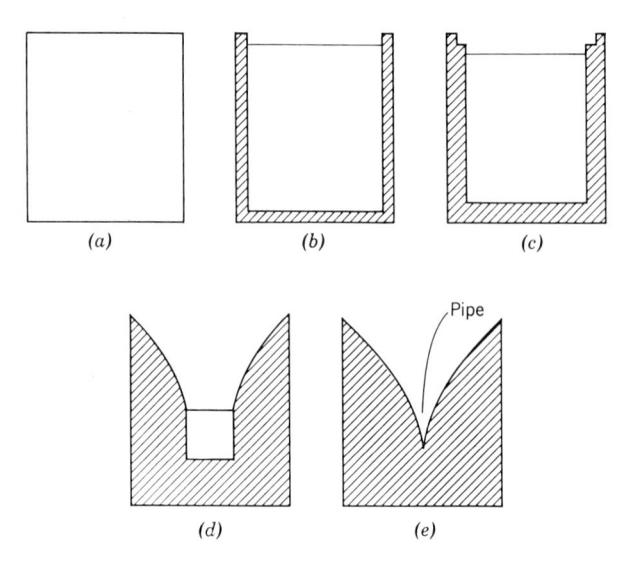

Fig. 6.49. Formation of pipe by shrinkage.

Fig. 6.50. Totally enclosed pipe.

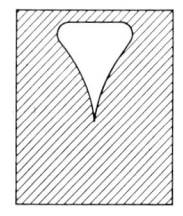

6.50. A combination of the two types of pipe is shown in Fig. 6.51. The cavities A and B formed because the liquid in these regions became isolated from the remaining liquid before it froze; it was therefore not possible for liquid to feed A and B, which consequently became internal cavities. In general, any pool of liquid that is completely surrounded by solid will develop a void as it freezes; this applies whether the mold has the simple shape shown in Figs. 6.49–6.51 or whether it is more complicated.

When an equiaxed zone is formed in the interior, the whole of this zone may be filled with a dendritic skeleton, resulting from the interpenetration and interlocking of the dendritic growth that originated at each center of nucleation. As pointed out above, the last stage of freezing takes place as a result of conduction of heat outwards. This process may leave islands of liquid with high solute content, and therefore lower liquidus temperature; these eventually freeze and cause shrinkage pores.

A special case of shrinkage porosity is found when the initial chilled layer has remelted as a result of the liquid having a very high temperature. In this case, the liquid cools past its liquidus temperature much more slowly than when it was first poured into the cold mold, and consequently far fewer crystals nucleate than in the initial chill zone; these grow dendritically and the space between them contains liquid of higher solute content, and therefore lower liquidus temperature. In this situation the temperature gradient is small and the interdendritic liquid does not freeze until the dendrites have progressed a long way into the interior. The negative pressure due to shrinkage may suck

Fig. 6.51. Pipe in ingot.

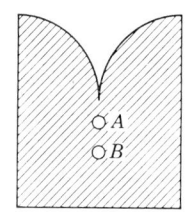

the interdendritic liquid inwards from the surface, leaving a very porous surface layer with a distinct dendritic structure.

Effects of gas evolution. There are many cases in which gases are much more soluble in a liquid metal than in the solid. The gas therefore behaves like a solute for which the distribution coefficient k is much less than unity. The gas that is rejected accumulates in front of the advancing interface, where it may reach a concentration that causes the nucleation of gas bubbles. It is unlikely that the interface is a good heterogeneous nucleation catalyst for bubble formation, because contact between the bubble and the solid surface would create a solid-gas interface, which has higher energy than the liquid-gas interface. Gas bubbles nevertheless frequently adhere to the interface, perhaps at specially favorable points, such as grain boundaries. The subsequent behavior of a bubble depends upon the rate of advance of the interface. It may grow, by diffusion of gas into it from the surrounding liquid, to such an extent that it floats away from the interface; it may rise to the surface of the melt and escape, or it may be trapped by the already formed solid and form a blowhole; blowholes are also formed by gas originating from a suitable chemical reaction in the liquid metal. Alternatively, bubbles may adhere to the surface of the solid, which continues to grow except where the bubble is actually in contact with it. The bubble itself also grows by the diffusion of gas towards it from the surrounding supersaturated liquid. It may grow at a speed equal to that of the advance of the interface, when it becomes a cylindrical hole, with its axis normal to the interface. An example of an ingot containing such holes is shown in Fig. 6.52a. Figure 6.52b shows similar holes in ice grown from water containing dissolved air. The third possibility is that the interface advances so rapidly that it closes off the bubble, which then remains as a spherical hole.

Interaction of shrinkage and gas evolution. The concentration of gas required for the nucleation of a bubble is reduced if a triaxial tension (negative hydrostatic pressure) is applied to the liquid. Such a negative pressure arises in any "sealed-off" part of the liquid; thus it is to be expected that if any dissolved gas is present, a bubble will nucleate and grow in liquid which has a tendency to form a shrinkage cavity as a result of being sealed off. Such sealed-off regions may be interdendritic liquid, or they may be regions that have remained liquid because of the geometry of the solidifying mass, as in the case of the holes A and B of Fig. 6.51. A pore may therefore be found to contain gas even if there was not sufficient gas content to have caused gas bubbles to form in the absence of triaxial tension.

An effect of the nucleation of a bubble in a region of sealed-off liquid

CHANGE OF STATE 281

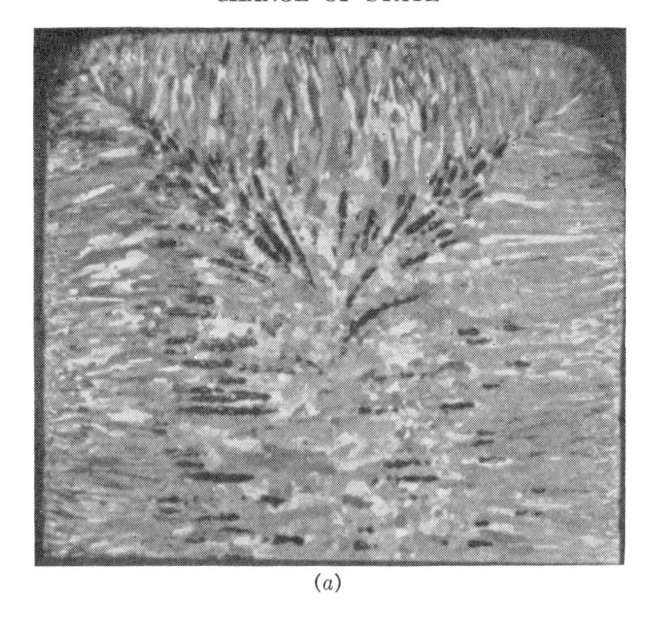

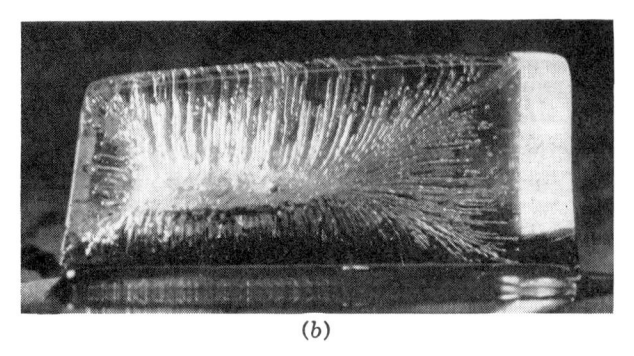

(a)

(b)

Fig. 6.52. Elongated gas pores in (a) metal and (b) ice.

is to cause the gas in the liquid to diffuse into the bubble; no more bubbles are likely to be formed, and the result is usually a single pore in each sealed-off region. The same amount of shrinkage in the absence of gas is more likely to produce a large number of much smaller pores.

6.9 Segregation

It can usually be assumed that the composition of a molten alloy is homogeneous; the resulting solid, however, varies considerably in

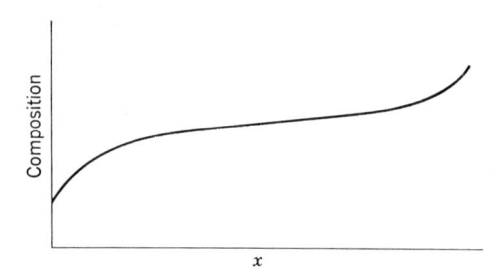

Fig. 6.53. Curve showing normal segregation.

composition from one place to another. These differences in composition are described as "segregation." There are several different kinds of segregation; more than one of these may occur in the same sample of metal, but they will be discussed separately.

Normal segregation. Normal segregation is the direct result of the rejection of solute at an advancing interface. It corresponds to the increase in concentration of a solute (for which $k < 1$) as the center of the mold is approached. An example of the simplest possible cases of normal segregation has been discussed on p. 253. A schematic example of the redistribution of a solute by normal segregation is given in Fig. 6.53. Normal segregation may be expected to occur when the direction of growth is inwards, as in the columnar zone.

Coring. A more localized type of segregation occurs during the growth of the equiaxed crystals in the central zone; the crystals begin to grow with a composition corresponding to the initial concentration of

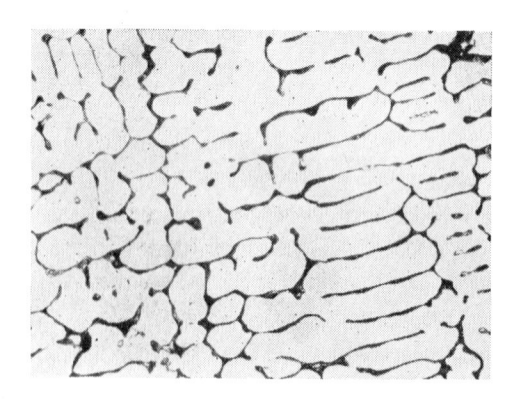

Fig. 6.54. Interdendritic segregation in iron-nickel-lanthanum alloy ($\times 335$) (Courtesy J. L. Walker, General Electric Research Laboratory.)

the liquid, but as the liquid changes in composition, as a result of the rejection of solute, the solid that is formed has a gradually increasing solute content. Such crystals, therefore, vary in composition from center to periphery, and are said to be "cored."

Interdendritic segregation. When the crystals, whether columnar or equiaxed, grow dendritically, coring will still occur but its geometry corresponds to its dendritic structure; the last liquid to freeze, between the dendrites, will have a higher concentration than the parts that were formed earlier. An example of this type of segregation is shown in Fig. 6.54. This is interdendritic segregation, and is a special case of coring.

Inverse segregation. It is sometimes found that a solute has a maximum concentration at or near the outside of a casting, its distribution being as shown in Fig. 6.55. This may be due to the outward movement of solute-enriched interdendritic liquid, which is probably caused by the contraction that takes place in regions such as A in Fig. 6.56. The only liquid that can flow into such regions is that which has already been enriched by the rejection of solute by the growing dendrites. This process can only occur when the columnar crystals grow dendritically.

Another type of inverse segregation is that which occurs when drops of liquid metal of high solute concentration are exuded into the space that sometimes develops between the mold and the metal surface. The space appears because of differential contraction of the mold and the metal in it, and the exuded drops, which are found on the ingot surface, are sucked out as the metal moves away from the mold. They originate as interdendritic liquid, which accounts for their high concentration of solute.

Gravity segregation. It is sometimes found that the composition is not the same in the upper and lower parts of a casting, apparently as a result of differences of density of the various constituents. No segregation is possible in the liquid before freezing begins, but there are two

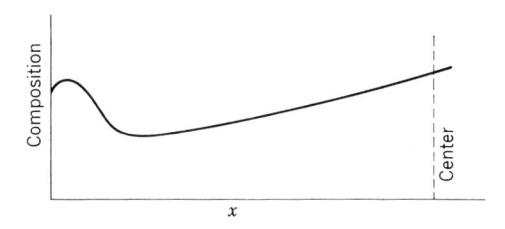

Fig. 6.55. Curve showing inverse segregation.

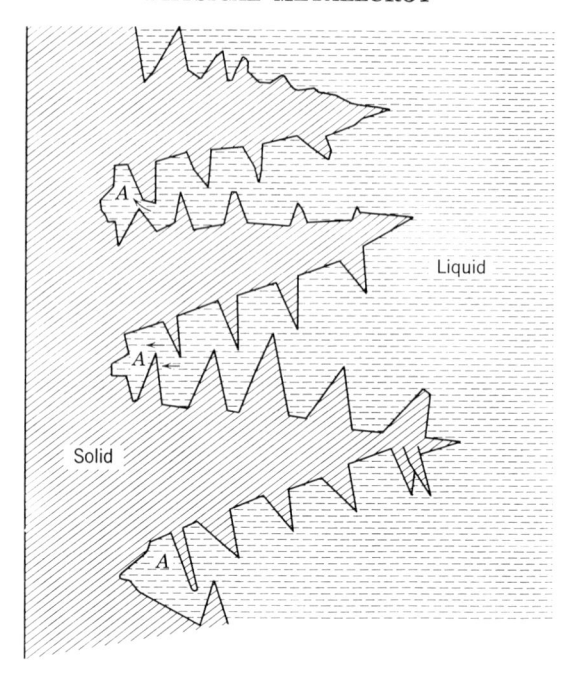

Fig. 6.56. Schematic illustration of origin of inverse segregation.

mechanisms which may contribute during freezing: (a) vertical movement of the enriched layer ahead of the interface, as a result of a difference of density due to the change of composition that is caused by rejection of solute, and (b) floating or sinking of the equiaxed crystals as a result of a difference in density between them and the surrounding liquid.

6.10 Heat Flow Considerations

It has been shown that the structure that is produced when a mass of metal solidifies depends upon its chemical composition, the rate of freezing, and the temperature gradient. The second and third of these parameters are determined by the heat flow conditions; the heat flow depends upon the size, geometry, temperature, and thermal properties of the metal and of the mold, and on the thermal resistance at the mold-metal interface. The equations relating these quantities have not been derived in the general case, but some indication of the relationships can be given.

Pure metals. The simplest case is that in which liquid metal is infinite in extent and is poured into contact with the plane surface of the mold material. Heat flow is perpendicular to the mold wall and it can be shown that the distance of the interface from the wall increases in proportion to $t^{1/2}$, where t is the time after the start of freezing (Fig. 6.57, curve A). If the mold is of circular section, the decrease of radius as the interface approaches the center causes the rate to increase, and curve B of Fig. 6.57 is obtained. The actual rates depend upon the temperatures and properties; the rate is reduced by raising the temperature of the liquid (pouring temperature) or the mold temperature, or by using a mold material of lower thermal conductivity. The analysis becomes more complicated if a gap develops between the mold and the metal; this often occurs as a result of differential contraction; the rate of freezing is slower after the formation of such a gap than it would otherwise be.

Alloys. The problem is more complicated and more important in the case of an alloy, because the nucleation of equiaxed crystals depends upon the supercooling in the remaining liquid, and this depends upon the temperature of the advancing dendritic interface and on the temperature gradient in the liquid. Qualitatively it may be expected that the same conditions that produce slow advance of the interface should also produce a long columnar zone.

6.11 Applications of Solidification

There are three main categories under which the applications of freezing processes may be considered; they are: 1. the production of ingots or billets, which are to be fabricated by plastic deformation; 2. castings, in which the solidified metal is used without subsequent de-

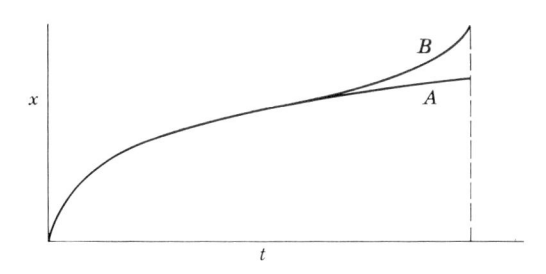

Fig. 6.57. Position of interface as a function of time; A, semi-infinite liquid, B, cylindrical mold.

formation (this does not exclude machining operations); 3. special purposes.

Ingots. Ingots are made either by pouring the metal into a mold and then allowing it to freeze, or by a "continuous" process in which metal is added continuously while freezing is taking place.

Since an ingot is intended for subsequent fabrication by deformation, the "as-cast" structure is no longer to be found in the finished product; the structure is therefore important in relation to the processes of fabrication rather than to the properties of the final product. The most important aspects are continuity and segregation; large pores are detrimental and may cause cracking during fabrication; segregation may limit the temperature at which hot working can be carried out because of the presence of material with a low melting point, and may also give rise to heterogeneity in the properties of the final product.

A large pipe in an ingot is undesirable, because the material surrounding it must be cut off before deformation processes can be conducted; this is uneconomic as it may require a substantial amount of metal to be rejected. Piping can be controlled by imposing conditions which cause solidification to be directional, i.e., from the bottom upwards. This can be achieved to a useful extent by using a "hot-top," which is a mass of large thermal capacity at the top of the mold. The hot-top is insulated to reduce its rate of cooling. This prevents cooling of the mold cavity proper from above and so increases the tendency for directional cooling. A second function of the hot-top is to contain liquid metal that prevents the formation of a pipe by feeding liquid metal into the region of shrinkage.

CONTINUOUS CASTING. If the formation of a pipe is avoided, there may still be porosity in the equiaxed region as a result of some liquid being sealed off, and therefore not being "fed" during the final stages of freezing. The only certain way of avoiding this is to eliminate the equiaxed region; this can be achieved by continuous casting, which also tends to reduce segregation.

In continuous casting, the molten metal is poured continuously into a very shallow mold, which is water-cooled, and the ingot, itself water-cooled, is withdrawn continuously downwards. The already formed ingot constitutes the bottom of the mold, Fig. 6.58. In this way, a steady state is maintained in which segregation is nearly eliminated, piping cannot occur, and the structure is usually columnar throughout. Adhesion of the ingot to the mold is prevented by the contraction of the ingot as it cools. The ingot is periodically cut off while it is moving continuously downward. Because the mold and the ingot are cooled from the outside, the interface is roughly parabolic in shape, as

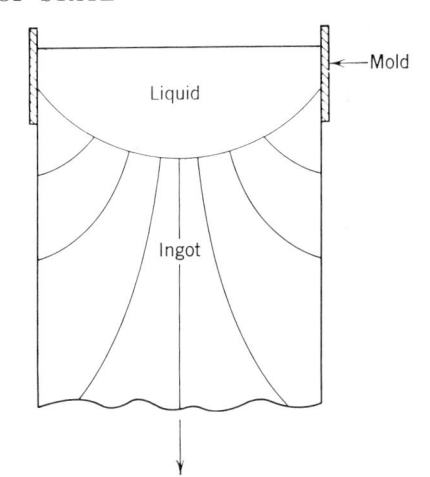

Fig. 6.58. Continuous casting (schematic).

shown in Fig. 6.58. When the process is conducted correctly, the temperature gradient in the liquid is so steep that there is no constitutional supercooling, and the structure is therefore columnar. Since the grain boundaries tend to form perpendicularly to the interface, the columnar structure is as shown in Fig. 6.58. Under ideal conditions, there should be no porosity, and the only segregation should be lateral, the outermost regions containing the least solute. If the process is speeded up, the interface becomes more sharply curved and liquid may exist in the interior well below the bottom of the mold. In extreme cases this may lead to the entrapment of liquid metal, resulting in porosity, or, in extreme cases, breakout of liquid through the "skin." Precise control of temperature and other conditions becomes increasingly necessary at higher speeds.

ARC MELTING. The processes discussed above are suitable for those metals and alloys for which there are suitable refractories in which the metal can be melted and held while it is heated to the pouring temperature. There are some metals and alloys for which no satisfactory refractory has been found; these include titanium and its alloys. The preparation of an ingot of such metals is usually achieved by melting the metal in an arc and allowing it to drip into a water-cooled mold, usually of copper on account of its high thermal conductivity. The arc has also the function of keeping molten some of the metal in the mold so that the liquid that "drips" in is immediately united with it. The arc may be between a rod of the metal that is to be cast and the metal in the mold (consumable electrode) or it may be between a tungsten or carbon electrode and the melt (inert electrode). The whole operation is usually conducted in a vacuum or an inert gas because the metals for which this process is necessary react with oxygen or with nitrogen.

VACUUM MELTING. It has been shown that the presence, in solution or as a compound, of oxygen in many alloys has a detrimental effect on their properties. Consequently, alloys for many special purposes are melted and cast into ingot form in a vacuum, so that no oxygen is added and any oxygen that is already present may be removed.

Castings. Castings are made by pouring molten metal into a mold so shaped that when the solid metal is extracted it has the required shape and dimensions. The dimensions of the mold must be greater than those required for the finished casting by the amount that the metal contracts *after* freezing. In the case of copper this amounts to about 2%. In addition to the cavity in which the finished casting will be formed, the mold must also contain a way for the metal to enter the mold (sprue) and a reservoir to hold molten metal to feed the casting when shrinkage takes place (riser). The sprue and the riser may be the same channel through the mold or, alternatively, the sprue may lead into the bottom of the mold cavity while the riser enters it from above. In a large or complicated mold, the *sprue* may divide into several *runners* that enter the mold at various points. The top of the sprue is usually widened out to form the *pouring basin*. The *gate* is a term that is applied to the whole of the pouring basin, sprue, and runners.

In addition to introducing the molten metal into the mold, it is also necessary to provide for the escape of the air or other gas that is originally present in the mold or is rejected by the metal during freezing. The gas may escape through the mold wall (if it is porous), or through a riser.

The design of castings to avoid porosity and shrinkage cavities depends upon the factors discussed on p. 277; the complete avoidance of these defects can only be achieved by causing freezing to occur progressively towards the riser. This is sometimes difficult because a thin section freezes completely sooner than a thicker section that started at the same temperature. Some control can be exercised by the use of chills, which are metal inserts in the mold. They remove heat more quickly than the mold material itself would, and therefore increase the speed of freezing locally.

A second purpose in the design of castings is to avoid the application of a stress to a metal when it is nearly solid but still contains a small amount of liquid; deformation at this stage may cause a "hot tear" by separating parts of the solid to such an extent that the limited amount of remaining liquid cannot rejoin them.

Different alloys vary greatly in their suitability for casting. The freezing range (i.e., difference between liquidus and solidus temperatures) is important; an alloy with a large freezing range is subject to severe segregation and porosity. The "fluidity" of an alloy is another important property; it determines the ease with which the liquid metal can flow into a complicated mold. It depends in part on the freezing range and in part on the viscosity of the liquid metal.

The mold may be either expendable, a new one being made for each casting, or permanent. Expendable molds are either sand molds or investments.

SAND MOLD CASTING. A sand mold is made in two parts, each part being molded on a pattern which represents the appropriate part of the required shape. When the two parts are put together the complete mold cavity is formed. A limitation is that it must be possible to withdraw the pattern section from the half mold. Cores, also of sand, are used for features that cannot be molded in this way.

INVESTMENT CASTING. In investment casting, the pattern, including gates, is made in one piece in wax or other readily fusible material. The wax is then immersed in a ceramic slurry that sets, forming a mold that completely surrounds the pattern or patterns including gate. The wax or other pattern material is then melted out, and the mold cavity is prepared to receive the molten metal by a suitable baking treatment. In this case the pattern, as well as the mold, is destroyed, and it is therefore necessary to make patterns accurately and easily. This is achieved by injecting the wax into a permanent steel die that can be opened into two halves. This imposes essentially the same limitation as the necessity for extracting the half patterns from a sand mold, except that wax components can readily be joined together to form a more complicated pattern. The investment method, sometimes referred to as "lost wax casting" or "precision casting," can produce castings with high dimensional precision (0.1%) and good surface finish. With suitable ceramic mold materials, it can be used for alloys with very high melting points. The air originally in the mold escapes mainly by diffusion through the investment, which must therefore be porous.

SHELL MOLDING. An alternative method for making an expendable ceramic mold is to coat the pattern with a number of thin layers of a ceramic, by repeatedly dipping it into a slurry and drying it. The subsequent procedure is similar to that for investment casting.

PERMANENT MOLD CASTING. Permanent molds are of metal, usually steel, and are used to produce castings when the quantity required is sufficient to justify the expense of a permanent mold. The higher thermal conductivity of the metal mold, as compared with sand, causes more rapid cooling and hence produces a finer grain size and better mechanical properties. Better surface finish and higher precision are attainable in permanent mold castings than in sand castings, and it is possible to use much thinner sections if the metal is forced in under pressure, as in pressure die casting. Another advantage of die casting is the high speed of production that is possible.

Special-purpose casting operations

JOINING. All fusion welding, brazing, and soldering operations depend upon the production of a pool of molten metal, its junction with the two pieces to be joined, and its subsequent freezing. In welding, the molten metal is either identical with or similar to the metals to be joined; it may be part of one or both or it may be added as a filler; it may be melted by electrically generated heat, either by resistance or by an arc, or it may be melted by externally applied heat in the form of a flame.

Subsequent freezing takes place as a result of the extraction of heat by the cooler surrounding metal; it usually takes place very rapidly, but it is sometimes found to be desirable to slow it down by maintaining the supply of heat to a controlled extent. Many of the problems that arise in welding are a result of the high stresses that result from heating and cooling localized regions in a piece of metal. Rapid freezing of the weld sometimes produces material with insufficient ductility to accommodate the cooling stresses, and a cracked weld is the result. Welding techniques have been developed that minimize both of these contributory factors.

Brazing and soldering are joining processes in which the joint is made by means of a different metal or alloy, which has a lower melting point than the materials to be joined. The process is called brazing if the joint is made with an alloy whose melting point is above about 425°C; when the joining alloy melts below 425°C the process is described as soldering. Since the filler differs from the stock, the problem of achieving a junction is not automatically solved, as it is in welding whenever the stock melts at the region of the weld. In many cases, as for example soft-soldering steel with a tin-lead alloy, an intermediate phase of the tin-iron system must form at the surface of the steel for a bond to be formed. This will not occur unless the surface of the steel is free from oxide. The main purpose of the flux that is used in soldering is to dissolve the oxide and to prevent it from re-forming.

COATING. The commonly practiced process of coating steel with tin, zinc, or aluminum by dipping it in the molten metal is called hot dipping. Its effectiveness depends on achieving a substantially continuous thin layer of the molten metal, and causing it to freeze uniformly. The success of the process depends upon the temperatures being adjusted so that freezing takes place when the quantity of liquid metal on the surface is correct and when it is uniformly distributed. In the manufacture of tin plate, the amount and distribution of the tin are controlled by means of rollers through which the coated sheet of steel is passed while the tin is still molten, i.e., just after the sheet emerges from the bath of tin.

Another method of coating that depends on freezing is metal spraying. In this process, a shower of drops of molten metal is directed at the surface to be coated. The drops remain molten until they hit, and then freeze as a reasonably continuous coating. It is remarkable that metals that oxidize rapidly in air can nevertheless be sprayed successfully in air.

6.12 Electroplating

The process of electroplating is the formation of solid metal on a cathode from ions in a solution, the solution being replenished by metal dissolving from an anode. The metal to be plated exists in the solution as the anion or positive ion; when the positive ion reaches the cathode, it is neutralized by one or more electrons and the metallic atom so formed either occupies a site on an existing crystal structure or takes part in the nucleation of a new crystal. The potential difference existing between the cathode and the adjacent anode during deposition is the "single-deposition potential." It is higher for the nucleation of a new crystal than for the growth of an existing one, and it also depends upon the current density as shown in Fig. 6.59. The amount of metal that is deposited is related to the amount of current that flows; under ideal conditions this relationship is defined by Faraday's law (1 gram equivalent is deposited for every 96,000 coulombs), but in most cases some of the current is used in liberating hydrogen at the cathode, and the amount of metal deposited is rather less than would be predicted from Faraday's law.

Two important variables are recognized in electroplating. One is the "throwing power," which determines the uniformity of deposit that is formed when various parts of the cathode differ in distance or inclination with respect to the anode. Plating solutions are known empirically to vary in throwing power. The other variable is the uniformity or smoothness of the deposits on a much smaller scale, that which determines whether it is smooth and bright or rough and dull.

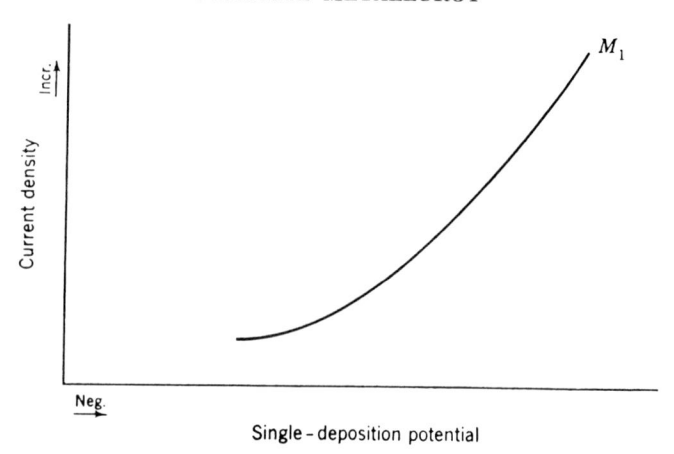

Fig. 6.59. Single-deposition potential as a function of current density. From *Modern Electroplating*, A. G. Gray, Ed., John Wiley & Sons, New York, 1953, Fig. 1, p. 74.

It is found empirically that a bright deposit can often be obtained when an organic colloid, such as a glue, is added to the bath. In extreme cases a rough deposit may develop dendritic characteristics.

When a plating bath contains two different metals, it is sometimes possible to cause them to deposit simultaneously, forming an alloy layer on the cathode. It appears that a solid solution is normally formed, although there is some evidence of the direct formation of intermediate phases.

6.13 Growth of Crystals from Vapor

It has been shown theoretically that the growth of a crystal when the neighboring phase is the vapor or a dilute solution may be extremely structure-sensitive. When the surface of the crystal is a perfect close-packed plane of atoms, the nucleation of a new layer requires a high density of atoms on the surface. This can only be achieved if the adjacent phase is one of high density (as in a molten metal) or if there is considerable supersaturation (for vapor or solution).

The theoretical rate of growth of the crystal, in a direction normal to such a face, should be related to the supersaturation as shown in Fig. 6.60, curve a. It is often found experimentally that the relationship is actually that of curve b, Fig. 6.60. The discrepancy between theory and experiment is resolved as follows: The rate of growth on

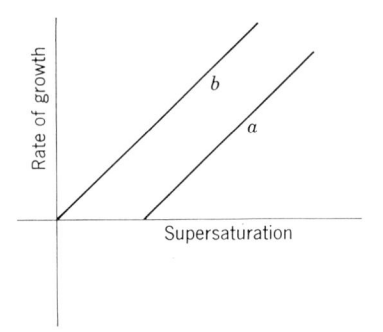

Fig. 6.60. Rate of growth of crystals from vapor.

a surface on which a growth step (Fig. 6.61) exists should correspond to curve b of Fig. 6.60, because single atoms can join the crystal at points such as A in Fig. 6.61 without the necessity for nucleating a new layer. If the crystal were perfect, the step would soon reach the edge of the crystal and growth would cease. It was pointed out by Frank that a *permanent* step must exist if a screw dislocation emerges at the surface in question. This is shown schematically in Fig. 6.62, in which the upper surface $ABCD$ is the crystal face under consideration. The screw dislocation RS produces the step STU. The addition of atoms to this step cannot eliminate it; each atom that is added causes it to move. It is "anchored" at S, and if all points (except S) on the step move at equal speeds, the step becomes a spiral. The observation of spirals on the surfaces of crystals that have grown from vapors and solutions provides convincing support for the theory. An example is shown in Fig. 6.63.

The requirement of screw dislocations does not appear to cause any difficulty in the practical processes in which metals are formed from the vapor phase. The production of solid metal by condensation of a vapor is useful for the production of very thin films for reflectors; it is also used as part of the process of purification of certain metals by distillation. The process, when used for the production of thin films of high melting point metals, is usually carried out in a good vacuum. The first requirement is that the metal be vaporized. This is often done by supporting a short length of wire on a tungsten filament that

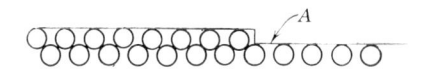

Fig. 6.61. Stepwise growth of crystals from vapor.

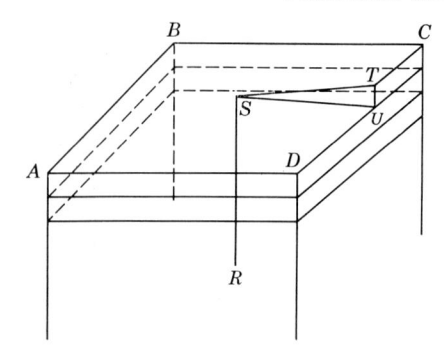

Fig. 6.62. Step due to emergence of screw dislocation.

is heated to a high temperature. The vapor then deposits on any available surface. Under these conditions, a smooth bright surface is produced.

When the process is one of distillation, an inert gas may be present. The form of the deposit may be smooth or it may be dendritic; this apparently depends upon the rate of deposition and the pressure of the inert gas. The production of dendrites is evidently a result of the establishment of a positive gradient of supersaturation, analogous to the case of freezing.

Metals in solid form are also produced by the decomposition of a

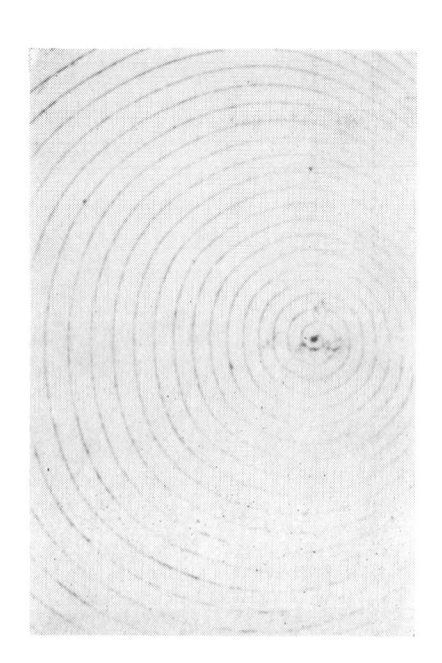

Fig. 6.63. Growth spiral on silicon carbide ($\times 60$). From S. Amelinckx, *Studie van Kristaloppervlakken*, Gent, 1955, Plate III, Fig. 12.

compound which is present in the form of vapor. This process can be used for the purification of titanium and other metals; the Van Arkel process depends on the decomposition of, for example, titanium iodide at a heated surface, on which the titanium is deposited in the metallic form.

The growth of some types of metal "whiskers" is also due to the decomposition of a suitable compound. Condensation occurs selectively on the "tip" of the whiskers for reasons that may be related to the dislocation mechanism discussed above.

6.14 Melting

The process of melting is not as difficult to understand as the process of freezing for two reasons. In the first place, there is usually no nucleation problem in melting, because the existing solid-vapor (or solid-oxide) surface can be replaced by solid-liquid plus liquid-vapor (or liquid-oxide) interfaces without an increase of free energy. Therefore melting can start at the surface of the solid as soon as the melting point is reached. Some superheating may be necessary for melting to start within a crystal, but this is a situation of little practical importance. The second reason that melting is a relatively simple process is that the product of the process (the liquid) has a structure that is independent of the conditions under which it was formed; in addition, there is little change in the distribution of the chemical species in the solid before melting takes place, and so most of the complications of the freezing process are absent.

6.15 Solution

The removal of metal from the surface of a solid metal below the solidus temperature takes place by a process of solution, as a result of which the ions that are removed go into solution in the adjacent phase, which must be a liquid. The driving force for a solution reaction may be either chemical, in which case the free energy of the system is reduced by the solution of some of the solid, or electrolytic, when the work required to raise the free energy is supplied by means of an externally applied electric current. It should be noted that even when the actual mechanism of removal of the metal is electrolytic, as it often is, the driving force is chemical, if the reaction proceeds in the absence of an external supply of energy. The processes to be considered can be conveniently divided into three groups:

1. Solution in liquid metals.
2. Solution in aqueous media without external driving force (corrosion).
3. Solution in aqueous media resulting from external driving force (electropolishing).

Solution in liquid metals. This problem, which is of increasing practical significance because of growing interest in liquid metals as coolants for nuclear reactors, can be subdivided into systems in which an intermediate phase does not form at the interface and those in which such a phase is formed.

No INTERMEDIATE PHASE. Typical of such systems is the one shown in Fig. 6.64, in which solid lead is brought into contact with liquid of composition S at the temperature T. It has been demonstrated that, in the absence of convection or other movement of the liquid, the process is controlled by the rate at which the lead, which dissolves (or melts) at the interface, diffuses away into the more remote liquid. It is also found that if the liquid is stirred sufficiently violently the process is limited by the absolute net rate of the melting process at the temperature of the reaction. The driving force for the reaction depends upon

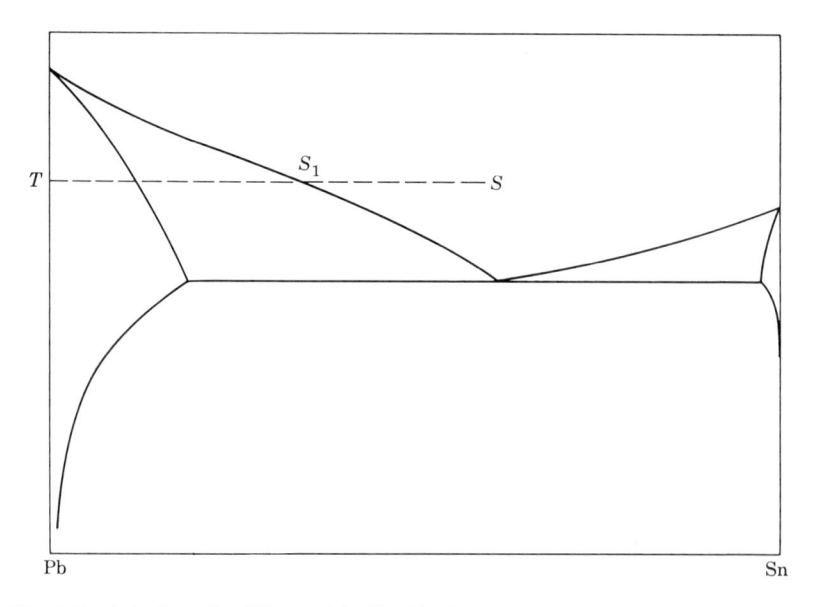

Fig. 6.64. Solution of solid metal in liquid alloy. From *Metals Handbook*, ASM, p. 1238.

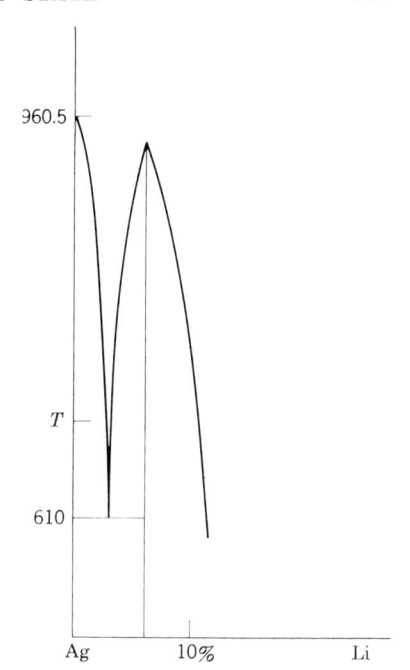

Fig. 6.65. Interaction of solid metal and liquid alloy. From *Metals Handbook*, ASM, p. 1150.

the difference between the actual composition and the composition corresponding to the liquidus, S_1.

There are some cases in which the attack by a liquid metal occurs preferentially at the grain boundaries. It is not clear whether this occurs in the complete absence of internal stress or whether it is a special case of stress corrosion (see p. 304). Examples are the intergranular attack of brass and of aluminum by mercury and of certain steels by liquid tin.

WITH INTERMEDIATE PHASE. The driving force for the reaction would not be altered if an intermediate phase, insoluble in the liquid phase, were formed. An example of such a system is that of silver-lithium, shown in Fig. 6.65. If solid silver is placed in contact with liquid lithium at the temperature T, a layer of the compound silver-lithium will immediately form. This does not represent an equilibrium situation, which would not be reached until the whole of the "minority" component was used up in forming the compound. The rate of formation of the compound, however, is limited by the rate at which silver and lithium atoms can come into contact with each other by the diffusion of one or the other through the compound layer. Simultaneously, the compound is dissolved by the liquid because equilibrium does not exist between them. This rate of solution is probably controlled by the diffusion of silver in the liquid lithium.

More complicated situations occur as a result of compound formation with an element that is present in smaller amount than in the case considered so far. Such compound layers may be very stable and cause a marked decrease in the rate of solution.

If the temperature of a solid-liquid metal system were uniform and constant, no more solution would occur once equilibrium had been established. However, the use of liquid metals as coolants requires that different parts of the system should be at different temperatures, and it is impossible for equilibrium to be achieved at all points in such a system because the liquidus line always slopes. There is, therefore, always a tendency for the solid to go into solution at the higher temperature and for it to redeposit in parts of the system that are at a lower temperature. The success of such a system may depend more on the suppression of deposition than on the suppression of solution. This is inherently probable because the deposition process is usually dependent on nucleation while the process of solution is not.

Solution in aqueous media. It has been established that the direct solution of metals in most liquids is not an important process. The process in which local electrolytic cells play an important part is much more significant, and consequently attention here will be concentrated mainly on such cases.

ELECTROCHEMICAL POTENTIAL. If any metal or alloy is in contact with a conducting aqueous or other solution, positive ions go into the solution until the resulting electromotive force (emf) is sufficient to prevent any further solution. In order to measure this emf, it is necessary to use a second electrode (also in the solution). The resulting measurement gives the difference between the solution potentials of the two electrode materials with respect to the solution that is used. It is usual to use a "hydrogen electrode" as the standard, and Table 6.5 shows the electrochemical potentials of a number of metals in an aqueous solution, based on the hydrogen electrode as zero. If two metals are immersed in the same electrolyte, a difference of potential equal to the difference between their respective values will exist between them.

If two dissimilar metals, such as zinc and copper, are dipped into water containing sufficient impurities to be reasonably conducting, a potential difference exists between them but very little chemical action takes place unless the two metals are in electrical contact. If they are connected, a current flows through the circuit. The strength of this current is determined by Ohm's law, and therefore depends upon (a) the emf and (b) the resistance in the circuit. The emf is that due

CHANGE OF STATE 299

Table 6.5. Electrochemical Potentials of Some Metals

Metal	Normal Electrode Potential (volts)	Metal	Normal Electrode Potential (volts)
Cesium	−3.0	Iron	−0.44
Lithium	−3.0	Cadmium	−0.40
Potassium	−2.9	Nickel	−0.25
Calcium	−2.9	Molybdenum	−0.2
Sodium	−2.7	Tin	−0.14
Magnesium	−2.38	Lead	−0.13
Titanium	−1.75	Hydrogen	0.000
Beryllium	−1.70	Copper	+0.52
Aluminum	−1.67	Mercury	+0.80
Uranium	−1.40	Platinum	+1.2
Zinc	−0.76	Gold	+1.4
Chromium	−0.56		

to the difference in the solution potential of the two metals, minus any "back emf" due to hydrogen, and the resistance is the total resistance of the whole circuit. Initially there is no back emf and the flow of current through the electrolyte takes place as follows.

In Fig. 6.66 a cell is shown in which zinc goes into solution as position ions, leaving an excess of electrons at the zinc electrode; the excess positive charge in the solution is neutralized by the arrival of positively

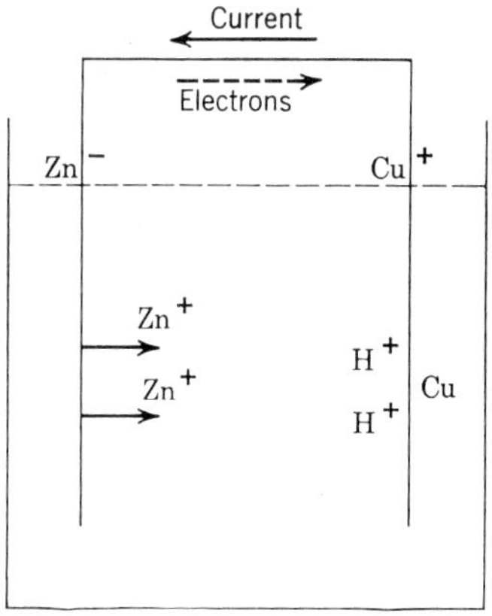

Fig. 6.66. Electrolytic cell.

charged hydrogen ions at the copper electrode. The hydrogen ions gather sufficient electrons from the copper electrode to become electrically neutral. The *external* current is therefore a flow of *electrons* from zinc to copper, or a *current* from copper to zinc. Any hydrogen ions that exist at the surface of the copper contribute to an emf in the reverse direction, i.e. back emf, or decrease the net emf of the cell. The second effect of hydrogen present on the surface of the copper is to increase the resistance of the circuit. The "polarization" of the cell therefore reduces the current and consequently it reduces the rate at which zinc leaves the zinc electrode. The amount of zinc that goes into solution is given by Faraday's second law, which states that one electrochemical equivalent is transported by 96,688 coulombs of electricity. The loss of zinc by the zinc electrode is an example of the type of corrosion which can occur wherever two metals with different electrochemical potentials are in contact with each other and with an electrolyte.

The type of electrolytic corrosion discussed so far is the easiest to understand; however, there are other sources for the emf besides a gross difference between the chemical compositions of the two electrodes. The emf may be present as a result of:

1. Differences in chemical composition, either due to segregation or to the presence of different phases.
2. Differences in crystal orientation.
3. Differences in strain, either applied or residual.
4. Differences in the composition of the electrolyte.

It follows that different parts of the same piece of metal may develop an emf which could cause one part to corrode by the mechanism

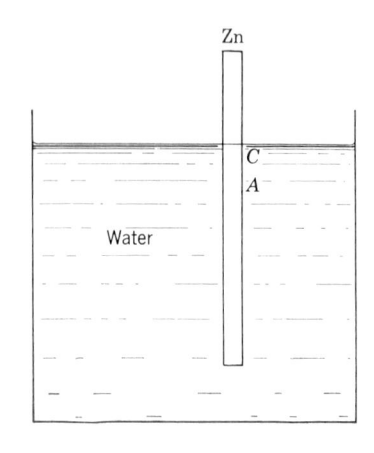

Fig. 6.67. Differential aeration cell.

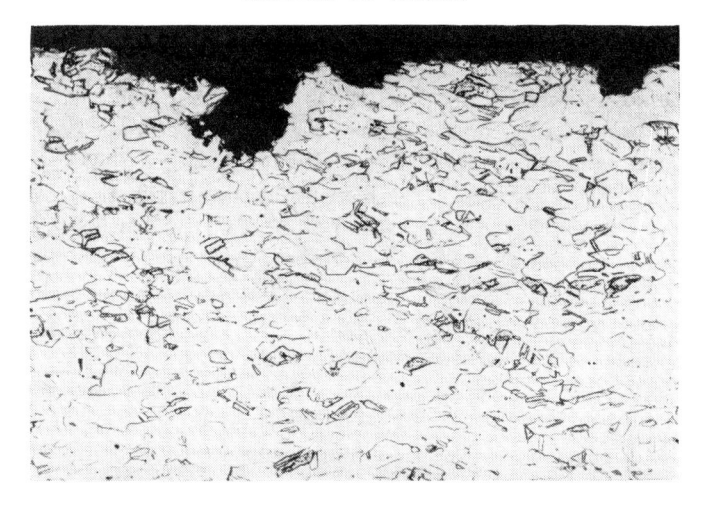

Fig. 6.68. Photomicrograph showing cross section of corrosion pit (brass, ×146). (Courtesy Chase Brass and Copper Company.)

discussed above. Another important case is that in which the oxygen content of the electrolyte is not uniform. Consider a zinc plate partly immersed in water (Fig. 6.67). The oxygen content of the water near the surface is higher than elsewhere, and this causes the zinc in this region to be cathodic compared with the zinc that is further away from the surface. The zinc at A is therefore anodic and a corrosion current flows; the metal at A goes into solution, and hydrogen is released at C. The plentiful supply of oxygen at C oxidizes the hydrogen, and this keeps the cell depolarized, which allows the corrosion to proceed relatively rapidly, although only one metal is present.

A similar process encourages the formation of corrosion pits (Fig. 6.68). As the depth of a pit increases, the difference in oxygen concentration of the electrolyte increases because the supply of oxygen depends on diffusion in the electrolyte.

INHIBITORS. Since electrolytic corrosion requires the action of both an *anode* and a *cathode*, it should be possible to prevent corrosion by suppressing either the anode reaction or the cathode reaction. The anode reaction can be reduced or stopped by the use of an inhibitor that produces an insoluble corrosion product. Chromates, for example, are sometimes used to produce insoluble chromate films on iron, magnesium, etc. A danger that arises is that if the area of anode is reduced, but not eliminated, corrosion is accelerated in the region that remains. Cathode inhibitors are intended to prevent access of the oxygen that causes depolarization.

SACRIFICIAL PROTECTION. The protection of a functionally important metal component by a second metal that is anodic to it is widely practiced; "galvanized iron," which is zinc-coated steel, is corrosion-resistant because the zinc is anodic to the steel and corrodes sacrificially by the mechanism discussed above. It may be noted that tin plate is corrosion-resistant only because the tin forms a very nearly continuous coating on the steel. Tin is cathodic to steel, which would therefore corrode preferentially if the two metals formed an electrolytic cell.

PROTECTION BY "INSULATION." The most obvious method of preventing corrosion is to prevent access of the electrolyte to the metal, and this is achieved, with varying success, by means of paint and other nonmetallic coatings. A second method is to protect a metal or alloy that is inherently vulnerable by one which is much less so. For example, many of the stronger aluminum alloys are very readily corroded, perhaps because of the inherent heterogeneity of an alloy consisting of more than one phase. They are often protected by a thin surface layer of a much purer aluminum.

There are many metals and alloys in which the naturally forming oxide skin affords good protection from corrosion so long as it is maintained. For example, aluminum is very corrosion-resistant except in alkaline solutions which dissolve the oxide; stainless steel (18–8) is very corrosion-resistant except in chloride and certain solutions which either dissolve, or do not restore, the chromium oxide film which is formed under oxidizing conditions.

TYPES OF CORROSION. The importance of a corrosion process may depend to a large extent on the geometry of the attack: it may be uniform over the whole surface of a component, it may occur as distinct pits, or it may occur preferentially at grain boundaries, welds, etc. For this reason it is often dangerous to measure corrosion rates by studying the rate at which metal is lost; the loss of strength or the rate of growth of corrosion pits may be more realistic criteria.

Uniform Attack. If the whole of the exposed surface of a metal undergoes corrosion simultaneously, it is evident that the process cannot be purely electrolytic; it may be purely chemical, in the same way that the solution of salt in water is chemical. The rate of this kind of attack depends greatly on the chemical nature of the metal and on the composition of the electrolyte, and may be controlled by the formation of a layer of a compound between them. Such layers may be formed deliberately by the use of a suitable reagent. Examples are the "passivation" of iron by the formation of an oxide film by contact with concentrated nitric acid, and the protection of magnesium from aqueous corrosion by the formation of a film of chromate on the surface.

The process of uniform corrosion is often the formation of an oxide or hydrated oxide, which may dissolve in the adjacent liquid. If so, then corrosion will proceed indefinitely. An example of this type of corrosion is that of aluminum or stainless steel in water containing chlorides, such as sea water.

Since the metals at the "lower" or positive end of the electrochemical series have lower solution potentials than hydrogen, they do not tend to replace hydrogen in the electrolyte, and therefore they exhibit less tendency to form oxides. Hence they are inherently more corrosion-resistant in the absence of electrolytic action, as well as being cathodic and therefore protected, in an electrolytic cell. Inherent corrosion resistance therefore follows much the same order as the electrochemical series.

Pit Formation. Pit formation is electrolytic in origin, as explained in the section on "Electrochemical Potential"; however, a sufficiently dispersed multiple pit formation can produce a result that looks similar to general corrosion. An example is that of magnesium at the level of purity at which it contains finely dispersed iron particles. Since iron is cathodic to magnesium, the magnesium surrounding each iron particle is removed electrolytically; the immediate result is shown schematically in Fig. 6.69. If the iron particles are sufficiently far apart, electrolytic action ceases when all the exposed iron particles are separated from the magnesium, either by falling out or by the formation of an insulating layer of corrosion product. If each pit exposes further iron particles, the electrolytic corrosion continues. This explains the extreme sensitivity of magnesium to small traces of iron. The effect of iron on the corrosion rate of magnesium is shown in Fig. 6.70.

Intergranular Corrosion. The third type of corrosion is that which is concentrated at crystal boundaries. It is frequently difficult to detect except by its effects on the mechanical properties, which are to reduce the ductility very drastically and consequently to reduce the effective strength. This is due to the stress concentrations that exist at the limits of the attack and to the reduction in the effective cross section of the specimen. The mechanism is presumably electrolytic, the material at the boundary being out of equilibrium with the remainder of

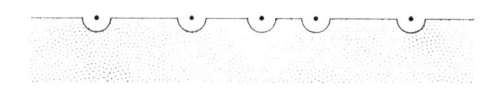

Fig. 6.69. Corrosion due to particles of a second phase.

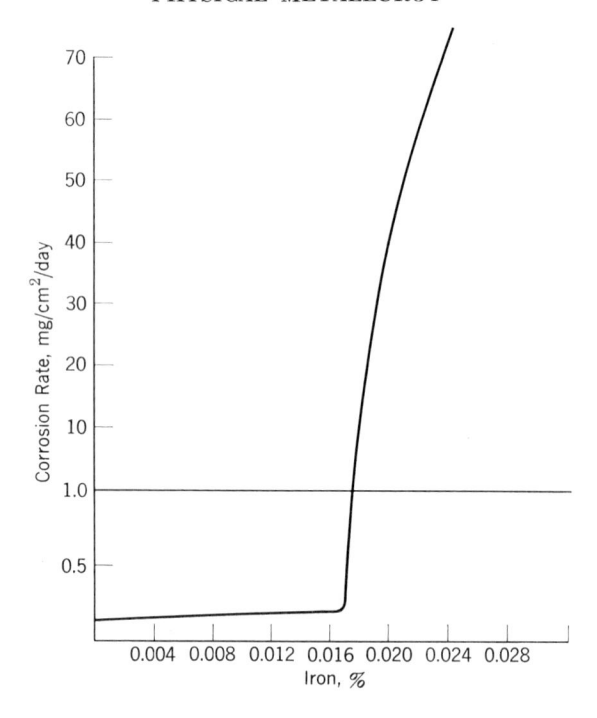

Fig. 6.70. Effect of iron on the corrosion rate of magnesium. After Hanawalt et al., Trans. AIME, Vol. 147, 1942, p. 299.

the surface. An example of intergranular attack is that of incorrectly heat-treated stainless steel; a typical case is shown in Fig. 6.71.

STRESS CORROSION. Corrosion is in some important cases greatly aggravated by the application of a stress. This may be either an externally applied stress, or it may be a residual stress resulting from fabrication, heat treatment, plastic deformation in use, or a change of temperature in metal that is anisotropic in thermal expansion.

An example of the combined effect of stress and corrosion is shown in Fig. 6.72. It is believed that the attack is electrolytic and that the stress concentration due to the existence of some localized attack leads to more rapid propagation of the cracks. It is not clear whether the plastic strain to be expected ahead of the cracks contributes to the process by causing the metal to be locally more anodic than the surrounding metal. It is found that many solutions will cause general corrosion but not stress corrosion, which therefore depends on the presence of specific chemical species. Stress corrosion cracking is sometimes transgranular and sometimes intergranular. The classical case

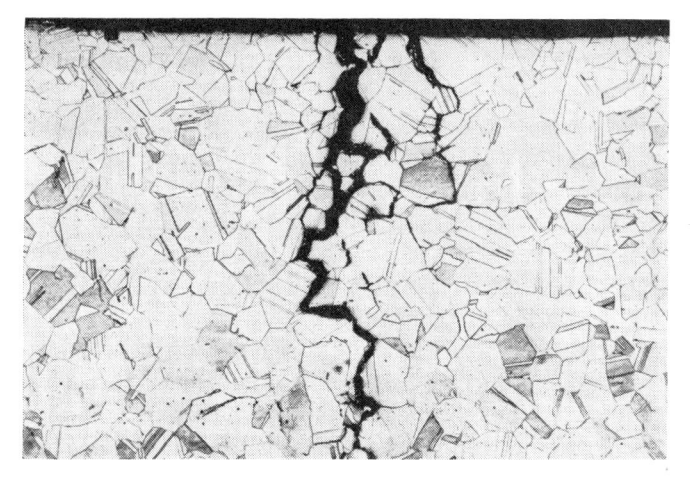

Fig. 6.71. Photomicrograph showing intergranular corrosion (brass, ×146). (Courtesy Chase Brass and Copper Company.)

of intergranular stress corrosion is the "season cracking" of brass. Brass with some internal stress will eventually crack if exposed to ammonia or to mercury. If the brass is given a stress-relieving treatment (see p. 326), it will not crack. Other cases of stress corrosion occur in high-strength steels and some aluminum alloys when exposed to sea water while under stress.

CORROSION FATIGUE. Another combined effect of stress and corrosion is corrosion fatigue, in which the presence of a corroding medium during the application of alternating or fluctuating stresses greatly accelerates failure, which is generally intergranular. The mechanism is not understood, but it seems to be at least partly electrolytic, since the process does not occur if the sample is the cathode of an electrolytic cell.

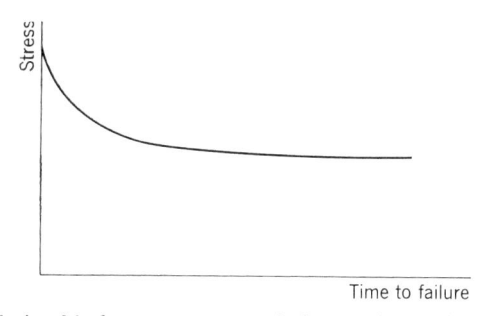

Fig. 6.72. Relationship between stress and time to failure in stress corrosion.

EFFECT OF TEMPERATURE ON CORROSION. In general, the effect of raising the temperature is to increase the rate of corrosion. This is because chemical processes such as solution proceed more rapidly at high temperatures. In addition, solubilities generally also increase. However, there are some cases in which an increase in temperature may drastically change the whole process; for example; in some exceptional natural waters, magnesium actually becomes cathodic to iron at temperatures approaching the boiling point of water. It is not, therefore, necessarily possible to accelerate a corrosion test by conducting it at a higher temperature.

Electropolishing. The electrolytic removal of metal can be controlled by passing a current from an external source through the cell instead of allowing the cell to generate its own current as in the typical electrolytic corrosion case. It has been found that the anode of such a cell may be *smoothed*, by eliminating irregularities on a relatively large scale, i.e., above a micron, and *brightened*, by removing irregularities down to about a hundredth of a micron in size. These two processes, together termed electropolishing, only occur when a suitable electrolyte is used in combination with a carefully controlled current density. It is thought that the smoothing action is due to the formation near the anode of a thick layer of a viscous liquid that has a relatively high electrical resistance. More current passes through the thinner parts of the film, allowing more rapid dissolution of the "hills" than the "valleys." The origin of the "brightening" action is not clear. The details of the electrolyte, current density, etc., vary from one metal to another and have been determined empirically. The process is used as a laboratory technique for preparing undeformed metal surfaces, and industrially as a method of polishing, and as a method of removing carefully controlled amounts of metal (electromachining).

Chemical polishing. Somewhat similar results can in some cases be obtained without the use of an external emf simply by immersing the metal in a suitable solution; for example, aluminum can be chemically polished by a solution of 70% (by volume) orthophosphoric acid.

6.16 Further Reading

J. H. Hollomon and D. Turnbull, "Nucleation," Chapter 7 in *Progress in Metal Physics 4*, Bruce Chalmers, Ed., Pergamon Press, London, 1953.

U. M. Martius, "Solidification of Metals," Chapter 5 in *Progress in Metal Physics 5*, Bruce Chalmers and R. King, Eds., Pergamon Press, London, 1954.

Liquid Metals and Solidification, American Society for Metals, Cleveland, 1958.

7

Deformation, Radiation Damage, and Recovery Processes

It has been shown in Chapter 6 that a process involving change of state can be used to produce almost any required shape in a sample of metal, and that some control of the structure, and therefore of the properties, is also possible. The major limitation of these processes is that the structure can only be controlled to a very limited extent. In particular, the crystals are necessarily completely free from work hardening, and their size is often not as small as would be desired.

The purpose of this chapter is to discuss the changes that can occur in metals without a change of state or of phase. There are three: changes due to plastic deformation; those due to irradiation by neutrons or other energetic particles; and the return towards the equilibrium structure which may occur after the structure has undergone plastic deformation or radiation "damage."

7.1 The Structure of Deformed Metals

One of the properties that contributes most to the engineering usefulness of metals is high strength combined with the capacity to undergo substantial permanent deformation without fracture. This property of ductility allows metals to withstand abnormal stresses in service without catastrophic failure, and it also makes it possible to produce useful shapes by plastic deformation.

As shown on p. 216, plastic deformation is usually limited either by the type of instability that results in the formation of a "neck," or by fracture, which occurs when the force which must be applied to cause further deformation exceeds that which causes fracture. The limitation of ductility due to "necking" is closely related to the geometry of the specimen and the stress system, and is not an inherent property of the material; the intersection of the flow and fracture curves is inherent, and can only be modified by varying the conditions, usually temperature or pressure, or the type of stress (triaxial instead of uniaxial). In general, therefore, the amount of deformation that can be achieved by the application of a given type of stress is a property of the material; it is a structure-sensitive property. In addition to changing the shape, plastic deformation modifies the structure and this in turn affects the properties. The mechanisms of plastic deformation and the resulting changes in structure and properties have been discussed in Chapter 5. Further deformation almost always becomes more difficult as a result of plastic deformation.

Dislocation content. The changes in structure may be summarized as follows. In the "as-cast" condition, a metal or alloy contains a number of dislocations that is usually between 10^4 and 10^7/cm^2. The number may depend on the amount of solute that is present, and on the conditions that existed during freezing. A plastically deformed metal has a greatly increased content of dislocations, ranging from 10^{11}/cm^2 for a moderately deformed sample to 10^{13}/cm^2 in extreme cases. These are the dislocations that remain in the metal when the deforming force is no longer applied; there is probably a relatively small decrease concurrent with the release of the force. The dislocation loops resulting from deformation are arrayed in their slip planes; they are prevented from retreating into their sources by one or more of the following mechanisms: they may have jogs as a result of intersection with other dislocations, they may be "locked" by interaction with other dislocations, or their source may have moved out of the slip plane in which it originated the loops, or parts of the loops may have emerged at a free surface or into a crystal boundary. In all cases in which the work hardening is severe there are arrays of dislocation on several intersecting sets of slip planes, although some work hardening occurs when only a single set of slip planes is activated. In addition to a high content of dislocations, there is also a considerable excess of vacancies, as a result of the movement of jogs in directions that do not correspond to slip.

Diffraction effects. The accumulation of large numbers of dislocations on parallel slip planes causes local curvature of the crystal lattice as shown in Fig. 7.1. It follows that there must be some spread in the orientation of what was previously a crystal with a uniform orientation. This spread in orientation is apparent in X-ray diffrac-

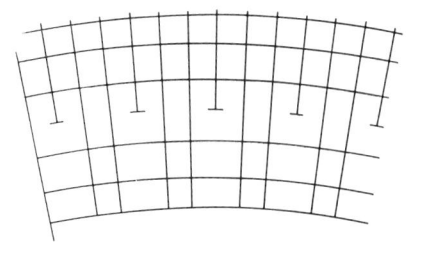

DEFORMATION, RADIATION DAMAGE, AND RECOVERY 309

Fig. 7.1. Lattice curvature due to dislocation pile-up.

tion studies as a broadening of the rings, or in a powder pattern, or as "asterism" of the Laue spots. Figure 7.2 shows an example of asterism resulting from deformation.

Shapes of crystals. In most examples of plastic deformation, the final shape of the sample differs from the original shape, and it follows that the individual crystals must change in shape in a generally similar manner. It is not to be expected that the change in shape of each individual crystal will be geometrically similar to that of the sample, because of the anisotropic character of the deformation process. Unless deformation is extremely severe, the original crystals are still recognizable if the sample is examined metallographically after deformation. An example is shown in Fig. 7.3.

Two-phase alloys. When the sample consists of more than one phase, the changes may be more complicated; if a second phase consists of hard particles, they may remain unchanged in shape, being redistributed in position within the surrounding softer phase that is deformed. If, however, a second phase is soft or weak, its shape may

Fig. 7.2. Asterism resulting from plastic deformation. From *Elements of X-Ray Diffraction,* B. D. Cullity, Addison-Wesley Publishing Co., Reading, Mass., 1956, Fig. 8-28(a), p. 246.

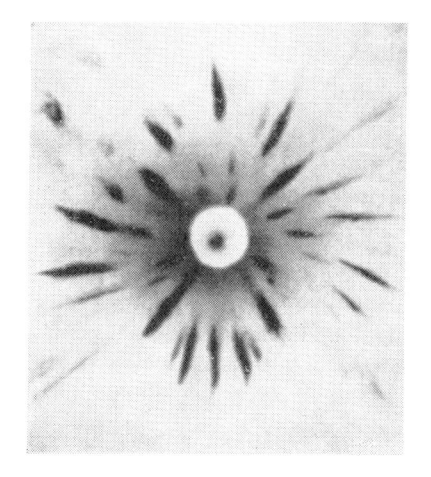

(a)

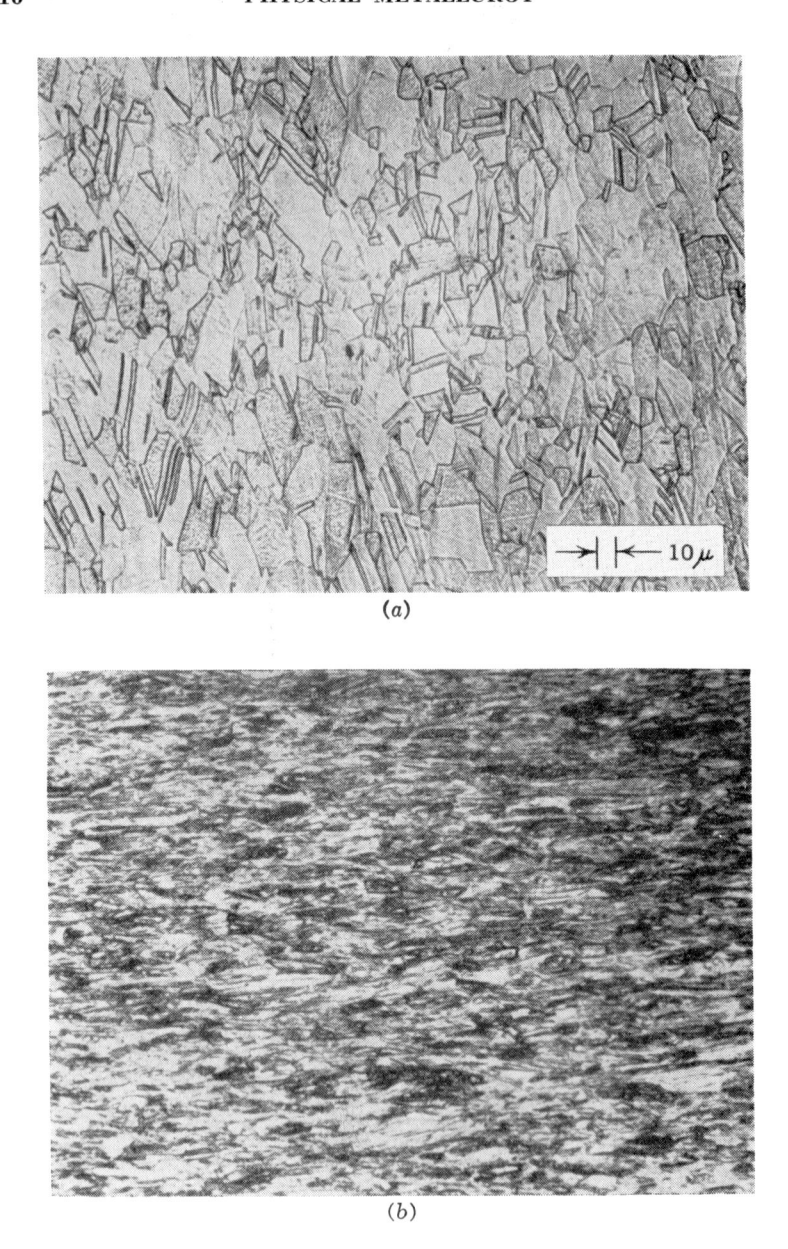

(b)

Fig. 7.3. Photomicrographs showing change of crystal shape of α brass by deformation. (a) Before deformation ($\times$230). (b) After deformation by rolling ($\times$230). (Courtesy P. Doherty and H. McP. Brown, Harvard University.)

change drastically, for example, a sphere of a soft inclusion being drawn out into a "thread" by severe deformation. The influence on properties of the resulting distribution of phases is discussed on p. 348.

Preferred orientation, deformation texture. A major effect of deformation is on the orientations of the crystals. It can readily be seen that the orientation of a single crystal must in general change if it undergoes deformation by slip. If slip is in a single direction in a single plane, as would be the case for a crystal stressed so that the axis of tension is T in Fig. 7.4, the orientation would initially change so that T moves along the great circle towards P; however, when point S is reached, a second system is equally stressed, and it would be expected that it would tend to cause T to move towards P'. The resultant of the two slip processes causes T to move along CH until the $\langle 121 \rangle$ position is reached. These purely geometrical considerations do not represent or account for the observed facts. In many cases the second slip system becomes active long before point S has been reached; however, in some cases, notably α brass, the onset of double slip is delayed beyond the point S.

It was pointed out in Chapter 5 that the interaction between adjacent crystals causes slip systems to operate that would not be sufficiently stressed by the applied stress alone. It follows that the changes of orientation that occur during deformation cannot be predicted completely from the behavior of single crystals. However, it is found that the deformation of a randomly oriented polycrystalline aggregate results in material with a preferred orientation, the crystals tending to approximate progressively to one or more well-defined orientations, which can usually be represented by simple indices. A more complete representation of the extent and kind of preferred orientation requires representation as a pole figure or a series of pole figures.

The orientation that is approached during deformation depends upon the nature of the applied stress system and on the crystal structure of the metal. Table 7.1 shows the most prominent preferred orientations for several stress systems and metals.

Fig. 7.4. Change of orientation during slip on one system.

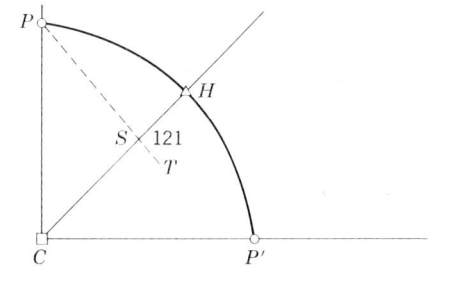

Table 7.1. Some Deformation Textures

	Uniaxial Tension or Compression	Drawn Wire	Rolled Sheet
Iron	[111] and [110]	[110]	[110] (001)
Copper	[110] normal to compression axis	[100] and [111]	[112] (110)
Aluminum	[110] normal to compression axis	[111]	[112] (110)

The oversimplification introduced by the representation of preferred orientations as in Table 7.1 is shown by the pole figures for the texture of a silicon-iron alloy reduced by 95% by rolling, Fig. 7.5. The spread of orientation about the "ideal" texture is shown by the extent of the variously shaded regions.

Work hardening. It has already been shown that an effect of plastic deformation is to make subsequent deformation more difficult; this effect, called work hardening, can be measured and represented by means of the stress-strain curve. It can also be measured (much more conveniently) by means of the hardness as a function of the amount of strain. This method can be applied even when the method of straining is such that the stress cannot be measured during deforma-

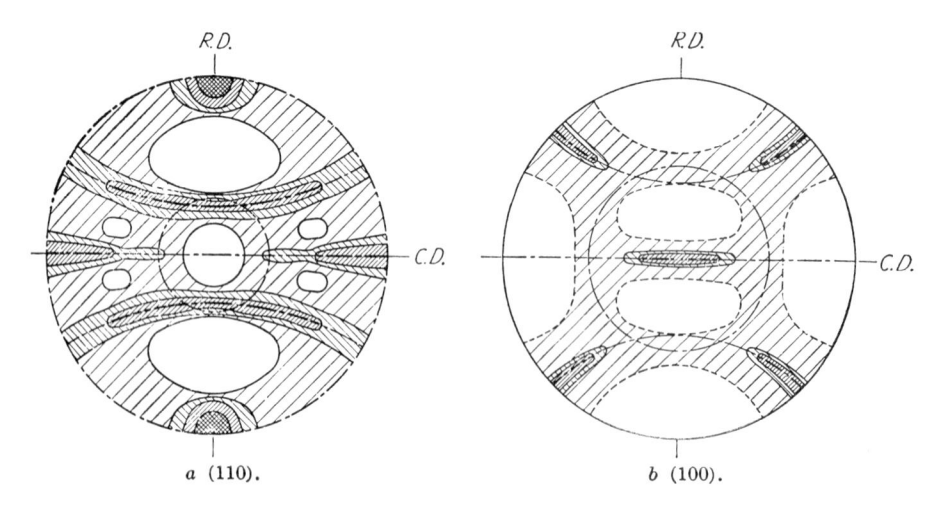

Fig. 7.5. Deformation texture (rolling) in copper. By permission from *Structure of Metals,* by C. S. Barrett, Fig. 29, p. 467. Copyright 1952. McGraw-Hill Book Co.

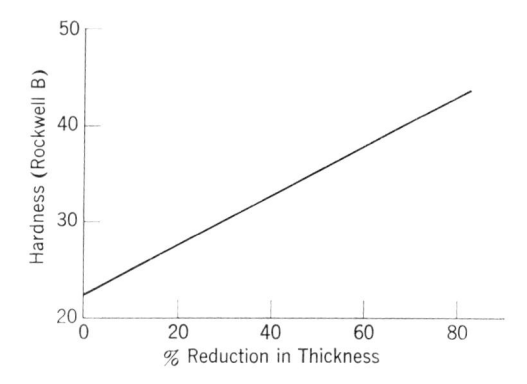

Fig. 7.6. Work hardening of copper by rolling.

tion; most of the methods used for fabrication are in this category. The criterion for the amount of strain also depends upon the geometry of the process; the percentage decrease in thickness or cross section is usually a good parameter for this purpose. Figure 7.6 gives an example of a work-hardening curve.

7.2 Radiation "Damage"

The advent of nuclear engineering has led to the existence of a new way of disturbing the structure of a metal. This is by the impingement of radiation of high enough energy to produce effects that are not accounted for by an increase in thermal energy. These effects are usually referred to as "radiation damage," although the term "damage" is not an apt description of the effects found in metals. The types of radiation that are usually considered are:

1. Electromagnetic radiation in the form of X rays or γ rays.
2. Light charged particles (electrons).
3. Heavy charged particles (protons, α particles, etc.).
4. Heavy neutral particles (neutrons).

This ignores cosmic radiation, which has not been extensively studied in relation to its effects on metals. The effect of radiation depends upon the energy, mass, and charge of the individual particle, and on the number of particles passing into or through a unit amount of the material under consideration. The energy is usually expressed in electron volts (ev, kev, or Mev); the charge in electron units, and the mass in terms of the proton (for heavy particles) or the electron.

Types of "damage." There are three major effects which radiation may have on the structure of matter: ionization, collision, and fission.

IONIZATION EFFECTS. The approach of a charged particle to an atom may cause electrons to be removed from their usual energy levels into higher ones. This may result in ionization of the atoms that are affected. This is of importance in materials in which electrons are not free to move, but in metals the electrons so displaced do not remain in nonequilibrium positions long enough for any consequences to be observable. Ionization effects in metals will therefore be ignored, except insofar as a particle is slowed down gradually by the energy it expends in displacing electrons as it goes by.

COLLISION EFFECTS. A particle moving through a crystal exerts forces not only on the electrons but also on the atoms (or ions) themselves. These forces depend in a very complicated way on the distance between the ion and the particle, but it is a reasonable and useful approximation to regard each ion in the crystal as having a *cross section*. If the moving particle passes within this cross section, there is a collision; if not, the path of the moving particle is not affected except by the energy dissipated by ionization. The cross section depends upon the speed, charge, and mass of the approaching particle, being small for fast particles and increasing as the particle slows down. The cross section is also larger for charged than for neutral particles.

Temperature Spikes. When an atom in a crystal is hit by an energetic particle, it immediately acquires an energy that is extremely high compared with its thermal energy. It is therefore not in equilibrium with its surroundings and its excess energy is quickly shared with other atoms in the vicinity. The region that is disturbed in this way is called a "temperature spike." There are two kinds of temperature spikes; namely, the "displacement spike," in which the energy is sufficient to move many atoms from their lattice sites into interstitial positions, and the "thermal spike," in which few or no atoms leave their lattice sites.

Thermal Spikes. It is of interest to consider a thermal spike from the quantitative point of view. It is assumed that the amount of energy transferred to the atom in question is 300 ev. If this amount of energy were suddenly liberated at a point in a medium with the thermal properties of copper, the outward conduction of heat would be as shown in Fig. 7.7, where curves are shown for various times after the release of the energy. It is seen that the temperatures at the outset are extremely high but that they persist for very short times.

Another aspect is that the energy assumed would be sufficient to raise the temperature of a sphere of copper containing about 1000

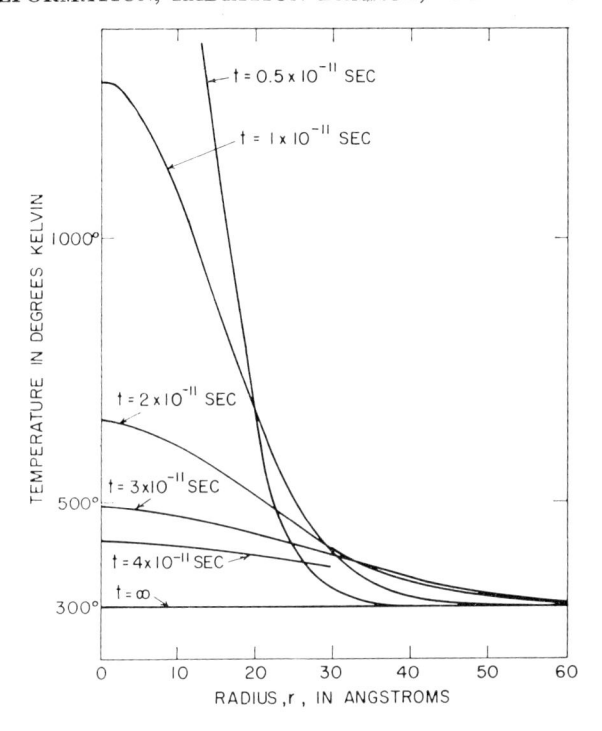

Fig. 7.7. Temperature distribution in a thermal spike as a function of time. From *Radiation Effects in Solids,* G. J. Dienes and G. H. Vineyard, Interscience Publishers, New York, 1957, Fig. 2.5, p. 35.

atoms from room temperature to the melting point. The same heat would be sufficient to *melt* about 700 atoms of copper. However, it is very doubtful if melting actually occurs, because the time during which any of the material should be above the melting point is only about 5×10^{-12} sec, and the nucleation of melting must take a finite time. As shown by Fig. 7.7, spreading and cooling is extremely rapid, the temperature being nowhere above 150°C after 2×10^{-11} sec.

The discussion so far has been in terms of spherical spikes. If, however, an energetic particle has a series of collisions that are fairly close together, the individual spherical spikes may overlap, giving a cylindrical spike. A fast neutron moving through a crystal of a heavy element may produce "knocked-on" atoms that collide with other atoms every few angstroms, and dissipate a few hundred electron volts of energy per angstrom of their path. The resulting cylindrical spike would be of the order of 1000 A in length and would have a diameter of about 25 A after 10^{-12} sec.

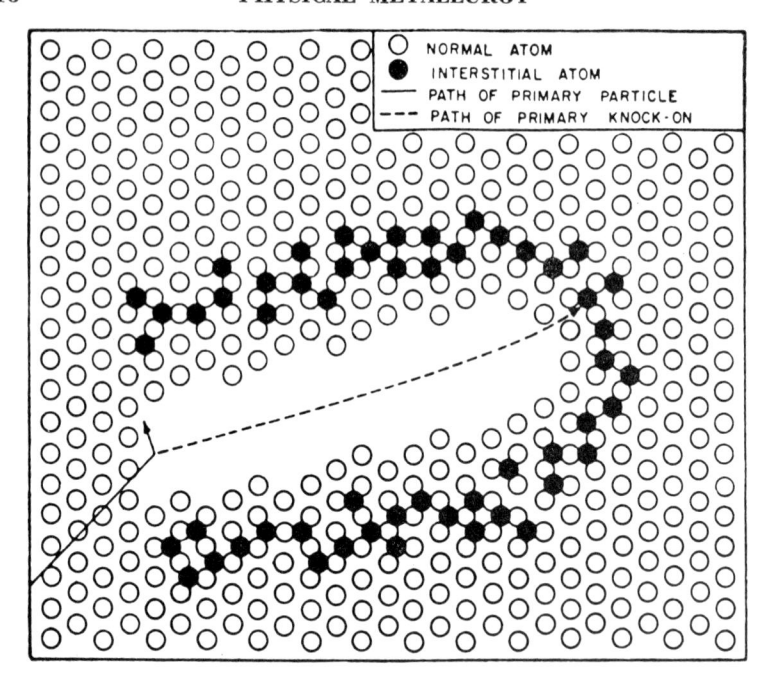

Fig. 7.8. Schematic drawing of a displacement spike (after Brinkman). From *Radiation Effects,* Dienes and Vineyard, Fig. 2.6, p. 42.

Whether melting takes place in a thermal spike or not, there must be thermal expansion, with the result that the surrounding material will be subjected to a very high stress. This may cause dislocations to be nucleated, but only in the very first stages of the formation of the spike, while the temperature gradient is still very steep. The resulting dislocation loops could disappear when the spike dies away, but it is more probable that the intersection of loops would prevent this.

Displacement Spikes. The displacement spike is postulated on the theory that the primary knocked-on atom is energetic enough to knock a considerable number of other atoms into interstitial positions. A schematic representation of the displacement spike is shown in Fig. 7.8. This type of spike would correspond to local volatilization, which would be followed by condensation into the momentarily hollow or partly hollow core. The result for a pure metal would be a region containing some dislocation loops and perhaps some interstitials and vacancies; for an ordered alloy or an intermediate phase there would also be some disturbance of the distribution of the different sorts of atoms. A displacement spike could also cause crystals of intermediate phases to be taken into solution in the surrounding material.

FISSION EFFECTS. In the types of collision discussed above, the number of atoms and their chemical nature remains unchanged. There are, however, important cases in which fission of the struck atom into two lighter atoms may occur. This is accompanied by the release of a large amount of energy, and at least in some instances, by the ejection of more fast neutrons.

Effects of radiation on properties

WITHOUT FISSION. A number of the properties of metals are changed by irradiation in which there is no fission and the effects are presumably due to temperature spikes.

Electrical Resistivity. The resistivity of metals is invariably increased by irradiation by fast particles such as neutrons, but is unaffected by ionizing radiation. The residual (structure-sensitive) resistivity (see p. 90) is the part which is affected by radiation. Since the effects of irradiation disappear spontaneously at suitable temperatures (see p. 322), it is necessary to perform the irradiation and the measurement at low temperatures in order to obtain the maximum

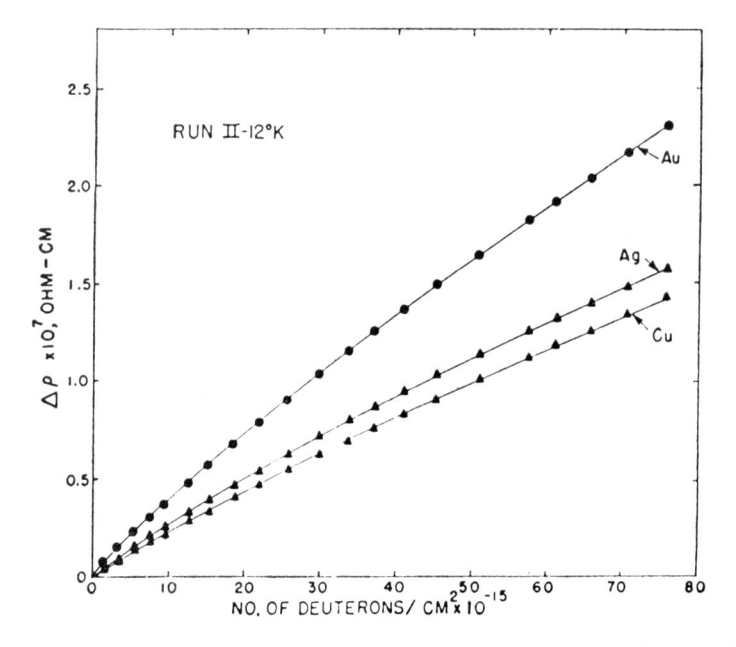

Fig. 7.9. Effect of irradiation on the resistivity of metals. From *Radiation Effects*, Dienes and Vineyard, Fig. 3.2, p. 63.

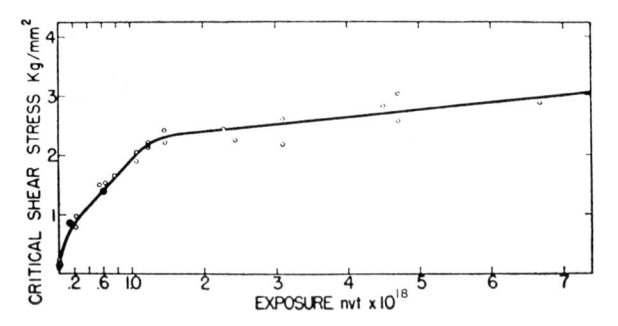

Fig. 7.10. Effect of irradiation on the critical resolved shear stress for copper. From *Radiation Effects,* Dienes and Vineyard, Fig. 6.2, p. 183.

information. The increase of resistivity under these conditions can be quite large, as shown in Fig. 7.9.

Elastic Properties. Quite large increases, 15 to 20%, in Young's modulus for copper have been reported; this change was accompanied by a large decrease in internal friction, and is thought to be due to the pinning of dislocations by point defects.

Plastic Deformation. Neutron irradiation also has large effects on plastic deformation; the critical resolved shear stress for the single crystal is raised by a very large amount in some metals. Figure 7.10 shows this for copper. The later parts of the stress-strain curve are less affected, Fig. 7.11. This is compatible with the theory that the existing dislocations are "pinned" by the condensation on them of vacancies or interstitials.

In polycrystalline samples of alloys of less pure metals, however,

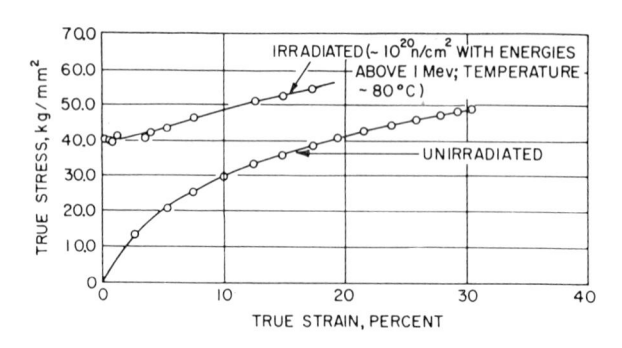

Fig. 7.11. Effect of irradiation on the stress-strain curve of nickel. From *Radiation Effects,* Dienes and Vineyard, Fig. 6.3, p. 184.

the effects of neutron irradiation on the stress-strain curve are sometimes quite large; in general, the rate of work hardening is increased, compared with the unirradiated metal, and the ductility is accordingly decreased. These effects can all be associated with the existence of obstacles to dislocation movement, probably as a result of the condensation of vacancies. Figure 7.12 shows the effect of neutron irradiation on the tensile properties of aluminum of 99% purity. The corresponding curves for a stainless steel (18% chromium, 8% nickel, niobium stabilized) are shown in Fig. 7.13. The main effect of irradiation in this material is the appearance of a sharp yield point, a phenomenon usually associated with the pinning of dislocations by means of a high local concentration of a solute. In this case the pinning is probably the result of the condensation of vacancies on the dislocations.

Other mechanical properties are changed by irradiation; for example, the impact properties of steels are greatly affected; the impact strength is decreased and the transition temperature between ductile and brittle behavior is raised. An example is shown in Fig. 7.14.

Diffraction Effects. Radiation damage can be detected by means of two X-ray diffraction effects: the Bragg reflections are usually displaced and become less sharp, and diffuse scattering is increased.

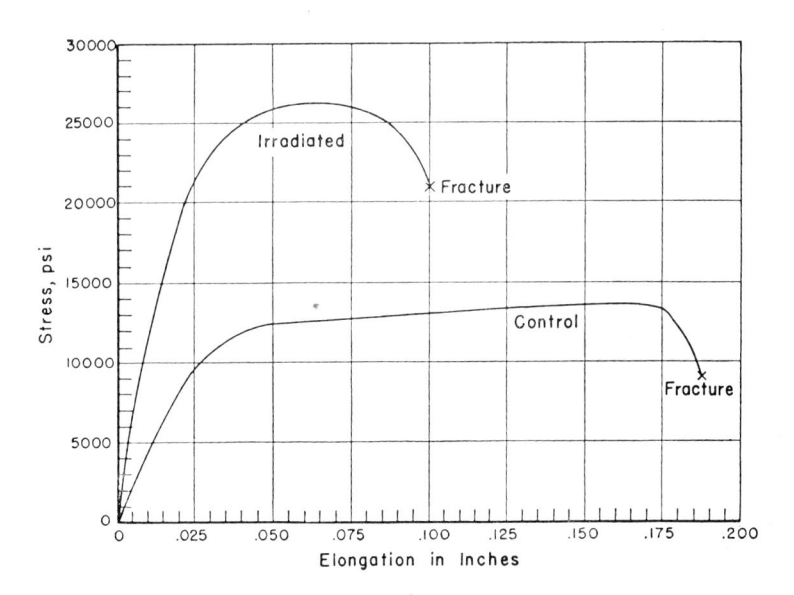

Fig. 7.12. Effect of neutron irradiation on the tensile properties of aluminum. From: Walter D. Wilkinson and William F. Murphy, *Nuclear Reactor Metallurgy*, Fig. 19-1, p. 318. Copyright 1958, D. Van Nostrand Company, Princeton, N. J.

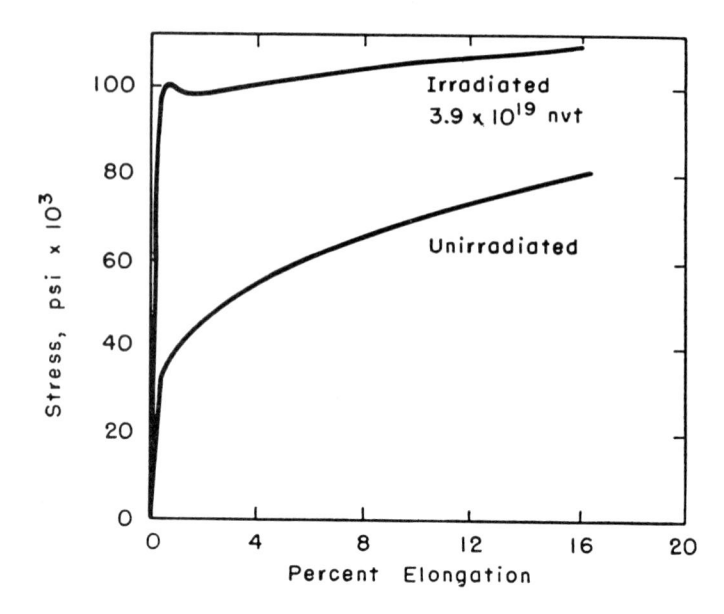

Fig. 7.13. Effect of irradiation on the tensile properties of stainless steel. From *Nuclear Reactor Metallurgy*, Wilkinson and Murphy, Fig. 19-9, p. 328.

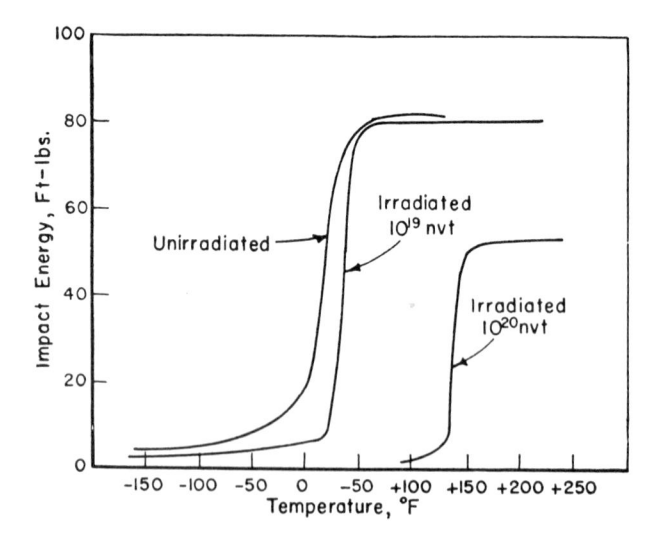

Fig. 7.14. Effect of irradiation on impact properties of a carbon steel. From *Nuclear Reactor Metallurgy*, Wilkinson and Murphy, Fig. 19-8, p. 325.

Irradiation at very low temperatures is apparently necessary to produce these effects in metals; a change of lattice parameter of 10^{-4} has been produced in copper by irradiation at 12°K.

EFFECTS PRODUCED BY IRRADIATION ACCOMPANIED BY FISSION. When fission takes place, the effects are even more complicated, because (a) the number of atoms is increased, (b) the chemical nature is changed, to the extent to which fission has occurred, and (c) fast particles (neutrons) originate within the material.

In uranium, in which metal these effects have been studied, the complicated crystal structure introduces anisotropy of behavior. The most important effect is the change of shape that takes place during irradiation and fission; the dimensions increase in the [010] direction, decrease in the [100] direction, and remain constant in the [001] direction of the orthorhombic structure. The effects are very large, increasing or decreasing the dimensions by a factor of 2 by the time 0.2% of the atoms have undergone fission. The change of density is relatively quite small. The effect on a polycrystalline specimen depends critically on the crystal size and on the departure from random orientation; a sample with fine grain size and random orientation shows a slight uniform increase in dimensions as irradiation proceeds, while coarse-grained uranium with a strongly preferred orientation may change its dimensions in a very anisotropic manner.

7.3 Recovery Processes

The changes in structure caused by deformation or radiation correspond to a departure from the equilibrium condition of the metal; the free energy is increased, mainly by the strain energy corresponding to the introduction of large numbers of dislocations, but also by the formation of vacancies and interstitials. There is therefore a tendency for the structure to change in the direction of restoring equilibrium conditions. The true equilibrium condition would be a single crystal of a shape that might be spherical, or it might be geometrically related to the crystallographic planes of the crystal, depending on the degree of anisotropy in the surface free energy of the crystal. Part of this change takes place as the forces causing deformation are released; the major part can only occur as a result of thermally activated processes; these are described in a general sense as "recovery processes." They may take place either simultaneously with or subsequently to the process of deformation. Since the recovery processes are thermally activated, their rates are temperature-dependent.

There are six processes by which a deformed or irradiated metal can approach the equilibrium state:

1. The decrease in vacancy and interstitial content towards the equilibrium number.
2. The mutual annihilation of dislocations of opposite sign, or the shrinkage of dislocation loops until they disappear.
3. The rearrangement of dislocations in more stable arrays.
4. The absorption of dislocations by crystal boundaries that move through the deformed crystal.
5. A reduction in the total area of crystal boundaries.
6. Decrease of surface free energy by change of shape or orientation.

Annealing out of point defects. Point defects in excess of the equilibrium number can be introduced into a metal in three ways: by lowering the temperature, and so decreasing the equilibrium number; by plastic deformation; and by irradiation. It has been shown that in all cases the point defects disappear spontaneously at relatively low temperatures. Positive identification of the types of defect that become mobile at the various temperatures is not yet possible; it is found, however, that a large part of the resistivity introduced by irradiation disappears at temperatures below about 50°K, while the greater part of the hardening produced by irradiating copper disappears at about 300°K (i.e., room temperature). Measurements of the release of stored energy have been made by comparing the heat input required to raise the temperature of a deformed or irradiated sample with that required to raise the temperature of a "standard" sample through the same intervals. It is found that a significant amount of stored energy is released from copper at about room temperature both after deformation and after irradiation. This has been interpreted as due to the disappearance of excess vacancies. The vacancies must move by the process discussed on p. 370. The disappearance of excess point defects appears to have little effect on the mechanical properties of metals except when large concentrations of vacancies are present, as a result of either irradiation or quenching from a high temperature. In these circumstances, the hardening produced by the vacancies disappears on annealing.

Recovery (polygonization). The first process that causes marked changes in structure when a deformed metal is heated is a rearrangement of the dislocations into a distribution of lower energy, without a very marked decrease in the total number of dislocations. This process is sometimes called *polygonization,* although this term was devised to describe the following particularly simple case. If a suitably

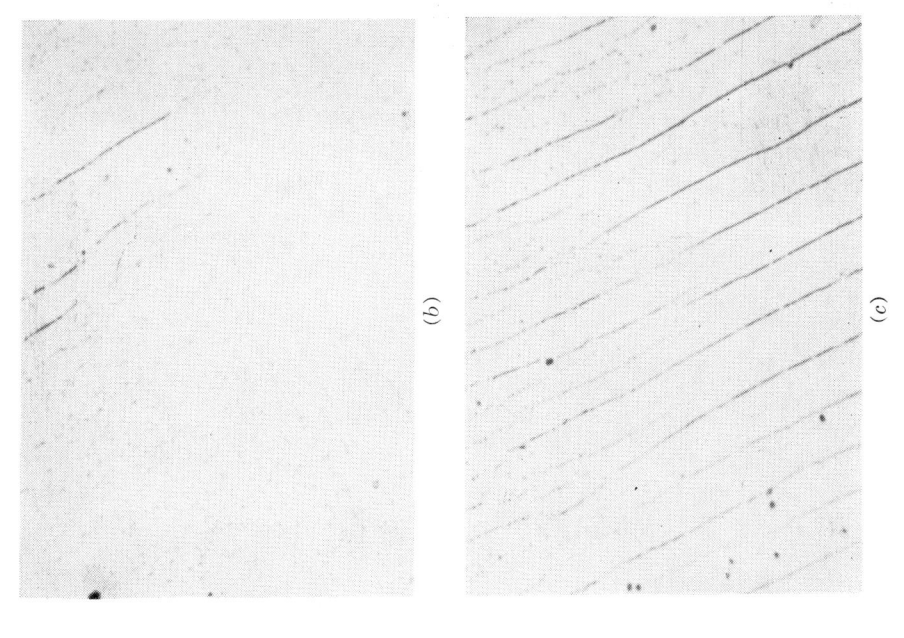

Fig. 7.15. Photomicrographs of etch pits in silicon iron, showing polygonization: (a, above) as deformed; (b, upper right) after heating at 700°C for 1 hr; (c, right) after heating at 875°C for 1 hr. From W. R. Hibbard, Jr., and C. G. Dunn, *Acta Metallurgica*, Vol. 4, No. 3, May 1956, Fig. 2, p. 308, Figs. 11 and 14, p. 311.

324 PHYSICAL METALLURGY

oriented single crystal of, say, a silicon-iron alloy is bent, it deforms by slip on a single system; the dislocation content can be revealed, in this case, by an etching technique, and the distribution of dislocations is shown in Fig. 7.15a. The dislocations are arrayed in their slip planes. Their energy is that due to the individual dislocations with the addition of the energy resulting from the repulsion of dislocations

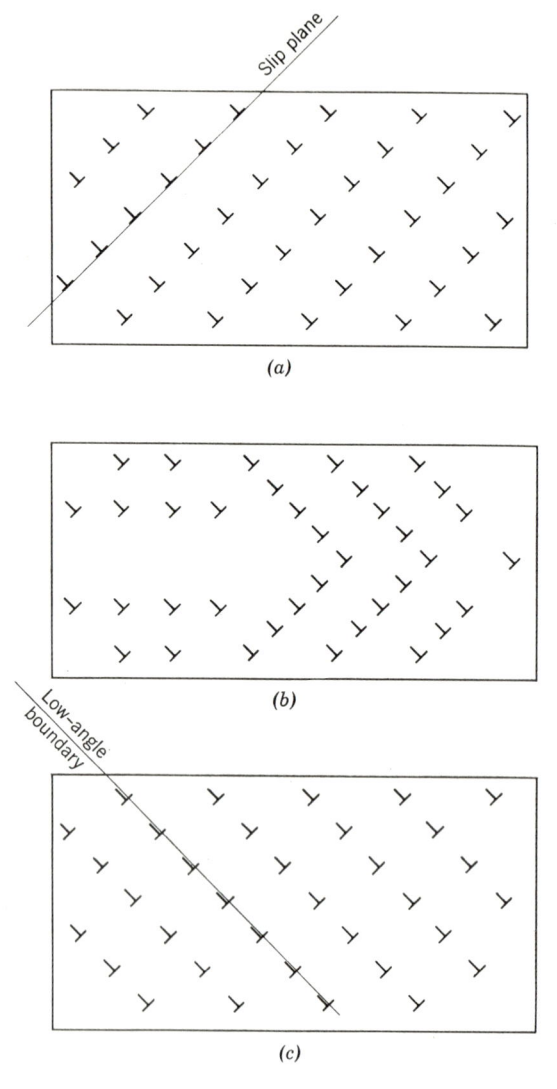

Fig. 7.16. Schematic diagrams corresponding to Figs. 7.15a, b, c.

Fig. 7.17. Photomicrograph showing polygonization in aluminum. From R. Cahn, *Progress in Metal Physics 2*, Bruce Chalmers, Ed., Butterworths Scientific Publications, London, 1950, Fig. 5, facing p. 155.

in their common slip planes. Figure 7.15b shows the result of heating the sample to 700°C for 1 hr; some of the dislocations have moved away from their slip planes and they are now much more randomly distributed; their energy is now close to the sum of the energies of the individual dislocations. Figure 7.15c shows the result of heating to 875°C for 1 hr; the dislocations, still essentially the same in number, are now arrayed in low-angle boundaries; the energy has been reduced by the interaction of the dislocations. The crystal lattice, instead of being uniformly distorted and bent, now consists of sections of "good" crystal separated by low-angle boundaries. The schematic diagrams of Fig. 7.16 correspond to the photomicrographs of Fig. 7.15; the final structure is said to be "polygonized." In the case described above, only a single type of dislocation is present. In more complicated cases, where several slip systems are invoked, equivalent but more complicated patterns of low-angle boundaries are formed. An example is shown in Fig. 7.17.

Polygonization occurs as a result of climb of dislocations out of their slip planes, followed by their motion by slip into the lower energy array. The process therefore takes place at an observable rate only at temperatures at which vacancies are sufficient in number and have adequate mobility. It is evident that a substantially higher temperature is needed for polygonization than for the elimination of excess vacancies. In a cold-worked metal, the Laue diffraction pattern consists of radially elongated spots (see Fig. 7.2). In a polygonized sample, the elongated spots are replaced by a series of sharp fine spots which represent about the same angular spread as the elongated spot.

This indicates that each section of the polygonized structure has nearly the same orientation as the mean of the region in which it formed.

RESIDUAL STRESSES AND STRESS RELIEF. The most important effects of recovery on the properties are caused by the relief of residual stresses. Residual stresses arise during inhomogeneous deformation in the following way. As an example of inhomogeneous deformation, consider the bending of a bar of rectangular cross section. This is represented in Fig. 7.18. The original section $ABCD$ is deformed to $A'B'C'D'$ by the application of a suitable stress system. The outer layer BC is stretched by an amount a, while the inner layer is compressed by an amount b which may be different. Part of the strain is plastic in nature. When the applied stress is released, the two extreme layers will change in length by the elastic deformation that corresponds to the stress that had been applied. This causes a decrease in the curvature, a phenomenon known as "springback." If springback is prevented by any kind of constraint, as for example if the sample is bent into a complete circle and the two ends joined together, the elastic stresses will be retained in the system. Even if springback is allowed, there will be some residual stress because the *strain* in any section is directly proportional to the curvature and the distance from the neutral axis; the springback is proportional to the stress and is related to the strain not by direct proportionality but by the stress-strain curve. Thus all sections would not spring back by amounts that are proportional to their distance from the neutral axis, and so some elastic stresses remain. In the absence of an externally applied stress, these residual stresses must balance, and there must therefore be both tensional and compressive stresses. Internal residual stresses also arise, on a microscopic scale of distance, because of the anisotropy of elastic constants and of the plastic behavior of most metals. Neighboring crystals, having been plastically deformed by the same stress, do not necessarily have the same elastic recovery when the stress is released.

Residual stress can also be caused by a change of temperature. If the temperature gradients are at any time large enough for the resulting stresses to cause local plastic deformation, elastic stresses develop as the temperature becomes uniform.

When the temperature of a deformed metal is raised to such an

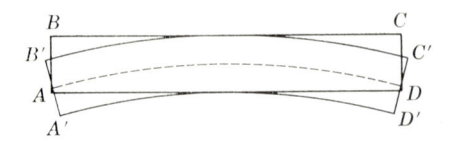

Fig. 7.18. Origin of residual stresses.

extent that dislocations are able to climb out of the slip planes in which they were generated, the residual stress is relieved by the movement of these dislocations, which are now free to move under quite small stresses. The residual stresses therefore disappear, with only minor changes of external dimensions. The treatment that is used for this purpose is called stress relief; it can be carried out at temperatures at which there is not sufficient movement of dislocations to cause noticeable polygonization or changes in properties.

There is in general very little change in hardness during the recovery process; this indicates that severe deformation does not occur much more readily in a polygonized sample than in the work-hardened state; however, there is evidence that the yield stress is reduced by polygonization.

GROWTH OF URANIUM DURING CYCLIC TEMPERATURE CHANGES. A remarkable example of stress relief has frequently been observed when polycrystalline uranium has been subjected to cyclical temperature variations, which are commonly encountered in nuclear reactors. The typical behavior is an increase in length and decrease in cross section of a cylindrical sample. The growth may be very large, an increase in length of 50% in 700 cycles being typical.

The explanation is as follows: Uranium is extremely anisotropic as regards thermal expansion, the coefficients in the three principal crystallographic directions being $+33 \times 10^{-6}$, -6.5×10^{-6}, and $+30 \times 10^{-6}$ in the a, b, and c directions respectively. Consequently, two neighboring crystals with different orientations expand and contract to different extents whenever the temperature changes. This causes one or both to deform plastically. The slip system that operates at low temperature (below 400°C) in uranium is (010) [100] and in a sample of the type under consideration some crystals are forced to deform on that system as the temperature is lowered. On the other hand, at higher temperatures, thermally activated (creep) deformation becomes predominant, and this occurs mainly on planes that do not operate at the lower temperatures. If the texture of the original sample contains several components, one of which consists of crystals so oriented that they have a large coefficient of expansion in the longitudinal direction, and deform relatively easily at low temperatures under the stress that develops during cooling, while the other component has its low or negative expansion coefficient in the longitudinal direction and deforms easily at high temperatures under the stress characteristic of heating, the sample will increase in length during each cycle.

It has been found possible to prevent growth by developing a random orientation in the uranium.

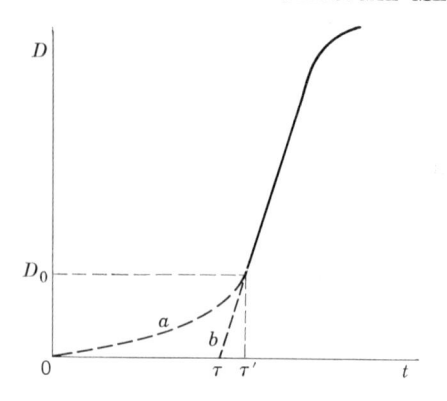

Fig. 7.19. Increase of size of recrystallized grain with time.

Recrystallization. If the amount of plastic deformation is small, polygonization may be the only recovery process to occur when a sample is heated to progressively higher temperatures. Under more usual conditions, however, a second process, called "recrystallization," * takes place in a temperature range above that of recovery. Recrystallization is the rearrangement of the atoms of the solid into an entirely new set of crystals. These crystals grow from very small "nuclei" until they impinge on each other, absorbing the intervening unrecrystallized material. The new crystals have a much lower dislocation content than the deformed material, and the driving force for recrystallization is the decrease in free energy resulting from the decrease in dislocation content.

KINETICS OF RECRYSTALLIZATION. The process of recrystallization can be described as follows: One of the results of plastic deformation is to produce regions in the material which can either support the nucleation of new crystals or which can themselves grow into new crystals under suitable conditions. The number of such "growth sites" increases as the amount of deformation is increased. On heating to a suitable temperature, each growth site becomes a new crystal, its "radius" increasing with time according to the law $D = G(t - \tau)$, as shown in Fig. 7.19.

It is evident that there is a nucleation problem in recrystallization in the sense that a "new" crystal could be below its critical size for growth. The critical size can be estimated as follows: The decrease in free energy per unit volume of metal resulting from recrystallization has been measured (see p. 336) and is about 0.2 cal/gram for severely

* The term "recrystallization" is also sometimes used in transistor technology to represent the process of melting followed by freezing. Its significance in chemical parlance is also different from the metallurgical usage.

deformed copper. This is equivalent to about 6.8×10^7 ergs/cc. This can be taken as the value of ΔF_v for the calculation of the critical size. The surface between the recrystallizing grain and the deformed matrix is usually a grain boundary, and has an energy of about 300 ergs/cm². If these values are substituted in the expression $r^* = 2\sigma/(\Delta Fv)$ (see p. 238), the value of r^* is found to be about 10^{-5} cm. It is evident that the critical size is larger for smaller values of ΔF_v, and therefore, presumably, for less severely deformed material. This case departs in one respect from the standard nucleation considerations. In the case, for example, of the nucleation of a crystal in a supercooled melt, an embryo that forms but does not reach critical size can disappear completely. A recrystallization embryo, on the other hand, can grow, by "absorbing" surrounding dislocations into its boundary, but it may not restore the surrounding material to its condition of high free energy by shrinking. In other words, a subcritical embryo may be metastable. Its growth, therefore, may be expected to be irreversible, and to continue wherever a suitable thermal fluctuation occurs. When the critical size has been reached by this process, which is necessarily intermittent and slow, *continuous* growth should occur.

The rate of growth is linear from this point until the growth of the new crystals is limited by their impingement on each other. There are, therefore, several possible interpretations of the part of the curve for which $D < D_0$, where D_0 is the smallest size at which the new crystal can be observed.

Curves a and b in Fig. 7.19 correspond to (1) continuous growth from time $t = 0$, from an existing nucleus, and (2) nucleation at time $t = \tau$. A third possibility is that growth of a pre-existing subcritical embryo takes place, starting at time $t = 0$, the embryo reaching critical size at $t = \tau'$. In this case, the embryo existing at $t = 0$ need not be as free from dislocations as the parts of the new crystals that grow later. It must, however, have lower free energy than its surroundings. It is not at present possible to distinguish between these possibilities; it is convenient for descriptive purposes to represent the process by a nucleation rate N and a growth rate G.

Numerous experiments suggest, but do not prove, that the rate of nucleation N is zero at zero time and increases rapidly with time; it is more firmly established that G, the rate of growth of the crystals that have nucleated, is independent of time. Both these parameters increase with a rise of temperature, and N is increased by greater plastic deformation and by smaller grain size in the original material. This may be related to the observed tendency for the new crystals to originate at or near crystal boundaries, surfaces, and other regions of inhomogeneity.

MECHANISMS OF RECRYSTALLIZATION. The mechanism of recrystallization is the outward movement of crystal boundaries into material that contains a high concentration of dislocations. The boundary increases in area as the crystal grows larger, but the energy of the material is decreased because the newly formed crystal has less dislocation energy per unit volume than the deformed crystals. If the boundary between the new crystal and the matrix behaves as a large-angle crystal boundary, it can "absorb" dislocations without any increase in energy per unit area. However, if it is a low-angle boundary, its energy may be either increased or decreased by the addition of dislocations. The detailed process whereby atoms of the deformed matrix enter the boundary at one side, while other atoms leave the boundary on the other to become part of the new crystal, has not been clarified. It is likely, however, that the unit process is the movement of an atom from the boundary onto the new crystal lattice, or from the deformed crystal lattice into the boundary, by a process of changing places with a vacancy. It is not clear, either, why the new crystal that is formed always contains dislocations apparently in about the same number as crystals grown from the melt.

RECRYSTALLIZATION TEMPERATURE. The rate at which recrystallization occurs can be represented by Fig. 7.20, which shows how the proportion of material that has recrystallized increases with time. It may be regarded as the average curve for many new crystals, each of which individually behaves in the manner of Fig. 7.19 but with different values of τ. The process may be characterized by the time, t, that it takes for, say, 95% of the material to have recrystallized. Owing to

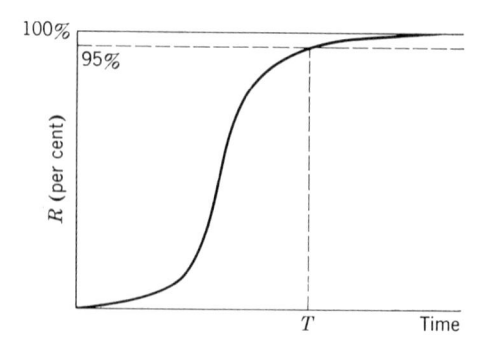

Fig. 7.20. Per cent recrystallized as a function of time.

DEFORMATION, RADIATION DAMAGE, AND RECOVERY

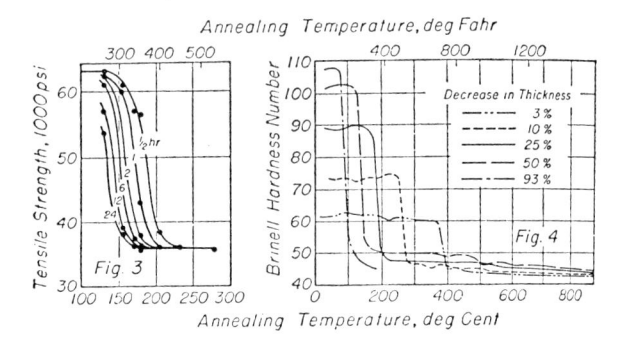

Fig. 7.21 (left). Effect of time on the recrystallization temperature. From *Metals Handbook,* American Society for Metals, Cleveland, 1948, Fig. 3, p. 260.

Fig. 7.22 (right). Effect of extent of deformation on the recrystallization temperature. From *Metals Handbook,* ASM, Fig. 4, p. 260.

the asymptotic approach to the condition of 100% recrystallization, it is not possible to measure how long it takes to reach completion. The time required for the process to reach a specified stage (e.g. 95%) depends upon the following factors: amount and kind of prior deformation, temperature, and grain size. The recrystallization temperature for a material is defined as the temperature at which recrystallization will be substantially complete in a standard time and after a standard amount of deformation.

The influence of time and amount of prior deformation on the recrystallization temperature are shown in Figs. 7.21 and 7.22. The recrystallization temperatures that are usually quoted are often not obtained under well-standardized conditions, and they therefore only serve as an approximate guide to the temperature at which recrystallization will in fact occur under any given set of conditions. Some values are quoted in Table 7.2. It is of some significance that the temperatures shown in the table are all in the general vicinity of 0.4 of the melting point.

The effect of impurities on the recrystallization temperature is known to be large; very pure silver (prepared by zone refining) is said to recrystallize below 0°C, for example, and in less extreme cases it is known that most impurities raise the recrystallization temperature. The effect of impurity atoms on recrystallization temperature is probably associated with their tendency to segregate at the boundaries, the movement of which would be retarded by the necessity for solute to diffuse with them.

Table 7.2. Table of Recrystallization Temperatures

	Recrystallization Temperature (°C)	Melting Point (°C)	$\dfrac{\text{RT (°K)}}{\text{MP (°K)}}$
Iron	450	1593	0.40
Nickel	600	1455	0.51
Gold	200	1063	0.36
Silver	200	960	0.39
Copper	200	1083	0.35
Aluminum	150	660	0.45
Platinum	450	1773	0.36
Magnesium	150	650	0.46
Tantalum	1000	2996	0.39
Tungsten	1200	3410	0.40
Molybdenum	900	2625	0.41
Zinc	206	419	0.69
Lead	<20	327	<0.49
Tin	<20	232	<0.58
Cadmium	20	321	0.49

RECRYSTALLIZATION TEXTURES. The new crystals that form during recrystallization are in general not randomly oriented. The textures that have been observed are bewildering in their variety; there are, however, several generalizations that can be made. It is often found that either some or all of the new crystals have orientations that belong to the deformation texture. The nuclei for these crystals may be polygonization cells of the deformed crystals, which for some reason reach the critical size for continued growth. This process is sometimes called "recrystallization *in situ.*" The second type of texture or texture component is that in which the new and deformed lattices are related by rotation about a common axis, a {111} axis in the case of face-centered cubic structures. It has been shown that a boundary that exists between crystals related by a 22° rotation about a common ⟨111⟩ axis contains many sites that are common to both crystals. The positions of a layer of atoms before and after the passage of such a boundary are shown in Fig. 7.23. It has also been found that such boundaries are much more mobile than those corresponding to other orientation differences, particularly in impure metals. There have been many observations of recrystallization textures in which an important component is related in this way to the deformation texture.

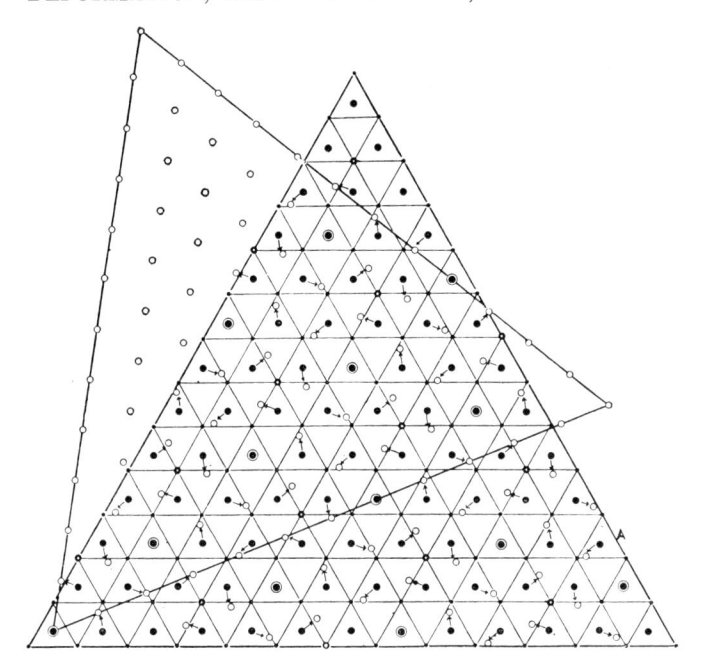

Fig. 7.23. Atom movements in recrystallization according to Kronberg and Wilson. From *Structure of Metals,* Barrett, Fig. 19, p. 506.

In addition to the orientations that can be described in this way, there are others that do not bear as simple a relationship to the deformation texture, and the complete recrystallization texture may be very complicated, containing several components. For example, Fig. 7.24 shows pole figures for rolled iron. The amount of spread of the various orientation components depends upon the sharpness of the deformation texture and on the amount of deformation and the temperature at which recrystallization is carried out. It is possible, in many cases, to produce a substantially random texture by proper selection of the amount of deformation and the annealing temperature; the composition of the metal or alloy also has an important influence. In general, a relatively small amount of cold reduction followed by a relatively high temperature anneal appears to be conducive to randomness. The amount of cold deformation that can be tolerated depends upon the prior grain size, there apparently being a minimum extent to which the grains should be reduced in order to produce a random texture. It seems that the production of either a random orientation or a desirable preferred orientation is still a problem that can best be solved by empirical methods.

a (110)   b (100)

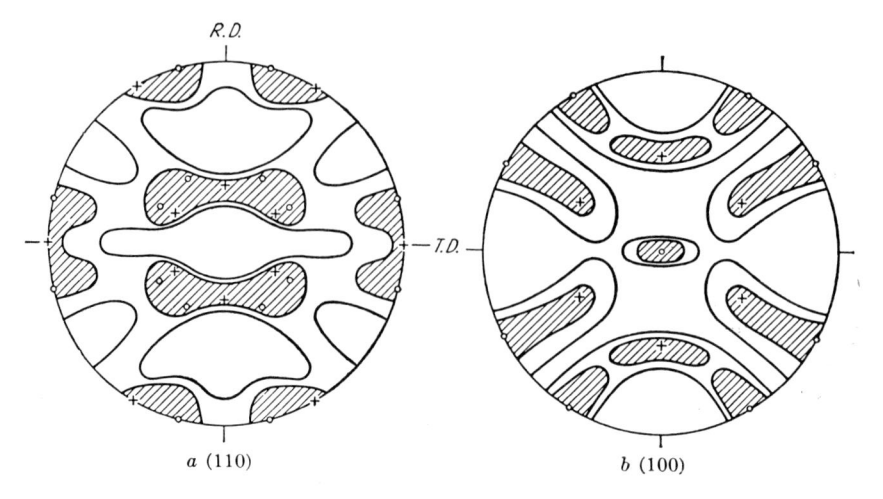

Fig. 7.24. Recrystallization texture for rolled iron. From *Structure of Metals,* Barrett, Fig. 14, p. 500.

One problem in the understanding of recrystallization textures is that of the origin of the "nuclei"; a second problem is that of the influence of the difference of orientation between crystal and matrix on the rate of growth of the crystal. The relative importance of these two problems is a matter of controversy, but the probable outcome is that nuclei occur, after deformation, in certain preferred orientations, probably with a wide spread around these orientations and, perhaps, randomly with the exception of certain missing orientations, and that the rate of growth of these nuclei is more sharply dependent on the orientation relationship. The crystals which are finally present in the recrystallized material are therefore those which are present as nuclei and which grow fast. It is necessary to assume both the "oriented nucleation" and the "selective growth" hypotheses in order to account for the fact that sharp textures are formed, but that all equivalent rotations about the $\langle 111 \rangle$ directions of the matrix, for example, are not found.

RECRYSTALLIZATION TWINS. The grains that form during the recrystallization of many face-centered cubic metals contain twinned regions. Figure 7.25 shows an example. The origin of recrystallization twins is probably as follows: If the boundary between a growing crystal and the matrix is a {111} plane of the new crystal, a new layer may form either at the correct position (see p. 126) for the growth of the original crystal, or it may form so that its atoms are at the "faulted" position. The faulted sites correspond to the positions of the atoms

DEFORMATION, RADIATION DAMAGE, AND RECOVERY

in a crystal that is twinned with respect to the original crystal. It is quite possible that each new {111} layer nucleates at many points, some normal and some faulted. The survival of the twin depends on an energetic advantage over its competitor, the normal crystal. An energetic advantage occurs if the boundaries between the twin and its neighboring crystals are of relatively low energy. This is possible because the twin has a different orientation from the original crystal.

Figure 7.26 illustrates this point. The surface $ABCD$ in Fig. 7.26a is the position of a recrystallization interface, which is moving to the left. Some impingement of new crystals has already occurred. PB and PC are grain boundaries, which can have substantially lower energy if the crystal PBC changes to its twin orientation. This condition may be reversed when a new grain edge is formed, as in Fig. 7.26b; the twin then terminates at Q, as shown in the figure.

STORED ENERGY. It has been found by means of calorimetric studies that a substantial release of energy occurs while recrystallization is taking place. This energy can be regarded as that of the dislocations that were introduced by the process of deformation. It has already

Fig. 7.25. Photomicrograph ($\times 230$) showing annealing twins. (Courtesy P. Doherty and H. McP. Brown, Harvard University.)

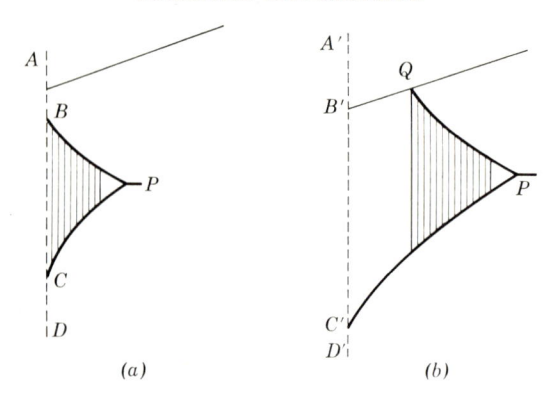

Fig. 7.26. Illustrating the formation of annealing twins.

been shown that the point defects resulting from deformation are removed at a lower temperature. The amount of stored energy that is released on recrystallization depends upon the material and on the amount of deformation. Some values are given in Table 7.3. The release of 0.2 cal/gram corresponds to the destruction of about 10^{13} dislocations/cm^2.

PROPERTIES OF RECRYSTALLIZED METALS. The changes brought about by plastic deformation are completely reversed by recrystallization,

Table 7.3. Some Value of the Stored Energy due to Cold Work *

Metal	Type of Deformation	Amount of Strain	Stored Energy Cal/g	Percentage of Work of Deformation
Copper	Tension	0.20	0.07	5
	Torsion	0.26	0.08	7
	Wire drawing	1–5	0.3–1.5	2
Copper	Compression	0.8	0.34	5
Nickel	Torsion	2.3	0.7	2
Silver	Compression	0.5	0.06	3
Aluminum	Compression	1.0	0.84	10
Iron	Torsion	0.6	0.66	8

* From A. L. Titchener and M. B. Bever, *Progress in Metal Physics* 7, Bruce Chalmers and R. King, Eds., Pergamon Press, London, 1958, p. 247.

except for the effects on grain size and preferred orientation. The properties of the individual crystals appear, therefore, to have been restored to the values characteristic of the cast material. This would imply that the dislocation content and distribution are the same in the as-cast and the recrystallized conditions. There does not appear to be any evidence that this is to be regarded as more than a very general similarity; it should not be concluded from existing evidence that there is a standard degree of imperfection that a crystal may attain by a variety of different paths.

A recrystallized metal is, in general, soft and ductile; i.e., it has a low yield stress and a high work-hardening rate.

Grain growth. If the temperature of a recrystallized metal is raised, further changes occur. Some of the grains grow at the expense of others, which disappear, with the result that the average grain size increases. This may occur in three ways:

1. Most of the grains, at any one time, may grow, only a few grains decreasing in size at any one time. The grain size at any time during this process is nearly uniform, a few small grains, however, always being present. This is normal grain growth.

2. Grains may grow selectively with orientations that are not representative of all the grains present. The new grains may or may not be of orientations corresponding to a major component of the recrystallization texture. The result of this process, called secondary recrystallization, is the simultaneous development of a new texture (the secondary recrystallization texture) and a much larger grain size.

3. In some cases, a very small number of primary grains grow to an abnormal extent, resulting in a very large grain size. The grain size is very far from uniform while this is taking place.

NORMAL GRAIN GROWTH. The driving force for grain growth is the decrease in free energy that results from a reduction in the total area of the grain boundaries. It has been pointed out, however, that an aggregate of geometrically similar crystals has no path of continuously decreasing energy by means of which the grain boundaries can move, while an aggregate containing crystals that do not all have the same number of sides can follow such a path.

This can be illustrated for the analogous two-dimensional case as follows: Consider an array of hexagonal cells, as in Fig. 7.27a. The cell walls, which correspond to the crystal boundaries, have energy, and therefore tend towards minimum area; this is equivalent to the statement that they behave as if they are under tension. The surface tension may be considered to be equal for all the walls, although in the

(a) (b)

Fig. 7.27. Stable (a) and unstable (b) cell structures.

case of a crystal aggregate a few per cent of them would have significantly lower values because there would be some low-angle boundaries in a random texture. The stable configuration is that in which all the angles are 120°. Any movement of a point of contact of three cell walls will increase the total area of the walls, and so there is no way of decreasing the number of cells by a continuous reduction of energy.

On the other hand, if the number of sides per cell is only statistically constant, as in Fig. 7.27b, there will be some cells with more than six sides, and some with fewer. A cell with other than six sides cannot have 120° angles and straight sides; if the angles are 120°, the sides must be curved as in Fig. 7.28. All the angles are 120°, but, because of the curvature of the sides, a reduction of total area can be achieved by moving the points of intersection outwards. Such a cell, or crystal, will grow at a rate that is controlled by the rate of movement of the boundary. This is a relatively slow process except when a moving point approaches a point that has no tendency to move, as in Fig. 7.29. The area of boundaries decreases substantially by achieving the positions shown by the thick lines, and the cell now has one more side than before; there are also two crystals fewer than before. An extension of this discussion would show that, as the number of sides goes on increasing, the

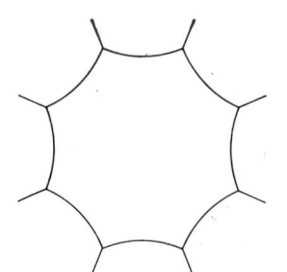

Fig. 7.28. Curvature of sides of eight-sided cell.

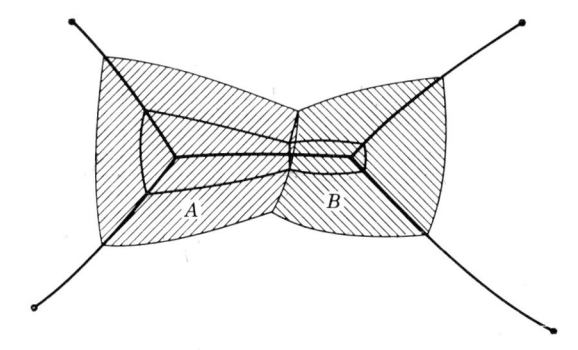

DEFORMATION, RADIATION DAMAGE, AND RECOVERY 339

Fig. 7.29. Showing the disappearance of grains A and B by the inward movement of curved boundaries.

sides become more curved and the possible energy decrease grows as the process continues. It must be pointed out that the condition for a cell to grow is that it must have more than the average number of sides, not that it must be abnormally large.

The final result may depend on other factors, such as the influence of the orientation difference on the rate of boundary movement, and this may be a cause of the very strong textures that are sometimes found as a result of secondary recrystallization.

SECONDARY RECRYSTALLIZATION. The process of secondary recrystallization has a similar time dependence to primary recrystallization; an incubation period, during which no observable change takes place, is followed by a period of uniform growth rate, which is terminated by impingement of the secondary grains.

The driving force for secondary recrystallization has not been clearly identified; it is clear that secondary recrystallization will occur if a small number of primary grains that conform to the new texture develop a greater than average number of sides very early in the process (i.e., in the incubation period). It is possible that the secondary grains already have this advantage by the time primary recrystallization is complete. The possible types of driving force are related to the volume free energy, the grain boundary free energy, and the surface free energy. The volume free energy is apparently not reduced by the process of secondary recrystallization; the new grains do not appear to be any more perfect than the primaries. The boundary energy is decreased by secondary recrystallization, but not to any greater extent than would be the case for normal grain growth. A decrease in the surface energy may be a significant driving force; it appears that secondary recrystallization has only been detected in sheet material, in which the surface free energy should make a relatively important contribution to the total free energy. The evidence derived from thermal etching shows that the surface free energy of a crystal sometimes depends strongly on the crystallographic orientation of its surface.

The best known example of secondary recrystallization is the production of the so-called "cube on edge" texture in silicon iron (see p. 225). The texture has nearly all the grains oriented with one $\langle 001 \rangle$ direction within a few degrees of being parallel to a direction in the plane of the sheet (the rolling direction) and a $\{110\}$ plane closely parallel to the plane of the sheet. This texture is produced by subjecting the alloy to an empirically determined sequence of deformation and thermal treatments.

FACTORS THAT AFFECT GRAIN GROWTH. It is frequently stated or implied that the grain size resulting from grain growth is a function of temperature only, and not of time. This is not strictly true, but under ordinary practical conditions it is sufficiently accurate for most purposes. Grain growth is influenced by the geometry of the sample, the presence of soluble impurities, and the presence of particles of a second phase.

The geometry becomes most significant when the material is in sheet form; normal grain growth usually ceases when the grain size becomes comparable with the thickness of the sheet; at this point, the boundaries are normal to the sheet surface and grain growth becomes a two-dimensional rather than a three-dimensional process. The movement of a boundary that reaches the surface is retarded by the formation of a groove by thermal etching. The presence of solutes appears to retard the process of grain growth, but little is known about the details.

Insoluble particles tend to anchor the boundaries because the strain around such a particle is reduced if it is in a boundary and the energy of the boundary itself is reduced if part of it is replaced by a crystal of some other phase. The energy of the system is therefore at a minimum when such particles are at grain boundaries. It follows that work must be done to pull boundaries away from inclusion particles. This fact is utilized in the control of the austenite grain size in steels. The grain size of the austenite grains is a function of the temperature to which the material has been heated; this relationship is shown in Fig. 7.30, curve A. This is the normal relationship. If, however, aluminum is used for deoxidation in the steel-making process, the grain-coarsening curve is as shown in Fig. 7.30, curve B. There is a rather sharp temperature below which the grain boundaries are prevented from moving

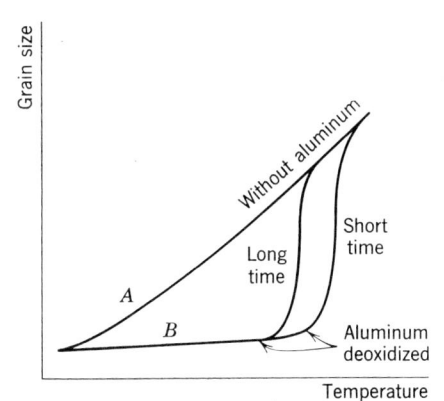

Fig. 7.30. Effect of aluminum on grain growth in austenite.

by the alumina particles in the steel, while at higher temperatures the boundaries are able to move. This is probably due to the solution of the alumina particles.

ABNORMAL GRAIN GROWTH. It is sometimes found that annealing produces an extremely large grain size. This occurs in regions in which the deformation has been relatively small, often around 2 to 4%. It is likely that this amount of deformation produces very few nuclei, and these are able to grow into primary grains of large size. The phenomenon under discussion is used for producing large crystals for experimental purposes by the "strain-anneal" technique. This is quite distinct from secondary recrystallization, which occurs after, and not instead of, recrystallization.

GRAIN SHAPES IN TWO-PHASE ALLOYS. The minimization of grain boundary energy is the driving force for grain growth in single-phase metals and alloys; it also dictates the equilibrium configuration of the crystals in two-phase alloys. The problem can be considered in terms of the shape of the crystals of a "minority" phase, the "majority" phase forming a continuous polycrystalline matrix in which the minority phase is embedded. The typical situation is the occurrence of a "minority" crystal at a line where three grains meet.

Three possible cases are shown in Fig. 7.31. In (c), the minority phase spreads all over the boundaries of the matrix, and may cause important changes in the mechanical properties, especially if it is brittle or has a very low melting point. The difference between the three cases illustrated is related to the angle θ, which is 180°, 60°, and 0° respectively. Under equilibrium conditions, θ has a value such that the three boundaries meeting at a point exert zero resultant force. This is equivalent to the statement that the value of θ is such that any change causes

(a) $\theta = 180°$ (b) $\theta = 60°$ (c) $\theta = 0°$

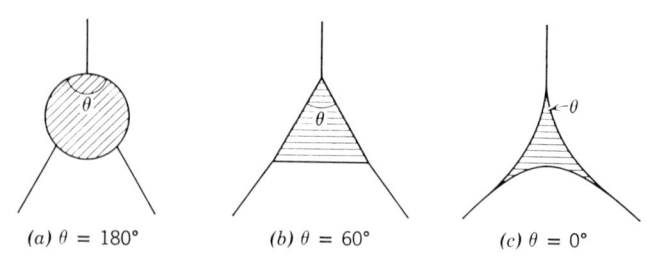

Fig. 7.31. Equilibrium shape of second-phase particle as affected by the interfacial angle θ.

an increase in energy. The value of θ is related to the interfacial tensions σ_{mi} and σ_{mm} (Fig. 7.32) by the relationship

$$\frac{\sigma_{mm}}{\sin \theta_1} = \frac{\sigma_{mi}}{\sin \theta_2} = \frac{\sigma_{mi}}{\sin \theta_3}$$

where σ_{mm} is the interfacial tension of the crystal boundary between two crystals of the matrix m, and σ_{mi} is the interfacial tension of the interface between the matrix and the inclusion i. It is assumed that both the interfacial tensions are independent of the orientation of the interface and of the adjoining crystals, and that the interfacial tension and interfacial energy are equivalent. The latter assumption is true in the case of liquids and is valid if equilibrium is reached at a temperature at which the interface can move under vanishingly small stresses, even if much motion is very slow. $\theta = 0$ if $\sigma_{mi} < \frac{1}{2}\sigma_{mm}$; this is the condition for the minority phase to spread over the boundaries. It has been shown that small changes of composition can alter the values of the surface tensions sufficiently to prevent spreading. For example, bismuth in copper satisfies the condition $\sigma_{mi} < \frac{1}{2}\sigma_{mm}$. It spreads, causing the copper to lose its strength at the melting point of bismuth. The addition of lead increases σ_{mi} to such an extent that $\sigma_{mi} > \frac{1}{2}\sigma_{mm}$; spreading no longer occurs, and although the bismuth phase still melts

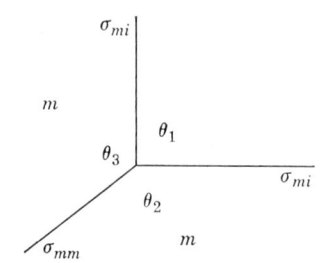

Fig. 7.32. Equilibrium of three boundaries.

DEFORMATION, RADIATION DAMAGE, AND RECOVERY 343

at a very low temperature (even lower because of the presence of lead) it does not seriously weaken the copper.

It should be emphasized that the kinetics of formation of the second phase usually control the shapes of the crystals, unless adequate steps have been taken to achieve equilibrium.

7.4 The Shaping of Metals

The purpose of this section is to discuss in general terms the processes by which a metal or alloy is changed in shape by the application of force. This stage of the fabrication sequence follows the solidification process: the main objective is to produce the required geometry, but the simultaneous development of desirable properties is by no means neglected. The choice of a process for a particular purpose depends partly on tradition, partly on economics, and partly on the properties required in the product.

The shaping process can be divided into three stages: 1. Changing the shape of the whole of a billet or other starting material. 2. Removing, by cutting, any part which is not required. 3. Producing the desired surface finish. All three stages can be carried out either by mechanical or other means. In this chapter, attention is concentrated on the mechanical methods.

Deformation processes

COLD DEFORMATION AND HOT DEFORMATION. There are in general two kinds of procedure that may be used when a metal or alloy is to undergo drastic changes of shape during fabrication. One is cold working, which takes place at a temperature below that at which the recovery processes are effective; the flow curve rises with increasing deformation, as a result of which a progressively increasing force must be used and deformation becomes more and more difficult; the fracture curve may be reached, in which case fracture occurs instead of further plastic deformation. There is often, therefore, a limit to the amount of cold deformation that can be achieved, either because of limited available force, or because of the incidence of cracking. Further deformation is possible only after the material has been restored to a softer condition by the recovery processes. This is achieved by subjecting it to a temperature at which the recovery processes occur at a useful rate; such a process is called "annealing."

The alternative procedure is hot working, which is conducted at a temperature at which recovery processes take place simultaneously with deformation. Deformation can be continued without limit because a condition is reached in which the recovery processes keep pace with the deformation.

A common feature in almost all the plastic working procedures is that the force is applied to the material locally and that the deformation takes place progressively. It is therefore not usually necessary to apply sufficient force to deform the whole of the piece simultaneously.

TYPES OF PROCESSES. There are various ways of classifying the processes of deformation; a convenient one is in terms of the symmetry of the operation. There may be no symmetry, a plane of symmetry, or a line of symmetry.

Forging. The basic no-symmetry process is that of forging, in which the metal is shaped, either on an anvil or in a die, by the application of a force, which may be steadily increased in a press, or applied impulsively by a hammer. Forging may be done from an ingot or billet, or the material may be partly formed first by one of the other processes. No tensile stresses are applied during forging, the stress being mainly one of uniaxial compression, although some triaxial compression arises from the reaction of the die or anvil. These conditions tend to cause plastic deformation rather than cracking; however, forging is always a hot-working operation unless the amount of deformation is to be quite small, when the process is called "coining."

Rolling. In rolling processes, in which the metal is shaped by passing it between pairs of rotating rolls, the resulting shape has a plane of symmetry. The forces that cause deformation are as follows: Figure 7.33 represents a pair of rolls into which material of thickness t_1 is fed; the thickness is reduced to t_2 by the rolls. The only force that causes the material to pass through the rolls is the friction force between the roll surfaces and the metal. The friction force arises all along the arc of contact. There is also a normal component of force between the

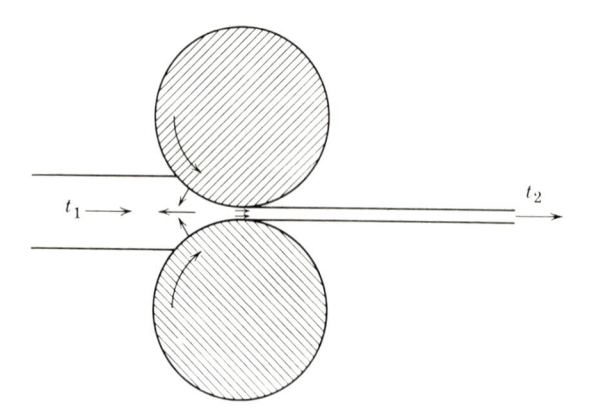

Fig. 7.33. Illustrating passage of metal through rolls.

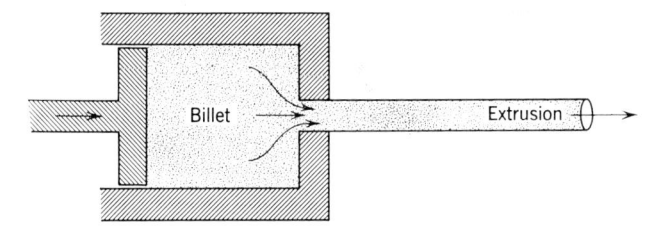

DEFORMATION, RADIATION DAMAGE, AND RECOVERY 345

Fig. 7.34. Schematic diagram of extrusion press.

rolls and the metal; the friction force, tending to move the metal to the right, is opposed by the resultant of the horizontal components of the normal force. The metal is therefore subjected to a longitudinal tension force, which, combined with the normal forces, produces the shear stresses that cause deformation. It should be noted that there must be some slipping between the metal and the rolls, as the metal enters at a relatively low speed and leaves at a higher speed since the volume of metal leaving the rolls is equal to the volume entering. Rolling is conducted either as a cold-working or as a hot-working process. It is common practice to produce sheet, in steel and in aluminum alloys, by a combination of the two, in which the major part of the reduction in thickness is achieved by hot rolling. The final process is cold rolling, in order to produce a better surface finish than is possible with hot rolling (in which surface oxidation is rapid, because of the high temperatures), and in order to finish with sheet that has the greater strength and hardness that results from cold working. Other shapes than sheet or strip are also produced by rolling; these are relatively simple shapes such as rods, I beams, and rail sections.

Extrusion. The process of extrusion may, but need not, result in a shape that is strictly symmetrical about a line; it consists in forcing the metal through a die of the desired shape. The process is shown schematically in Fig. 7.34. In this process the metal mainly is subjected

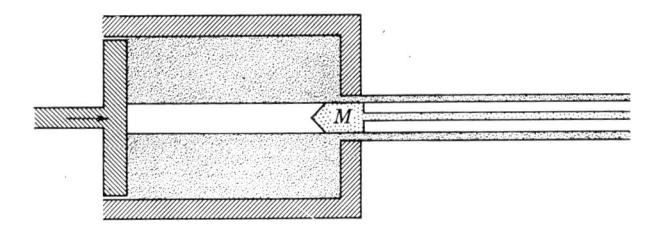

Fig. 7.35. Tube extrusion press, showing mandrel M.

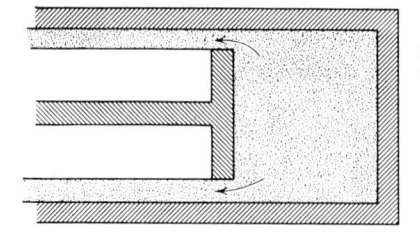

Fig. 7.36. Back extrusion.

to hydrostatic pressure, but the stresses in the neighborhood of the die orifice must contain large components of shear stress. Shapes of considerable complication can be extruded by using suitably shaped dies. Tubes, either of circular or of less simple shapes, can be extruded by using a hollow extrusion billet and extruding it round a mandrel as shown schematically in Fig. 7.35. The mandrel may be fixed, as shown in the diagram, or it may "float."

A variation of the usual extrusion process is back extrusion, used for making tubes, by forcing the metal through an annular space between a piston and a cylinder; this is illustrated in Fig. 7.36. In this process the force is often applied by impact.

Wire Drawing. A second process of circular symmetry is wire drawing, in which the material is drawn through a series of dies by means of a tension force. The typical process is represented in Fig. 7.37. Wire drawing is nearly always a cold-deformation process, intermediate annealing between stages being used when necessary. In the case of copper, for example, it is found that it is desirable to anneal after reduction in thickness by about 95%, although slower drawing, under less tension stress, allows greater reduction before the danger of fracture becomes important.

Pressing. In pressing, a sheet, originally flat, is placed between mating dies which are forced together, the sheet being stretched and deformed into the required shape. Because a high proportion of the metal is deformed simultaneously (rather than sequentially) and because the

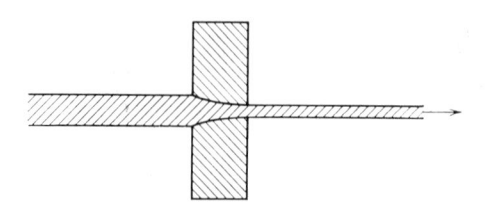

Fig. 7.37. Wire drawing.

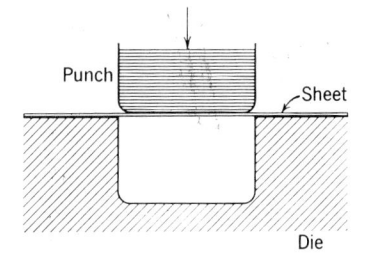

Fig. 7.38. Deep drawing, first stage.

process can be used for forming very large sheets, extremely large forces are often required.

Deep Drawing. Deep drawing is an extension of the pressing process. It is usually carried out on a much smaller scale, and consists of a sequence of pressing operations, each of which requires a separate die, with intermediate anneals. A stage of the process is shown schematically in Fig. 7.38.

Spinning. An alternative method for converting sheet to nonplanar, circularly symmetrical shapes is by spinning. In this operation, a disk of sheet metal is rotated about its center, and it is forced against a mandrel by means of a tool that applies pressure locally. The process is illustrated in Fig. 7.39, in which a partly completed cupping operation is shown by the full lines, and the finished shape by the broken line.

Properties of metals fabricated by deformation. Although the processes of plastic deformation were devised in order to produce metal components of required shapes, they can often be used to produce desirable properties as well as shape; at the same time, undesirable properties sometimes develop, and it is therefore necessary to consider those aspects of the structure of deformed metals that contribute to the properties. This discussion can be conveniently divided into two parts, dealing with hot-deformation and cold-deformation processes respectively.

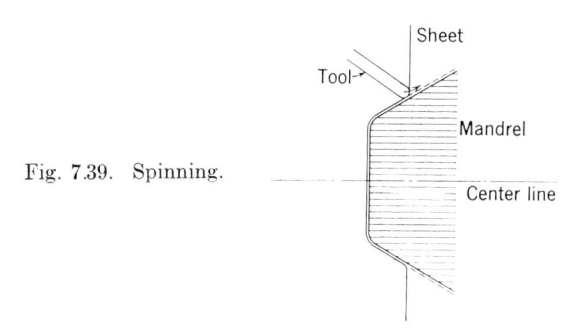

Fig. 7.39. Spinning.

HOT-DEFORMATION PROCESSES. It was pointed out above that, in hot deformation, recovery processes occur simultaneously with deformation; when deformation ceases, however, the recovery processes continue while the metal is at a high enough temperature; in particular, recrystallization is likely to become complete and grain growth may occur. The extent of grain growth, and hence the coarseness of the resulting grain size, depends upon the temperature at which deformation finishes, and on how long it takes for the metal to cool down to below the recrystallization temperature. This in turn depends upon the size of the component. In order to produce a fine grain size, which is desirable for most purposes, it is necessary to continue deformation to the lowest possible temperature while the metal is cooling down. This is possible in forging operations, but not in the extrusion process. However, the cross section of an extrusion is usually fairly small and it normally cools quickly.

A second structural characteristic of a forging is the distribution of nonmetallic inclusions, the presence of which is inevitable in most industrial processes. As pointed out above, some inclusions are drawn out into "strings," and the metal is weaker in directions transverse to these strings than parallel to them. The strings of inclusions delineate the "flow lines" of a forging, and indicate the directions and regions of weakness. A forging containing inclusions revealed by the flow lines is shown in Fig. 7.40. It is sometimes very important to design the forging operation so that the flow lines are parallel rather than per-

Fig. 7.40. Macrophotograph showing flow lines in a section of a forging. From *Metals Handbook*, ASM, Fig. 2, p. 44.

Table 7.4. Effect of Cold Rolling on the Properties of 2S Aluminum Sheet

Reduction in Thickness (%)	Yield Strength (psi)	Tensile Strength (psi)	Elongation (%)	Brinell Hardness
0	5,000	13,000	35	23
20	13,000	15,000	12	28
40	14,000	17,000	9	32
60	17,000	20,000	6	38
80	21,000	24,000	5	44

pendicular to the directions of the tensile stresses that will be encountered in service.

COLD-DEFORMATION PROCESSES. The amount of work hardening since the last anneal has a great influence on the yield stress, tensile strength, ductility, and hardness; this has been discussed in terms of the tensile test (p. 214), but it must be pointed out that much larger deformation and therefore much more work hardening is possible when deformation takes place under a suitable combination of biaxial or triaxial pressure and tension, as in rolling for example, than under tension alone, as in the tensile test. The effect of various amounts of cold working on the properties of aluminum sheet are given in Table 7.4.

The preferred orientation that may result from cold working followed by annealing may be either desirable or undesirable. A case in which it is desirable is that of iron-silicon alloy sheet used for transformer cores, referred to on p. 225. An example of an undesirable preferred orientation is in sheet that is to be used for deep drawing. The tensile properties are a function of direction in a sheet which has a preferred orientation; an example is shown in Fig. 7.41. The operations of deep drawing depend mainly on the tensile properties of the material in the radial direction; these properties vary from one direction to another and consequently the drawing process is not uniform around the circumference of a circular cup. The result is that the final cup may exhibit "ears," as illustrated in Fig. 7.42.

Cutting processes

MACHINING. Most metal-forming sequences include one or more operations in which metal is separated by a cutting action. These operations include sawing, filing, drilling, milling, and turning. The process of cutting in general may be one of cleavage (or brittle fracture)

350   PHYSICAL METALLURGY

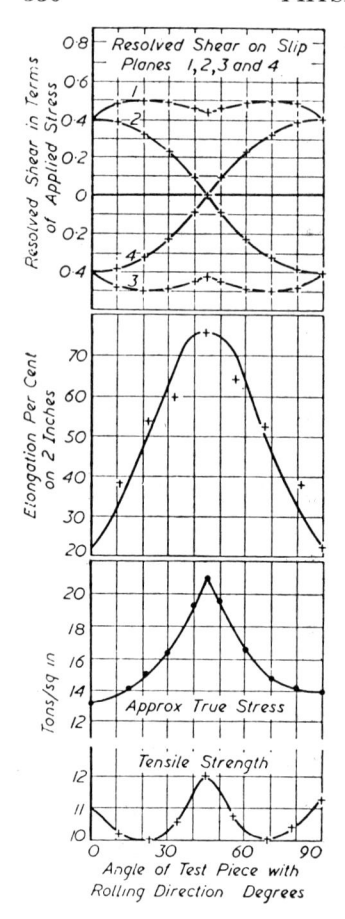

Fig. 7.41. Variation of properties with direction in a sheet with preferred orientation. From T. Ll. Richards, Chapter 6 in *Progress in Metal Physics 1,* Bruce Chalmers, Ed., Butterworths Scientific Publications, London, 1949, Fig. 6, p. 303.

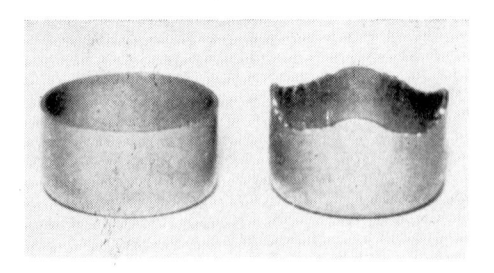

Fig. 7.42. Cylindrical cups made from isotropic and anisotropic copper strip, the latter with a (100) [001] recrystallization texture. From Richards, *Progress in Metal Physics 1,* Plate XXXVII, facing p. 296.

Fig. 7.43. Cutting by shear deformation.

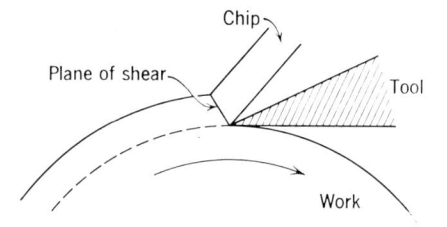

or it may consist largely of plastic deformation. The fundamental process of plastic cutting is illustrated in Fig. 7.43. The tool, either directly or through a built-up layer of the material being cut, deforms the material, by shear, into a "chip" that is separate from the "work." A substantial part of the work done in machining is expended in shear deformation. Some work is also done against friction, both on the lower surface of the tool, where it may be in sliding contact with the "work," and on the upper surface, where the chip may slide over it.

Four factors determine the efficiency of a cutting operation: 1. the time taken, 2. the resulting surface finish, 3. tool life, 4. the power expended per unit amount of metal removed. These parameters are interdependent, and they also depend, in complicated ways, on the tool shape, the cutting fluid used for lubrication and cooling, and the structure of the metal being cut. In general, higher machining speed improves the surface finish by decreasing the extent of the "built-up edge," but it also results in higher temperatures and this may necessitate the use of tool materials that retain their strength at high temperatures. The structure of the metal has an important influence on its machinability; steels are most readily cut when hot-worked or annealed rather than when cold-worked; steels containing either sulfur or lead are easily machined, apparently because there are particles or layers that break up the chip as it forms. This is said to promote long tool life and a good surface finish. The graphite in gray cast iron, similarly, breaks up the chip and also acts as a lubricant. No completely satisfactory criterion or test for the machinability of a metal has yet been devised.

ABRASION AND WEAR. Abrasion is the process of removing material from a surface by means of a series of miniature cutting operations usually conducted by sharp hard particles that are rubbed against the surface. The process is more effective if the hard particles are prevented from rotating, either by sticking them to paper or cloth, or by embedding them in a soft surface, such as a lead lap. It is commonly found that when two surfaces are in sliding contact it is the harder one that wears; this is because the foreign particles that cause abrasion become embedded in the softer material and abrade the harder one.

Polishing and burnishing. Mechanical polishing is the process in which the material near the surface is redistributed in the form of a smooth layer, by a process of flow. It is thought that in some cases the surface material is actually melted by friction with the particles of the polishing medium, which always has a very high melting point. The alternative explanation is that plastic flow takes place locally under a combination of high shear stress and high hydrostatic pressure. The structure of the polished surface is either amorphous or else crystalline but with such a small crystal size that it cannot be distinguished from amorphous material. It is the structure that would be expected to result from melting followed by cooling that is sufficiently rapid to prevent the growth of any crystal above some extremely small size, which is perhaps the maximum expected size of crystal embryo at the lowest temperature at which the atoms are effectively mobile. The same structure would be produced by extremely severe plastic deformation, at a fairly high temperature, as a result of the work done during deformation, followed by a very rapid cooling which would prevent any crystals growing above the extremely small size they may be considered to have during the incubation period of the recrystallization process. Such a structure could equally well be described as a supercooled liquid or a very fine grained crystalline solid.

Burnishing is the production of a substantial layer of severely cold-worked metal by friction with a tool that leaves a smooth surface; this can be achieved if the tool is extremely smooth and does not weld to the surface.

7.5 Solid-State Welding

It has been shown (see p. 134) that the free energy of a sample of metal contains a surface component. The free energy can therefore decrease by means of any change that reduces the surface area, except in the case, discussed on p. 135, in which specific low energy planes are developed. The amount of free energy that can be gained in this way is small; surface energies are of the order of 10^3 ergs/cm^2, while the stored energy of cold work can amount to 0.2 cal/gram or about 10^7 ergs/cm^3; and while the whole of the latter is released by recrystallization, only a part of the former is available because the surface can never be eliminated completely. In consequence, the processes that depend upon this driving force are in general slow, and only take place at high temperatures, at which plastic deformation can occur under small stresses.

Nevertheless, there are several important fabrication techniques that depend upon this driving force. They are all based upon the fact that, if two clean metal surfaces are maintained in contact with each other at a high temperature, a bond is formed between them. The bond starts to form at the few points where atoms in the two surfaces come within the range of interatomic faces (10^{-7} cm) of each other. The bond, which is a crystal boundary, grows outwards, with a consequent reduction of the area of the two surfaces. The process is represented in Fig. 7.44a, in which a region of initial contact A is shown surrounded by regions BB where there is no contact. The formation of the bond, as shown in Fig. 7.44b, replaces two surfaces by one grain boundary. The mechanism by which the spaces at B are filled by atoms depends on the motion of vacancies. A vacancy originally at V_1 (Fig. 7.44a) can exchange places with an atom, which helps to form the bond. The vacancy continues to migrate until it emerges from the crystal either at a surface or at the boundary that has already been formed at A. In either case there is a net transfer of atoms from the boundary, or the surface, to the regions BB. Regions where the surface is concave act as "sources" for vacancies, or "sinks" for atoms. The transport of atoms away from the boundary allows the two pieces of metal to move bodily towards each other. It has been clearly demonstrated that this process occurs in silver, at temperatures at which the oxide is not stable; it has also been shown that the junction formed in

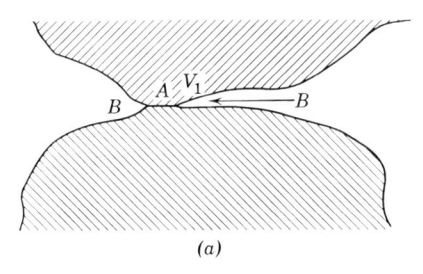

Fig. 7.44. Formation of sintered joint between two crystals.

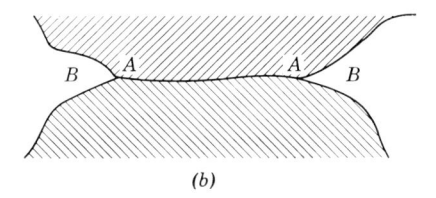

this way between two crystals has the energy and the mobility of an ordinary crystal boundary. It also occurs in other metals and alloys, but is complicated by the presence of surface layers such as oxides, which can prevent metal-to-metal contact.

The principles discussed above are used in powder metallurgy, and in various processes for joining one piece of metal to another without fusion.

Powder metallurgy. The process known as powder metallurgy consists of two steps: a powder is pressed into a die, and it is sintered. These two processes may be carried out in sequence, the pressing being conducted at room temperature (cold pressing), or simultaneously, by hot pressing.

The compression of the powder into the die causes the particles to be deformed so that intimate contact is established between adjacent particles, and it simultaneously increases the surface area of each particle so that they are no longer completely covered with oxide. This allows metal-to-metal contact to be made at numerous isolated points, with the result that the compressed powder has sufficient strength to withstand normal handling.

The second stage is that of sintering,* in which the compact is heated to an elevated temperature, at which it is held for a predetermined time. This temperature is well below the solidus temperature; consequently no fusion occurs. However, the compact becomes progressively denser and stronger, and may finally have properties comparable with those obtained by the traditional methods. Alternatively, conditions may be adjusted so that the compact remains porous even after considerable densification and strengthening have taken place. In this form, powder metallurgy products are widely used for so-called "oilless bearings"; lubrication is supplied by oil that is forced into the interconnecting pores of the sintered compact.

The process of sintering takes place in the following stages:

1. Establishment of metal-to-metal contact. It is to be expected that, if two completely clean metal surfaces are brought into contact with each other, the metallic bond should be set up between the atoms of one surface and those of the other. The area of contact is a crystal boundary, and if the temperature is high enough to allow some rearrangement of the atoms it takes up the structure characteristic of a crystal boundary of whatever difference of orientation there happens to be.

* Not to be confused with the sintering of ores, which depends on the formation of a liquid phase.

Fig. 7.45. Change of shape of void formed between particles.

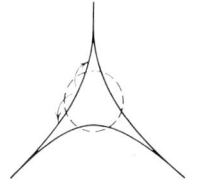

2. Reduction of surface area. The surface area is reduced by motion of the particles towards each other and the resulting increase in the area of the grain boundary between them. The atoms move from convex and less concave to the more concave parts of the surfaces; a region of contact of three grains, as shown in Fig. 7.45, first changes shape until it is spherical and then decreases in size if it is associated with a grain boundary that can act as a sink for the vacancies that must be eliminated if the void is to be filled with atoms.

Fabrication by powder metallurgy has become an important process for metals and alloys for which there are no satisfactory crucible materials, and which have very high melting points. Very high strength can be achieved when the conditions are suitably adjusted, and standard fabrication techniques can be used on sintered objects. The production of controlled porosity is another important application. For materials that are brittle at ordinary temperatures, it is preferable to combine the pressing and sintering operations, carrying out the pressing at a high temperature. Sintering is probably much faster under these conditions because of the high concentration of vacancies and dislocations that exists during deformation.

Other solid-state bonding techniques. The same principles apply to a variety of other methods for joining metals that do not depend on fusion. The simplest is the "blacksmith's weld," in which the two components to be joined are hammered into close contact while hot. Considerable deformation while the two surfaces are in contact is necessary, probably to cause metal-to-metal contact at enough points to allow sintering to proceed at a fast enough rate.

"Cold welding," similarly, is a joining process that results from very severe deformation of the two surfaces while they are in contact. No heating is necessary in this operation, which works best on relatively low melting point metals such as aluminum. A very large amount of deformation is necessary to produce sufficient metal-to-metal contact for a strong bond to be obtained. The bond is formed rapidly, and therefore must depend only on establishing contact and not on diffusion.

Bonding of dissimilar metals can also be achieved by deforming them while in contact, as for example by rolling, and either sintering subsequently, or conducting the rolling operation at an elevated temperature. In such operations, the disposal of the oxide or other nonmetallic films is important. Care is taken to start with surfaces that are as clean as possible, but air-formed oxide films cannot be completely avoided. The oxide either goes into solution in one or other of the metals, or it remains as oxide but spheroidizes into isolated particles of much reduced surface area. This requires the operation of a diffusion process, which necessarily takes time and can occur only at an elevated temperature.

7.6 Further Reading

Cold Working of Metals, American Society for Metals, Cleveland, 1949.
Radiation Effects in Solids, G. J. Dienes and G. H. Vineyard, Interscience Publishers, New York, 1957.
"Recrystallization and Grain Growth," Chapter 7 in *Progress in Metal Physics 3,* Bruce Chalmers, Ed., Pergamon Press, London, 1952.

8

Solid-State Transformations

8.1 Classification of Solid-State Transformations

In the processes of solidification and plastic deformation, the main objective is to produce a metal component of a required shape; some control of the structure, and therefore of the properties, is possible in these processes. However, most of the mechanically useful alloys have been developed on the basis of subsequent control of some aspects of the structure without further change of geometry of the piece as a whole, by causing phase transformations to occur. This is usually achieved in several stages. The first stage is to hold the alloy at an elevated temperature for a long enough time for phase equilibrium to be closely approached. The temperature is then lowered, so that the existing phases are no longer in equilibrium, with the result that a phase change takes place. The most desirable structures are usually obtained when the approach to a new equilibrium condition is interrupted before equilibrium is achieved. The technical operations that are conducted for this purpose are described as "heat treatment," a term that also includes the annealing processes discussed in Chapter 7.

It will be apparent from Chapter 2 that there are many alloy systems in which a change of temperature causes a change in the relative free energies of the possible phases. The result is that in such cases equilibrium can be maintained during a change of temperature only by the partial or complete replacement of existing phases by new ones. Such rearrangements of the atoms are called phase transformations. They not only occur reversibly, as implied above, but also as an approach to equilibrium in a system in which the temperature has been changed too rapidly for equilibrium to be maintained while the temperature is changing. The new phase that is formed by such a transformation need not be the equilibrium phase; its formation must reduce the free energy of the whole system, but the criterion that determines which of several possible phases is actually formed is the rate at which the phase forms as a result of nucleation followed by growth, rather than the extent to which the free energy of the system is reduced. It is therefore possible to have competing transformations; the detailed thermal history determines the relative proportions. Examples of competing transformations will be discussed later.

There are several distinct types of phase transformations; they may be classified according to the number of new phases that form, and whether the new phases differ in composition from the original phase. The following types must be distinguished:

1. One phase transforms into one new phase without change of composition; this is referred to as the diffusionless or martensitic transformation.

2. One new phase is formed, differing in composition and structure from the original phase, which remains in existence although undergoing some change in composition. This is typified by proeutectoid and precipitation reactions.

3. One phase transforms into two new phases, each differing in composition from the original phase. This includes the eutectoid transformation (during cooling) and the peritectoid transformation (during heating), in which both the new phases differ in structure and composition from the original phase, and it also includes the much less common type in which the two new phases differ in composition but have structures that are the same as that of the original phase.

4. An ordered structure is formed by rearrangement of the atoms without any other change in the over-all structure of the crystal.

5. A solid and a gas react to form a new phase or to change the composition of the solid phase.

There is a tendency to concentrate attention on the transformations that occur while the temperature of an alloy is being reduced; this is because most alloys are used at temperatures below those at which reactions take place at significant rates, and therefore the structure of the alloy as used is the result of descending temperature transformations. The properties depend on the phases that are present and on their structure, shape, size, distribution, and the interactions associated with their interfaces. These depend to a large extent on the kinetics of nucleation and growth of the new phases, and therefore on the transformations that take place when temperatures are lowered; but the transformations also depend upon the detailed structure of the antecedent phase or phases, which have usually resulted from an ascending transformation, to which some attention must also be given. In addition, there are important cases in which the final result is reached by a controlled exposure to a higher temperature following a lower one. Consequently, it is necessary to include a consideration of the inverse reactions, as well as those categorized above.

A useful distinction may be drawn between those transformations in which no atom moves, during the transformation, through a distance that is more, relatively to its neighbors, than a fraction of the interatomic distance, and those in which atoms move through many times this distance. The latter type of transformation depends upon diffusion, the process by which atoms move about within the crystal.

8.2 Diffusionless, or Martensitic, Transformations

General. A diffusionless transformation is one in which the atoms of the existing phase move into positions on the crystal lattice of a new, more stable phase by a co-operative type of process in which each atom is enabled to move into its new position by the strain energy resulting from the similar movement of neighboring atoms; the process is not thermally activated, but takes place at a speed that is probably that of the propagation of an elastic disturbance, i.e., the velocity of sound. One of the consequences of the co-operative character of the atom movements is that the shape of the newly formed crystal is usually different from that of the space previously occupied by the same atoms. Such a change of shape of part of a crystal sets up strains that oppose further transformation; consequently, the transformation may fail to go to completion under the conditions which cause it to start. A lower temperature (or more driving force) is required to produce more transformation.

Three distinct temperatures characterize a diffusionless transformation, as follows:

1. M_D is the temperature above which the new (martensitic) phase cannot form because the free energy of the system would not be reduced thereby.

2. M_S is the temperature at which the martensitic phase begins to form if the temperature is lowered. M_S is lower than M_D because some supercooling is necessary to cause nucleation of the new phase.

3. M_F is the temperature at which the whole of the parent phase has been transformed into the martensitic phase.

Crystallography of diffusionless transformations. Because of the co-operative nature of this type of transformation, it is necessary that the interface between the parent and martensitic phases should be coherent, because otherwise it is not possible for each atom to undergo similar movements. The equivalence of the movements of all the atoms is demonstrated experimentally by the fact that an ordered arrangement in the parent phase yields an ordered arrangement in the martensitic phase.

The coherency of the interface that must exist when the transformation is actually occurring may be lost immediately afterwards by plastic deformations of one or both phases, but a definite relationship between the orientations of the two phases must persist, and the interface of the transformation (habit plane) must also have a definite crystallographic significance. The crystallographic data for some diffusionless transformations are given in Table 8.1.

Various studies of the crystallographic relationship between parent and martensitic phases have shown that, in most cases, the new structure cannot be reached by a single shear process of the parent crystal, but that it can be described in terms of two consecutive simple deformations, the first of which could usually be shear parallel to the habit plane. It is not known whether the individual atoms undergo two successive changes in position, or whether an atom makes a single composite jump as the interface passes it.

Reversibility of diffusionless transformations. In the strict thermodynamic sense these transformations are irreversible, since they do not occur at the temperature of equilibrium of the two phases; from a crystallographic point of view, however, they are reversible, in the sense that if a single crystal of a parent phase is transformed into the "martensitic" phase by lowering the temperature, and is then caused to transform back by raising the temperature, a single crystal of the original orientation is formed. This is found to apply even when the

Table 8.1. **Crystallographic Data for Some Diffusionless Transformations** *

System	Composition	Transformation on Cooling	Relationship between Crystallographic Planes	Relationship between Crystallographic Directions
FeC	0.5–1.4% C	F.C.C. → B.C. Tetragonal	$\{111\}\gamma \parallel \{110\}\alpha$	$\langle 110\rangle\gamma \parallel \langle 111\rangle\alpha$
FeNi	27–34% Ni	F.C.C. → B.C.C.	$\{111\}\gamma \parallel \{110\}\alpha$	$\langle 211\rangle\gamma \parallel \langle 110\rangle\alpha$
Li		B.C.C. → C.P. Hex	$\{110\} \parallel \{0001\}$	$\langle 111\rangle$ 3° from $\langle 11\bar{2}0\rangle$
AuCd	47.5 atomic % Cd	B.C.C. → orthorhombic β'	$\{001\}\beta \parallel \{001\}\beta'$	$[111]\beta \parallel [110]\beta'$
Co		F.C.C. → C.P. Hex	$\{111\} \parallel \{0001\}$	$\langle 110\rangle \parallel \langle 11\bar{2}0\rangle$

* Data from J. S. Bowles and C. S. Barrett, Chapter 1 in *Progress in Metal Physics 3*, Bruce Chalmers, Ed., Pergamon Press, London, 1952, p. 6.

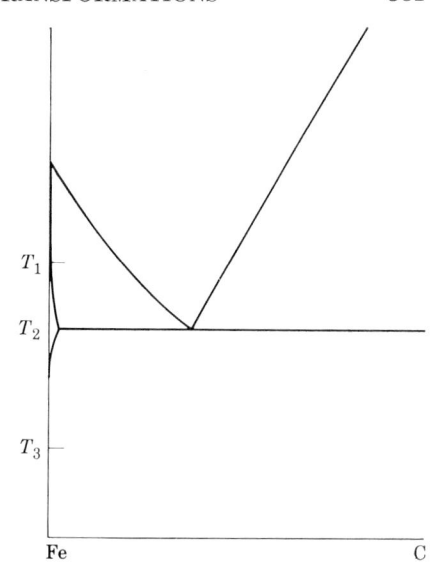

Fig. 8.1. Part of the phase diagram for iron-carbon alloys. From *Metals Handbook*, American Society for Metals, Cleveland, 1948, p. 1182.

martensitic crystals have a number of different orientations, all of which are, of course, similarly related crystallographically to the parent crystal. This reversion to the original form may be prevented by the occurrence of a competing transformation.

Effect of plastic deformation. At any temperature below M_D, the parent phase is metastable, its transformation being prevented either by the lack of nuclei or by the presence of strain energy that would be increased by further transformation. Plastic deformation of such a metastable parent phase causes the martensitic transformation to take place, evidently by contributing to the nucleation of martensitic crystals. Further aspects of martensitic transformations will be considered in relation to specific cases.

The martensitic transformation in steel. The phase diagram for iron-carbon alloys (Fig. 8.1) indicates that austenite containing, for example, 0.6% carbon transforms, under equilibrium conditions, to ferrite and cementite by the time the temperature falls to 720°C. In this condition, practically the whole of the carbon in the steel is concentrated in the cementite, which constitutes only a small fraction of the material. It is therefore necessary for the carbon to undergo considerable redistribution by diffusion, a process that requires time (see p. 369). If cooling is too rapid for this to occur, the austenite instead remains in existence until the temperature is reached at which the free energy can be reduced by a diffusionless transformation resulting in the formation of martensite.

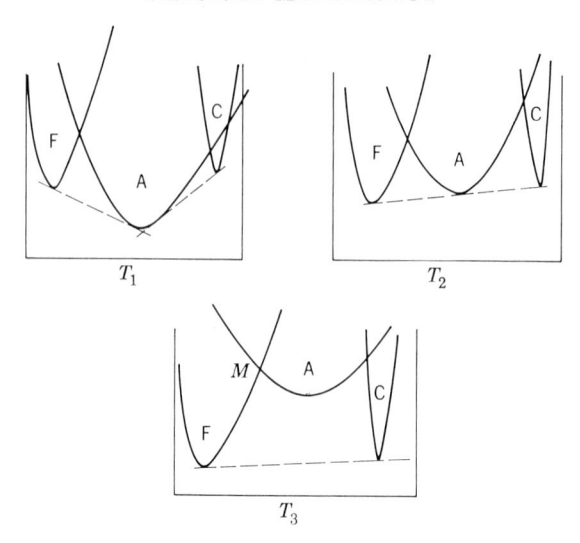

Fig. 8.2. Free energy relationships for ferrite, austenite, and cementite (schematic).

The free energies of the different phases are illustrated for various temperatures in Fig. 8.2. At T_1 (Fig. 8.1) austenite is a stable phase, and it may be in equilibrium with either ferrite or with cementite. At T_2, the eutectoid temperature, austenite is in equilibrium with both ferrite and cementite. At T_3 austenite is no longer a stable phase, but if no change of composition can occur (owing to lack of time for diffusion) the only transformations that can take place are to ferrite or to cementite. These transformations represent a decrease in free energy for all compositions at which the free energy curve for austenite is above that for ferrite or cementite. This applies to all carbon concentrations below that of point M in Fig. 8.2. The ferrite curve at point M corresponds to ferrite containing far more carbon than is usually to be found in ferrite. It is not stable, as the free energy would be reduced if the supersaturated ferrite, called martensite, decomposed into equilibrium ferrite and cementite.

Thus the transformation to supersaturated ferrite is energetically possible when the temperature is below some value M_D that depends on the composition. Since this transformation can take place without change of composition, it is a diffusionless or martensitic transformation.

THE STRUCTURE OF MARTENSITE. As stated above, martensite may be regarded as ferrite that is supersaturated with carbon. The extent

Fig. 8.3. Distortion of ferrite by interstitial carbon.

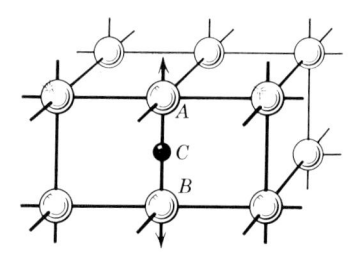

of the supersaturation depends upon the carbon content, which may be as high as 1.4%. The *equilibrium* solubility limit for carbon in ferrite is about 0.02%. Structurally, as well as thermodynamically, martensite is supersaturated ferrite. A carbon atom in interstitial solution in ferrite causes the lattice in its immediate vicinity to become tetragonal, instead of being cubic as it is in the absence of carbon. The tetragonality is due to the fact that a carbon atom takes up a position such as C in Fig. 8.3. This causes an *increase* in the distance between the atoms A and B, and a *decrease* in the spacing in the transverse directions. If carbon atoms are distributed randomly on sites of the three types shown in Fig. 8.4, there is no tetragonality of the lattice as a whole, the local directions of tetragonality being randomly distributed. If, however, the carbon atoms are relatively close together, they tend to co-operate to produce tetragonality that is constant in direction, as this corresponds to less elastic strain. It is probable that "long-range tetragonality" occurs whenever there is more than about 0.2% carbon (by weight) present. This corresponds to about 1 atom of carbon to every 50 unit cells. The interstitial sites occupied by carbon in austenite do not cause tetragonality, and it is evident that the ordering of the carbon must take place during the transformation. A crystal of martensite is shown in Fig. 8.5, in which the spacing of the carbon atoms corresponds very roughly to 0.8% carbon. The tetragonality is such that the c spacing is increased

Fig. 8.4. Alternative interstitial sites in ferrite.

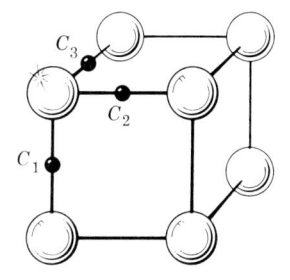

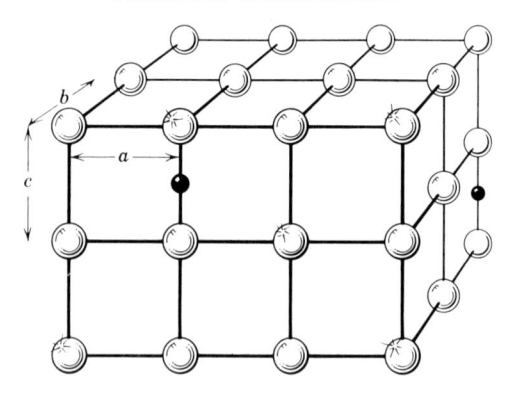

Fig. 8.5. Carbon sites in a ferrite crystal.

and the a and b spacings are decreased. The dependence of these spacings on the carbon content is shown in Fig. 8.6.

THE MORPHOLOGY OF MARTENSITE. Photomicrographs of martensite, an example of which is shown in Fig. 8.7, suggest that the individual crystals are needle-shaped. This is an example of the misleading impression that can be given by studying a two-dimensional section of a three-dimensional object. If the martensite crystals were needle-shaped, many of them would be seen as transverse sections, which would be roughly circular, which is not the case. The conclusion is that they must be platelike in form. The "plates" are elongated along one axis, and are thicker in the center than at the edges.

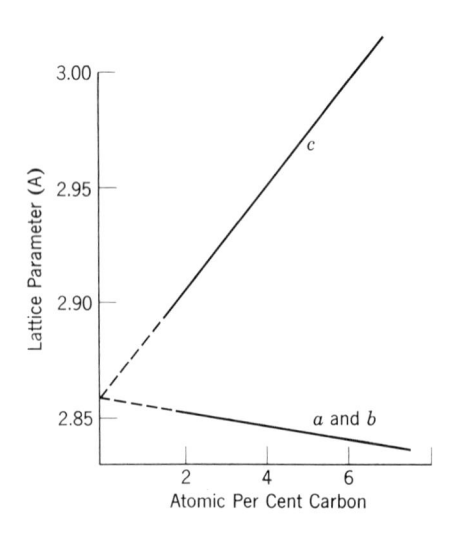

Fig. 8.6. Variation of lattice parameters of martensite with carbon content.

SOLID-STATE TRANSFORMATIONS 365

Fig. 8.7. Photomicrograph of martensite (×1660). (Courtesy U. S. Steel Corporation.)

THE PROPERTIES OF MARTENSITE. The most important and obvious property of martensite is its extremely high hardness. It is difficult to measure the properties of individual martensite crystals directly, but the hardness of austenite containing various proportions of martensite gives a clear indication that the hardness of martensite itself increases with the carbon content and reaches very high values when the carbon content is high. Figure 8.8 shows these relationships.

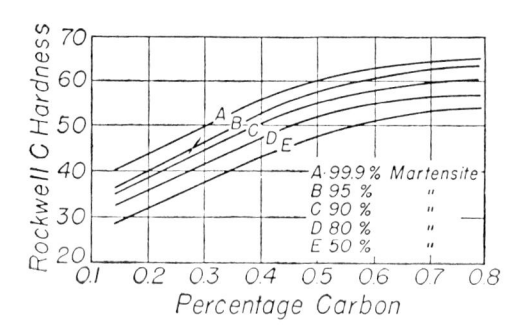

Fig. 8.8. Variation of hardness with carbon content and amount of martensite. From *Metals Handbook*, ASM, Fig. 7, p. 497.

A steel containing a high proportion of high carbon martensite has high strength, high hardness, and extremely low ductility. It is concluded that the martensite plates themselves also have high strength and low ductility. This case evidently differs from that of the random substitutional solid solution discussed on p. 166, where it was shown that there should be no strengthening effect due to the solute. Presumably it is very difficult to drive dislocations through a crystal that is as severely strained locally as martensite.

VOLUME CHANGE IN MARTENSITE TRANSFORMATION. The body-centered cubic structure is less close-packed than the face-centered cubic structure, and consequently the volume increases when austenite changes to ferrite. The volume increase is about 4%. There is a greater volume change when austenite is transformed to martensite.

This increase in volume is sufficient to set up large stresses when one part of a piece of steel has transformed while the remainder has not. Martensite is usually formed while the steel is being rapidly cooled from the outside; the stress distribution in a cylindrical bar during rapid cooling is that of an expanded martensite shell surrounding an untransformed austenite core. Circumferential cracking may occur as a result. In less regularly shaped specimens, more complicated stress patterns arise, and cracking often occurs at regions of rapid changes of section.

NUCLEATION AND GROWTH OF MARTENSITE. The martensitic transformation in steel begins when a temperature M_S is reached which is significantly below the temperature M_D at which martensite becomes more stable than austenite. There is therefore a nucleation barrier that requires some supercooling below the temperature M_D of quasi equilibrium between martensite and austenite. The two temperatures, M_D and M_S, vary with carbon content; as would be expected from Fig. 8.2, the temperature M_D is higher for low than for high carbon contents. The variation of M_S with carbon content is shown in Fig. 8.9 for plain carbon steels.

The nature of the nuclei for martensite formation has not been elucidated; however, it is found that the formation of martensite in steel is only very slightly time-dependent, practically all the martensite that will form at any given temperature being formed immediately that temperature is reached. Consequently, it is probable that all the nuclei are present while the austenite is being cooled, and that they do not form subsequently as a result of fluctuations of composition due to diffusion. It is possible that the nuclei are regions of abnormally low carbon content, as a result of statistical fluctuation, in the austenite; dislocations or other imperfections may also contribute.

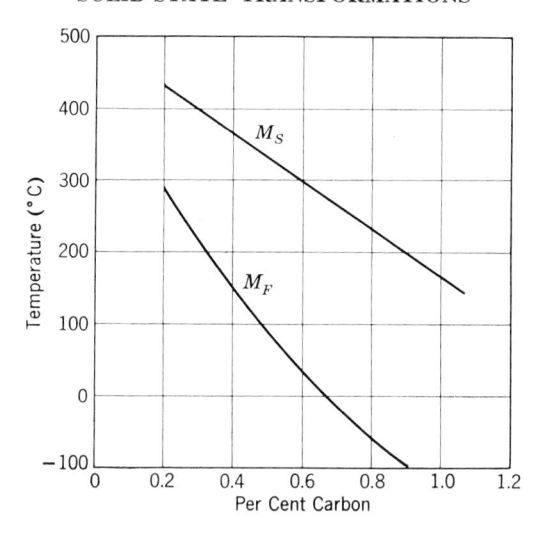

Fig. 8.9. M_S and M_F temperatures for carbon steel.

It is also found that the amount of martensite increases progressively as the temperature is lowered below the M_S temperature, until the transformation is virtually complete, at the M_F temperature. The reasons why the transformation does not go to the completion once the M_S temperature has been passed are that each individual martensite plate is only able to grow to a limited size, and that at any given temperature below M_S there are only a limited number of nuclei that can operate. The amount of transformation that occurs at any given temperature is less for a smaller grain size, probably because the amount of transformation corresponding to each nucleus is decreased. At any temperature below M_D, some transformation occurs if the material is plastically deformed. A substantial amount of transformation is produced in 18% chromium 8% nickel stainless steel, for example, if it is cold-worked at room temperature, which is between M_D and M_S. A similar situation exists with Hadfield's steel (13% manganese), which remains austenitic indefinitely at room temperature unless it is plastically deformed, in which case martensite is formed at the regions where deformation actually occurs. This steel therefore has a very high effective rate of work hardening, since the martensite is extremely hard.

STABILIZATION OF AUSTENITE. An effect of holding austenite at a temperature below M_D is to reduce the amount of austenite that will form at any lower temperature; the M_S temperature is also lowered.

This is presumably due to the disappearance of nuclei which are subcritical in size at such temperatures. The phenomenon is called stabilization.

Diffusionless transformations in titanium alloys. Many other examples are known of diffusionless transformations; all occur for the reason discussed above, i.e., the formation of a lower free energy phase even when no diffusion is possible.

Alloys of titanium with many of the transition metals undergo a diffusionless transformation on rapid cooling. The phase diagram for a typical case, that of titanium-manganese alloys, is shown in Fig. 8.10. The transformation is characterized by an M_S temperature, which decreases with increasing concentration of solute, exactly as in the case of the iron-carbon alloys. The M_S temperatures for a number of systems are shown in Fig. 8.11.

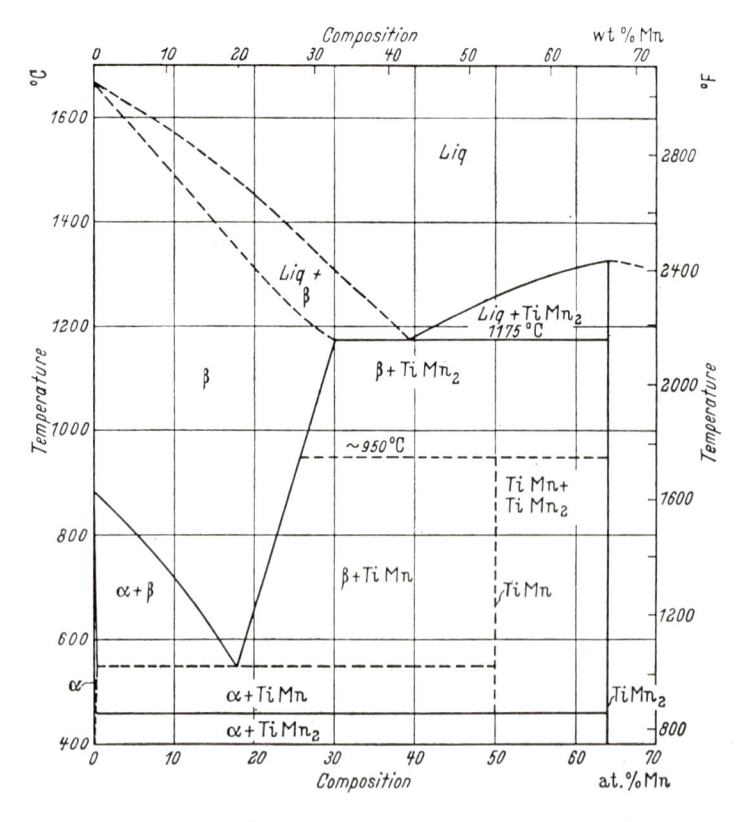

Fig. 8.10. Phase diagram of titanium-manganese system. From *Titanium*, A. D. McQuillan and M. K. McQuillan, Academic Press, New York, 1956, Fig. 78, p. 227.

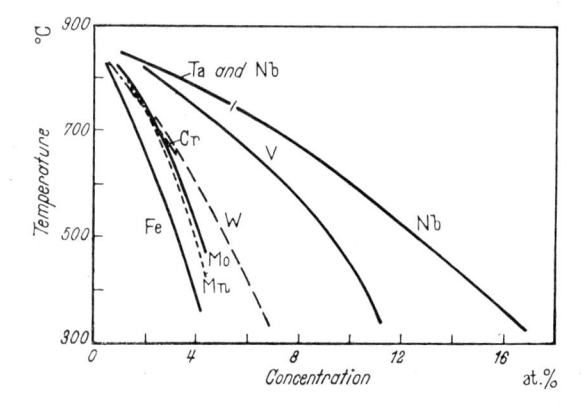

Fig. 8.11. M_S temperature for titanium alloys. From *Titanium*, McQuillan and McQuillan, Fig. 125, p. 315.

The supersaturated α-titanium solid solutions produced by the "martensitic" transformation do not have the mechanical properties characteristic of martensite. This is because the solute in this case is substitutional instead of interstitial, and far less lattice distortion is produced by a substitutional solute atom (e.g., of manganese in the hexagonal close-packed α-titanium structure) than by an interstitial carbon atom in the body-centered cubic structure of α iron. The diffusionless transformations in the alloys based on titanium and zirconium do not appear to have any practical value.

8.3 Diffusion

All the remaining types of solid-state transformations require the movement of atoms with respect to the crystal as a frame of reference by distances that are at least as great as the distance between atoms. It is therefore necessary to discuss the process, diffusion, by which these movements occur.

The term *diffusion* is used here to describe any changes in the relative positions of atoms or molecules in a medium which is stationary. This definition excludes atom movement that is brought about by the flow or plastic deformation of the medium. Diffusion can take place in gases and liquids, and, under suitable conditions, in solids. Diffusion in gases is not relevant to the present study, and consideration will be confined to diffusion in liquid metals and solid metals. Diffusion in solid metals can again be subdivided into diffusion in crystals, diffusion along dislocations, diffusion in crystal boundaries, and diffusion on surfaces. The best understood of these types of diffusion is diffusion in crystals; this will be discussed first.

Diffusion in crystals

THE UNIT PROCESS OF DIFFUSION. Diffusion in crystals, as defined above, takes place by the movement of individual atoms in relation to the crystal, the position of which is defined by the positions of all those atoms which do not change their relative positions. The unit process is, therefore, the motion of an individual atom from the site that it occupied previously into another site. All the atoms vibrate continuously about their mean positions, and the unit diffusion process or "jump" is an event in which the mean position of an atom changes. This can happen only if the atom in question has a sufficiently high energy and if there is a site into which it can go.

JUMP FREQUENCY. The process of diffusion, whether discussed theoretically or studied experimentally, is a result of all the diffusion jumps that take place in a crystal that contains a very large number of atoms in a time period that is very long compared with the time characteristic of thermal vibration. Consequently, the process must be considered statistically rather than in terms of individual atoms, and the most important quantity is the average time τ that elapses between consecutive jumps made by any atom in the aggregate. It is impossible to make any predictions about the time at which an individual atom will jump; it is, however, possible to predict the average interval; it is given by

$$\frac{1}{\tau} = P\nu \exp\left(-\Delta H/RT\right)$$

where $\exp\left(-\Delta H/RT\right)$ is the probability that an atom has an energy ΔH due to thermal motion, during a single vibration; ΔH is the energy of activation for the jump; ν is the number of thermal vibrations per second and is given approximately by $\nu = kT/h$; and P is a factor which corresponds to the probability that an atom with high enough energy can reach an available site. The factor P contains three components:

1. The number of equivalent sites that an atom can reach by a single jump. This is 4 for a body-centered and 12 for a face-centered cubic structure if the atom in question is in an interstitial site and can only move into another interstitial site; if the atom is in a *lattice* site and moves into a lattice site, this factor is 12 for a face-centered cubic and 8 for a body-centered cubic structure.

2. The probability p that one of these sites is available. This is unity for an interstitial atom because neighboring interstitial sites are never filled; but for an atom in a lattice site, it is the probability of a

neighboring site being vacant, and is consequently equal to β, the fractional concentration of vacancies in the crystal.

3. The probability of the gap through which the atom must pass being large enough, as a result of the thermal vibrations of the surrounding atoms, at the instant its thermal energy is high enough. This probability corresponds to an entropy of activation ΔS for the jump. The three components are combined as $P = p\beta \exp(\Delta S/R)$. In order to calculate the jump frequency, the fractional concentration of vacancies β, if applicable, must be known; this is given by $\beta = \exp(-\Delta G_1/RT)$, where ΔG_1 is the free energy associated with the formation of a vacancy, a quantity for which good estimates have been made; the value of ΔS must also be known; it can be found from the variation of elastic properties with temperatures. Table 8.2 gives some typical values of jump frequencies for metallic crystals.

THE DIFFUSION COEFFICIENT. It can be concluded from the considerations discussed above that atoms move about in crystals at temperatures that are commonly encountered. As has been shown, the movement of vacancies, which could not occur except by diffusion jumps of the atoms, becomes sufficient at temperatures of about 0.4 of the melting point to allow dislocation climb, which leads to stress relief and polygonization. The jump time for vacancies is, however, shorter than for atoms, by the factor β, which in a typical case may be 10^{-6}. Thus processes that depend on the movement of specific atoms, rather than of vacancies, do not become significant until much higher temperatures are reached.

An important aspect of the diffusion process is that, at any rate in cubic crystals, the jumps are in random directions. In noncubic crys-

Table 8.2. Average Diffusion Jump Times τ

Solvent	Solute	Temperature (°C)	τ (sec)
Ag	self-diffusion	800	1.8×10^{-7}
Ag	Cd	600	2.0×10^{-9}
Ag	Pd	444	5.5×10^{-5}
Ag	Zn	700	1.5×10^{-8}
Au	Ni	550	1.0×10^{-4}
Fe	C	20	1

tals the activation energies are not necessarily the same for jumps in nonequivalent directions, and consequently different jump frequencies exist in relation to different directions. The effects of this kind of anisotropy have been measured, but will not be discussed further here.

In a cubic crystal, a particular atom makes a series of jumps, at an *average* time interval that can be calculated in the manner discussed above. Each jump is in a direction determined by chance, without any relation to the direction of the previous jump. This behavior is an example of the "random walk" process, which has been examined as a statistical problem. It can be shown that an atom starting at a given site at zero time, and making n jumps each of length a, will on the average have moved a distance proportional to $\sqrt{n}\,a$ in a direction determined by chance. It follows that, if some of the atoms in a crystal are recognizable (in practice either because they are chemically different or because they are radioactive) and if these atoms are initially not randomly distributed, the result of their random motion is towards a random distribution. Thus if the initial distribution of the "observable" species A is as shown in Fig. 8.12, curve a, the distribution at two later times will be as shown in curves b and c. The curve after an infinite time is the horizontal line d. If the concentration of the marked species A is represented by C, then the change of distribution can be considered as follows. Figure 8.13 represents a crystal in which the concentration C varies linearly from a maximum at the plane AB to a minimum at the plane CD. The plane PQ is a plane which separates the atomic planes 1 and 2. The simplest quantitative representation of the diffusion process is the net flow of the species A across such a plane. The process must be looked at as the average of a very large number of individual jumps, and it is not satisfactory just to consider individual jumps across

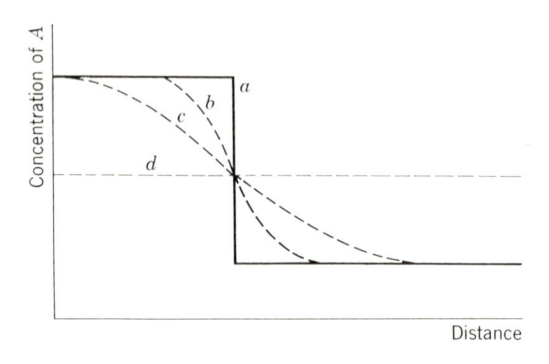

Fig. 8.12. Effect of diffusion on the distribution of A in a crystal.

Fig. 8.13. Schematic diagram of diffusion system.

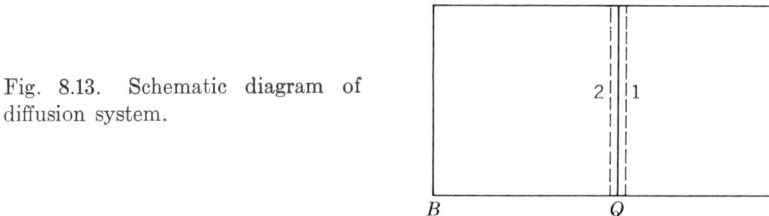

one plane because this does not allow appropriate averaging; instead, the rates of "flow" in the two directions must be considered. Since the distance traveled, on the average, is proportional to the square root of the time, it follows that the time required for atoms, on the average, to travel a given distance is proportional to the square of that distance; each atom, on the average, makes a jump every τ seconds, but only a fraction ρ of these carries it across the plane under consideration. The rate of movement of the atoms is therefore $(\rho C_1 \alpha^2)/\tau$. The number which cross from plane 1 to plane 2 is therefore $(\rho C_1 \alpha^2)/\tau/\alpha$ if α is the separation of the two planes under consideration. The number per unit area for unit time, therefore, is $(\rho C_1 \alpha)/\tau$ crossing in one direction and $(\rho C_2 \alpha)/\tau$ in the other. Hence the net flow is $[\rho(C_1 - C_2)\alpha]/\tau$. $C_1 - C_2$ is the concentration difference over a distance α; it is equal to $\alpha(dc/dx)$. Thus the net flow is $(\rho \alpha^2/\tau)(dc/dx)$.

The diffusion constant D is defined as the net flow per unit time per unit cross section under unit concentration gradient; therefore $D = \rho \alpha^2/\tau$, where ρ is a geometric factor depending on how many directions there are in which an atom can jump. ρ has the values $\frac{1}{24}$ for body-centered and $\frac{1}{12}$ for face-centered cubic crystals for interstitial diffusion, and $\frac{1}{8}$ and $\frac{1}{12}$ respectively for substitutional diffusion; α is the jump distance and τ is the mean jump time. As would be expected from the temperature dependence of jump frequency $1/\tau$, it is found experimentally that $D = D_0 \exp(-Q/RT)$, where the activation energy Q is equivalent to the sum of ΔH and ΔG, if vacancies are required for diffusion, and D_0 is the temperature-independent part of the expression. D_0 is given by $D_0 = p\alpha a^2 \nu \exp(\Delta S/R)$.

All the values of D_0 predicted from the expression fall within a small factor of 1 cm²/sec. Experimental results on diffusion constants consist of a series of values of D obtained at various temperatures. The Arrhenius equation shows that there should be a linear relationship between $\log D$ and $1/T$ if a single activated process is involved. This has been confirmed in many experimental investigations. The values of Q and D_0 are obtained from the slope and intercept of the straight

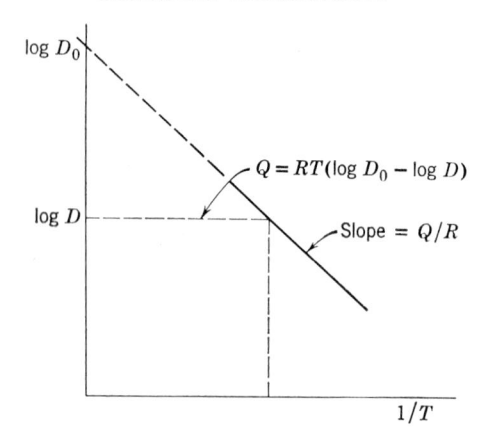

Fig. 8.14. Variation of diffusion constant with temperature.

line (see Fig. 8.14), since $\log D = \log D_0 - (Q/RT)$ or $Q = RT(\log D_0 - \log D)$. An example, the diffusion of carbon in α iron, is shown in Fig. 8.15.

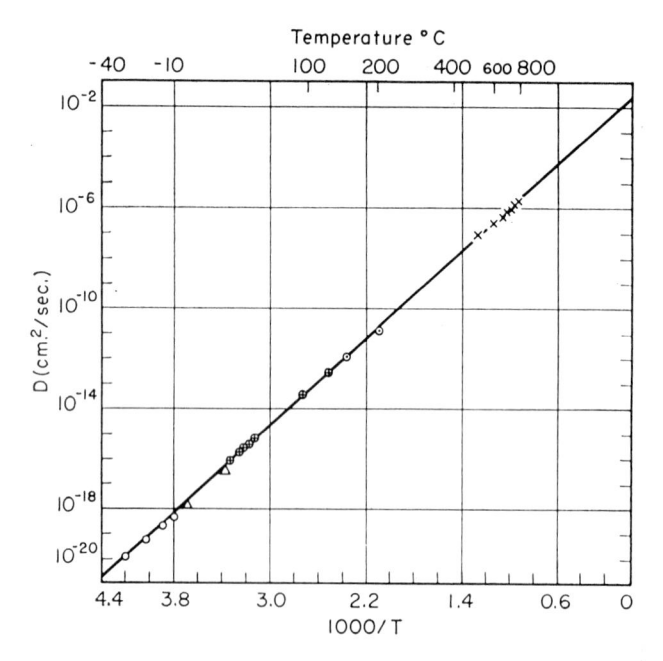

Fig. 8.15. Diffusion of carbon in α iron. From *Modern Research Techniques in Physical Metallurgy*, American Society for Metals, Cleveland, 1953, Fig. 7, p. 238.

Until recently, most diffusion measurements were reliable over such a small temperature range that the slope and intercept could not be determined accurately. Recent work, however, has shown that the values of D_0 in all cases fall close to the predicted value, and that the values of Q for various solute elements are all close together and in the case of substitutional diffusion close to the value for self-diffusion of the solvent. Some values of Q and D_0 are given in Table 8.3.

The general expression for a diffusion coefficient D is given by Fick's law, which can be expressed as $S = -AD(\partial c/\partial x)$, where S is the amount of material diffusing through the area A in unit time, and $\partial c/\partial x$ is the concentration gradient in the direction along which S is measured.

This expression implies that the concentration gradient is the "driving force" for diffusion, the end point of which should be complete uniformity of composition. This implication is not correct when diffusion of metallic systems is under consideration, because the equilibrium state of an alloy is not necessarily that in which the composition is the same everywhere; as shown in Chapter 2, the coexistence of two or more

Table 8.3. **Some Data for Diffusion Metals** *

Solvent	Solute	D_0 (cm²/sec)	Q (kcal/mole)
Ag	self-diffusion	0.89	45.9
Au	" "	0.16	53.0
Co	" "	0.37	67.0
Cu	" "	11	57.2
α Fe	" "	2300	73.2
γ Fe	" "	5.8	74.2
Pb	" "	6.6	27.9
Pt	" "	0.048	55.7
Zn { ∥ to c axis	" "	0.046	20.4
Zn { ⊥ to c axis	" "	91	31.0
Ag	Au	1.1×10^{-4}	26.6
Ag	Cd	4.8	22.3
Ag	Cu	6.0×10^{-5}	24.8
Ag	Pb	6.4×10^{-6}	20.2
Ag	Sb	7.8×10^{-5}	21.4
α Fe	C	2×10^{-2}	20.1

* Data from *Metals Reference Book*, 2nd ed., C. J. Smithells, Ed., Interscience Publishers, New York, 1955.

phases is often the stable situation; the criterion is the equality at all points of the chemical potential of each component of the system. The proper expression for the driving force for diffusion is therefore the gradient of chemical potential. This point of view explains why so-called "uphill diffusion" can take place. This is diffusion of a component from a region of low to a region of higher concentration. Many cases will be referred to later in which this takes place; for example, in the eutectoid reaction discussed on p. 415, each component diffuses towards the phase in which it has a high concentration. The same consideration applies in all cases of precipitation and pre-precipitation, discussed on p. 385.

For diffusion within a single phase, however, the gradient of chemical potential is nearly always quite similar to that of concentration, and it is usually permissible to measure and define the diffusion coefficient in terms of the concentration gradient; this is the form in which experimental results are usually given.

A second difficulty in the application of Fick's law to metals is their lack of homogeneity and isotropy. As will be shown on p. 381, diffusion takes place differently and with different temperature dependence in the grain boundaries and within the crystals of a polycrystalline metal; the value of D varies with direction in any crystal that has a noncubic structure. For fundamental studies, therefore, the single crystal is the ideal sample, although even in a single crystal the presence of dislocations can be significant.

A third difficulty is a more subtle one; if the gradient of concentration or of chemical potential is the driving force for diffusion, it would be expected that there would be no diffusion when there is no gradient; however, measurements show that *self-diffusion* takes place; this is the spontaneous rearrangement of the atoms in a crystal, without any mass transfer and without any thermodynamic driving force. Fick's law does not apply to *self-diffusion;* it only applies to the *result* of *chemical* diffusion. Nevertheless, a diffusion coefficient can be attributed to a self-diffusion process; these coefficients are comparable with those for chemical diffusion, and it is concluded that the two processes are basically identical. In the ideal case of a crystal of cubic structure, the diffusion coefficient can be calculated from the jump frequency. It is assumed that each jump takes place in a direction that is independent of the direction of the previous jump, i.e., that the jumps are randomly distributed between the ρ directions that are possible in the crystal. This assumption is not quite true, because an atom that has just changed places with a vacancy is slightly more likely to take its next jump back to its previous position than to some other position; this is due to the fact that the vacancy will stay in its new position for a finite time, which is much shorter than the jump time for an atom because it can change places with any of its surrounding atoms. The error due to this "correlation" effect can be ignored for most purposes.

The discussion so far only applies strictly to self-diffusion and to diffusion in dilute solutions; when the solution is more concentrated, so that the energy of a solute atom is changed by the proximity of other solute atoms, then the basic assumption that an atom is equally likely to jump in any direction is no longer valid. In these circumstances, the redistribution does not take place purely as an increase in randomness. There may in addition be a driving force which increases the probability of an atom moving towards either the region of low, or the region of high, concentration. The direction depends on whether the solute atoms have their energy increased or decreased by proximity to each other. The expression of the diffusion coefficient in terms of the gradient of chemical activity, instead of concentration, takes account of these effects.

There are two possible effects of solute concentration. The *driving force* for diffusion may be changed; this is equivalent to a jump being more probable in one direction than in another, owing to different free energy maxima existing to the higher concentration and lower concentration sides of a solute atom. The second effect is on the mobility, without regard to direction. This is equivalent to changing the free energy of the atom in its equilibrium site, and thereby changing the free energy of activation for a jump, and, therefore, the jump frequency.

It follows that the diffusion coefficient may be expected to depend on the concentration. This is verified experimentally. It is also found that a gradient of pressure or of temperature can cause diffusion in the absence of a concentration gradient.

Discussion of diffusion mechanisms. The assumption has been made in the preceding sections that the mechanisms of diffusion are not in any doubt. It is necessary to discuss the evidence for the mechanisms that have been assumed.

The diffusion of interstitials does not present a problem, because there does not appear to be any possible alternative to the thermally activated movement of the atoms from one interstitial site to an adjacent one. The situation has not always been as clear for the diffusion of substitutional solutes, and self-diffusion. A priori, there are three geometrically possible ways in which the atoms of a crystal could make diffusive movements:

378   PHYSICAL METALLURGY

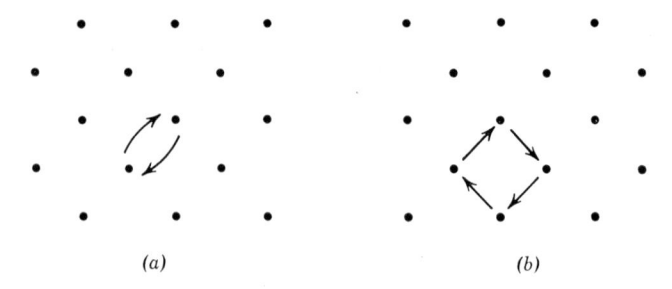

(a)   (b)

Fig. 8.16. Direct interchange and ring mechanisms.

1. By *direct interchange*, either of a pair of adjacent atoms or by the simultaneous movement of each of the atoms constituting a "ring" of atoms. These two types of movement are illustrated in Fig. 8.16.

2. By movement *as interstitial atoms*, or by the formation of a "crowdion" (see p. 95). An atom could leave a "crowdion" at a different point from that at which an atom entered it, giving a diffusion movement. These two interstitial-type mechanisms are represented in Fig. 8.17.

3. *The vacancy mechanism discussed above.* The evidence that supports the vacancy mechanism is of two kinds. One is that the theoretical calculations have shown clearly, at least in the case of face-centered cubic crystals, that it is energetically much more probable that diffusion should take place in this way, and that reasonably good predictions of the value of D_0 have been made on this basis. The other kind of evidence is based on the Kirkendall effect.

KIRKENDALL EFFECT. When a pure metal and an alloy, or two different alloys of the same two metals, are put into contact and held at a high temperature, the interface between them moves while diffusion is taking place. The typical experiment is to form two diffusion couples with, for example, copper and an alloy of copper with 30% zinc, and to mark the positions of the interfaces with very fine tungsten wires. A

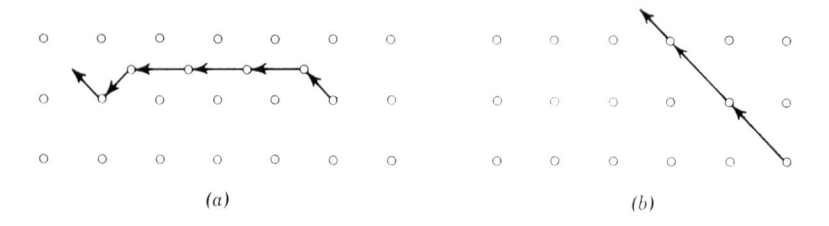

(a)   (b)

Fig. 8.17. Interstitialcy and crowdion mechanisms.

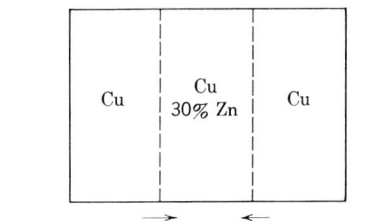

Fig. 8.18. Kirkendall experiment.

sample is shown diagrammatically in Fig. 8.18. During diffusion, the markers move towards each other, showing that zinc diffuses out of the alloy more rapidly than copper diffuses in. This experiment and many similar ones cannot be explained by a direct exchange or a ring mechanism, since in each of these any movement of an atom in one direction must be accompanied by the movement of another atom in the opposite direction, and so there cannot be a difference of rate of the two components. The Kirkendall effect has been confirmed in both face-centered cubic and body-centered cubic structures. It is observed in many experiments with diffusion couples that marked porosity develops in the region traversed by the interface. The Kirkendall effect is explained in terms of the vacancy mechanism as follows: In the case discussed above, it is postulated that it is easier for zinc atoms than copper atoms to change places with vacancies. This causes zinc to diffuse more rapidly than copper with the result that the interface moves, and it also causes a net flow of vacancies in the opposite direction to the movement of the zinc atoms. The vacancies that move towards the copper in the diffusion zone reach a high enough concentration to agglomerate into the pores which are observed.

A somewhat similar explanation of the effect could be made in terms of interstitial diffusion. It should be emphasized that the preference for the vacancy mechanism to the interstitial one depends on the much higher energy believed to be associated with an interstitial atom than with a vacancy in a metal, when the interstitial is comparable in size with the atoms of the crystal; the activation energy necessary for an atom to leave a lattice site and become interstitial is so high that the observed diffusion rates could not be accounted for on this basis.

These considerations apply to the close-packed and nearly close-packed structures; there is evidence that this may not be the case in less closely packed structures. In silicon, for example, which has the diamond-cubic structure, copper is able to diffuse interstitially. The interstices in silicon are much larger than in the more closely packed structures.

The activation energy for diffusion by the vacancy mechanism consists, as shown above, of two parts; these correspond to the variation of vacancy content with temperature, and the activation energy for the jump process. The variation of vacancy content with temperature has been studied by measuring diffusion immediately after rapid quenching from a higher temperature. The vacancy content at that time has the value characteristic of the higher temperature, and the diffusion coefficient is, accordingly, higher than it is later, when the vacancies have come to equilibrium at the new temperature. It is found that about one half of the measured activation energy for diffusion is due to the variation of vacancy content with temperature.

EFFECT OF IRRADIATION ON DIFFUSION. Since one of the effects of irradiation by energetic particles is to produce vacancies, it would be expected that the diffusion constant should be higher while a material is under irradiation than when it is not. Direct evidence of this effect has not been obtained, and in fact it has been shown theoretically that the effect should be too small to observe at temperatures at which direct measurements of diffusion constants can be made. Indirect methods of measuring diffusion effects, such as the rate at which ordering or disordering occurs, give results that can be explained on the basis that a higher vacancy content exists during irradiation.

Dislocation diffusion. The discussion so far has been limited to the ideal case of diffusion in crystals that are perfect except for vacancies. It has been shown, however, that diffusion takes place with lower activation energy along edge dislocations, so that this may be an important process at relatively low temperatures, where the vacancy process becomes very slow. It is possible that certain of the transformations to be discussed take place by dislocation or "pipe" diffusion. The mechanism for diffusion along an edge dislocation may be the movement of atoms from a position such as A (Fig. 8.19) to the corresponding position in the next plane. The activation energy should be low because of the relatively large space available for movement of the atom, which can become "semi-interstitial" in the relevant region.

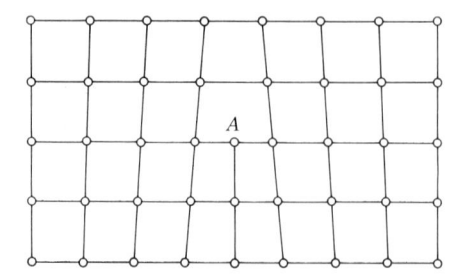

Fig. 8.19. Diffusion along an edge dislocation.

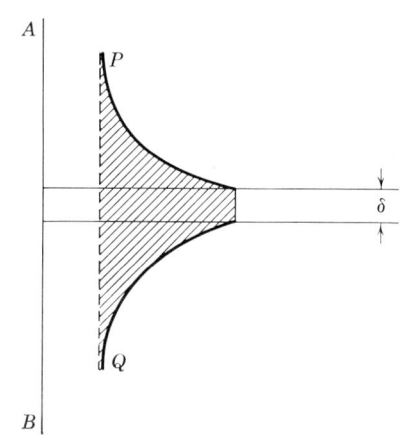

Fig. 8.20. Grain boundary diffusion.

Grain boundary diffusion. Because the arrangement of the atoms in a grain boundary must be different from that in a crystal, it is to be expected that the diffusion constants will also be different. In order to analyze the results of diffusion experiments on specimens containing grain boundaries, we adopt the following procedure, which follows that of Fisher. Assume that a grain boundary is a uniform slab of thickness δ, with a diffusion coefficient D_B. The crystal has a diffusion constant D_L. The idealized geometry is represented in Fig. 8.20. Diffusion takes place from left to right, and it is assumed that atoms diffuse according to D_B while in the boundary and that they diffuse away from it laterally according to D_L. The result of diffusion of solute from the surface AB, where the concentration of solute is maintained constant, is a distribution of solute represented by the "iso-concentration line" PQ. The amount of diffusion that is measured in an experiment is the area between PQ and AB, and this clearly depends on both D_B and D_L. The increase in the amount of diffusion attributable to the boundary is the shaded area in Fig. 8.20. The variation of concentration with distance from AB and from the boundary has been analyzed; the resulting expressions are complicated and not subject to direct experimental verification. Certain consequences of the results are, however, of importance in the present discussion.

It has sometimes been stated that, if the measured diffusion coefficient is the same for a single crystal and for a polycrystalline sample of the same metal, it may be concluded that the values of D_B and D_L are equal. It has been demonstrated that this is not correct, as follows. Assume that all the grain boundaries are normal to the "starting" surface AB, and that the average linear dimension of the grains, normal to the diffusion direction, is R. The amount of material reaching, by diffusion, a plane parallel to AB consists of two parts: that which has reached this plane by lattice diffusion and that which has reached it by boundary diffusion followed by lattice diffusion (it is assumed that the volume of the boundary is so small that the amount of solute *in* it can be ignored). It was shown by Turnbull that the grain size at which the two contributions to the total diffusion are equal depends on the ratio of D_B to D_L, as shown in Table 8.4 under a specified set of conditions. It can be concluded approximately that, for example, the contribution of grain boundary diffusion would be one-tenth that of volume diffusion if the grain size were 2.6 cm and the value of D_B/D_L were 10^6. Similarly, if D_B/D_L is 10^5, the grain size above which the boundary contribution is less than 10% is 0.014 cm. The absence of any observed effect on D of grain size, in the range 0.014 to 2.6 cm, would mean that $D_B/D_L < 10^5$, which is by no means equivalent to the earlier conclusion that it corresponded to $D_B \approx D_L$.

The grain boundary diffusion coefficient D_B can, however, be determined from measurements by radioactive tracer methods of the amount of solute that has diffused along and out of the boundary at distances from the surface of origin that are such that there is virtually no contribution from pure lattice diffusion. The results that are obtained can only be expressed as diffusion coefficients if the thickness δ of the boundary is known; since the cross section of the diffusion path enters into the definition of D, it is customary to assume that this is 5×10^{-8} cm; but it is in general better to express the result as $D_B\delta$. The variation with temperature of the boundary diffusion coefficient does not present this difficulty, unless δ varies with temperature. On the assumption that it does not, values of activation energies may be calculated. A typical result is an activation energy of 20 kcal/mole for the boundary

Table 8.4. **Relationship between Ratio of Diffusion Coefficients for Crystal D_L and Boundary D_B and Grain Size at Which Equal Amounts of Diffusion Take Place by the Two Processes** *

D_B/D_L	Grain Size (cm)
10^6	2×10^{-1}
10^5	1.4×10^{-3}
10^4	2.1×10^{-5}
10^3	$<10^{-6}$

* Data from A. D. Le Claire, Chapter 6 in *Progress in Metal Physics 4*, Bruce Chalmers, Ed., Pergamon Press, London, 1953, p. 283.

SOLID-STATE TRANSFORMATIONS

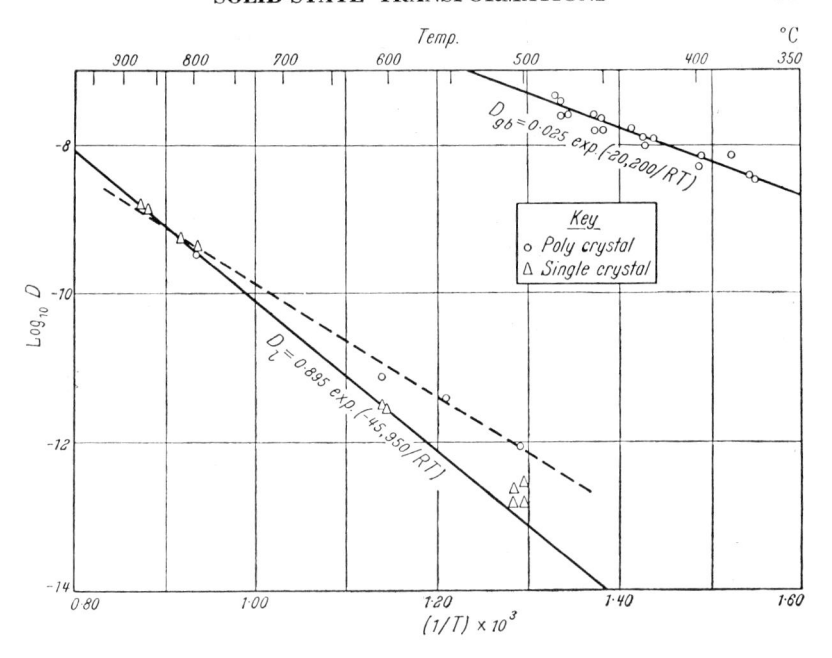

Fig. 8.21. Lattice and grain boundary diffusion in silver. From Le Claire, *Progress in Metal Physics* 4, Bruce Chalmers, Ed., Pergamon Press, London, 1953, Fig. 6, p. 285.

self-diffusion of silver. If the thickness is taken as 5×10^{-8} cm, the value of D_0 is 0.03 cm²/sec, and at 500°C $D_B/D_L = 10^5$. At lower temperatures, the ratio of D_B/D_L decreases, because D_B changes less rapidly with temperature than does D_L. Thus at lower temperatures the effect of grain boundaries on the measured diffusion constant should be greater than at the higher temperatures at which most measurements are made. Figure 8.21 shows the results of Hoffman and Turnbull on silver.

It has been shown by Hoffman and Turnbull that the diffusion coefficient D_B may, at least in simple cases, be replaced by the "pipe diffusion coefficient" D_P, which represents the diffusion along edge dislocations. They have also shown that diffusion in a tilt boundary at right angles to the dislocation direction is much slower than that along the dislocation direction. An interesting feature of their results is that this anisotropy is still observed when the boundary corresponds to a large angle (>30°) at which the dislocation model of the boundary does not seem to be physically possible because the dislocation cores would overlap each other.

Surface diffusion. The diffusion of metallic atoms over metal surfaces appears to be another process of relatively low activation energy. This is consistent with the view that atoms striking a surface are able to migrate randomly on it for a finite time before being trapped by the crystal or evaporating. The diffusing atoms could undergo the same sort of migration; in this case they would originate at the surface rather than arriving as condensing vapor.

Diffusion in liquid metals. The diffusion of solute atoms in liquid metals appears to be a thermally activated process and to have the same activation energy as that required for viscous transfer of momentum. Diffusion constants for liquid metals appear to vary much less from one metal to another than the corresponding values for solids. Some values are given in Table 8.5.

Table 8.5. Diffusion Data for Liquid Metals

Solvent	Solute	Temperature (°C)	D (cm²/sec × 10⁵)
Hg	self-diffusion	room temp.	0.007
Hg	Ag	" "	1.1
Hg	Au	" "	0.8
Hg	Bi	" "	1.5
Hg	Cd	" "	1.8
Hg	Na	" "	0.9
Hg	Pb	" "	1.58
Al	Cu	700	7.2
Al	Mn	700	0.6
Al	Fe	700	1.4
Al	Ni	700	1.5
Al	Si	700	81
Bi	Au	500	5.2
Fe	Si	1428	2.4
Pb	Au	490	3.5
Pb	Pt	490	2.0
Sn	Pb	500	3.7
Al	Cu	700	7.2
Al	Cu	800	11.0
Al	Cu	900	14.0
Al	Cu	1000	15

8.4 Precipitation and Solution

Definition. For the present purpose, precipitation will be defined as the formation of a new phase as a result of supersaturation produced by a change in the temperature of a solid phase. The reverse process, the disappearance of a phase when the temperature has been changed so that it is no longer stable, will be referred to as solution. The allotropic transformation of a pure metal could be considered as a precipitation phenomenon but is excluded by the above definition, which implies that the new phase has a different chemical composition from that of the existing phase. The thermodynamic condition for precipitation is that an alloy system should move, as a result of a change of temperature, from a single-phase to a two-phase region of the phase diagram, for a binary alloy, or from an n-phase to an $(n + 1)$-phase region in the general case. The discussion will be limited to binary systems, in which the main cases of interest are shown in Fig. 8.22. It should perhaps be pointed out that precipitation in a solid system differs from the more familiar precipitation in a liquid solution in that the precipitate remains where it forms in the solid, instead of sinking to the bottom, as it does in a liquid.

Reversible and irreversible precipitation. If the change of temperature that causes precipitation is extremely slow and if there is no "nucleation barrier," then equilibrium can be maintained, and the compositions and quantities of the phases present at any time can be predicted directly from the phase diagram. However, the time scale for equilibrium, or reversible, precipitation is set by the necessity for diffusion in the matrix to maintain a uniform composition in spite of the fact that growth of precipitate changes the composition of the matrix at its interface with the precipitate particle. The diffusion in question

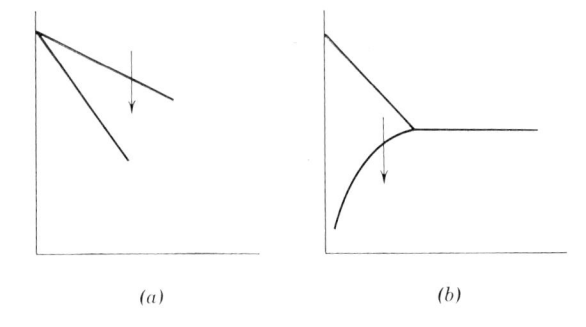

Fig. 8.22. Precipitating systems.

is solid diffusion, which is even slower than the liquid diffusion that exerts an analogous controlling effect on the solidification of an alloy. It is therefore reasonable to consider only the case in which the process is not reversible, but is controlled by the rate of diffusion as well as by the temperature. An additional reason why equilibrium is not maintained during precipitation is that the new phase must nucleate, and this inherently requires some supersaturation.

Nucleation of precipitate. The process of precipitation consists of two stages: nucleation of precipitate particles, and their subsequent growth. The nucleation of precipitates is more complicated than the nucleation of liquids from vapors, or of crystals from liquids, for three distinct reasons:

1. The interfacial energy depends upon the geometrical match or mismatch between the precipitate nucleus and the matrix.
2. The precipitate, having a different crystal structure, may differ in volume or shape from the matrix from which it formed; this causes volume strains or shear strains in the matrix and in the nucleus. The energy associated with such strains must be added to the interfacial energy in constructing the free energy equation for the critical size of nucleus.
3. Imperfections in the matrix itself may serve as preferred nucleation sites, requiring less supersaturation than would be required for truly homogeneous nucleation, which would occur in a region containing no imperfections. The reason for this is that, if the energy of an imperfection is reduced by the formation of a precipitation nucleus, this energy must be subtracted from the surface energy term of the nucleation equation.

The first of these aspects can be considered in more detail as follows.

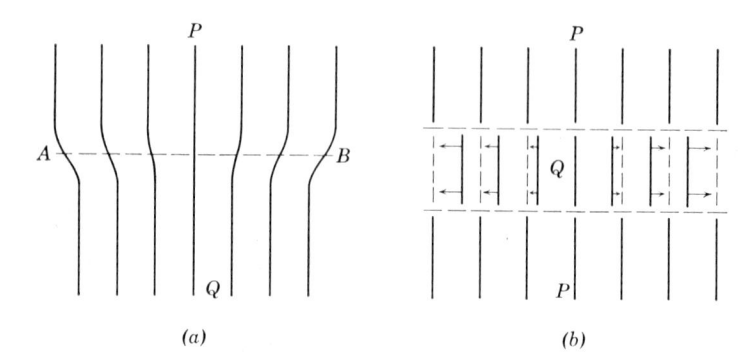

Fig. 8.23. Distortion near coherent interface.

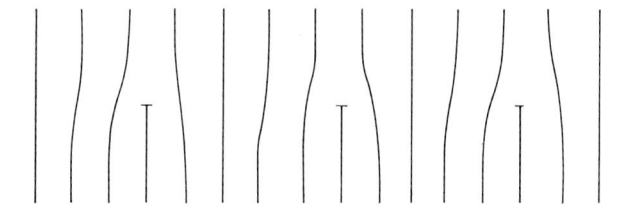

Fig. 8.24. Semicoherent interface.

The interface between two crystals may be coherent, in which case the structures of the two crystals match at the interface. If the two lattices do not naturally match exactly, they may be elastically deformed to a limited extent to make them do so. An example of this is shown in Fig. 8.23a, in which the interface is represented by the broken line AB. The two crystals P and Q differ in lattice parameter (in regions that are unaffected by the interface) by a few per cent. Near the interface, the crystal with the larger spacing (P) is therefore contracted, and in a state of compression, while Q is expanded. The energy corresponding to these distortions is elastic in nature, since it would disappear if the two crystals were separated. The amount of elastic energy that would be required depends on the amount of misfit, the appropriate elastic constants for the two crystals, and the geometry of the system. In a discussion of nucleation it would be unrealistic to assume that the volume of the new phase was infinite, as implied in Fig. 8.23a. It is more realistic to assume a limited amount of Q as in Fig. 8.23b. The strain energy is much less in this case, especially if the crystal Q has the lower elastic moduli. The elastic strain energy is much less, therefore, for a precipitate particle in the form of a thin plate than for a particle of equal volume in the form of a sphere, if the interface is coherent in each case. It should be emphasized that it is assumed here that the whole of the energy of a coherent interface is the strain energy required to achieve coherency.

The other case is that in which there is no coherency, when the energy of the interface is analogous to that of a high-angle crystal boundary. The surface energy term is now proportional to the area of the surface, and this leads to nucleation of spherical particles. An intermediate case is that in which the degree of misfit is such that the strain energy will be too high for complete coherency to be possible; in this case, local regions of coherency, separated by dislocations, may occur, producing a result shown schematically in Fig. 8.24. The question of whether it is valid to regard a noncoherent interface as a more extreme case of a dislocation array is similar to the problem of the most appropriate description of the large-angle crystal boundary.

The elastic strain energy discussed above arises from shear deformation of the precipitate particle and the matrix near the interface. Elastic strain energy also arises if the unstrained volume of the precipitate particle differs from that previously occupied by the same atoms. The strain in this case is triaxial tension or compression. This component of the strain energy is also minimized if the precipitate particle assumes a platelike form.

DISTRIBUTION OF NUCLEATION SITES. In any alloy in which precipitation can occur, it should be possible to define the following temperatures: 1. the equilibrium temperature; 2. the temperature for nucleation of precipitate on the most favorable sites, i.e., at crystal boundaries; 3. temperatures for nucleation at other preferred sites, such as dislocation nodes; and 4. the temperature for nucleation in regions of perfect crystal. These are in order of descending temperature. It follows that the temperature at which precipitation takes place determines whether it occurs at the most preferred sites, at less preferred sites, or at non-preferred sites. Specifically, at higher temperatures, nucleation is predominantly at the grain boundaries and the surface, while at lower temperatures it is distributed throughout the crystals as well as at the boundaries.

Growth of precipitates

KINETICS. It will be assumed that precipitation takes place isothermally, after an instantaneous change of temperature (quench). Once a precipitate has nucleated, its rate of growth is controlled by the rate at which the necessary change of composition can take place. This usually requires the replacement of some solvent atoms by solute atoms, and the rearrangement of the atoms into the new structure. The rate at which this can occur is determined by diffusion, usually of the solute. The diffusion rate depends on temperature, in the way described on p. 374 and represented in Fig. 8.14, and it also depends upon the con-

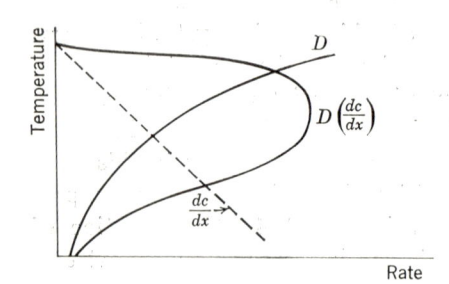

Fig. 8.25. Effect of temperature on rate of growth of precipitate particle.

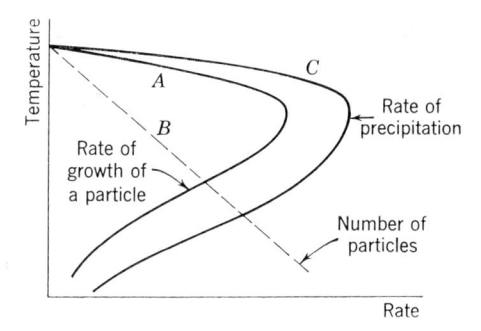

Fig. 8.26.

centration gradient. The concentration gradient in the matrix depends upon the difference between the original composition of the matrix and the composition that exists at the interface with the precipitate particle, i.e., very close to the equilibrium composition, and on the geometry of the particle. The concentration gradient, therefore, is proportional to the degree of supersaturation of the solution (for a given geometry), and this increases with decreasing temperature, as shown in Fig. 8.25. The diffusion *rate* is proportional to the product of the concentration gradient (dc/dx) and the diffusion constant D, also shown in Fig. 8.25.

The resultant curve, which is characteristic of many solid-state reactions, is due to the opposing effects of the decrease of diffusion constant with temperature, and the increase of driving force with departure from the equilibrium temperature. When the departure from equilibrium is in the direction of increasing temperature, as in the solution of a precipitate, both factors work in the same direction, and the rate increases continuously with rising temperature.

The discussion so far has been about the rate of growth of a single particle after it has nucleated. The over-all rate of growth of the precipitate phase depends also on the number of particles that are growing. This increases with increasing departure from equilibrium, and therefore the resultant curve of Fig. 8.25, shown as curve A of Fig. 8.26, must be combined with curve B, representing the number of particles growing, to give the *rate of precipitation* curve C. The rate must vary with time, for several reasons. When a particle nucleates, the concentration gradient in the surrounding matrix is very steep; as growth proceeds, the diffusion gradient decreases in steepness and an increasing volume of the matrix supplies solute atoms. The rate therefore decreases for this reason. It decreases more abruptly when the precipitate particles begin to compete for the same solute atoms; the time dependence of nucleation may also cause a change of rate. If the nucleation process is "spread out" over a substantial time, the rate will increase during this time, because of the increase in the number of growing particles. A measure of the rate of precipitation is the time taken for some property to undergo a fixed proportion (e.g., one-half) of the total change that corresponds to complete precipitation. An example of the variation of precipitation rate with temperature and with composition is given in Fig. 8.27.

Attempts to apply diffusion data quantitatively to precipitation phenomena have led to anomalous results; a conclusion is that diffusion along dislocations may play a very important part in the process.

MORPHOLOGY. The growth of precipitate particles is related to the same free energy considerations as nucleation. If a precipitate particle is not coherent with the matrix, it grows either as a sphere (if its surface free energy is not sensitive to the crystallographic orientation of the surface) or with a shape that is bounded by low energy planes. In very many cases, the precipitate occurs along close-packed planes of the matrix. This is often due to coherent nucleation followed by growth

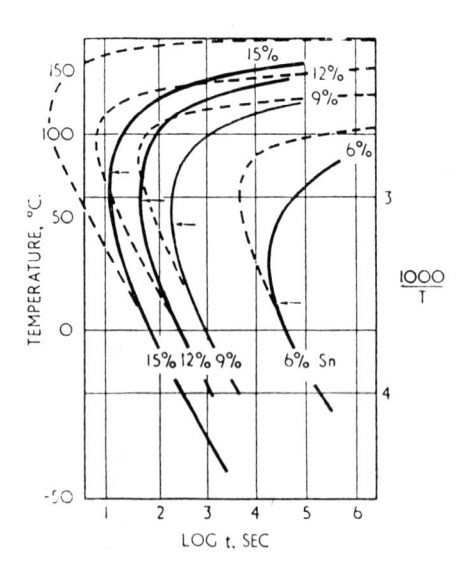

Fig. 8.27. Aging of tin-lead alloys; time for one-half of the total resistance change. From H. K. Hardy and T. J. Heal, Chapter 4 in *Progress in Metal Physics 5*, Bruce Chalmers and R. King, Eds., Pergamon Press, London, 1954, Fig. 109, p. 256.

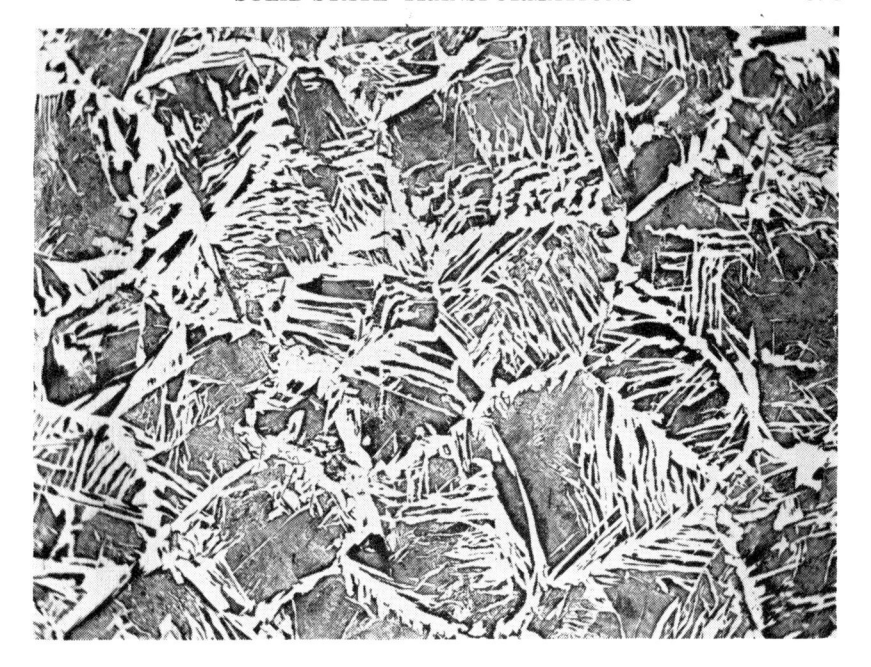

Fig. 8.28. Photomicrograph of typical Widmanstatten structure (×176). (Courtesy U. S. Steel Corporation.)

determined by low energy faces. This type of structure, shown in Fig. 8.28, is known as a Widmanstatten structure.

It is also found that under conditions of very high supersaturation the precipitate forms with a dendritic structure. An example, that of cementite precipitating in austenite, is shown in Fig. 8.29.

EFFECT OF QUENCHING RATE. It has been assumed in the preceding discussion that the alloy is cooled instantaneously from the "solution" temperature to the "precipitation" temperature. This can never be the case in practice, but it should be recognized that the desirable results of precipitation are almost always obtained when a very large number of precipitate particles or pre-precipitation zones are formed. This only occurs when conditions correspond to the lower part of the rate curve of Fig. 8.26. The alloy must therefore be quenched quickly through the temperature range in which coarse precipitation would occur quickly and into the region of fine, slow precipitation.

However, a very drastic quench may set up large stresses because the outer layers are cooled sooner than the interior, which sets up large tensile stresses near the surface, and cracking may result. If cracking

(a) 150°C, 170 hr (b) 260°C, 3 min (c) 260°C, 10 min

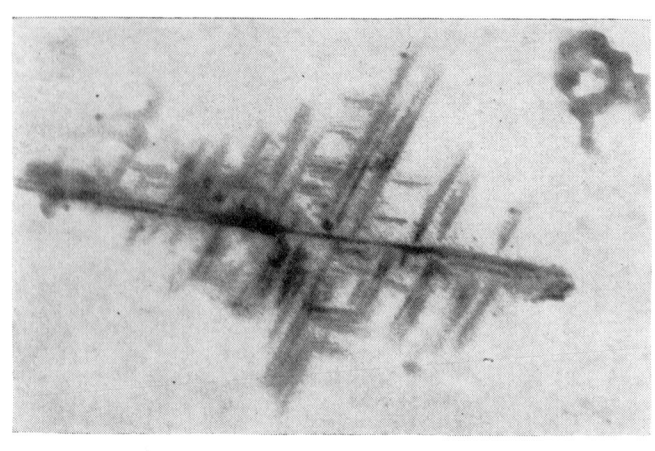

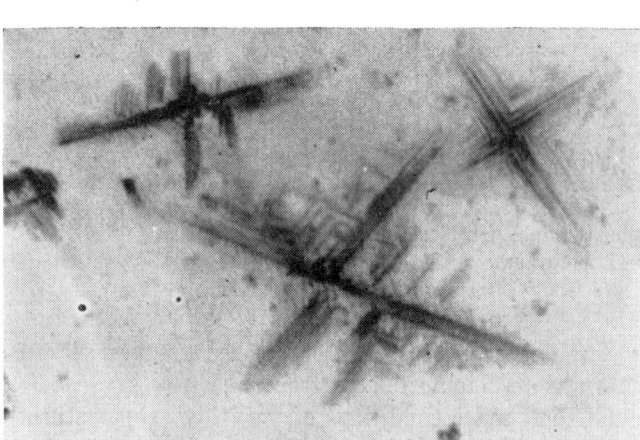

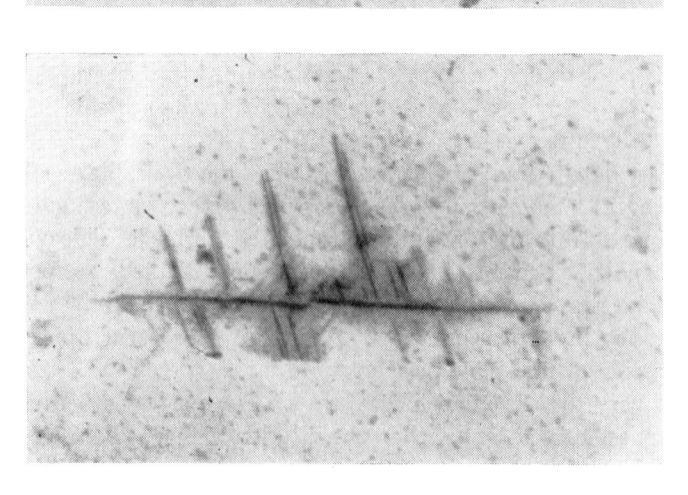

Fig. 8.29. Dendritic carbides precipitated during aging of iron–0.014% carbon alloy brine quenched from 740°C, aged as noted. Extraction replicas. ×38,500. (Courtesy W. C. Leslie, U. S. Steel Corporation.)

has not occurred previously, stress relief may take place during aging. This is sometimes accompanied by marked dimensional changes, which may be troublesome in, for example, instrument components. The tendency for cracking to occur can often be reduced by decreasing the severity of the quench, by, for instance, quenching into hot water instead of cold.

EFFECT OF STRAIN ON PRECIPITATION. It is found that plastic deformation of a quenched alloy increases the rate of precipitation; this may be related to the fact that when the precipitate becomes visible in such cases it occurs along lines that correspond to the slip planes of the crystals. It follows that nucleation or growth of the precipitate occurs more readily in regions of high dislocation content than in regions of lower dislocation content; a greater nucleation rate could be accounted for in terms of the strain energy associated with an array of dislocations, while an enhanced growth rate would be expected as a result of rapid diffusion along dislocations.

Crystal structures of precipitates. It has been clearly demonstrated that in many alloy systems the final equilibrium phase is not formed at the beginning of precipitation, but after a complicated sequence of events. A very important case that has been subjected to detailed study is the aluminum-copper alloy system. The relevant portion of the phase diagram is shown in Fig. 8.30.

It has been shown that, under appropriate conditions, four distinct stages of precipitation can be observed, when an alloy of aluminum with a few per cent of copper is quenched from a temperature of 500°C and allowed to age at lower temperatures, such as 200°C. The first stage is the formation of zones, known as Guinier Preston zones, or as GP[1] zones, in which an increased concentration of copper occurs in some {001} planes of the aluminum matrix. The zones are extremely small (of the order of 50 atoms across), and can only be observed by very refined X-ray techniques. The second stage, known as θ'' or GP[2], differs from the first in that the structure of the zones is now apparently ordered. The third stage is the formation of the θ' phase, which is a phase of composition $CuAl_2$, that is coherent with the matrix. The final stage is the formation of massive $CuAl_2$, the equilibrium phase θ, which is not coherent with the matrix.

It appears that the formation of each of the three later precipitates is not dependent on the prior formation of the earlier ones, since θ', for example, is apparently the first precipitate phase formed when a 2% alloy is aged at 220°C; also, GP[2] is the first formed phase when a 4% alloy is aged at 190°. A 4½% alloy aged at 110°C forms GP[1]. It is also believed that θ can either nucleate independently or be formed by the transformation of θ'. It may be concluded that the sequence is due to kinetic reasons rather than to the nucleation of each stage by the previous one.

The case discussed above is an unusually complicated one, but it is not unusual for a two-stage process to occur, in which the first stage is the formation of a coherent precipitate, which loses coherency when the particles reach a critical size.

Effect of precipitation on properties. The mechanical, and in some cases, magnetic, properties of alloys are drastically changed by precipitation. It is therefore of interest to discuss several specific cases.

BINARY ALUMINUM-COPPER ALLOYS. Although the binary alloys of aluminum and copper are of less practical interest than some of the more complex aluminum alloys (to be discussed below), these alloys form a useful starting point for discussing the effect of precipitation on the mechanical properties. The most convenient measure of the mechanical properties is the hardness; the variation of hardness with time of various binary aluminum-copper alloys, aged at 130°C after

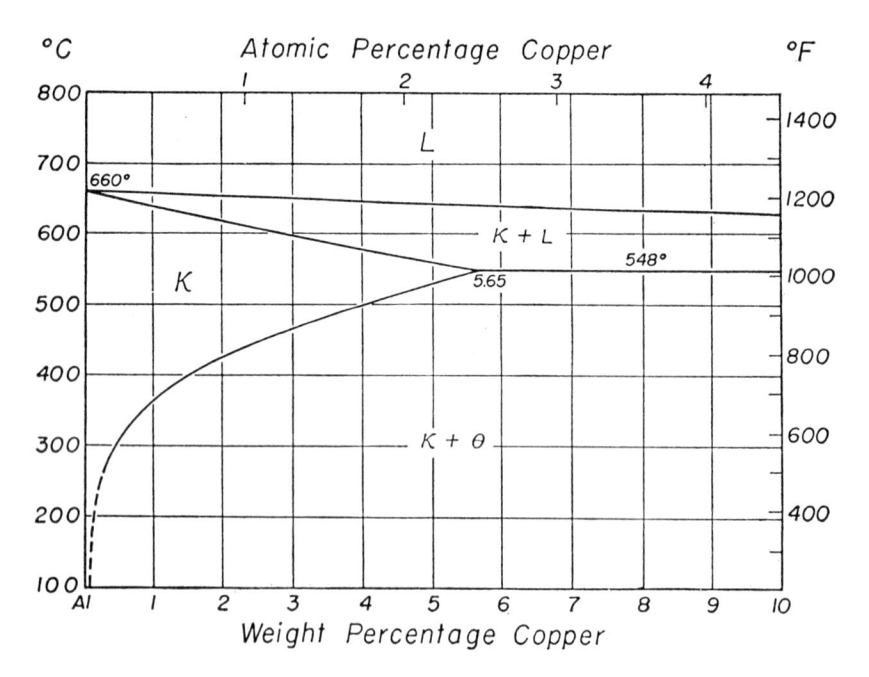

Fig. 8.30. Phase diagram for aluminum-rich end of the aluminum-copper system. From *Metals Handbook,* ASM, p. 1160.

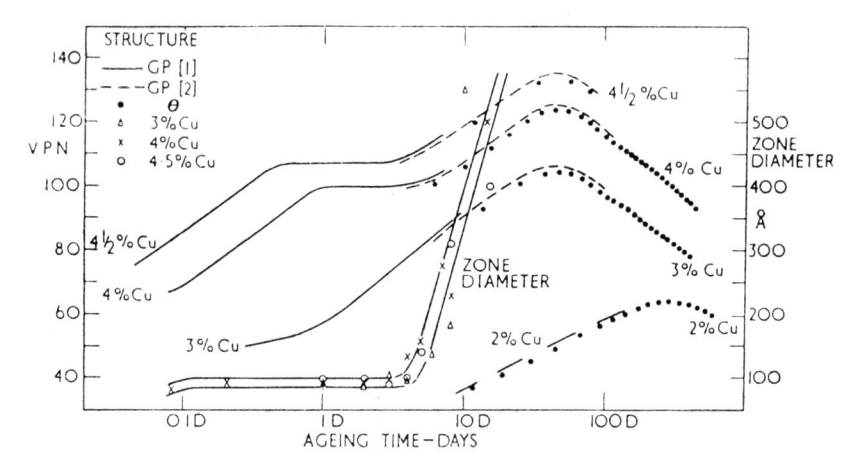

Fig. 8.31. Variation of hardness with time for various aluminum-copper alloys aged at 130°C. From Hardy and Heal, *Progress in Metal Physics 5*, Fig. 63, p. 195.

quenching, is shown in Fig. 8.31. It is seen that the hardness increases to a maximum in two stages and then decreases. The maximum hardness is reached when the amount of GP[2] has reached its maximum. The hardness decreases when GP[2] is replaced by θ'. However, when GP[2] has not been formed (high copper content combined with high aging temperature), then a maximum is reached when θ' is at its maximum. The formation of the θ phase causes a reduction of hardness.

The increase in hardness, and the corresponding change in the other mechanical properties, can be accounted for on the following basis. A precipitate particle (this includes the so-called "pre-precipitation" zones) has different elastic constants from the matrix and this causes a stress field to be set up when a dislocation approaches it. This may have several effects. One is that it should be more difficult to drive a dislocation through a crystal that contains a large number of such regions, and so the stress required to cause deformation would be higher than for the homogeneous solid solution. This would account for a higher yield stress but not for a greater rate of work hardening. Secondly, if the particles offered sufficient resistance to the passage of dislocations, the dislocations would not pass through them but would pass between them and leave closed loops around the particles. This is illustrated in Fig. 8.32. As each successive dislocation from a source moves across the crystal, an additional loop is formed around each obstacle. The spaces between the obstacles decrease, and this makes it progressively more difficult for more dislocations to be forced through. The

(a) (b) (c) (d)

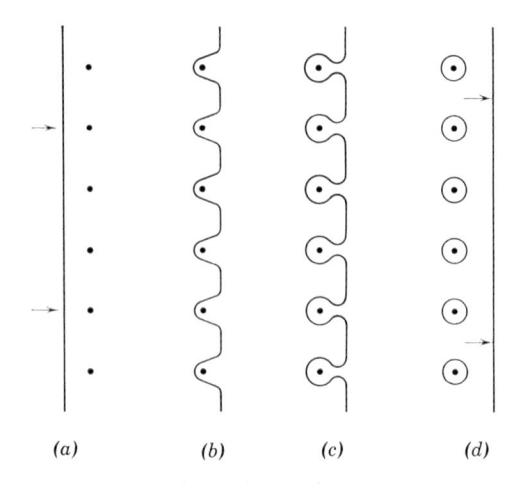

Fig. 8.32. Motion of dislocation past precipitate particles.

maximum effect would be achieved when each particle is large enough to hold up the dislocations and when the particles are as numerous, and therefore as close together, as possible. This theory shows good agreement with many of the experimental observations.

Figure 8.31 shows that the hardness decreases if aging is carried beyond the optimum point. This stage is called overaging; it can be prevented only by lowering the temperature when the maximum hardness (and strength) have been reached; however, in many cases, overaging is so slow that it does not present a problem in practice.

As a reference point for considering other aluminum alloys, the tensile properties of aluminum-copper alloys are shown in Fig. 8.33, in which the effects of heat treatment and copper content are given, both for cast and for wrought alloys. The wrought alloy differs from the cast alloy in having been very severely deformed and recrystallized; it may be expected to have a much finer grain size and much more uniform distribution of copper.

OTHER ALUMINUM ALLOYS. Higher strengths are obtained, after heat treatment, by using an alloy that contains, besides copper, smaller amounts of silicon, manganese, and magnesium, a typical composition being Cu 4.4%, Si 0.8%, Mn 0.8%, Mg 0.4%. The U.T.S. and yield stress are, after heat treatment, 70,000 psi and 60,000 psi respectively, with an elongation of 13%. The greater strength is probably due to the precipitation of other phases, such as Mg_2Si, as well as $CuAl_2$. The precipitation treatment takes 10–20 hr at a temperature around 160°C. Even higher strengths are obtained in aluminum alloys containing zinc

SOLID-STATE TRANSFORMATIONS

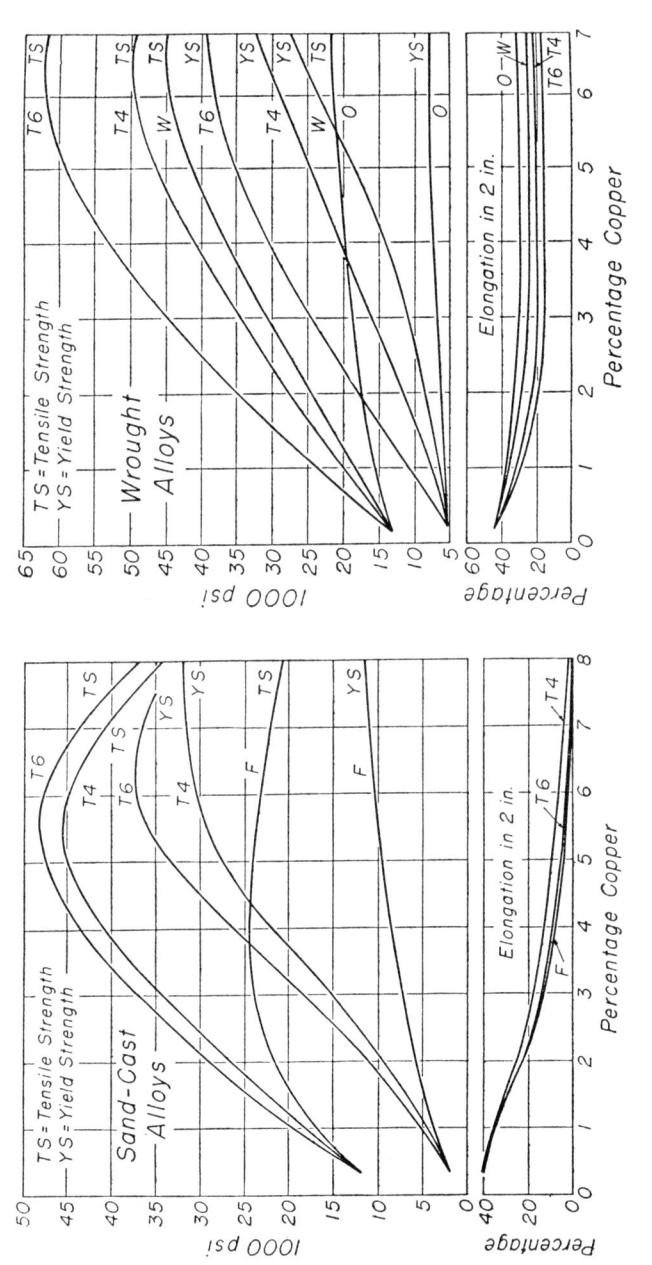

Fig. 8.33. Effect of copper content on properties of aluminum-copper alloys. (F, as cast; W, solution treated and quenched; O, annealed; $T4$, solution treated, quenched, and aged at room temperature; $T6$, solution treated, quenched, and aged at elevated temperature.) From *Metals Handbook*, ASM, Figs. 1 and 2, p. 804.

(5%) in addition to magnesium (2.5%), copper (1.5%), chromium (0.3%), and manganese (0.2%). The maximum strength properties are U.T.S. 88,000 psi, yield stress 80,000 psi, elongation 10%, obtained after aging at about 120°C for about 20 hr. There are probably several precipitating phases in this system.

The development of these alloys has necessarily been largely empirical, because there is no theory that can make detailed predictions on the variation of solubility with temperature in complicated alloy systems; the objective has usually been to achieve the highest strength that is compatible with sufficient ductility (sometimes arbitrarily specified as 8%), adequate corrosion resistance, and other less easily defined properties that contribute to the suitability of the alloy for industrial use.

BERYLLIUM COPPER. One of the most interesting of the precipitation hardening alloys is beryllium copper. Part of the phase diagram for the copper-beryllium system is shown in Fig. 8.34; it will be seen that 2% beryllium (by weight) should be in solid solution at about 850°C,

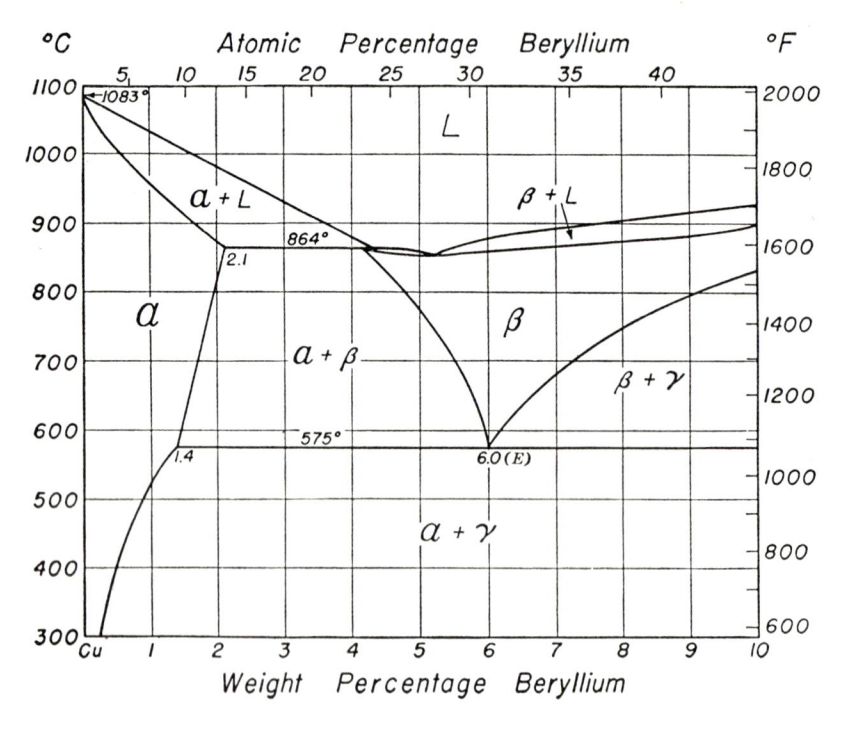

Fig. 8.34. Part of the copper-beryllium system. From *Metals Handbook*, ASM, p. 1176.

and that it should precipitate the β phase above 575°C and the γ phase at lower temperatures. In practice, the alloy that is used contains 2% beryllium and 0.25% cobalt, and precipitation is carried out at about 600°C, for a period of 2–3 hr. The tensile strength attainable by this treatment is 200,000 psi, combined with 2.8% elongation.

CREEP RESISTANT ALLOYS. Most of the alloys that have been developed for creep resistance at high temperatures appear to depend, at least to a limited extent, on the presence of a precipitate. The problem of the instability of a precipitate of optimum size can only be solved partially, and it is apparently necessary to compromise to the extent of using an alloy in the overaged condition.

TEMPERING OF MARTENSITE. In the cases discussed so far, precipitation is used to increase the resistance to deformation of an alloy by producing the optimum dispersion of the solute or the second phase. The effect on the matrix is, in these alloys, relatively unimportant.

In the case of the tempering of martensite, the reverse is true. Tempering of martensite is the treatment by which the supersaturation is reduced by precipitation of carbon as a carbide phase. Consider a steel containing, for example, 0.8% carbon which has been quenched from 800°C (at which temperature it is austenite) so that the martensite transformation takes place; some austenite will be retained unless the temperature reaches the M_F line. The resulting material is too hard and brittle for most applications; its ductility must be increased, and its hardness decreased, by controlled precipitation, which decreases the carbon content of the martensite.

The equilibrium structure would be ferrite and carbon (graphite), and the precipitation processes to be considered are the stages that are observed if martensite is heated to temperatures at which change can occur.

It is found that tempering of high carbon steel takes place in four stages, which may overlap, but which can be observed separately. They are:

1. 80–160°C; slight increase in hardness.
2. 230–280°C; slight softening.
3. 260–360°C; marked softening.
4. High temperature tempering.

These four stages are interpreted as follows: The first stage appears to correspond to decomposition of the high carbon martensite (α') into low carbon martensite (α'') and ϵ iron carbide. α'' is martensite containing about 0.25% carbon, and ϵ iron carbide is either Fe_2C, or a distorted, coherent form of Fe_3C. The increase in hardness is attributed to the formation of the precipitate. The second stage (230–280°C) appears to be the decomposition of the retained austenite, presumably into the final phases ferrite and cementite. The third stage is the decomposition of the α'' and ϵ into ferrite and cementite. This is accompanied by marked softening, presumably because the α'' is much harder than ferrite. Further softening (the fourth stage) occurs as a result of spheroidization, i.e., the growth of larger cementite particles and the solution of smaller ones. The total amount of cementite does not change appreciably, but the reduction in the number of particles reduces the hardness, in much the same way as in the overaging of precipitated alloys.

The process of tempering is both temperature- and time-dependent, and is modified by the presence of alloying elements in a steel. For plain carbon steel, the relationship between hardness and tempering temperature is given in Fig. 8.35. The time dependence is relatively unimportant, and can usually be ignored. An unexplained aspect of the tempering process is the embrittlement that occurs when tempering is carried out in a temperature range around 300°C.

HIGH-SPEED STEELS. Some steels have been developed in which the tempering temperature is extremely high. They are used in applications, such as machining, in which it is necessary to retain high hardness at high temperatures. They all contain tungsten and molybdenum, which are strong carbide-forming elements. Possibly the special properties are due to the formation of W_2C instead of the usual iron carbide. The resulting structure would be particularly stable, both because of the high stability of the carbide and because the relatively slow diffusion of tungsten (as compared with carbon) would retard the spheroidization of the carbide.

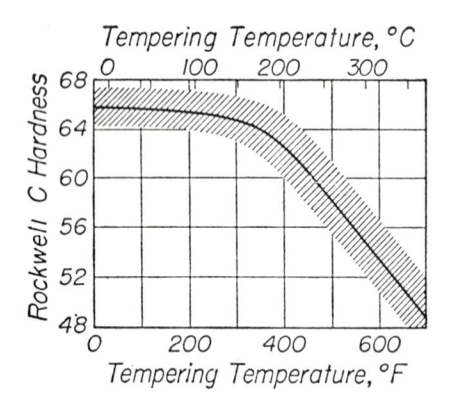

Fig. 8.35. Effect of tempering temperature on hardness for quenched 0.90% carbon steel. From *Metals Handbook,* ASM, Fig. 1, p. 660.

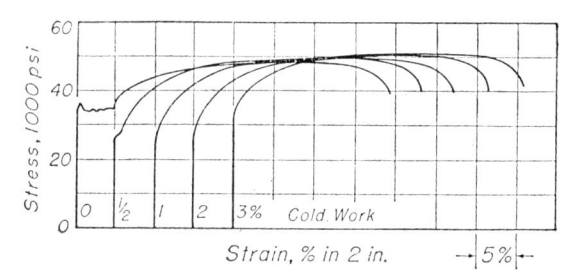

Fig. 8.36. Suppression of yield by cold work. Curves obtained without cold work and immediately after ½, 1, 2, and 3% cold work. From *Metals Handbook*, ASM, Fig. 6, p. 441.

Precipitation of solute on dislocations. It has been pointed out (p. 107) that the energy of a crystal containing solute atoms and dislocations is reduced if the solute atoms take up positions in the regions of high distortion associated with the dislocations. Interstitial solutes, for example, cause less lattice expansion when they are in the region of expanded lattice near an edge dislocation than when they are in ordinary interstitial sites; similarly, substitutional solute atoms tend to occupy positions on the compressive or tension sides of an edge dislocation if they are respectively smaller or larger than the solvent atoms. In a somewhat similar way, solute atoms also tend to associate with screw dislocations.

On the other hand, the entropy of a solid solution is reduced if the solute atoms are restricted to special positions; consequently, we should not expect all the atoms to segregate at the dislocations. The condition for minimum free energy requires that $C = C_0 \exp(-U/kT)$, where C_0 is the average concentration in the crystal, C is the concentration in the dislocation, and U is the binding energy between the dislocation and the solute atom. If U is much greater that kT, as is the case for carbon and nitrogen at room temperature, the dislocation becomes "saturated" by the condensation of a single line of solute atoms along the dislocation line.

THE YIELD POINT. If a specimen of steel containing 0.2% carbon is subjected to a tensile test, the load-elongation curve is as shown in Fig. 8.36, first curve. The first appreciable plastic deformation takes place at the upper yield point. The stress required to cause further deformation drops to the lower yield point, and the specimen deforms by the propagation of Lüders bands (see p. 157). Iron containing nitrogen instead of carbon shows a similar effect.

The existence of the yield point in low carbon steels and in certain other alloys is attributed to the fact that it should require a higher shear stress to activate a Frank-Read source when solute atoms are segregated along the dislocation line than when there is no solute present. The production of the first dislocation loop from a source allows the solute atoms to diffuse away and it is therefore easier to generate successive loops than it was to produce the first one. This accounts in general terms for the drop in stress after the upper yield point has been passed. The subsequent rise in stress is due to work hardening.

STRAIN AGING AND QUENCH AGING. If a specimen is retested immediately after having yielded, the load-elongation curve is as shown in Fig. 8.36. The yield point is not observed. If retesting is delayed, the specimen being held at room temperature, the yield point reappears. The reappearance of the yield point is called "strain aging." The process of strain aging is illustrated in Fig. 8.37.

The behavior of a steel that has been stressed beyond its yield point corresponds to the fact that the carbon atoms are at first distributed at random (having diffused away from the previous positions of the dislocation), at which stage there is no yield point if the specimen is retested. If the sample is held at room temperature for a time of the order of days, the yield point gradually reappears. The rate at which this takes place, and its variation with temperature, are in quantitative agreement with the theory that they are controlled by the resegregation of carbon at the dislocations. Internal friction measurements have shown, moreover, that the amount of carbon that is able to move freely decreases during the process of strain aging.

The existence of the "Cottrell atmosphere" explains why it requires a higher stress to generate the first dislocation of a Frank-Read source than to generate subsequent ones; the carbon atoms are unlikely to be in exactly the right positions to lock the second and subsequent dislocations as they move past the original position of the first one. The

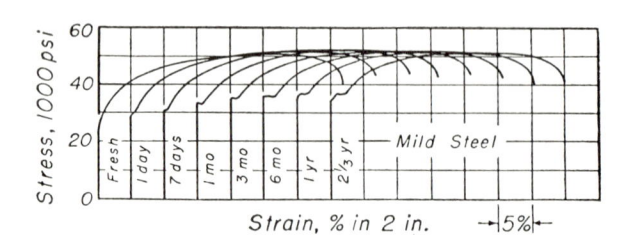

Fig. 8.37. Strain aging. Tests at various times after initial straining. From *Metals Handbook*, ASM, Fig. 7, p. 441.

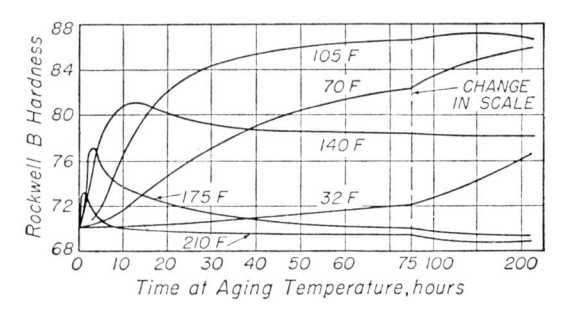

Fig. 8.38. Variation of hardness with time after quenching for a steel containing 0.06% carbon. From *Metals Handbook*, ASM, Fig. 2, p. 439.

observed yielding, which is very pronounced in polycrystalline samples, occurs when slip propagates from one crystal to the next throughout a zone that extends right across the crystal to form a Lüders band; slip in one crystal sets up additional stresses in neighboring crystals and they slip, producing a "chain reaction" right across the specimen. Once a Lüders band has formed, it can extend laterally at a relatively low stress (the lower yield point) because some of the crystals that have yielded (i.e., are in the band) apply stresses to crystals that are adjacent to the band which therefore do not require as high an externally applied stress as would otherwise be necessary.

Quench aging is the name given to the change in properties that occurs after a steel has been cooled from a high to a lower temperature. Figure 8.38 shows the variation of hardness with time at various temperatures for a steel containing 0.06% carbon. Quench aging does not occur in the absence of carbon, nitrogen, and oxygen. An important aspect of the problems of aging in steels is that the deoxidizing practice used in the manufacture of the steel has a very significant influence on its aging properties. In general, strongly deoxidized steels do not show either quench aging or strain aging. In an extreme case, in which the steel is deoxidized by means of a relatively large addition of titanium, the yield point is suppressed as well as the aging phenomena. It is highly probable that quench aging, like strain aging, depends on segregation of carbon, nitrogen, and perhaps other interstitials, at dislocations.

"Precipitation" of gases in metals. A different type of precipitation phenomenon is that in which a gas, usually hydrogen, comes out of solution within a metal, often with undesirable consequences. Steel may contain hydrogen either as a result of its reaction with water vapor at a high temperature, or as a result of electrolytic decomposition of water, whether by corrosion or during electroplating. The hydrogen, present in atomic form, may exist in such a high concentration that it nucleates "islands" of molecular hydrogen within the steel; sites for nucleation have not been identified, but it is probable that they are imperfections of some kind, as it does not seem likely that nucleation can arise in a region of perfect crystal. It is possible that the "condensation" of hydrogen can occur at regions of high stress such as are formed where two arrays of dislocations of opposite sign block each other; under conditions of an applied stress and some hydrogen content, there is a delay time before fracture takes place that increases as the stress decreases. This is consistent with the time taken for hydrogen to diffuse to an embryo crack to increase its size until it reaches the "Griffith critical size," after which catastrophic fracture will occur. It is found that large steel forgings which contain hydrogen may be prevented from forming internal cracks (known as "snowflakes") by sufficiently slow cooling; this allows hydrogen, which becomes supersaturated as the temperature falls, to diffuse to the external surface and so to escape.

"Blistering" of aluminum and magnesium alloys during heat treatment is also due to the formation of hydrogen bubbles; the hydrogen is a product of the reaction of the metal with water vapor. The partial pressure of *atomic* hydrogen in a blister is extremely low (because of the low dissociation of molecular hydrogen into atomic hydrogen at the temperatures involved) and so a very high pressure of *molecular* hydrogen would be required for equilibrium with the *atomic* hydrogen in the metal. The pressure of molecular hydrogen is sufficient to cause plastic deformation of the overlying metal, with the result that a blister is formed.

Graphitization. A phenomenon that may be included under the general heading of precipitation is graphitization. When a steel containing cementite is held for a long time at an elevated temperature (below the transformation temperature), the cementite is gradually replaced by graphite. Thermodynamically, graphite is more stable than cementite, but either the nucleation or the growth of graphite in steel is so slow that under ordinary conditions austenite transforms into ferrite and cementite, no graphite being formed. The problem of graphitization is not well understood, but it appears that relatively small amounts of alloying elements have a considerable effect. It is found that graphitization occurs first in the zones near welds that have approached, but not reached, the transformation temperature. This

SOLID-STATE TRANSFORMATIONS

may be due to partial spheroidization of the cementite, or it may be due to changes of composition resulting from the proximity of the transformation interface.

Solution of precipitates. When the temperature of an alloy containing a precipitate is raised, the precipitate tends to dissolve. Whether solution is complete or not depends upon the temperature, which determines whether the equilibrium state is a single-phase alloy, and on time, because the process of solution is controlled by the diffusion of the dissolving material into the matrix. No nucleation barrier exists in the solution of a precipitate in an existing phase. It is important in heat treatment operations to allow sufficient time for the complete solution of existing phases, as otherwise the alloy does not effectively have the expected composition.

Spheroidization. When a precipitated alloy is held at a temperature at which the precipitate is close to equilibrium with the matrix and

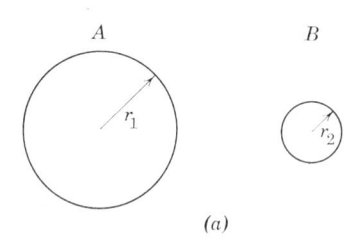

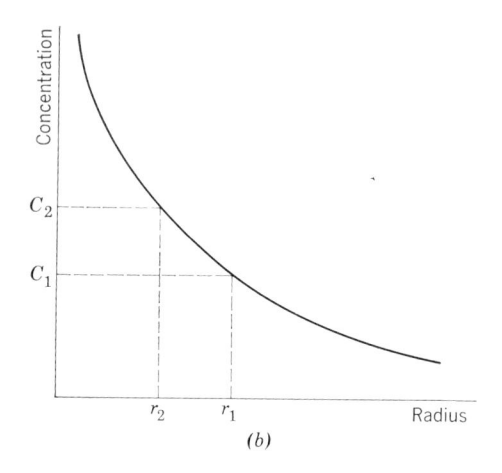

Fig. 8.39. Effect of radius on equilibrium concentrations.

at which diffusion can occur, the process of *spheroidization* takes place. The precipitate particles increase in size and decrease in number, and their shape becomes more nearly spherical. The driving force for these events is the resulting reduction in the surface area, and therefore the surface free energy, of the precipitate.

Consider spherical precipitate particles A and B with radii of curvature r_1 and r_2, Fig. 8.39a. By analogy with the variation of equilibrium temperature with radius discussed on p. 237, the composition C of the matrix that is in equilibrium with the particle varies with r in the manner shown in Fig. 8.39b. The matrix in contact with a has an equilibrium composition C_2, and in the neighborhood of b it has a lower concentration C_1. Solute therefore diffuses from the matrix near b towards a. This causes the matrix near b to become unsaturated; a dissolves in it in order to restore equilibrium. The matrix near b, similarly, becomes supersaturated, and this causes solute to deposit on b.

This discussion can be generalized to include nonspherical particles; the radius r now becomes the local radius of curvature. It follows that a nonspherical particle approaches a spherical shape because material moves by diffusion in the neighboring phase from the regions of small radius of curvature to regions of larger radius of curvature, as shown in Fig. 8.40.

These considerations only apply when the free energy associated with the interface between matrix and precipitate is not strongly dependent on the orientation of the interface in relation to either phase. When this condition is not satisfied, the precipitate particle may be in equilibrium when it is bounded by crystallographic faces.

Diffusion controlled "allotropic" transformation. When a pure metal undergoes an allotropic transformation, no diffusion is necessary, and the rate of the transformation is governed by the kinetics of the process itself; when a solid solution of another element in such a metal undergoes the corresponding transformation, it may be diffusionless, as in the martensitic transformations, or it takes place closer to equi-

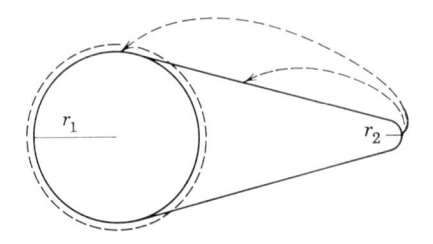

Fig. 8.40. Approach to equilibrium shape for nonspherical particle.

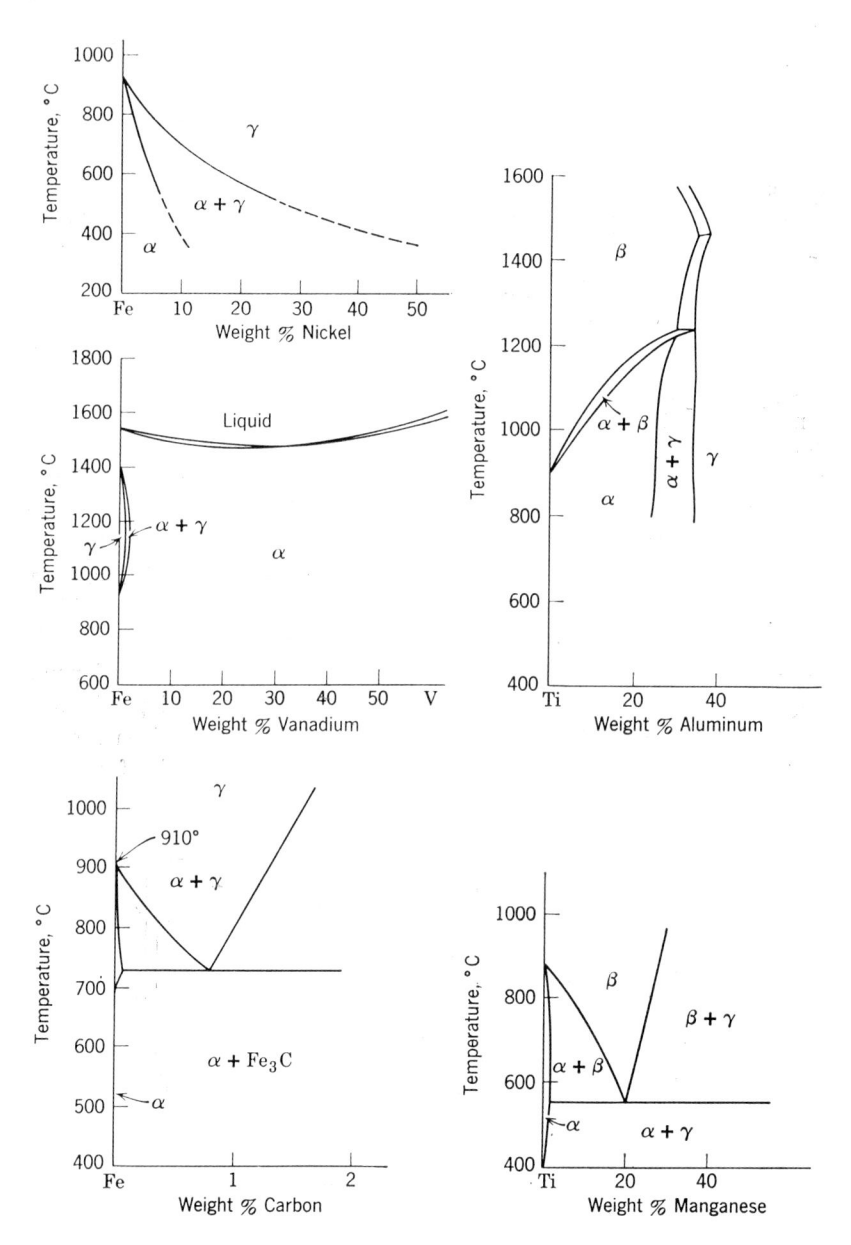

Fig. 8.41. Typical systems containing two solid solutions.

librium conditions at a rate that is controlled by diffusion. This type of transformation is important in alloys based on iron, titanium, zirconium, and uranium, each of which has an allotropic transformation in the temperature range in which diffusion can occur. A type of transformation that is in some respects intermediate between the martensitic and the equilibrium type is the bainite reaction, to be discussed on p. 425.

The part played by diffusion in these transformations is closely analogous to that in the solidification of alloys. The equilibrium solute content of one phase differs from that of the other; examples are shown in Fig. 8.41 for some systems based on iron and on titanium. In all systems like these, the transformation nucleates when the temperature is somewhat below the equilibrium temperature appropriate to the composition, and the new phase grows at any temperature between the nucleation temperature and the M_S temperature. Since the newly forming phase has a different composition from the matrix, the latter must change in composition at the interface where the transformation takes place. The change of composition is always in such a direction that the equilibrium temperature is lowered; if the equilibrium temperature reaches the actual temperature the reaction stops, but this is opposed by the diffusion of solute away from the interface into the matrix. The rate at which the new phase grows, therefore, depends upon the diffusion rate of the solute in the matrix; this depends upon the "driving force" (i.e., departure from equilibrium), and on the diffusion constant. The relationship between rate of growth of the new

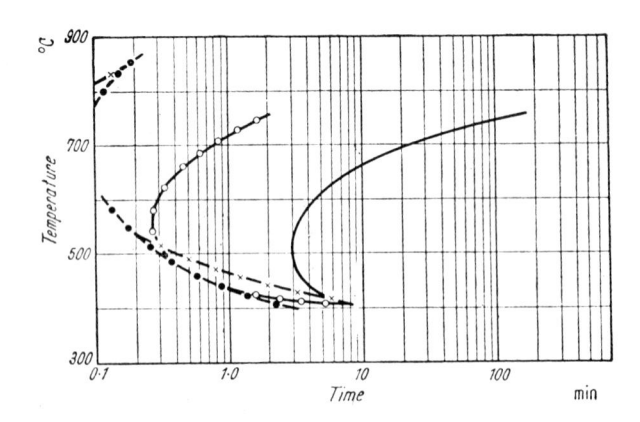

Fig. 8.42. T.T.T. curves showing the relationship between temperature and transformation time for titanium-molybdenum alloys containing oxygen. From *Titanium*, McQuillan and McQuillan, Fig. 131, p. 322.

Fig. 8.43. Photomicrograph showing proeutectoid ferrite (0.35% carbon steel slow-cooled) (×450). (Courtesy U. S. Steel Corporation.)

phase and the temperature is therefore similar to that discussed on p. 388. An example of the dependence of rate on temperature and composition for an alloy of titanium and molybdenum is shown in Fig. 8.42.

As in the previous discussion of the kinetics of precipitation (p. 389), the rate of transformation depends on the number of growing particles as well as on their individual growth rate. The nucleation of the new phase in the iron alloys and the titanium alloys under discussion usually occurs first at grain boundaries; this is most pronounced in hypoeutectoid steels, i.e., those containing less than the eutectoid content of carbon. An example of the formation of primary or proeutectoid ferrite at grain boundaries is shown in Fig. 8.43.

In some cases, the newly formed phase is coherent with the matrix, and its orientation is related to that of the matrix in a well-defined way. There is some uncertainty about the case of the austenite-ferrite transformation, but it is established that the $\beta \to \alpha$ transformation in titanium and its solid solutions is such that

$$(110)_\beta \parallel (0001)_\alpha$$

$$[111]_\beta \parallel (11\bar{2}0)_\alpha$$

a relationship discovered by Burgers. This type of relationship allows several different orientations of the α phase to form from a crystal of the β phase; it often results in a Widmanstatten type of structure, as shown in Fig. 8.44.

HEAT TREATMENT AND PROPERTIES OF TITANIUM ALLOYS

Interstitial Solutes. Since many titanium alloys are used after a heat treatment that produces controlled precipitation, it is convenient to discuss here the general subject of titanium alloys. One of the most significant facts about titanium is its extreme sensitivity to nitrogen, oxygen, carbon, and hydrogen, which form interstitial solid solutions. The effect of nitrogen is shown in Fig. 8.45; small amounts of nitrogen produce a very large increase in hardness and strength, but the ductility decreases seriously if the nitrogen content is too high. The situation is similar for oxygen, but the extent of the changes is limited, in the case of carbon, by the formation of a titanium carbide phase when the carbon concentration exceeds about 0.3 weight % (1.2 atomic %). The effect of hydrogen on the hardness and tensile properties is much smaller; however, the solubility of hydrogen decreases rapidly between 300°C (8%) and room temperature (0.1%) and the resulting precipitation of the brittle titanium hydride phase reduces the impact strength seriously. The decrease in impact strength continues for months after the alloy has been cooled, as shown in Fig. 8.46.

The useful alloys of titanium may be divided into three classes, those in which: 1. the α phase is strengthened by a solute and is the only stable phase at room temperature; 2. the β phase is present at room temperature; and 3. the α phase is strengthened and the β phase is stabilized. In general, the transition metals are more soluble in the β phase and stabilize it, while the other metals are more soluble in the α phase and raise the transformation temperature.

α Phase Alloys. The titanium-aluminum system is the most important example of a system in which a solute is added that stabilizes the

Fig. 8.44. Micrograph showing precipitate in titanium-3% molybdenum. From *Titanium*, McQuillan and McQuillan, Fig. 130, p. 321.

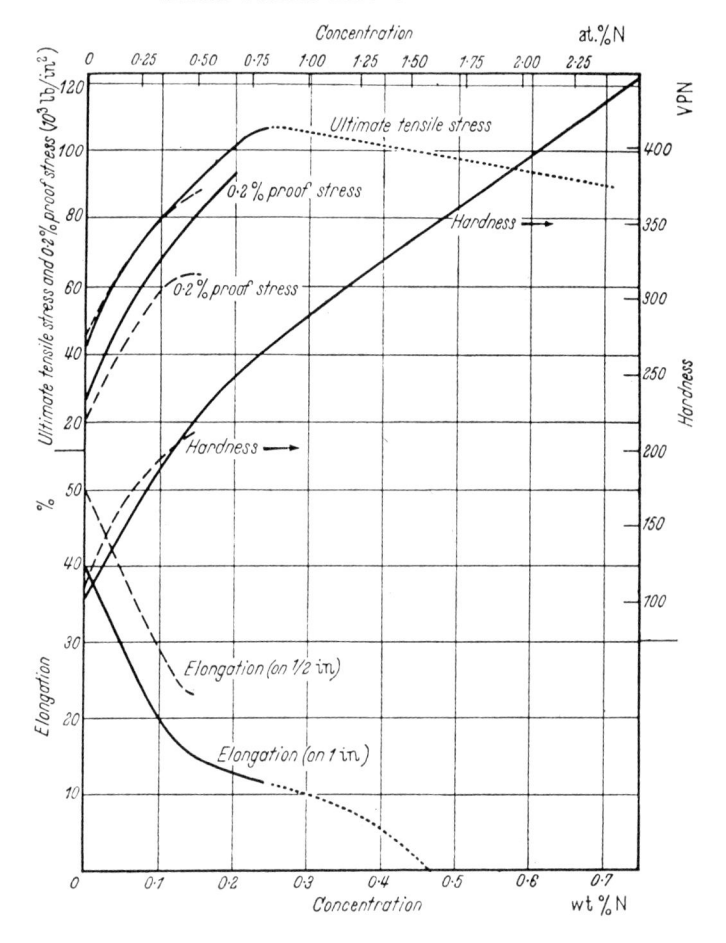

Fig. 8.45. Effect of nitrogen on properties of titanium. From *Titanium*, McQuillan and McQuillan, Fig. 135, p. 339.

α phase. The phase diagram is shown in Fig. 8.41. The effect of aluminum on the mechanical properties is entirely due to solid solution strengthening; it is shown in Fig. 8.47. Further strengthening can be obtained by the addition of tin, an alloy containing 5% aluminum and 2.5% tin giving a good combination of strength (115,000 psi) and ductility (10%) and good retention of strength at elevated temperatures, as shown in Fig. 8.48.

Titanium Alloys Containing β Phase. The transition metals, being more soluble in β titanium than in α titanium, produce alloy systems typified by that of titanium manganese. The β phase is greatly

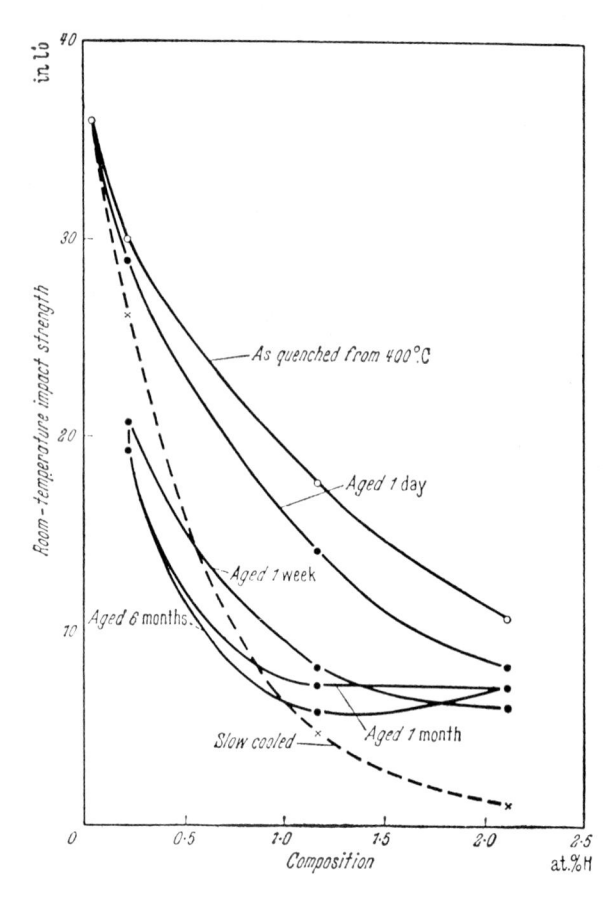

Fig. 8.46. Effect of hydrogen on impact properties of titanium. From *Titanium*, McQuillan and McQuillan, Fig. 139, p. 344.

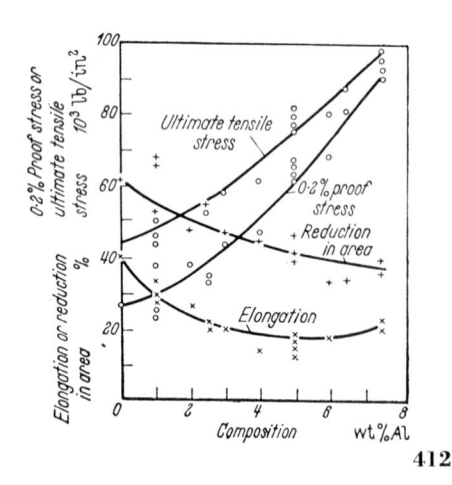

Fig. 8.47. Effect of aluminum on the properties of titanium. From *Titanium*, McQuillan and McQuillan, Fig. 156, p. 366.

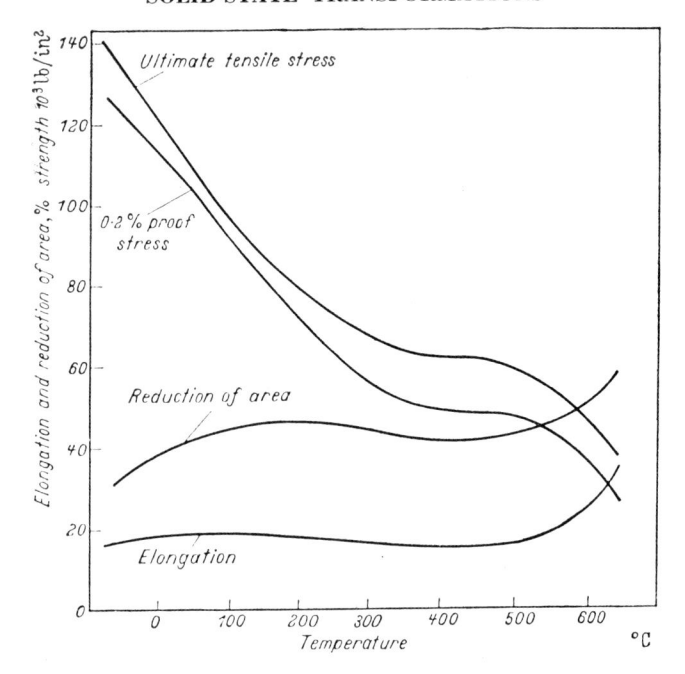

Fig. 8.48. Effect of temperature on the tensile properties of a titanium alloy. From *Titanium,* McQuillan and McQuillan, Fig. 160, p. 371.

strengthened by the manganese that is dissolved in it, while the α phase, being relatively free from manganese, is soft. The most useful properties for alloys of this kind are therefore obtained when the proportion of β phase is high. It is not possible in practice to use alloys that contain sufficient solute for the β phase alone to be stable at room temperature. The best properties are obtained when the alloy is worked or annealed at a high temperature, at which its equilibrium constitution is mainly β, and then cooled down to retain as much β as possible. The alternative is to quench from the "all β" condition and then to allow α to form by a controlled aging process. The effect of the annealing temperature is shown in Fig. 8.49. For commercial use, the optimum treatment may consist in hot-working the alloy in the α + β field and finishing the operation at the temperature at which the two phases are present in about equal quantities. The alloy is allowed to cool slowly. An example of the properties reported for alloys of this type is Ti, 3% Cu, 1.5% Fe. The U.T.S. is 150,000 psi and the yield stress is 120,000 psi, with an elongation of 15%.

Some alloys have been developed in which an "α strengthener" such

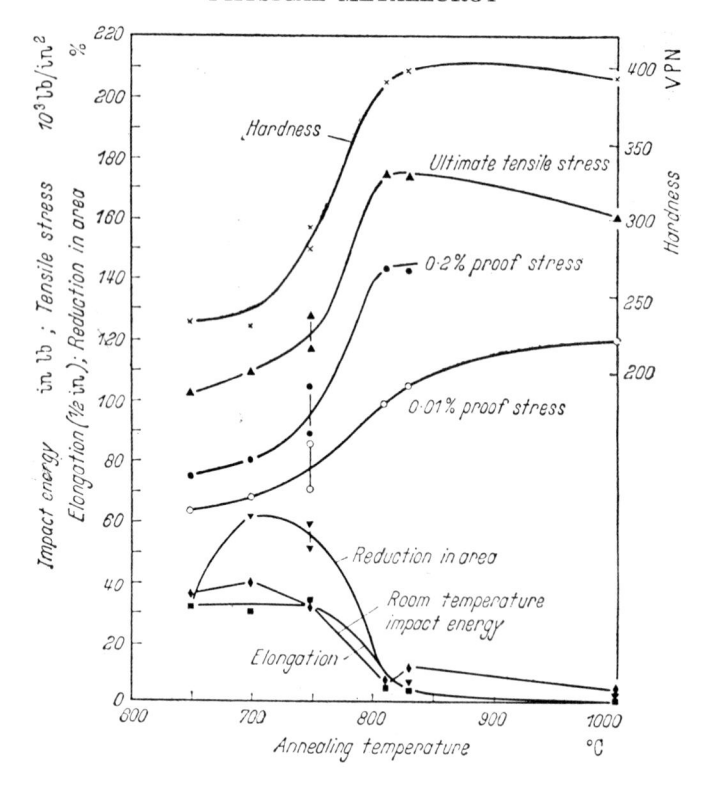

Fig. 8.49. Effect of annealing temperature on properties of titanium-4.4% manganese. From *Titanium*, McQuillan and McQuillan, Fig. 164, p. 380.

as aluminum is added in addition to a "β stabilizer" such as manganese. The transition metal content is usually reduced when aluminum is added in order to retain sufficient ductility; this results in a slight decrease in density, a very desirable characteristic for many applications. The properties obtainable with alloys of this type are slightly better than when no aluminum is present.

Recent work has shown that an "all β" alloy containing 2.5% aluminum and 16% vanadium can achieve an U.T.S. of 270,000 psi.

The substitution of smaller amounts of several transition metals for a larger amount of a single element appears to be beneficial, for reasons that are obscure, and the various possible alloying elements are not interchangeable; consequently, much empirical development work is required before the optimum compositions and heat treatments are found.

8.5 The Eutectoid Transformation

A eutectoid transformation is one in which a single solid phase transforms into two new phases on cooling. The two phases that are formed both have different compositions and structures from the pre-existing phase, and therefore diffusion is necessary for the transformation to occur. Because of the slowness of diffusion in solids, it is possible to produce the transformation under conditions that are far from equilibrium. There is, therefore, a wide range of temperatures at which a eutectoid reaction can occur. The resulting structures and properties show a correspondingly wide variation.

Although eutectoid reactions occur in many alloys, the only system that has been studied intensively is that of iron-carbon. The discussion will be confined to this system, although the principles, but not the details, apply to any eutectoid system.

Pearlite. The two phases that are formed by the eutectoid decomposition of austenite are ferrite and cementite, Fig. 8.50; the former contains 0.025% carbon if the transformation takes place at the equilibrium temperature, and less if the temperature is lower. Cementite, Fe_3C, contains 25 atomic % of carbon, which is about 6% by weight. Consequently, when austenite containing 0.8% carbon decomposes into ferrite and cementite, about one-eighth (by weight) is cementite and the remainder is ferrite. The fact that two phases are formed, one containing very little of the carbon and the other containing nearly all of it, allows all the diffusive movement of carbon to take place locally.

This is achieved by the simultaneous growth of lamellae of the two phases, a process exactly analogous to the freezing of a lamellar eutectic, discussed on p. 263. Figure 8.51a shows schematically how the carbon diffuses from regions where it is rejected by the advancing ferrite lamellae to regions where it is required for the formation of cementite. It will be seen that the diffusion of carbon takes place close to the interface and mainly parallel to it. Pearlite occurs in "colonies," each of which has common orientations for ferrite and for cementite, and derives from a single nucleus. A typical pearlite structure is shown in Fig. 8.51b.

There is strong evidence that the first stage of the formation of a colony is the nucleation of a cementite crystal. As the cementite crystal grows, it removes carbon from the surrounding austenite; the result is an increase in the tendency for ferrite to nucleate, because this happens most easily in a region of abnormally low carbon content. Eventually nucleation of ferrite occurs, at the surface of the cementite

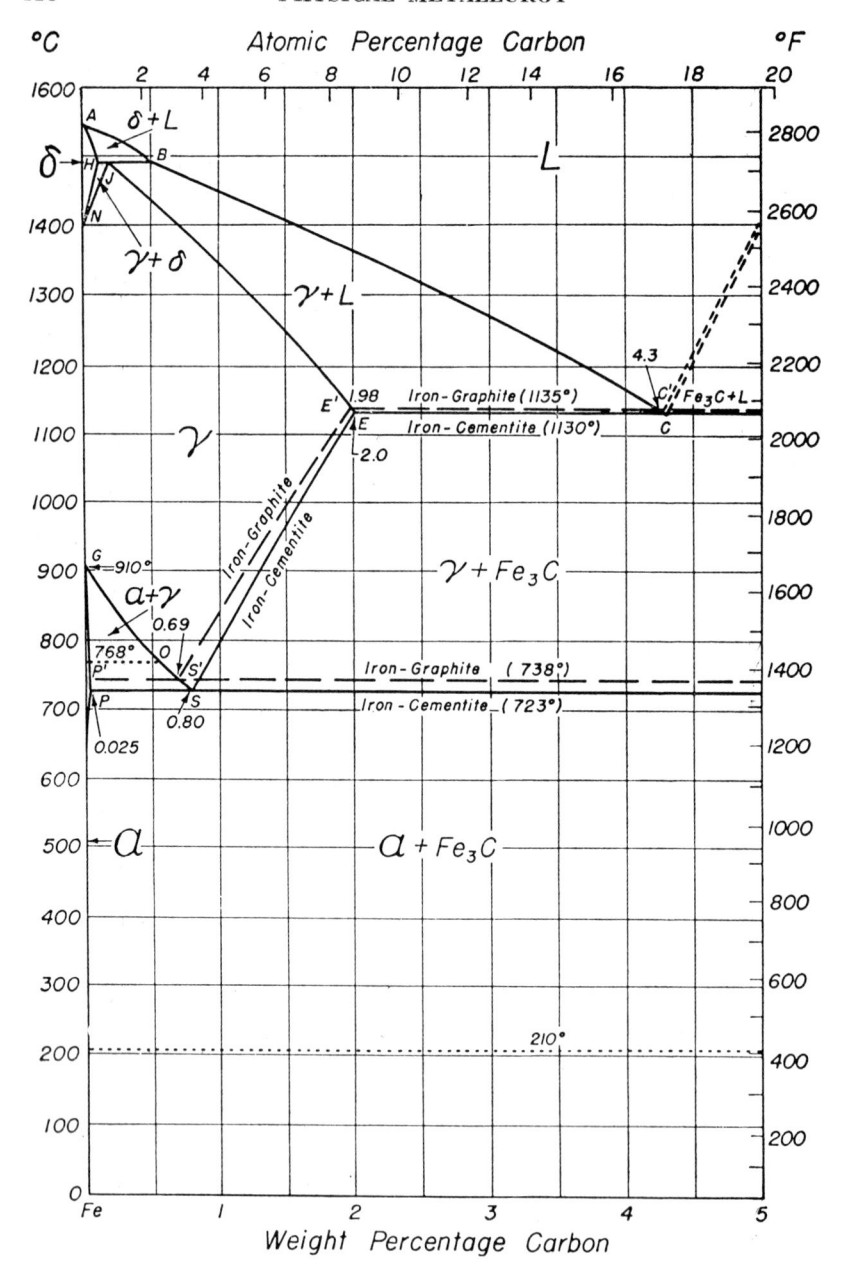

Fig. 8.50. Part of iron-carbon phase diagram. From *Metals Handbook*, ASM, p. 1182.

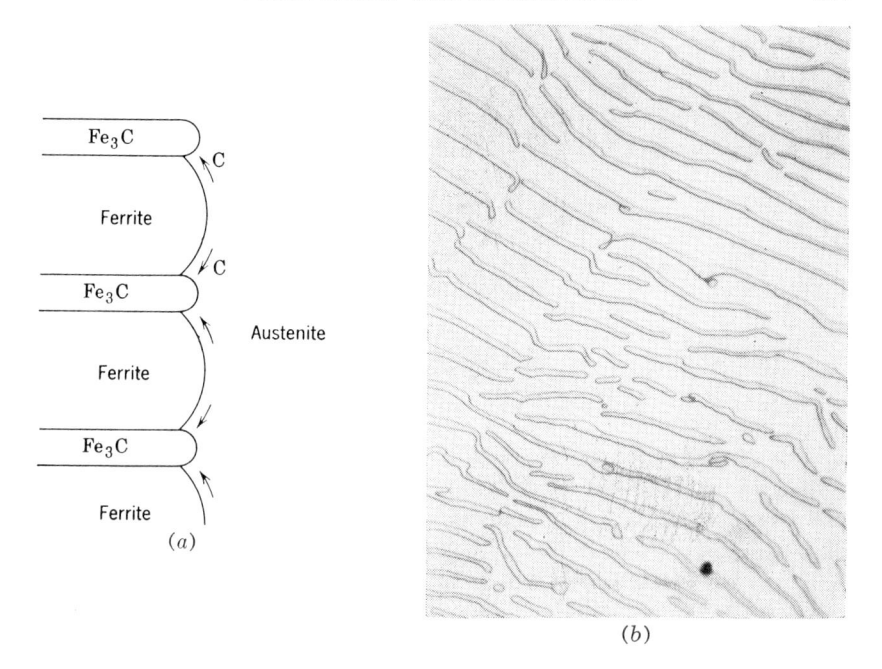

Fig. 8.51. Formation and appearance of pearlite. (a) Schematic. (b) Photomicrograph (×1250). (Courtesy U. S. Steel Corporation.)

and coherently with it; the two phases then grow simultaneously, the excess carbon from the ferrite being used up by the cementite, at regions where their surfaces are close together. At other parts of their surfaces either an excess or a deficit of carbon develops; cementite therefore spreads over the surface of ferrite and vice versa. This results in the formation of alternate layers of cementite and ferrite, each being continuous with its original nucleus. (It is possible in a three-dimensional system to have two interpenetrating continuous phases; this cannot be illustrated in a two-dimensional diagram.)

The lamellar structure of ferrite and cementite is called pearlite. The original nucleus of a pearlite colony is usually formed at a grain boundary of the austenite or at a pre-existing cementite particle; the colony then spreads rapidly along the grain boundary and, more slowly, normally to it.

It will be noted that the eutectoid transformation is not self-suppressing, as is the formation of proeuctectoid ferrite above the eutectoid temperature, and martensite; pearlite continues to form until the austenite

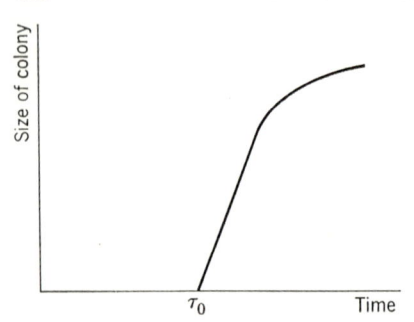

Fig. 8.52. Growth of pearlite colony.

is completely transformed, if the initial composition is 0.8% carbon and the temperature is below 740°C.

Kinetics of austenite-pearlite transformation. In most practical situations, pearlite is formed while the steel sample is cooling down; in order to understand the details, and to make useful generalizations, however, it is necessary first to study the changes that take place isothermally, i.e., when the temperature of austenite has been suddenly lowered to a specified temperature and then held there. Under these conditions, it would be expected that the growth of a single pearlite colony would increase the radius with time in the manner indicated in Fig. 8.52, which is analogous to the growth of a new crystal in recrystallization (see p. 328). Nucleation occurs at time τ_0, and growth takes place at a constant rate thereafter until impingement begins. In a sample of steel, many colonies will nucleate at various times and the increase in *amount* of pearlite as a function of time is shown in Fig. 8.53, which is a typical experimental curve.

The variation of this curve with temperature can be considered as follows: For reasons discussed on p. 239, more nuclei will form per unit time as the supercooling is increased, i.e., as the temperature drops

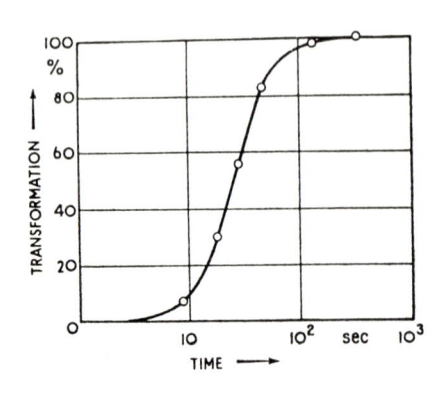

Fig. 8.53. Increase of amount of pearlite with time. From R. F. Mehl and W. C. Hagel, *Progress in Metal Physics 6*, Bruce Chalmers and R. King, Eds., Pergamon Press, London, 1956, Fig. 5, p. 76.

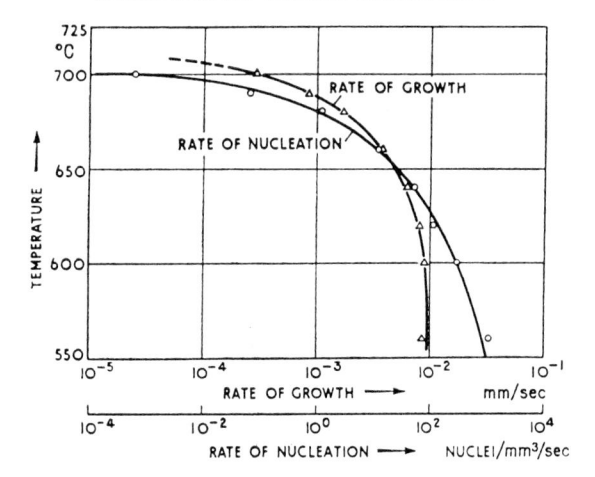

Fig. 8.54. Effect of temperature on rate of nucleation and growth of pearlite. After Mehl and Hagel, *Progress in Metal Physics 6*, Fig. 28, p. 102.

further below the eutectoid temperature. Also, the driving force for growth of pearlite *increases* with increased supercooling, while the diffusion constant *decreases* with a fall in temperature.

Figure 8.54 shows the results of measurements of rate of nucleation and rate of growth of pearlite colonies in the temperature range 725°C to 550°C. Both these rates would be expected to decrease as the temperature falls below 550°C. The rate of transformation is roughly proportional to the product of the number of nuclei and the growth rate, and must therefore follow a curve of the kind shown in Fig. 8.55.

Many experimental investigations have been made of the isothermal transformation of austenite to pearlite; the typical result is the time-temperature-transformation (T.T.T.) diagram shown in Fig. 8.56 for a eutectoid steel. The solid curves represent the time at which decomposition of pearlite is first detectable, and the time at which it is com-

Fig. 8.55. Effect of temperature on rate of eutectoid transformation.

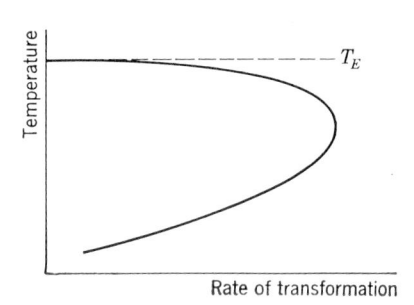

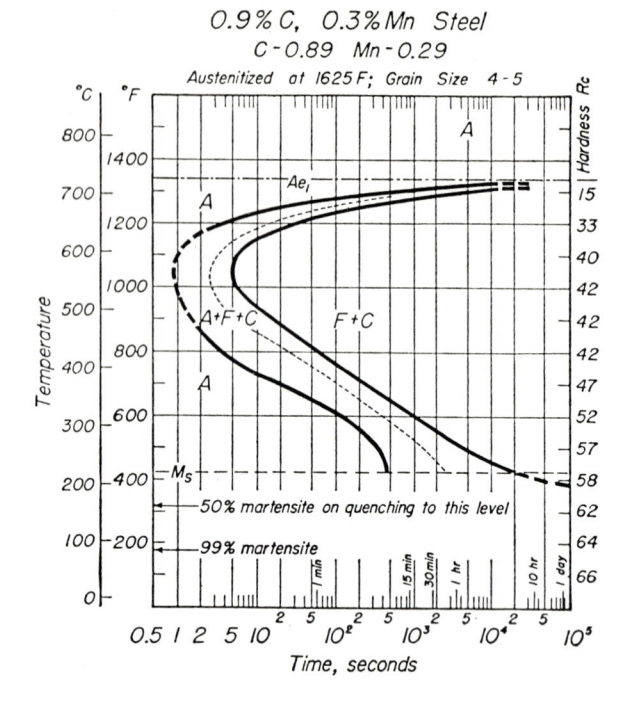

Fig. 8.56. T.T.T. curve for eutectoid steel. From *Metals Handbook*, ASM, Fig. 2, p. 608.

plete. If the steel is not of eutectoid composition, the T.T.T. diagram includes an extra branch, as shown in Fig. 8.57, which shows the start of the formation of proeutectoid ferrite.

Properties of pearlite. The properties of pearlite depend to an important extent on the thickness of the lamellae, and this in turn depends upon the speed of its formation, the thickness being limited by the distance through which carbon can diffuse during the time that is available. The variation of pearlite spacing with the temperature of the transformation is shown in Fig. 8.58. The effect of the pearlite lamellar thickness on its hardness is given in Fig. 8.59. The properties of steel containing ferrite and pearlite can be considerably modified by cold deformation, which causes work hardening of the ferrite. This is widely used for mild steel (0.1% to 0.2% carbon), frequently in the form of cold rolling. The variation of strength and ductility with amount of cold deformation in wire drawing is shown in Fig. 8.60 for several ranges of carbon content.

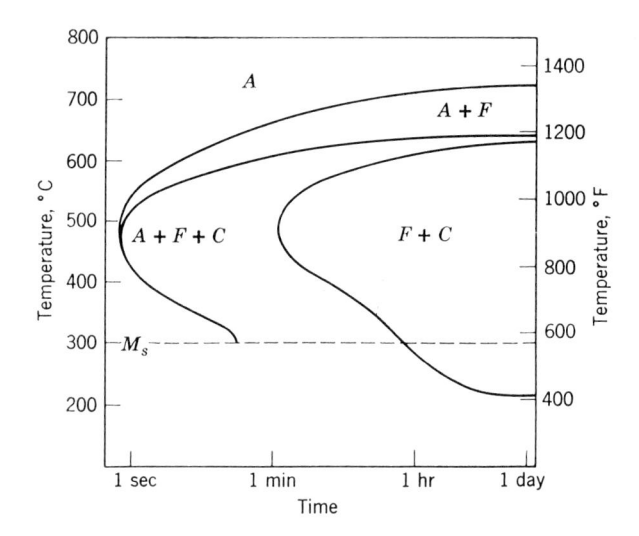

Fig. 8.57. T.T.T. curve for hypoeutectoid steel, showing region of proeutectoid ferrite. From *Metals Handbook*, ASM, Fig. 1, p. 607.

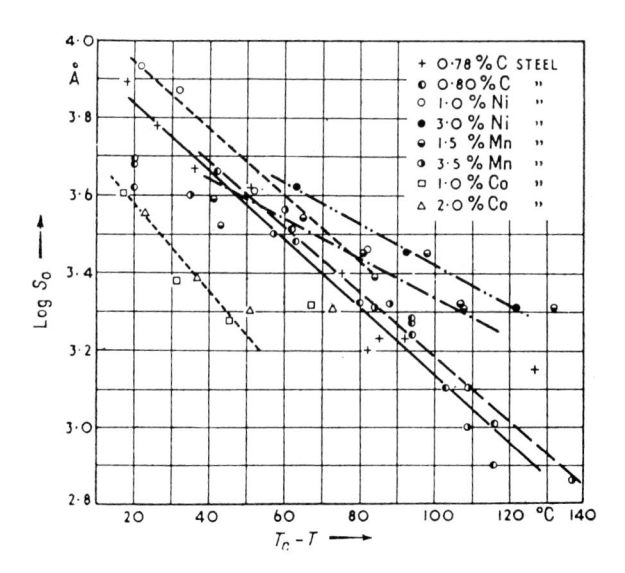

Fig. 8.58. Effect of transformation temperature on lamellar spacing; T_c and T are equilibrium temperature and transformation temperature, S_0 is the interlamellar spacing. From Mehl and Hagel, *Progress in Metal Physics 6*, Fig. 25, p. 98.

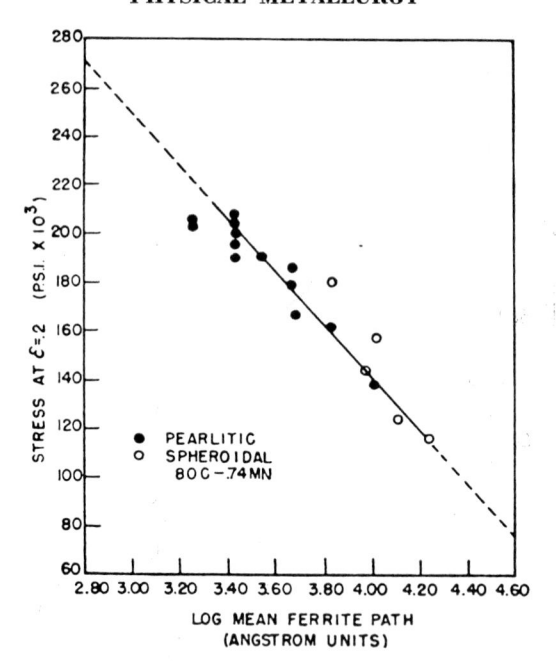

Fig. 8.59. Relationship between lamellar spacing in pearlite and hardness. From *Ferrous Metallurgical Design*, J. H. Hollomon and L. D. Jaffee, John Wiley & Sons, New York, 1947, Fig. 26, p. 83.

Metallography of pearlite. When the concentration of the carbon in austenite is less than 0.8% (hypoeutectoid), the result of slow cooling is the formation first of some ferrite (described as primary, or proeutectoid) until the austenite has reached the eutectoid composition,

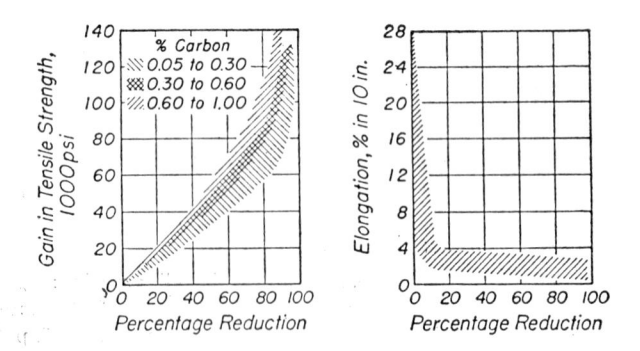

Fig. 8.60. Variation of strength and ductility with cold deformation in wire drawing. From *Metals Handbook*, ASM, Figs. 4 and 5, p. 360.

SOLID-STATE TRANSFORMATIONS   423

(a)

(b)

(c)

(d)

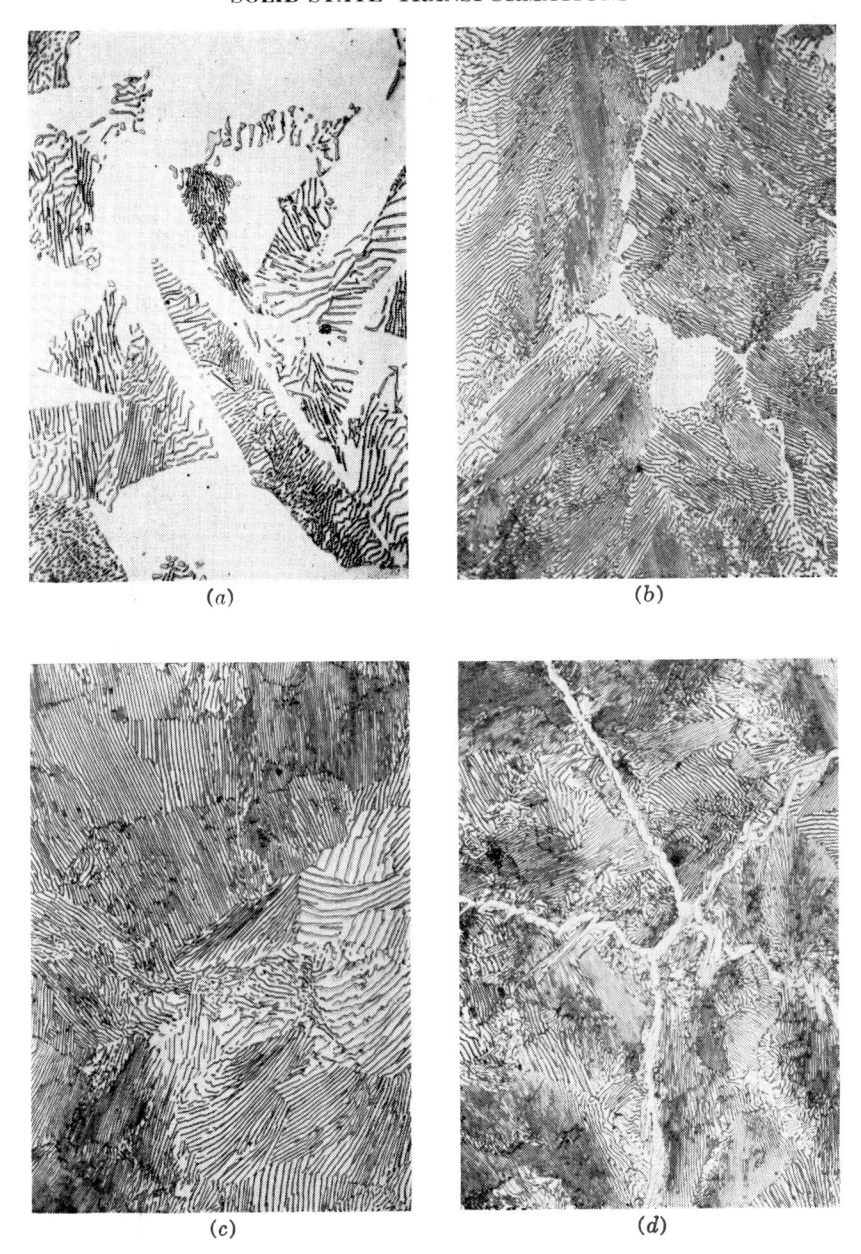

Fig. 8.61. Effect of carbon content on microstructure of slowly cooled carbon steels. (a) 0.35%C ($\times$450); (b) 0.65%C ($\times$500); (c) 0.87%C ($\times$500); (d) 1.2%C ($\times$500). (Courtesy U. S. Steel Corporation.)

after which the remaining austenite transforms into pearlite. The structure then contains predictable proportions of ferrite and pearlite. The distribution of the ferrite and the pearlite depends on the grain size of the austenite, because nucleation of the primary ferrite or (primary cementite if the carbon content exceeds 0.8%) occurs at the grain boundaries. The ferrite forms a "network" at the austenite grain boundaries and the pearlite is formed in the interior of the austenite grains. Examples of these structures are shown in Fig. 8.61.

If cooling is more rapid, so that a lower temperature is reached before nucleation of primary ferrite occurs, then pearlite may be formed even at carbon concentrations as low as 0.4%. This can be understood from Fig. 8.62. Consider a steel of carbon content given by S. Above the temperature T_A, the stable phase is austenite; from T_A to T_B, equilibrium is between austenite and ferrite. Between T_B and T_C, cementite is less stable than austenite, and therefore cannot nucleate from it, and so ferrite is formed until the austenite composition crosses the line EC', which defines the conditions for cementite to be in equilibrium with austenite. If the temperature is below T_C, cementite can nucleate immediately and pearlite is formed. Pearlite formed in this way has more ferrite and less cementite than eutectoid pearlite, and it is softer.

It should be mentioned that the description of the phase under discussion as pearlite, irrespective of its fineness, is not consistent with the classical terminology in which it was called troostite if it was not resolvable as a lamellar structure under the optical microscope.

The thickness of the lamellae should be consistent in a homogeneous sample of steel that has transformed isothermally; this frequently does not appear to be the case, because different colonies are sectioned at

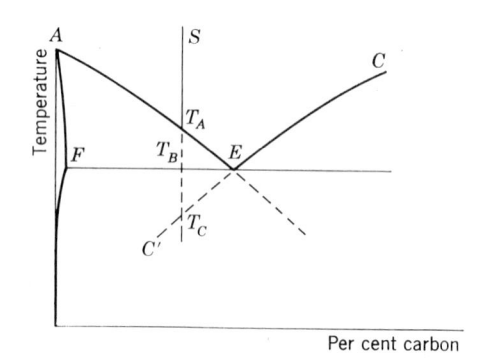

Fig. 8.62. Showing limits for pearlite formation.

(a)   (b)

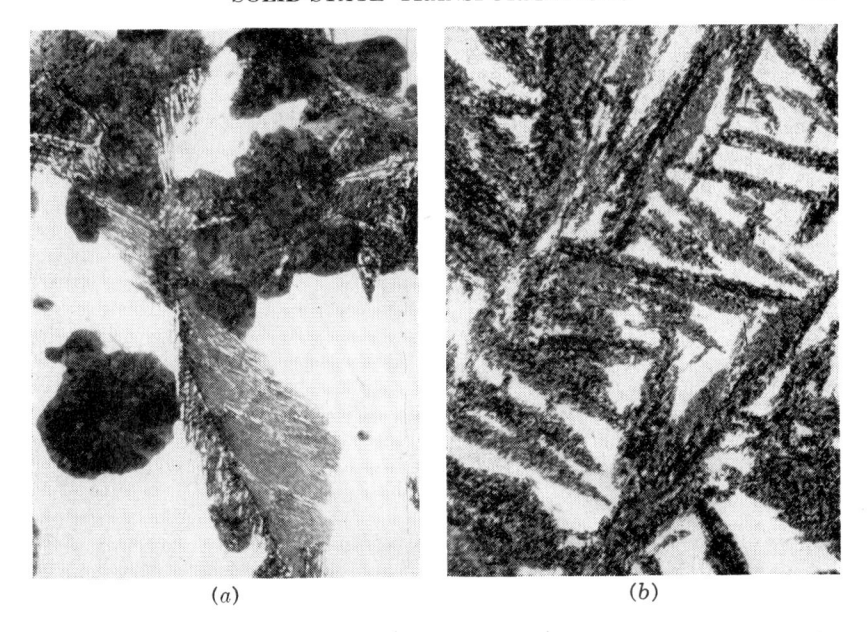

Fig. 8.63. Photomicrographs (×2500) of upper bainite (a) and lower bainite (b) (Courtesy U. S. Steel Corporation.)

different angles when a sample is prepared for microscopical examination.

The bainite transformation. When conditions are very far removed from equilibrium, so that nucleation of ferrite becomes rapid, an alternative transformation, the *bainite* transformation, may supersede the pearlite-forming mechanism that has been discussed. The alternative transformation, which produces crystals of supersaturated ferrite, and of cementite, results in structures similar to those shown in Fig. 8.63. Crystals of ferrite are probably formed coherently with the austenite matrix, but reject some carbon, with the result that the growth is controlled by the diffusion of carbon towards nuclei of cementite that form where the carbon concentration reaches a high enough value. Because the bainite reaction depends upon the frequent nucleation of ferrite and cementite, it is slower than the pearlite reaction at higher temperatures; but because it requires less diffusion of carbon, it is faster at lower temperatures. We may therefore schematically illustrate the two reactions on a single T.T.T. curve as shown in Fig. 8.64, which represents the conditions for a eutectoid plain carbon steel. Some alloy-

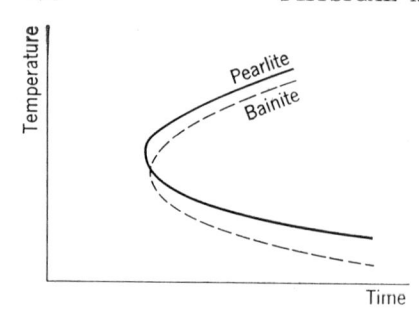

Fig. 8.64. Distinction between T.T.T. curves for pearlite and bainite.

ing elements have different effects on the two reactions, as will be discussed in the next section, with the result that the T.T.T. curve may have two "noses," as shown in Fig. 8.65.

Influence of alloying elements on T.T.T. curve. One of the main reasons for the addition of alloying elements to steel is to slow down or suppress the transformation from austenite to pearlite or bainite; this may be done in order to make it possible to form martensite even when cooling is not very rapid, as in the interior of a thick section, or in order to preserve the austenitic form at the temperature at which the steel will be used.

The usual alloying elements may (1) change the equilibrium conditions or (2) change the kinetics of the transformation; these will be discussed in turn.

EFFECT OF ALLOYING ELEMENTS ON EQUILIBRIUM CONDITIONS. The effects on the equilibrium conditions are most readily seen from the appropriate phase diagrams, shown in Fig. 8.66. It is seen that the only elements apart from carbon that have a large effect in reducing the

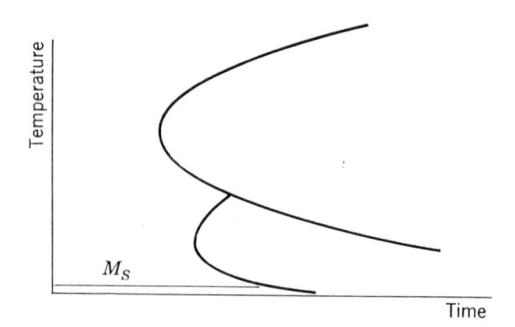

Fig. 8.65. T.T.T. curve showing two regions of fast transformation.

transformation temperature are nickel and manganese; chromium, silicon, etc., form the "γ loop" and so prevent the formation of austenite beyond a limited concentration of the alloying element.

EFFECT OF ALLOYING ELEMENTS ON KINETICS OF TRANSFORMATION. Most of the alloying elements have a noticeable effect on the kinetics of the transformation, causing it to occur more slowly. This may be due to the fact that the alloying elements are less soluble in ferrite than in austenite, and also form carbides, in competition with cementite. The other carbides probably do not act as efficient nuclei for pearlite formation, and so may slow down the nucleation part of the process. The slower diffusion of the substitutional elements, as compared with carbon, may slow down the growth part of the transformation process. Some typical T.T.T. curves showing the magnitude of the effect of alloying elements are given in Fig. 8.67.

One result of the retardation of the transformation is that a steel containing a relatively small amount of alloying elements undergoes the pearlite transformation at a lower temperature than for a plain carbon steel of the same carbon content; an example is shown in Fig. 8.68, in which a cooling curve is superimposed on the T.T.T. curves for a plain carbon steel and a low alloy steel. Less ferrite is formed, and the pearlite is harder as a result of its finer structure as well as, to a lesser extent, of the alloy content of the ferrite component. The effect on the hardness of ferrite of various alloying elements is shown in Fig. 8.69.

The alloy steels that are used in the "ferritic plus pearlitic" form are the low alloy structural steels. The use of steels of this type, in which a low carbon content (0.1%) is combined with copper and chromium (0.5% and 1%), has advantages especially if welding is contemplated; one of the results of welding a higher carbon steel is to produce martensite during the very rapid cooling that occurs as a result of heat flow from the weld into the adjacent cold metal. This martensite is not readily tempered, and the result is likely to be a hard, brittle weld.

Continuous cooling curves. The T.T.T. curve provides information on the time that elapses, at any selected temperature, before the transformation begins, and before it finishes. The most important practical problem that arises, however, is whether a given steel, cooled under given conditions, will consist of 100% martensite, or whether any ferrite, pearlite, or bainite will have formed during cooling.

It is necessary, therefore, to consider the relationship between the continuous cooling curve encountered in practice and the T.T.T. curve. It is possible to relate a continuous cooling curve to a T.T.T. curve by

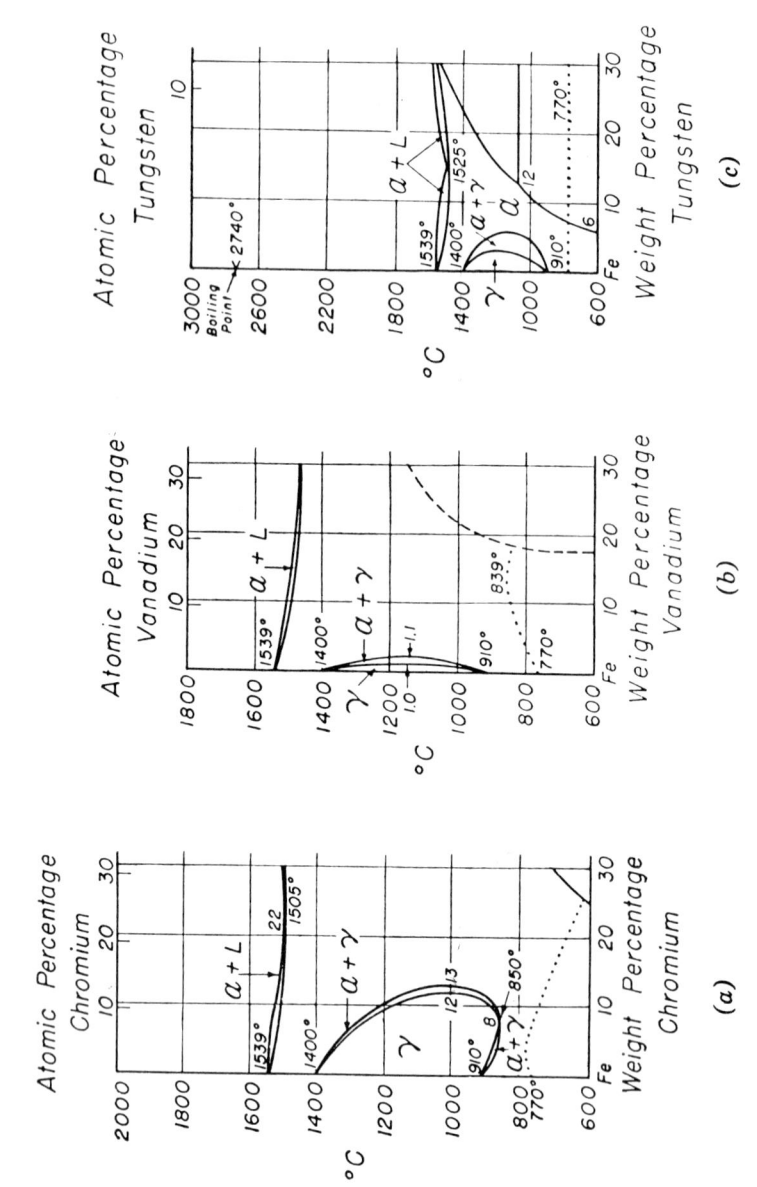

Fig. 8.66. Binary diagrams of iron with various alloying elements. From *Metals Handbook*, ASM, pp. 1194, 1219, 1220, 1211, 1217, and 1210.

SOLID-STATE TRANSFORMATIONS 429

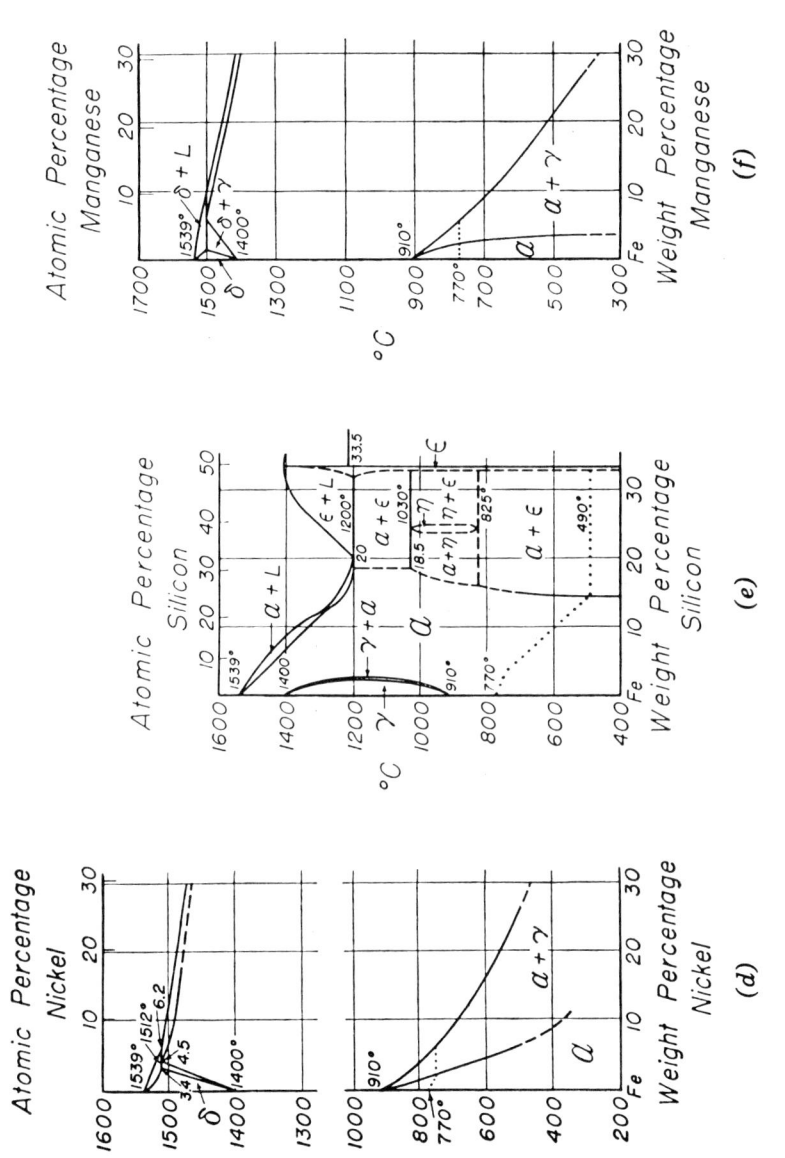

Fig. 8.66 (continued).

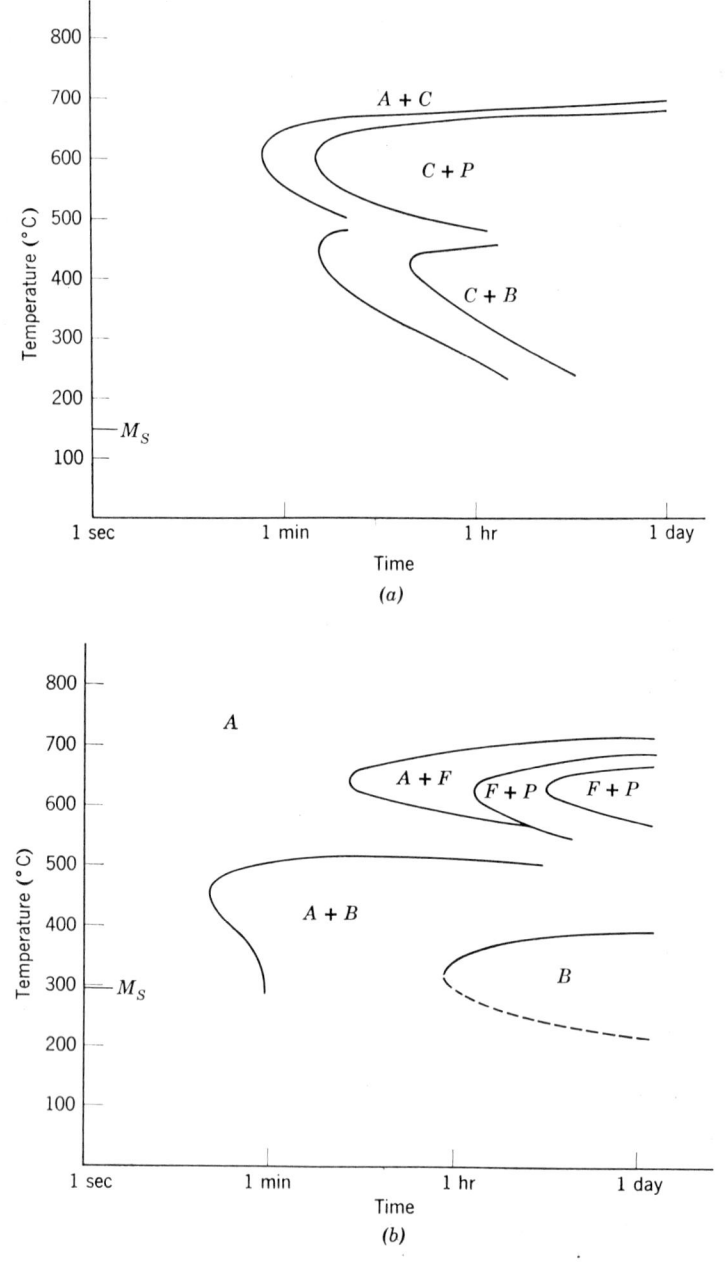

Fig. 8.67. T.T.T. curves for steel of composition (a) C 0.91%, Mn 0.65%, Ni 1.35%, Cr 0.60%; (b) C 0.42%, Mn 0.78%, Ni 1.79%, Cr 0.80%, Mo 0.33%. (Courtesy International Nickel Company.)

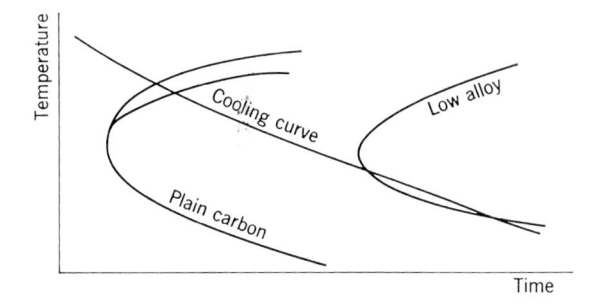

Fig. 8.68. Effect of alloying elements on transformation temperature.

the following procedure. Figure 8.70 shows a T.T.T. curve and a continuous cooling curve for the same steel. The problem is to determine whether any pearlite is formed. Consider a temperature interval ΔT. The time that elapses while the temperature falls by ΔT is Δt. This is a fraction $\Delta t/t'$ of the time t' before the transformation begins at the temperature T'. It may be assumed that the transformation begins when the sum of the fractions $\Delta t/t'$ for successive intervals reaches the value unity. Thus a small fraction of the time t' is "used up" at high temperatures where the transformation is slow and cooling is rapid, but as the temperature falls and the transformation time becomes less, the fraction used in each successive interval becomes larger. A new continuous cooling curve, shown as the broken line PQ, is drawn from the cumulative fraction of the time before the transformation begins. Pearlite begins to form when the derived curve PQ cuts the T.T.T. curve. The rate of cooling that just avoids the pearlite transforma-

Fig. 8.69. Effect of alloying elements on the strength of ferrite. From *Fundamentals of Physical Metallurgy*, Ralph Hultgren, Prentice-Hall, New York, 1952, Fig. 166, p. 247.

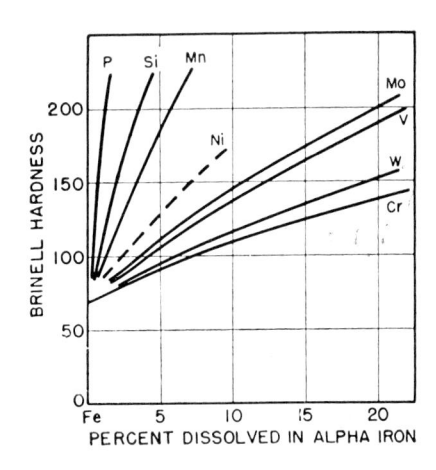

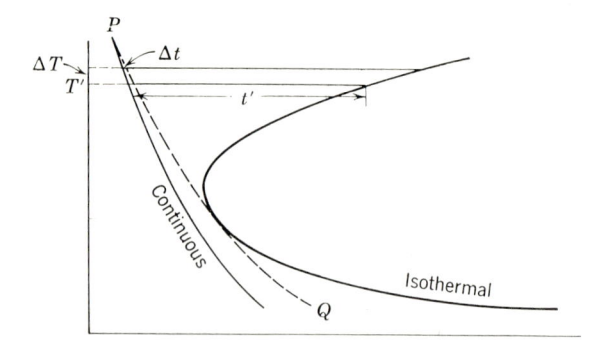

Fig. 8.70. Correction of continuous cooling curve by means of fractional transformation times.

tion is called the critical cooling rate. It is not strictly a single rate although it depends largely on how much time is spent in the temperature range just above the "nose" of the T.T.T. curve.

Quenching. It will be clear from the foregoing that for any given steel a fully martensitic structure can be obtained only if it is cooled through the critical range of temperature in less time than it takes for the pearlite or bainite transformation to start. It is therefore necessary to consider the rate of cooling in relation to the imposed conditions.

The only way in which a piece of metal can be cooled is by heat transfer from its surface. This is most efficiently achieved by immersing it in a liquid; because of its high latent heat of vaporization and specific heat, water or aqueous solutions are the most efficient. When a steel sample at, say, 900°C is quenched into cold water, heat is extracted in at least three distinct stages, as shown by the temperature-time curve of Fig. 8.71. In the first stage, A, a layer of vapor is formed

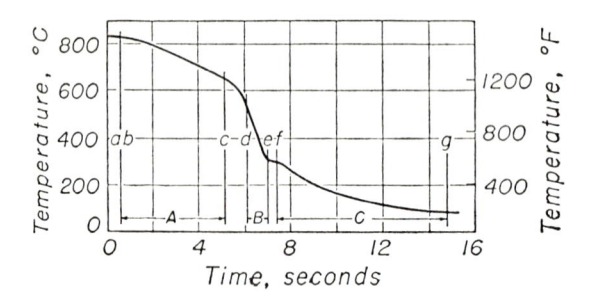

Fig. 8.71. Stages in the cooling of a quenched metal. From *Metals Handbook,* ASM, Fig. 3, p. 617.

on the surface of the steel and heat transfer through it is by conduction and radiation and is relatively slow. When the vapor layer reaches a critical thickness, it begins to break away as bubbles and since the vapor layer is no longer continuous the metal surface is wetted by the water. Heat is removed largely as the latent heat of vaporization of the water that boils in contact with the metal surface and cooling is much faster. This is stage B. The final stage, C, which is of little importance in the quenching of steel, is that in which the water does not boil but removes heat by convection. The duration of the vapor stage, A, is of importance in producing a rapid quench in the critical range, and it is found that aqueous solutions of sodium chloride, sodium hydroxide, or lithium chloride increase the cooling rate during the early stages by decreasing the time during which the continuous vapor layer persists. It is not clear why these solutes lend to earlier wetting of the steel by the water.

These considerations relate to the rate of extraction of heat at the surface. The actual rate of cooling depends upon the size of the sample and the position in it. Figure 8.72 shows how the time-temperature relationship varies from the surface to the center of a 2-in.-diameter steel cylinder quenched in water and in oil. It is evident that the rate of

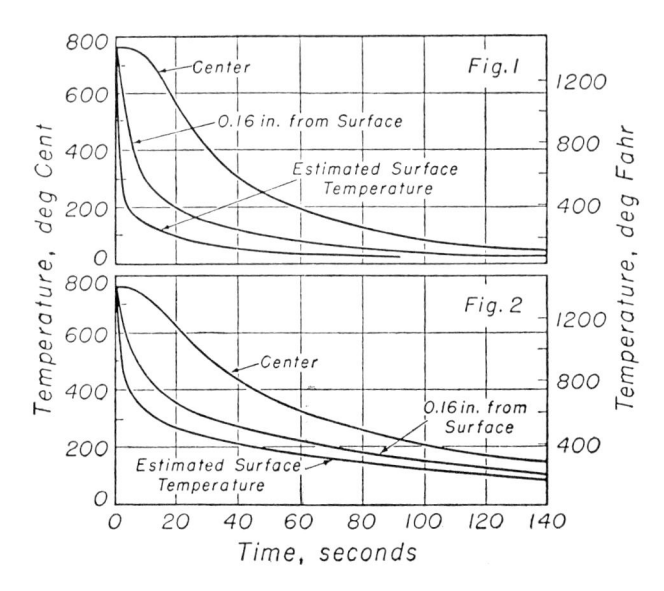

Fig. 8.72. Rate of cooling at surface, A, center, B, and intermediate point, C, quenched in water (Fig. 1) and oil (Fig. 2). From *Metals Handbook*, ASM, Fig. 1, p. 616.

cooling decreases from the surface inwards and that the surface may exceed the critical cooling rate while the center does not.

Hardenability. The term *hardenability* is used to denote the depth to which a fully martensitic structure can be obtained, or, alternatively, the radius of a cylinder that can just be fully hardened to its center by a standardized quenching technique. The hardenability bears **no** relationship to the hardness, which depends largely on the carbon content, but only to the depth to which the maximum hardness can be produced.

A conventional test for hardness is the Jominy end quench test, in which a cylindrical bar is quenched from one end by means of a water jet, and its hardness is measured as a function of distance from the quenched end. The distribution of hardness along the bar is shown in Fig. 8.73; the hardenability is the distance at which the hardness drops to some specified value.

EFFECT OF AUSTENITE GRAIN SIZE. The effect of austenite grain size on hardenability is a result of the fact that pearlite nucleates at the grain boundaries and grows from there into the crystals. The amount of pearlite that is formed, as a function of time, can be expressed as the thickness of the pearlite layer. This is shown in Fig. 8.74. The volume of pearlite that corresponds to this thickness depends upon the area of grain boundary in a unit volume of the steel. Two cases are represented in Fig. 8.75. It is seen that a far higher *proportion* has transformed in Fig. 8.75a than in b, although the thickness of pearlite is the same in both cases. This difference should not apply to the

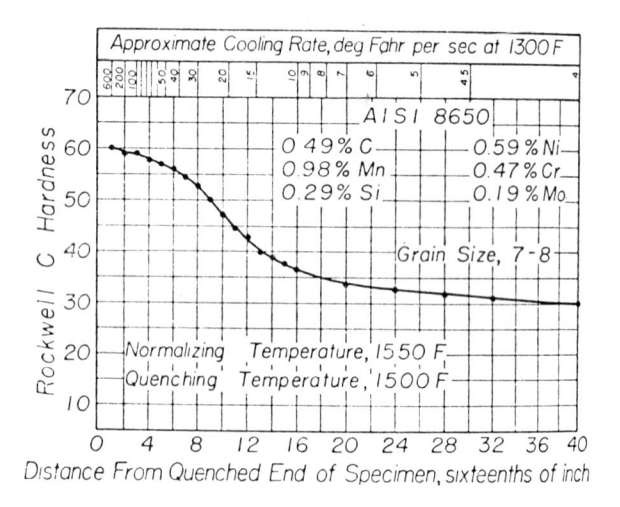

Fig. 8.73. Result of Jominy test. From *Metals Handbook,* ASM, Fig. 3, p. 490.

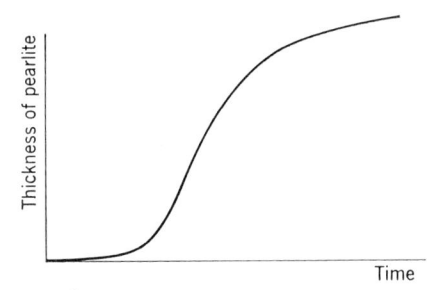

Fig. 8.74. Growth in thickness of pearlite layer at grain boundaries of austenite.

criterion of *zero* pearlite, but will have a marked influence when the more realistic criteria of 10% or 50% are adopted.

CALCULATION OF HARDENABILITY. Several proposals have been put forward for predicting the hardenability of a steel. The most successful of these is that due to Hollomon and Jaffee, who point out that the formation of either pearlite or bainite may limit the hardenability of a steel, and that the two are affected differently by some of the alloying elements and by grain size. It is therefore necessary to calculate both the "pearlite hardenability" and the "bainite hardenability" and to use the smaller value.

Although of doubtful theoretical validity, it is sufficiently accurate for most purposes to calculate the hardenability as the product of a number of factors, each of which represents the contribution of an alloying element or of the grain size. The assumption implicit in this method is that each alloying element has an effect that is linear with its concentration, and each factor is independent of the other elements that are present. The factors proposed by Hollomon and Jaffee are given in Table 8.6.

ADVANTAGES OF HIGH HARDENABILITY. There are two reasons for using steels of high hardenability. One is that a thicker section can

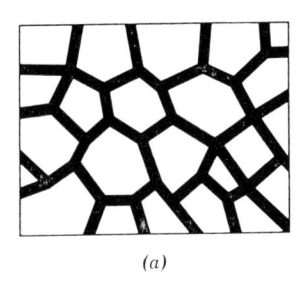

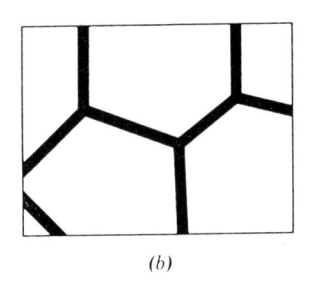

Fig. 8.75. Effect of austenitic grain size on hardenability.

Table 8.6. Hardenability Factors *

Element	Pearlitic Hardenability Factor	Bainitic Hardenability Factor
Carbon { 50% pearlite or bainite	$0.338 \times \sqrt{\% \text{ C}}$ in.	$0.494 \times \sqrt{\% \text{ C}}$ in.
Carbon { Almost no pearlite or bainite	$0.254 \times \sqrt{\% \text{ C}}$ in.	$0.272 \times \sqrt{\% \text{ C}}$ in.
Manganese	$1 + 4.10 \times \%$ Mn	$1 + 4.10 \times \%$ Mn
Phosphorus	$1 + 2.83 \times \%$ P	$1 + 2.83 \times \%$ P
Sulfur	$1 - 0.62 \times \%$ S	$1 - 0.62 \times \%$ S
Silicon	$1 + 0.64 \times \%$ Si	$1 + 0.64 \times \%$ Si
Chromium	$1 + 2.33 \times \%$ Cr	$1 + 1.16 \times \%$ Cr
Nickel	$1 + 0.52 \times \%$ Ni	$1 + 0.52 \times \%$ Ni
Molybdenum	$1 + 3.14 \times \%$ Mo	1
Copper	$1 + 0.27 \times \%$ Cu	$1 + 0.27 \times \%$ Cu

* From *Ferrous Metallurgical Design*, Holloman and Jaffee, Table 3, p. 200.

be fully transformed into martensite, and this is of importance for many applications in which maximum strength is required. The second reason is that full hardening to a moderate depth can be achieved with a less drastic quench if the hardenability is high; for example, a quench into oil instead of into cold water can be used for quenching a ½-in. bar with a hardenability of 1 in. The less drastic quench reduces the incidence of quenching cracks, which are caused by a combination of stresses resulting from the transformation and those caused by cooling different parts at different rates; it also decreases the distortion caused by quenching stresses.

The extreme case of high hardenability is found in those steels that are described as "air hardening"; these steels transform so slowly that full hardening can be attained by allowing the steel to cool in air, providing the section is not too thick. The Jominy test, which uses a water quench, is unsuitable for air-hardening steels, and the "hardenability" for these steels must be measured under conditions of slow cooling. For example, a steel of composition carbon 0.5%, manganese 0.5%, silicon 0.7%, chromium 7.4%, nickel 0.22%, tungsten 7.8%, and molybdenum 0.15% will air-harden if it is in the form of a bar of not more than 1-in. radius.

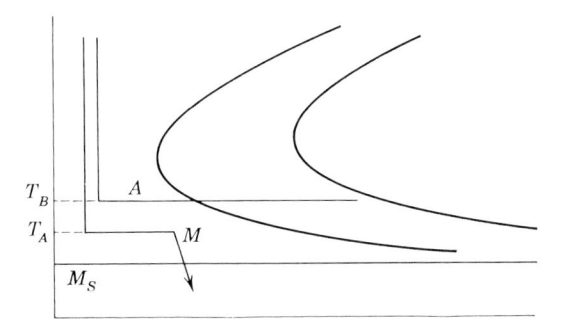

Fig. 8.76. Illustrating martempering, M, and austempering, A.

MARTEMPERING. A further step that can be taken to avoid quench cracking is to quench to the temperature T_A (Fig. 8.76), to hold the steel at this temperature long enough for the whole section to have reached essentially the same temperature, and then to allow it to cool relatively slowly to room temperature. The transformation takes place much more uniformly throughout the whole mass, and the stresses due to differential expansion are reduced.

This procedure is known as martempering. It depends upon the fact that at a temperature such as T_A there is ample time for temperature equalization before the transformation begins.

AUSTEMPERING. Another special treatment that is sometimes used is to allow the isothermal transformation to bainite to take place instead of producing martensite and then tempering it. This is done by quenching to a temperature such as T_B in Fig. 8.76 and holding at that temperature for a long enough time for completion of the bainite transformation. It is said that greater toughness is obtained at the same hardness by austempering as compared with the usual procedure.

Austenitic steels. It is seen from Fig. 8.66 that steels containing large amounts of nickel or manganese do not transform to pearlite in the usual temperature range. The depression of the temperature of the transformation also slows it down to such an extent that it is completely suppressed if cooling is fairly rapid.

The M_S temperature is also below room temperature, but in many cases the M_D temperature is not; such steels are stable in the austenitic form unless they are plastically deformed, when, if the carbon content is sufficient, the martensite transformation occurs in the deformed regions. This gives an extremely high work-hardening rate; steels containing 1.2% carbon and 12% manganese, called Hadfield's manganese steel, are remarkable for their resistance to wear, which is due to the local formation of martensite at regions that are subject to abrasion or repeated impact.

The steels in which transformation is suppressed by the addition of nickel, or nickel and chromium, also show the characteristic rapid work hardening due to strain-induced martensite. The nickel-chromium steels, containing at least 6% nickel and 16% chromium, are austenitic at room temperature and have remarkable corrosion resistance and oxidation resistance, hence their classification as *stainless steels*. In these steels, the pearlite transformation temperature is depressed by the nickel and retarded by the nickel and the chromium.

All the austenitic steels are nonmagnetic, because austenite does not become ferromagnetic until it contains about 40% of nickel, a composition that is outside the range of "austenitic steels." The nonmagnetic property is of great value in some applications, and is also useful for identification.

Annealing of steel. If a steel consisting of ferrite and pearlite is heated, spheroidization of the cementite takes place; this lowers the hardness of the steel, as would be expected from the reduction in the number of particles. If the steel has been cold-worked, recrystallization takes place below the transformation temperature when the carbon content is low; for higher carbon steels, it is necessary to transform to austenite to return the material to the soft condition. The properties on subsequent cooling depend only on the austenite grain size that is reached and on the cooling rate (for a steel of a given composition). The austenite grain size depends upon how far the temperature has risen above the transformation temperature, and on the presence or absence of "grain growth inhibitors," such as aluminum, in the steel. If the steel is allowed to cool very slowly, i.e., if it is left in the furnace and allowed to cool with it, it is said to be "fully annealed," and a very coarse pearlitic structure is obtained. If it is allowed to cool at a moderate rate (by being removed from the furnace and allowed to cool in air) it is said to be "normalized." These two methods of treatment are poorly defined but each produces a condition that is recognizable and useful for some purposes.

Typical microstructures of annealed and normalized 0.2% carbon steels are shown in Fig. 8.77.

8.6 Order-Disorder Changes

It was shown on p. 47 that there are a number of alloy systems in which a phase, usually a solid solution phase, is disordered at temperatures above some critical value, and has long-range order when it is in

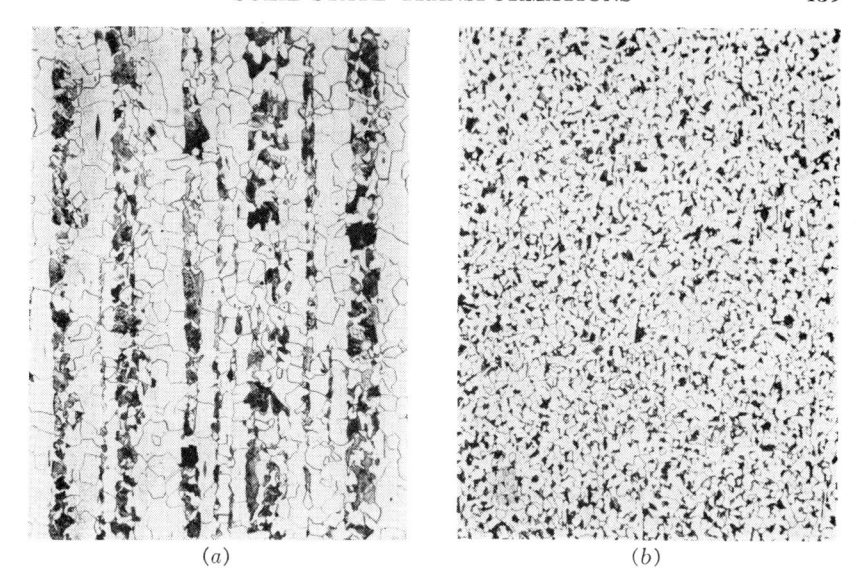

Fig. 8.77. Photomicrographs of annealed (a) and normalized (b) mild steel [(b) Courtesy U. S. Steel Corporation.]

equilibrium at lower temperatures. The change from the disordered to the ordered structure takes place by diffusion, and can therefore be suppressed if the temperature is lowered sufficiently fast. The rate of the ordering reaction follows the usual T.T.T. curve, depending, like most transformations, on nucleation followed by the growth of ordered regions. The change from an ordered to a disordered structure cannot be suppressed, because the temperature is high and there is no nucleation problem, the antiphase domain walls acting as nuclei for the growth of disorder.

Disordering can be produced below the critical temperature by high energy radiation. The extent of the radiation-induced disordering is too great to be accounted for by displacement spikes and is probably due to thermal spikes (see p. 314), in which the effects are partly due to the local high temperature and partly to the accompanying plastic deformation. The rate of cooling after a thermal spike has been formed is so high (see p. 315) that no ordering could occur during cooling.

Effect of ordering on properties. It is to be expected that the properties of an alloy should be different when it is ordered and when it is not; the properties that will be considered are density, mechanical, electrical, and magnetic.

DENSITY. The density of an alloy is always reduced by ordering; this may be explained as a result of the better geometrical packing of atoms of different sizes in an ordered array as compared with the same atoms distributed in the same structure in a random way. This increase in density is small, and usually of no practical importance.

MECHANICAL PROPERTIES. The major effects of ordering on mechanical properties arise when the ordered structure has lower symmetry than the disordered one, for example, when the disordered structure is cubic and the disordered one tetragonal. The ordered structure in such alloys consists of domains with any of three different tetragonal axes, and substantial strains are produced at the junctions of such domains. These regions of high strain have effects similar to the strains around pre-precipitation zones and substantial increases in yield stress and hardness are observed. The effect depends not only on the degree of order but also upon the sizes of the domains, a quantity that does not appear in the usual order parameters.

ELECTRICAL PROPERTIES. Since the electrical resistivity depends upon the scattering of electrons as they move through the crystal, it would be expected that an ordered crystal should have a lower resistance than a disordered one. A very good example of this effect is that of the alloys of gold and copper. The resistivity of gold-copper alloys, at room temperature, is shown in Fig. 8.78, in which the ordered alloys are represented by curve 2 and disordered alloys (retained by quench-

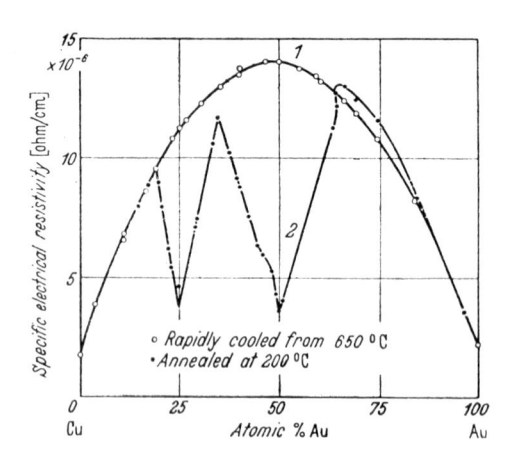

Fig. 8.78. Resistivity of copper-gold alloys, slowly cooled, curve 2, and quenched, curve 1. From H. Lipson, Chapter 1 in *Progress in Metal Physics 2*, Bruce Chalmers, Ed., Butterworths Scientific Publications, London, 1950, Fig. 34, p. 41.

ing) by curve 1. It will be observed that this effect is large and does not depend upon tetragonality, which occurs only in the 50 atomic % alloy (AuCu) and not in the AuCu$_3$ alloy.

MAGNETIC PROPERTIES. The most important effects of ordering are those in which the magnetic properties are changed. The effects to be considered are of two kinds.

In the first place, there are systems in which an alloy may be ferromagnetic or not, at a particular temperature, depending upon whether it is ordered or not. An example is the alloy Ni$_3$Mn, which is ferromagnetic if ordered, paramagnetic if not. This effect will not be considered further here.

The other way in which ordering affects the magnetic properties is by introducing or relieving stresses that change the ease with which ferromagnetic domain walls can pass through the crystal. An example of the increase in magnetic hardness by ordering is the alloy iron platinum, which orders with tetragonality. The highest coercivity is obtained with rapid cooling, which produces a small domain size. Slower cooling allows larger domains to be formed, with less internal strain and lower coercivity.

The converse is found with nickel-iron alloys containing 75–80 atomic % of iron. Annealing at 600° followed by a quench produces material of high permeability, apparently because the disordered structure does not impede the motion of domain walls. This is analogous to the relative ease of movement of dislocations through a crystal containing a randomly distributed solute as compared with the same crystal after the solute has formed pre-precipitation zones.

8.7 Reaction with Environment

Oxidation

THERMODYNAMIC CONSIDERATIONS. Most metals occur in nature in the form of compounds, often oxides or sulfides. This is because these metals combine with oxygen or sulfur, etc., with a decrease in free energy, or exothermically. The difference in free energy between a *mixture* of metal and oxygen and a *compound* of metal and oxygen is a good thermodynamic criterion for the tendency of the metal to oxidize. Some free energy differences and their variation with temperature are shown in Fig. 8.79, in which a *positive* value of ΔG corresponds to an *increase* in free energy when the oxide is formed. Thus any positive value corresponds to an *unstable* oxide. It is found that a few of the most noble metals, such as gold, do not form stable oxides at room

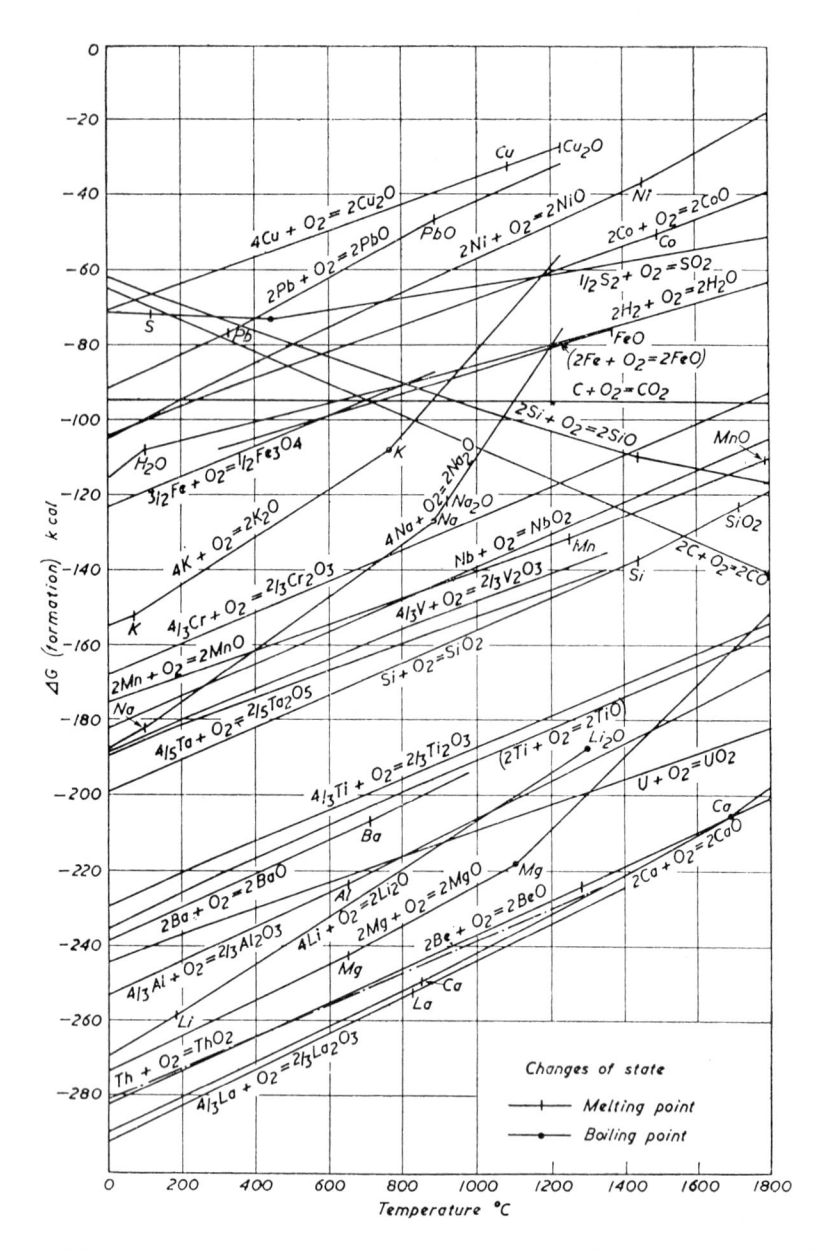

Fig. 8.79. Free energy of formation of oxides. From *Metals Reference Book*, Vol. II, 2nd ed., C. J. Smithells, Ed., Interscience Publishers, New York, 1955, p. 592.

temperature, because the free energy would be increased. Silver, however, forms an oxide at room temperature, but not at temperatures above about 300°C.

Figure 8.79 also indicates that, if a metal and the oxide of another metal are allowed to react, the oxygen will show a preference for the metal with the most negative ΔG; it follows, for example, that titanium oxide can be reduced by means of calcium, and that iron oxide can be reduced by carbon at high temperatures. In each case both oxides would be formed to the extent demanded by the law of mass action.

COMBUSTION OF METALS. It follows that, when oxidation takes place, the free energy is decreased and energy is liberated in the form of heat. In some cases this is sufficient to raise the temperature to a point where the reaction is fast enough to constitute combustion. Whether this happens or not depends upon:

1. The amount of heat released per unit amount of oxide formed.
2. The rate of the reaction.
3. The rate of loss of heat by conduction, radiation, and convection.

Of these factors, the first is characteristic of the system, and is, in principle, known. The second will be discussed in detail in the section entitled "Rate of Oxidation," and the third depends largely on the geometry of the system.

The following examples will illustrate the influence of the various factors:

Gold. Since gold does not form an oxide at any temperature, no combustion is possible.

Magnesium. Magnesium forms an oxide with considerable release of heat. It also forms an oxide film which causes the reaction rate to be so slow that, under most conditions, the temperature is not appreciably above that of the surroundings. However, if magnesium is heated in air to its melting point, the reaction rate is greatly increased, perhaps by the fact that the liquid metal is not protected much by the oxide film, and if the geometry is suitable, combustion may occur. Suitable geometry requires a relatively small volume per unit area of surface, so that the heat generated by the reaction raises the temperature quickly. A large mass of magnesium would presumably also burn if it were heated for a long enough time to raise the temperature everywhere to the melting point, but magnesium fires are usually caused by local heating of fragments such as turnings, which heat up quickly enough for the reaction to become self-sustaining.

Iron. Iron in massive form does not burn in air because, even at high temperatures, the supply of oxygen is not sufficient to sustain a high temperature. However, in very finely divided form, such as a fine powder, iron burns spontaneously on exposure to air, without being externally heated. This is because a particle of small size heats up very rapidly, as a result of its small thermal capacity in relation to its surface area, and its small size allows the available oxygen (the oxygen within a specific distance) to be sufficient to allow the speed of the reaction to be controlled by the temperature rather than by the supply of oxygen. Hence iron, and many other metal powders, are pyrophoric when finely divided.

RATE OF OXIDATION. The rate of oxidation of a metal surface exposed to air depends upon the rate of access of metal and oxygen atoms to each other, and this in turn depends upon the structure of the film that has already been formed. In general, the sequence of events is probably as follows:

Stage 1. Immediately contact between metal and oxygen is established, a film begins to form. Its structure is probably dictated by that of the metal surface, and its growth is controlled by the rate of diffusion through the film of the metal ion. The rate of growth is governed by the parabolic relationship

$$x^2 = at$$

where x is the film thickness, t is time, and a is a constant that depends upon the metal, the oxygen partial pressure, and the temperature.

Stage 2. If the film remains adherent and if it does not crack or buckle, it may be completely protective in the sense that the parabolic relationship persists. In such circumstances, growth essentially stops when the film is still thin enough to be transparent, and the metal apparently does not oxidize. Aluminum behaves in this way in air and is consequently quite resistant to oxidation at ordinary atmospheric temperatures. A number of other metals behave in the same way; lead, copper, nickel, chromium, and tin are examples.

Stage 3. There are, however, a number of metals such as uranium that are not completely protected by their oxide films at ordinary temperatures, and many that are not at high temperatures. This is associated with the fact that, in such cases, the specific volume of the oxide is either less than that of the metal or much greater. If the ratio of the specific volumes (known as the Pilling-Bedworth ratio) is less than 1, the oxide tends not to cover the metal completely and not to be adherent: oxidation then frequently proceeds linearly, i.e., at a constant rate. If the Pilling-Bedworth ratio is much greater than 1, the oxide is too bulky and tends to spall off, again leading to continuous oxidation. Table 8.7 shows the Pilling-Bedworth ratio for some of the metals.

EFFECT OF TEMPERATURE. Oxidation usually increases in rate at higher temperatures. This is to be expected because the rate-controlling process is diffusion, either of metal ions through the oxide, or of oxygen through cracks and discontinuities in the film. An example of the effects of temperature on oxidation rate is shown in Fig. 8.80. A second effect of temperature is noticed in the case of molybdenum, whose oxide is so volatile that it evaporates from the surface instead of protecting it.

OXIDATION OF ALLOYS. Two important cases arise in considering the oxidation of alloys: those in which the solvent metal oxidizes, and those in which it does not.

Protective Alloying. In the former case, the addition of a relatively small proportion of an alloying element that forms a protective coating may give very good protection to the alloy as a whole. A good example of this is in the stainless steels, in which, typically, 8% of chromium allows the formation of a very protective film of chromium oxide. It is important that the chromium is not in the form of a chromium carbide second phase, which can happen as a result of wrong heat treatment or near welds; chromium that is present as chromium carbide offers no protection from oxidation. This difficulty is avoided by the addition to the steel of a small amount of titanium or columbium

Table 8.7. Pilling-Bedworth Ratio for Some Metals

Protective		Not Protective	
Be	1.59	Li	0.57
Cu	1.68	Na	0.57
Al	1.28	K	0.45
Si	2.27	Ag	1.59
Cr	1.99	Cd	1.21
Mn	1.79	Ti	1.95
Fe	1.77	Mo	3.40
Co	1.99	Cb	2.61
Ni	1.52	Sb	2.35
Pd	1.60	W	3.40
Pb	1.40	Ta	2.33
Ce	1.16	U	3.05
		V	3.18

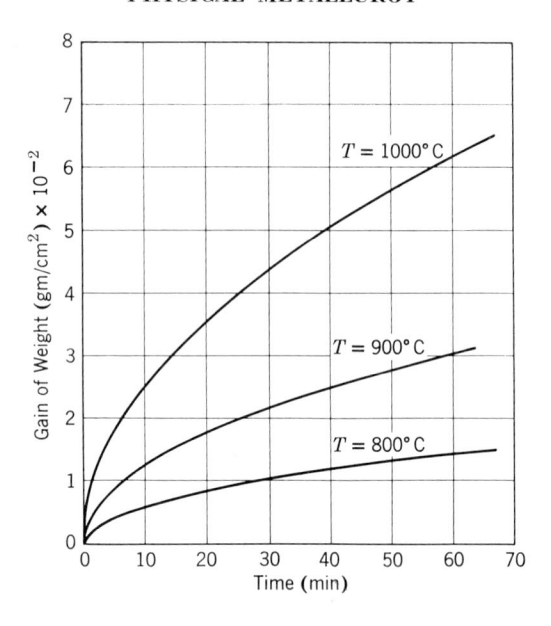

Fig. 8.80. Effect of temperature on oxidation rate of iron in oxygen.

which combines preferentially with the carbon that would otherwise form the undesirable chromium carbide. Such stainless steels are said to be "stabilized." The effect of chromium on the oxidation rate of steel at various temperatures is shown in Fig. 8.81.

Various other alloys developed for oxidation resistance at high temperatures show greatly reduced oxidation rates. These depend largely

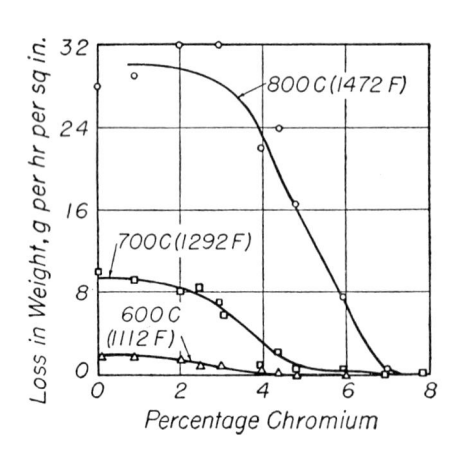

Fig. 8.81. Effect of chromium on the oxidation of steel. From *Metals Handbook*, ASM, Fig. 2, p. 225.

on protective films of chromium oxide or aluminum oxide, but the other components of the alloy are also very significant: for example, iron-nickel-chromium alloys are much more oxidation-resistant than iron-chromium alloys. It is also found that small amounts of impurities or minor additions can have quite large effects on the oxidation rate. Some or all of these results may be related to the effects of minor components on the diffusion characteristics of the oxide film, but these effects are not understood.

Internal Oxidation. Internal oxidation is the formation, within a matrix of one metal, of particles of oxide of a second metal that was originally in solid solution. It can take place only when no protective oxide film is formed. It occurs as a result of the inward diffusion of oxygen from the surface, and occurs to a depth that increases with the square root of time. This suggests that a definite oxygen concentration is necessary for the nucleation of the oxide crystals. They then grow as a result of diffusion of solute to the already nucleated crystals. A condition for the occurrence of internal oxidation is that the oxide of the solute should be more stable than that of the solvent metal.

It is found that the properties of an alloy are modified by internal oxidation; maximum resistance to deformation is produced by a distribution of a very large number of very small particles; this is analogous to the effect of precipitate size when second-phase particles are produced by a phase change. (See p. 172.)

Carburization and decarburization. An adjacent gas phase does not only react with a metal to form an oxide; it may also cause a change in the composition of an alloy. The most important case is that in which the carbon content of a steel is changed, near the surface, by interaction with the atmosphere in which it is heated. The change may be in either direction.

DECARBURIZATION. If the atmosphere does not contain carbon, but is either oxidizing or reducing, carbon at the surface combines with oxygen or hydrogen to form carbon monoxide or hydrocarbons, and, since these compounds are gases, it leaves the steel. Carbon diffuses outwards to reduce the concentration gradient set up by the loss of carbon, and a layer with low carbon content is formed. A layer that is deficient in carbon is softer and weaker than the remainder of the steel, especially after heat treatment, and many failures of steel components have been attributed to a decarburized layer.

CARBURIZATION. The addition of carbon to the surface layer of steel is a widely practiced method for producing a hard "case" on a tough "core." Carbon is supplied at the surface by the decomposition of a compound, often CO_2 or CH_4; the supply of these compounds is maintained by the presence of carbon in suitable form. Distribution of the carbon in the steel is by diffusion. The carbon content at and near the surface can be raised to a value such that a hard martensitic case is formed when the steel is heat-treated. The depth of the case depends upon the time allowed and on the temperature.

The effect of nitrogen on steel is similar to that of carbon and therefore nitriding and carbonitriding provide alternative methods for case hardening.

8.8 Further Reading

J. S. Bowles and C. S. Barrett, "Crystallography of Transformation," Chapter 1 in *Progress in Metal Physics 3*, Bruce Chalmers, Ed., Pergamon Press, London, 1952.

M. Cohen and L. Kaufman, "Martensitic Transformations," Chapter 3 in *Progress in Metal Physics 7*, Bruce Chalmers and R. King, Eds., Pergamon Press, London, 1958.

A. D. Le Claire, "Diffusion of Metals in Metals," Chapter 7 in *Progress in Metal Physics 1*, Bruce Chalmers, Ed., Butterworths Scientific Publications, London, 1949.

A. D. Le Claire, "Diffusion in Metals," Chapter 6 in *Progress in Metal Physics 4*, Bruce Chalmers, Ed., Pergamon Press, London, 1953.

Karl Hauffe, "The Mechanism of Oxidation of Metals and Alloys at High Temperatures," Chapter 2, *ibid.*

D. W. Jost, *Diffusion*, Academic Press, New York, 1952.

H. K. Hardy and T. J. Heal, "Report on Precipitation," Chapter 4 in *Progress in Metal Physics 5*, Bruce Chalmers and R. King, Eds., Pergamon Press, London, 1954.

R. F. Mehl and W. C. Hagel, "The Austenite:Pearlite Reaction," Chapter 3 in *Progress in Metal Physics 6*, Bruce Chalmers and R. King, Eds., Pergamon Press, London, 1956.

H. Lipson, "Order-Disorder Changes in Alloys," Chapter 1 in *Progress in Metal Physics 2*, Bruce Chalmers, Ed., Butterworths Scientific Publications, London, 1950.

E. W. Elcock, *Order-Disorder Phenomena*, Methuen & Co., London, 1956.

Appendix: Problems

Chapter 1

1. Define atomic number and atomic weight; are there any instances in which the atomic weight and the atomic number do not follow the same sequence? Explain.

2. (a) Which element has the highest atomic weight? (b) Which stable element has the highest atomic number?

3. What is the relationship between atomic number and X-ray absorption edge? Explain the origin of the K and L X-ray emission lines.

Chapter 2

1. Define unit cell, Miller index, and stereographic projection.

2. Calculate the interplanar spacing of the (331) planes in a copper crystal.

3. What is the angle between the (110) plane and the (231) plane in a cubic crystal? How can these planes be represented in a stereographic projection?

4. What is the mean thermal energy per atom of copper at room temperature?

5. Calculate the probability of an atom having thermal energy equal to (a) $2kT$; (b) $10kT$.

6. If an atom vibrates 10^{13} times/sec, at what temperature will it attain an energy of 20 kcal/mole once per second?

7. Discuss the equilibrium between a crystal and a vapor in terms of the transfer of atoms across the interface between them.

8. Give the co-ordinates of the positions of the atoms and the number of atoms per unit cell for the body-centered cubic, diamond-cubic, and close-packed hexagonal structures.

9. Calculate the volume per atom in terms of the distance of closest approach of atoms, for face-centered cubic, body-centered cubic, and close-packed hexagonal structures.

10. Find an alloy system in which a solid phase transforms on cooling into a solid phase and a liquid phase.

11. Using the van't Hoff expression, calculate the difference between the slopes of the initial liquidus and solidus lines for solutions of tin in lead and of lead in tin; compare your results with the values derived from the phase diagram.

12. Draw a series of free energy versus composition curves for the copper-lead system.

13. Describe the occurrence and structure of the sigma phase.

Chapter 3

1. Calculate the density of copper from the atomic mass, Avogadro's number, and the lattice parameter; compare with data for density determined experimentally.

2. Calculate the Young's modulus for an iron crystal in the [321] direction.

3. Would you expect any correlation between the bulk modulus and the thermal expansion for metals? Examine available data to determine whether there is any correlation.

4. Find a value for the saturation intensity of magnetization for iron, and calculate the magnetic moment per atom.

Chapter 4

1. Can a screw dislocation in a face-centered cubic crystal dissociate into partials? Describe the partials and stacking fault that would result.

2. What dissociation of dislocations can occur in a body-centered cubic crystal?

3. Discuss the criteria that determine whether two coplanar dislocations can combine to annihilate each other.

4. How can an extended dislocation climb?

5. Derive the relationship between dislocation spacing and boundary angle for a tilt boundary consisting of an array of edge dislocations. If the core of a dislocation has an effective radius of 2 atoms, what is the largest boundary angle which can be represented by a single array of separate dislocations?

6. Prepare a table showing the relationship between the energies of large-angle grain boundaries, surface energies, and solid-liquid interfacial energies for as many metals as possible.

Chapter 5

1. Calculate the theoretical yield stress in tension (based on a critical shear strain of $1/15$) for tin, aluminum, silver, iron, and tungsten.

2. Describe the action of a Frank-Read source consisting of a segment of screw dislocation.

3. Plot a graph for length of Frank-Read source versus stress required for operation.

4. Show, on a stereographic projection, the change of orientation due to slip on a single system.

5. What are the two most highly stressed slip systems when a tensile stress is applied in a [135] direction in a face-centered cubic crystal? Calculate the Schmid factors.

6. Outline the method used for finding the solubility of carbon in α iron by internal friction.

7. Calculate the theoretical fracture strength for iron and for tungsten. Discuss the possible significance of elastic anisotropy in this connection.

8. Does fatigue failure occur when the stress cycle is purely compressive?

9. What would be the yield stress of a copper alloy for which the elastic strain at yielding is equal to the plastic deformation?

Chapter 6

1. Calculate the free energy relationships, and angles of contact, for heterogeneous nucleation of lead at 320°C, and of nickel at 1355°C.
2. On the basis of the number of nearest neighbors, calculate the interfacial energy per atom of (a) a {111} surface of a face-centered cubic crystal, (b) a {100} face, and (c) a {120} face. How are these values changed if one-half of the sites corresponding to the next layer are filled?
3. What information is needed in order to predict the redistribution of solutes during the solidification of a ternary alloy?
4. How would you prepare a homogeneous crystal of a solid solution?
5. Describe continuous zone refining.
6. Calculate the proportion of a perfectly insulated melt of lead, supercooled by 10°C, that will freeze if nucleation occurs.
7. How would you change the thermal conductivity of the mold or the temperature of the melt, in order to reduce the extent of columnar freezing of an alloy melt?
8. Are dislocations necessary for the growth of metal crystals from the melt?
9. Sketch the phase diagram of an alloy system that would be ideal for a liquid metal coolant system.
10. What metals could be cathodically protected by (a) gold, (b) lead, (c) aluminum, (d) magnesium, (e) sodium, (f) tin?
11. If two phases of an alloy are in equilibrium with each other, would you expect them to have the same electrochemical potential?

Chapter 7

1. Sketch the pole figures for the 100, 111, and 110 poles for a texture oriented with its (100) planes within 5° of being parallel to the plane of projection, and with a spread of 10° from a given direction in the plane.
2. How can residual stresses be caused by tensile deformation?
3. What changes in shape and dimensions occur as a result of polygonization?
4. Discuss the temperature dependence of recrystallization in terms of activation energies.
5. Determine the relative curvatures of the sides of regular 4-, 5-, 6-, 7-, and 8-sided figures, of which the angles are 120° and the sides are arcs of circles. Express the curvature in terms of the radius of a circle passing through the corners.
6. Is recrystallization a necessary part of the process of solid-state welding?

Chapter 8

1. What is the effect of nickel on the M_S temperature of nickel-iron alloys?
2. Describe an example of reversible diffusionless transformation.
3. Draw a graph of the self-diffusion coefficient as a function of temperature for a metal with $D_0 = 1$, $Q = 3$. What units should these constants have in order to correspond to reality?

4. What is the driving force for self-diffusion?

5. What is known about the diffusion of inert gases in metals?

6. What is the jump frequency for (a) zinc atoms in solid solution, (b) vacancies, and (c) copper atoms, in copper at room temperature?

7. To what extent can the observed rates of precipitation be accounted for in terms of diffusion?

8. Calculate the hardenability of the steel SAE 4340 at various grain sizes. Suggest a composition for a steel that will have a hardenability of 2 in.

9. Under what conditions should silver burn? should aluminum burn?

10. Calculate the ratios of yield stress to shear modulus for the strongest alloys of iron, copper, titanium, magnesium, aluminum, lead, and nickel.

Index

Abnormal grain growth, 341
Abnormal thermal expansion, 87
Abrasion, 351
Accommodation coefficient, 19
Activated process, 17
Activated state, 16
Activation energy, 19
  for creep, 185
  for diffusion, 373
  for self-diffusion, 185
Aggregates of atoms, 10
Aggregation of vacancies, 199
Agitation, thermal, 17
Air-hardening steels, 436
Alkali metals, 3, 30, 95
Allotropic forms of plutonium, 25
Allotropic transformation, diffusion controlled, 406
Allotropism, 25
Alloying elements, effect on equilibrium conditions, 426
  effect on kinetics of transformation, 427
  effect on transformation temperature, 431
Alloys, density of, 80
Alpha iron, diffusion of carbon in, 374
Alpha particle, 313
Aluminum alloys, Young's modulus of, 84

Aluminum-copper system, precipitation in, 393
Aluminum-silicon eutectic, 266
Amelinckx, S., 110, 294
Amorphous layer theory, 114
Amplitude of thermal agitation, 87
Andrade, E. N. da C., 161, 184
Anelasticity, 139, 191
Angle of contact, 243
Anion, 291
Anisotropy, elastic, 81
  magnetic, 228
  magnetocrystalline, 136
Anisotropy of iron crystals, 83
Annealing, 181, 343
  of steel, 438
Annealing out of point defects, 322
Annealing twins, 335
Antiferromagnetism, 90, 136
Antiphase domain, 51
Aqueous media, solution of metals in, 298
Arc melting, 287
Asterism, 155, 309
ASTM grain size, 123
Atom, structure of, 1
Atom movements in recrystallization, 333
Atomic number, 1, 3
Atomic size, 33

454   INDEX

Atomic weight, 2
Atoms in metals, sizes of, 31
Aust, K. T., 120, 121
Austempering, 437
Austenite, grain growth in, 341
  stabilization of, 367
Austenite grain size, effect on hardenability, 434
Austenite-pearlite transformation, kinetics of, 418
Austenitic grain size, control of, 340
Austenitic steels, 437, 438
Axial ratio, 30, 80

Back extrusion, 346
Bainite, 425
Bainite hardenability, 435
Bainite transformation, 408, 425
Band, deformation, 158, 159, 160
  Lüders, 157
  Neumann, 158
Barrett, C. S., 312, 333, 334, 360
Bauschinger effect, 219, 220
Bend test, 222
Beryllium copper, precipitation in, 398
Bever, M. B., 336
Bicrystal, 176, 177
Binary system, 37
Blacksmith's weld, 355
Blistering, 404
Body-centered cubic structure, 28, 29
Bohr, N., 2
Boltzmann distribution, 16
Boltzmann's constant, 15
Bond, covalent, 13, 14, 28
  heteropolar, 12
  homopolar, 13
  ionic, 12
  metallic, 14
Boundary, crystal, 112
  lineage, 260
  twin, 112, 113
Boundary diffusion, anisotropy of, 383
Boundary groove, 136
Bowles, J. S., 360
Bragg condition, 65
Brass, season cracking of, 305
Brazing, 290
Bridgman, P. W., 205
Brinkman, J. A., 316

Brittle fracture, 212, 349
Brittle transition temperature, 206
Brittleness, notch, 203
  temper, 400
Broadening of diffraction rings, 309
Brown, H. McP., 310, 333
Brillouin zones, 64, 65, 66, 67, 71
Bulk modulus, 82
Burgers, W. G., 409
Burgers dislocation, 97
Burgers vector, 96, 97, 98
  for partial dislocation, 103
Burnishing, 352

Cahn, R. W., 325
Carbon steels, microstructure of, 423
Carbonitriding, 448
Carburization, 447
Carreker, R., 180
Casting, continuous, 286
  investment, 289
  lost wax, 289
  permanent mold, 289
  precision, 289
  sand mold, 289
Castings, 288
Catalyst, nucleation, 243
Cathode, 291
Cell, differential aeration, 300
Cementite, spheroidization of, 438
Chemical diffusion, 376
Chemical polishing, 306
Chill, 288
Chill zone, 273
Cleavage, 195, 369
Climb, 101
  forced, 154
Climb of dislocations, 199
Close-packed hexagonal structure, 28, 30
Clustering, 45, 46, 47
Coating, 290
Co-deposition, 292
Coercive force, 224, 230
  effect of nonmagnetic particles on, 226
Coercive force and grain size, 225
Coercive force and lattice mismatch, 229
Coherent interphase interface, 112, 128
Coherent twin boundary, 127
Cohesive force, 194
Coining, 344

INDEX

Cold deformation, 343, 349
Cold rolling, 345
  effect of, on properties, 349
Cold welding, 355
Cold work, suppression of yield by, 401
Collapse of vacancy disk, 249
Collision effects, 314
Columnar grain size, 275
Columnar zone, 274
Combustion of metals, 443
Complete dislocation, 103
Complex point defects, 93, 95
Compound, covalent, 69, 70
  interstitial, 74, 75
  ionic, 69, 70
Compression, load deformation curve in, 219
Compression loading, 218
Condensation, 19
Condition for homogeneous nucleation, 241
Conductivity, electrical, 66
  n-type, 67
  p-type, 67
Constant load creep testing, 184
Constant stress creep testing, 184
Constitutional supercooling, 258, 259, 261, 271, 273
Consumable electrode, 287
Continuous casting, 286
Continuous cooling curve, 427
Cooling rate, critical, 432
Cooling rate during quenching, 432, 433
Coordination number, 28, 34
Coring, 282
Correction, Goldschmidt, 34
Corrosion, electrolytic, 300
  intergranular, 303, 305
  types of, 302
  uniform, 302
Corrosion by pit formation, 303
Corrosion fatigue, 305
Corrosion pit, 301
Cosmic radiation, 313
Cottrell, A. H., 201
Cottrell atmosphere, 402
Cottrell-Lomer supersessile dislocation, 106, 155, 162
Cottrell mechanism, 166, 180
Cottrell theory of yield point, 107

Covalent bond, 13, 14, 28
Covalent compounds, 69, 70
Covalent phases, 75
Crack, critical size of, 198
  fatigue, 212, 213
  origin of, 201
Crack propagation, criterion for, 196
  critical speed for, 198
Creep, 139, 181, 183
  activation energy for, 185
  effect of stress and temperature on, 185
  effect of structure on, 186
  fracture resulting from, 190
  primary, 184
  quasi-viscous, 184
  secondary, 184
  steady-state, 184, 189
  tertiary, 184
  transient, 184
Creep resistant alloys, 399
Creep testing, constant load, 184
  constant stress, 184
Critical cooling rate, 432
Critical nucleus, 238, 245
Critical radius, 237
Critical resolved sheer stress, 152
Critical sheer stress, 149
Critical size of crack, 198
Critical speed for crack propagation, 198
Cross slip, 155
Crowdion, 95, 378
Cry of tin, 158
Crystal, definition of, 13
  growth of, from vapor, 292
  imperfect, 92
  imperfections in, 92
Crystal boundaries, effects of, 175
  equilibrium disposition of, 121
Crystal boundary, 112, 113
  dislocation model of, 118, 124
  mobility of, 120
Crystal growth, dislocation theory of, 293
Crystal lattice, 13
Crystal size, effect of, on plastic properties, 179, 180
Crystal size and specific surface energy, 244
Crystal structure, 14, 22

Crystalline form, 21
Crystalline solids, structure of, 21
Crystallography, influence of, on freezing, 232
Crystallography of dendrites, 235
Cube-on-edge texture, 340
Cube texture, 226
Cullity, B. D., 309
Cup and cone fracture, 210, 211
Cupping test, Erichsen, 222
Curie temperature, 89
Cutting processes, 349

Damage, radiation, 95
Dash, W. C., 110, 148
DeBlois, R. W., 137
Decarburization, 448
Decoration of dislocations, 107
Deep drawing, 347
Defect lattice, 50
Deformation, anelastic, 139
  cold, 343, 349
  diffraction effects of, 308
  effect on recrystallization temperature, 331
  elastic, 139
  heterogeneous, 157, 218
  hot, 343
  plastic, 93, 94, 139, 140, 307
    effect of size on, 161
  shear, 140
  simultaneous shear, 142
  stored energy of, 335, 336
Deformation bands, 158, 159, 160
Deformation texture, 311, 312
Deformed metal, dislocation content of, 308
  structure of, 307
Delayed fracture, 206
Demagnetizing field, 224
Dendrites, 260
  crystallography of, 235
Dendritic carbide precipitate, 392
Dendritic freezing, 234, 259, 262
Dendritic growth, 260
  direction of, 235
Dendritic structure of precipitate, 391
Density, effect of ordering on, 440
Density of alloys, 80
Density of elements, 78, 79

Density of interstitial solutions, 80
Density of states, 65
Depolarization, 301
Deviation from Hooke's law, 195
Diagram, equilibrium, 37
Diamond, 71
Dienes, G. J., 315, 316, 317, 318
Differential aeration cell, 300
Diffraction effects of deformation, 308
Diffraction effects of radiation, 319
Diffraction rings, broadening of, 309
Diffusion, 369
  activation energy for, 373
  chemical, 376
  direct interchange mechanism, 378
  driving force for, 377
  effect of irradiation on, 380
  jump frequency, 370, 371
  mechanism of, 377
  ring mechanism, 378
  solid state, 272
  surface, 384
  unit process of, 370
  uphill, 376
  vacancy mechanism, 378, 380
Diffusion coefficient, 371
  grain boundary, 382
  pipe, 383
Diffusion constant, variation with temperature, 374
Diffusion constants for liquid metals, 254
Diffusion controlled allotropic transformation, 406
Diffusion data for liquid metals, 384
Diffusion in crystals, 370
Diffusion in liquid metals, 384
Diffusion in metals, data for, 375
Diffusion layer, thickness of, 255
Diffusion of carbon in alpha iron, 374
Diffusion of interstitial atoms, 377
Diffusion of substitutional solutes, 377
Diffusion zone, 254
Diffusionless transformation, 358, 359
  crystallography of, 359, 360
  effect of plastic deformation on, 361
  reversibility of, 360
Diffusionless transformation in titanium alloys, 368
Direction of dendritic growth, 235

INDEX

Direction of easy magnetization, 136, 225
Discontinuous eutectic, 266
Dislocation, 95, 96, 248
  Burgers, 97
  climb of, 199
  complete, 103
  edge, 97
  energy of, 100
  flexibility of, 164
  glissile, 105
  motion of, 101
  partial, 103, 104, 105
  perfect, 103
  screw, 97
  segregation at, 106, 401
  supersessile, 106
  Taylor, 97
Dislocation content of crystals, 109
Dislocation content of plastically deformed metal, 308
Dislocation diffusion, 380
Dislocation loop, 99
Dislocation mechanism for twinning, 158
Dislocation model of crystal boundary, 118, 124
Dislocation network, 100
Dislocation node, 100
Dislocation segment, 100
Dislocation theory of crystal growth, 293
Dislocation theory of plastic deformation, 144
Dislocations, evidence for, 107
  forest of, 154
  pile-up of, 151, 155
Disordered structure, 47, 48
Disordering by high energy radiation, 439
Dispersion hardened system, 172
Displacement spike, 316
Distribution coefficient, 250
Distribution of embryo sizes, 240
Distribution of solute ahead of interface, 254
Divorced eutectic, 270
Doherty, P., 310, 335
Domain, antiphase, 51
  ferromagnetic, 137

Domain configuration, 223
Domain patterns and grain boundaries, 225
Domain structure, 137
Domain wall, 137
  ferromagnetic, 112, 136, 138
Drilling, 349
Ductile fracture, 196
Ductility, 214, 307
Dudzinski, N., 84
Dunn, C. S., 135, 323

Earing, 349
Easy glide, 153, 156
Easy magnetization, direction of, 136, 225
Edge dislocation, 97
$8 - N$ rule, 28, 29, 30, 67
Elastic behavior, 215
Elastic coefficient, 81
Elastic constants, 81
Elastic deformation, 139
Elastic limit, 215
Elastic modulus, 81
Elastic properties, 81, 214
  effect of radiation on, 318
Elastic properties of polycrystalline aggregates, 82
Elastic properties of polyphase alloys, 83
Elastic properties of some metals, 83
Elastic region, 153
Electrical conductivity, 66
Electrical properties, 90
  effect of ordering on, 440
Electrical resistance, 146
Electrical resistivity, effect of radiation on, 317
Electrochemical potential, 289, 298
Electrolytic cell, 299
Electrolytic corrosion, 300
Electron, 1, 11, 313
Electron configuration of the elements, 3, 4
Electron microscopy of dislocations, 107
Electron phases, 67, 68, 69, 76
Electronegative element, 12
Electrons, extranuclear, 2
Electrons in crystal, energy of, 64
Electrons per atom, 67, 69, 71, 169

458 INDEX

Electroplating, 291
Electropolishing, 306
Electropositive element, 12
Electrostatic forces, 11, 89
Element, electronegative, 12
  electropositive, 12
Elements, density of, 78, 79
Elongation, 215, 218
  practical importance of, 218
Embrittlement, hydrogen, 206, 404
  low temperature, 206
Embrittlement of titanium, 199
Embryo, 237
  recrystallization, 329
  two-dimensional, 246
Embryo size, distribution of, 240
End quench test, Jominy, 434, 436
Energy levels, 11
Energy of dislocation, 100
Energy of electrons in crystal, 64
Energy of interface, 110, 131
Energy of interphase boundary, 129
Energy of solid-liquid interface, 132, 133, 134
Energy of twin boundary, 159
Engineering stress, 215
Entropy, 10, 20
Equiaxed zone, 274
Equilibrium, 19, 20, 21
  effect of alloying elements on, 426
  thermodynamic criterion for, 63
  three-phase, 38
Equilibrium between phases, kinetic considerations, 17
  thermodynamic considerations, 20
Equilibrium between solid and liquid alloys, 54
Equilibrium diagram, 37, 85
Equilibrium distribution of crystal boundaries, 121
Equilibrium structure, 36
Equilibrium temperature, 19, 56, 232
Equilibrium vacancy content, 93, 94
Equivalent temperature, 182
Erichsen cupping test, 222
Etch pits, 124, 146
Etching of dislocations, 107
Eutectic, 38, 41, 62
  aluminum silicon, 266
  discontinuous, 266

Eutectic, divorced, 270
  iron carbon, 266
  modification of, 267
  nonlamellar, 265
  tin lead, 268
Eutectic liquid, freezing of, 262, 263, 269
Eutectic point, ternary, 45
Eutectoid, 41
Eutectoid steel, 419
  T.T.T. curve for, 420
Eutectoid temperature, 419
Eutectoid transformation, 358, 415, 417
  effect of temperature on rate of, 419
Evans, E. Ll., 75
Evaporation, 19
Excited state, 8
Exclusion principle, 3
Extended dislocation, Cottrell-Lomer, 106
Extranuclear electrons, 2
Extrusion, 345, 348

Face-centered cubic crystal, homogeneous nucleus for, 246
  structure of, 25, 26
Faraday's law, 291
Fatigue, 207
  corrosion, 305
  effect of mean stress, 208
Fatigue crack, 212, 213
Fatigue failure, 207, 212
Fatigue limit, 207, 208
Feather markings, 212
Fermi surface, 66
Ferromagnetic domain, 137
Ferromagnetic domain wall, 112, 136, 138
Ferromagnetic material, 136, 137
Ferromagnetic properties, 89
Ferromagnetism, 90
Fick's law, 375, 376
Fine grain size, production of, 276
Fisher, J. C., 381
Fisher mechanism, 167
Fission, irradiation accompanied by, 321
Fission effects, 317
Flake graphite, 199, 200
Flexibility of a dislocation, 164
Flow curve, 203

INDEX 459

Flow lines, 348
Flow stress, 172
Fluidity, 289
Forced climb, 154
Forces between atoms, 32
Forest of dislocations, 154
Forging, 344, 348
Fractional transformation times, 432
Fracture, 139, 151, 194, 215
  brittle, 212, 349
  characteristic appearance of, 210
  cup and cone, 210, 211
  ductile, 196
  effect of temperature on, 205
  intercrystalline, 196, 199, 213
  shear, 196
  transcrystalline, 199
Fracture by plastic deformation, 195
Fracture curve, 202, 203
Fracture resulting from creep, 190
Fracture strength, theoretical, 194
Frank, F. C., 293
Frank partial dislocation, 106
Frank-Read source, 146, 147, 209, 402
Free energy, 10, 15, 21, 59, 60, 62
Free energy of formation of oxides, 442
Free energy of solid solution, 63
Free energy relationship for martensite transformation, 362
Freezing, 19, 231, 232
  change of composition during, 251, 252
  change of volume on, 277
  dendritic, 234, 259, 262
  influence of crystallography on, 232
  interface, rate of advance of, 285
  nucleation of, 243
  rate of, 232
  smooth, 232
  unidirectional, 232
Freezing of alloy, steady-state condition for, 252
Freezing of alloys, single phase, 249
Freezing of eutectic at noneutectic composition, 269
Freezing of eutectic liquid, 262–263
Freezing of peritectic, 271
Freezing of two-phase alloys, 262
Freezing range, 289

Frequency-independent internal friction, 193
Frost, B. R. T., 79
Fully annealed steel, 438
Fundamental particles, properties of, 1
Fusible alloys, 277
Fusion, latent heat of, 35, 56, 85, 86, 130, 132, 232, 243
Fusion welding, 290

Gage length, 215, 218
Gamma loop, 427
Gamma rays, 313
Gas bubbles, formation during freezing, 280
Gas evolution, 280
  interaction of, with shrinkage, 280
Gases in metals, precipitation of, 403
Gate, 288
Geometrical hardening, 151
Geometrical softening, 151
Gibbs phase rule, 39
Gilman, J. J., 109
Glissile dislocation, 105
Goldman, J. E., 90
Goldschmidt correction, 34
Goodman diagram, 208
Graham, C. D., Jr., 137
Grain boundaries, domain patterns and, 225
Grain boundary diffusion, 381
  detection of, 382
Grain boundary diffusion coefficient, 382
Grain boundary energy, 133
Grain boundary slip, 194
Grain growth, 337, 340
  abnormal, 341
  mechanism of, 338
Grain growth in austenite, 341
Grain growth inhibitor, 438
Grain oriented texture, 225
Grain shapes in two-phase alloys, 341
Grain size, 123
  coercive force and, 225
  columnar, 275
  effect of vibration on, 277
Grain size number, 124
Graphite, flake, 199, 200
  nodular, 199, 200

460 INDEX

Graphite, spheroidal, 267, 268
Graphitization, 404
Gravity segregation, 283
Greenwood, J. N., 201
Griffith crack, 197, 201
Griffith theory, 196
Growth cell, 259
Growth of lamellar eutectic, 263, 265
Growth of uranium during temperature cycling, 327
Growth spiral, 294
Guinier Preston zones, 393

Hagel, W. C., 418, 419, 421
Halogens, 5
Hanawalt, J. D., 304
Hard atoms, 32
Hard testing machine, 216
Hardenability, 434, 436, 435
  advantages of high, 435
  calculation of, 435
Hardening, geometrical, 151
  precipitation, 395
  substructure, 180
Hardening by a second phase, 165, 172
Hardness, 179, 312
  correlation with U.T.S., 220–221
  definition of, 220
  lamellar spacing and, 422
Hardness testing, 220
Hardy, H. K., 390, 395
Heal, T. J., 390, 395
Heat flow considerations, 284
Heat of formation of intermediate phase, 74, 75
Heat of transformation, 19
Heat of vaporization, 133
Heat treatment, 357
Heterogeneous deformation, 157
Heterogeneous nucleation, 238, 241, 242, 243, 246
Heteropolar bond, 12
Hibbard, W. R., Jr., 169, 171, 180, 323
High energy radiation, disordering by, 439
High-speed steels, 400
Hirsch, P. B., 111
Hoffman, R., 383
Hollomon, J. H., 132, 178, 245, 422, 435, 436

Homogeneous nucleation, 238, 241, 243
  supercooling for, 245
Homogeneous nucleation temperature, 244
Homogeneous nucleus for F.C.C. crystal, 246
Homopolar bond, 13
Hooke's law, 81, 191
  deviation from, 195
Hot deformation, 343
  properties resulting from, 348
Hot-dipping, 290
Hot rolling, 345
Hot tear, 288
Hot-top, 286
Hultgren, R., 431
Hume-Rothery, W., 4, 7, 24, 33, 64, 71, 72, 73, 74
Hume-Rothery phases, 67, 68, 69, 75, 76
Hydrogen embrittlement, 206, 404
Hypoeutectoid steel, T.T.T. curve for, 421
Hysteresis loop, 224, 226
Hysteresis loss, 225

Impact test, 221, 222
Imperfect crystals, 92
Imperfections, line, 92, 95
  point, 92, 93, 96
  surface, 92, 110
  thermodynamically stable, 92
  thermodynamically unstable, 92
Imperfections in crystals, 77, 92
Imperfections resulting from freezing, 248
Impurities, effect on recrystallization temperature, 331
Inclusions, nonmetallic, 348
Incoherent interphase interface, 112
Induction, residual, 224
Inert electrode, 287
Inert gases, 5
Ingot, structure of, 275, 276
Ingot structure of pure metal, 276
Ingots, 286
Inhibitors, anode, 301
  cathode, 301
Initial permeability, 223, 228
  effect of composition on, 229
Insulation, protection by, 302

INDEX

Insulator, 66
Intensity of magnetization, 90
Interatomic distance, 32
Intercrystalline fracture, 196, 199, 213
Intercrystalline slip, 185, 189
Intercrystalline weakness, 213
Interdendritic segregation, 282, 283
Interface, energy of, 110, 131
  nucleation ahead of, 271
  smooth, 259
  solid vapor, 133, 134
Interface transition, 20
Interfacial energy, 243
Intergranular corrosion, 303, 305
Intermediate phase, 36, 63
  heat of formation of, 74, 75
Internal energy, 10, 18, 19, 20
Internal friction, 191, 194, 402
  frequency-independent, 193
Internal oxidation, 447
Interphase boundary, energy of, 129
Interphase interface, 112, 128
Interstitial atoms, 93, 94
  diffusion of, 377
Interstitial compound, 73, 75
Interstitial solid solution, 36, 53, 95
Interstitial solutions, density of, 80
Interstitialcy mechanism, 378
Invar, 89, 90
Inverse segregation, 283
Investment casting, 289
Ionic bond, 12
Ionic compound, 69, 70
Ionic phases, 75
Ionization effects, 314
Ionization potential, 8
Iron, effect of, on corrosion rate of magnesium, 204
  elastic anisotropy of, 83
Iron-carbon eutectic, 266
Iron-silicon alloy, 225, 227
Irradiation, effect of, on diffusion, 380
Irradiation accompanied by fission, 321
Isothermal section, 44
Isothermal transformation of austenite to pearlite, 419
Isotopes, 2

Jaffee, L. D., 178, 422, 435, 436
Jog, 153, 154

Johnston, W. G., 109
Joining, 290
Jominy end quench test, 434, 436
Jump frequency, diffusion, 370, 371

k space, 65
Kaidodekahedron, 245
Kinetics of growth of precipitates, 388
Kinetics of recrystallization, 328
Kinetics of transformation, effect of alloying elements on, 427
Kirkendall effect, 378
Kocks, U. F., 143, 160
Kronberg, M. L., 333
Kubaschewski, O., 75

Lamellar eutectic, growth of, 263, 265
  mechanical properties of, 268
  structure of, 265, 266
Lamellar spacing and hardness, 422
Lang, A. R., 111
Large-angle boundary, 118, 119, 121
Latent heat of fusion, 19, 35, 56, 85, 86, 130, 132, 232, 243
Latent heat of vaporization, 36, 86
Lattice, crystal, 13
Lattice mismatch and coercive force, 229
Lattice parameter of solid solutions, 80
Lattice vacancies, 93
Laves phase, 73, 74, 75
LeClaire, A. D., 382, 383
Leslie, W. C., 392
Lessells, J. M., 212
Lever rule, 37, 38
Limit of solubility, 53
Line imperfections, 92, 95
Lineage boundary, 260
Lineage structure, 259
Lipson, H., 440
Liquid, 15, 21
Liquid metals, diffusion constants for, 254
  diffusion in, 384
  solution in, 296
Liquid monolayer boundary, 119
Liquid solution, 36
Liquids, structure of, 34
Liquidus, 37, 43, 85
Liquidus surface, 45

Liss, B., 157
Livingston, J. D., 173
Load deformation curve in compression, 219
Load elongation curve, 214, 215, 216, 217
Local elongation, 217
Long range order, 46, 47, 438
Long-range-order parameter, 50, 52
Lost wax casting, 289
Low, J. R., Jr., 211, 213
Low-angle boundary, 121, 124, 180
Low temperature embrittlement, 206
Lower bainite, 425
Lower yield point, 216
Lüders band, 157, 401, 403
Ludwig's theory, 202

M_D temperature, 366, 367, 437
M_F temperature, 367
M_S temperature, 366, 367, 408, 437
Machinability, 351
Machining, 349
McQuillan, A. D., 368, 369, 408, 410, 411, 412, 413, 414
McQuillan, M. K., 368, 369, 408, 410, 411, 412, 413, 414
Macromosaic structure, 259
Macroscopic continuity condition, 175, 176
Magnesium, corrosion rate of, 304
Magnet, small particle, 228
Magnetic anisotropy, 228
Magnetic field, 89
Magnetic induction, 223
Magnetic properties, 89, 223
  effect of ordering on, 441
  effect of strain on, 230
Magnetically hard materials, 225, 228, 230
Magnetically soft materials, 225
Magnetization, direction of easy, 225
  intensity of, 90
  saturation, 223
  saturation intensity of, 90
Magnetization curve, 223
Magnetocrystalline anisotropy, 136
Magnetostriction, 193, 228
Martempering, 347
Martensite, lattice parameters of, 364

Martensite, morphology of, 364
  nucleation and growth of, 366
  properties of, 365
  structure of, 362, 363
  tempering of, 399
  transformation, volume change in, 366
  variation of hardness with carbon content, 365
Martensitic transformation, 358, 359
Martensitic transformation in steel, 361
Maximum embryo size, 240
Mechanical properties, 213
  effect of ordering on, 440
Mechanical properties of lamellar eutectic, 268
Mechanical testing, 213
Mehl, R. F., 418, 419, 421
Melting, 295
  change of volume of, 78, 79
Melting point, 56, 232
Melting point and thermal expansion, relation between, 88
Melting points of pure metals, 84
Metal, definition of, 14
Metallic bond, 14
Metalloid, 72
Meyer index, 221
Microhardness test, 221
Microscopic continuity condition, 175, 176
Microstructure of carbon steels, 423
Milling, 349
Mobility of crystal boundary, 120
Modification of eutectic, 267
Molecule, 13
Monotectic, 41
Moore, A. J. W., 135, 160
Motion of dislocation, 101
Moulding, shell, 289
Murphy, W. F., 319

Necking, 217, 218
Neumann band, 158
Neutrons, 1, 313
Nickel arsenide structure, 71, 72
Nickel arsenide type phases, 75
Nitriding, 448
Nodular graphite, 199, 200
Nondestructive testing, 223

Nonlamellar eutectic, 265
Nonmagnetic particles, effect on coercive force, 226
Nonmetallic inclusions, 348
Nonstoichiometric alloys, 52
Nonthermal energy, 19, 20
Nonuniform deformation, 218
Normal grain growth, 337
Normal segregation, 281
Normalized steel, 438
Notch brittleness, 203
Notch sensitivity, 222
n-type conductivity, 67
Nucleation, 21, 132, 235, 237
   heterogeneous, 238, 241, 242, 243, 246
   homongeneous, 238, 241, 243
   oriented, 334
   recrystallization, 328, 329
   surface, 246
Nucleation ahead of interface, 271
Nucleation and growth of martensite, 366
Nucleation catalyst, 243
Nucleation in supercooled melt, 261
Nucleation of freezing, 243
Nucleation of precipitation, 386
Nucleation temperature, homogeneous, 244
Nucleus, critical, 238, 245
Nucleus of atom, 1

Oilless bearing, 354
Order, long range, 46, 47
   short range, 46
Order-disorder changes, 438
Order parameter, long-range-, 50, 52
   short-range-, 46, 52
Ordered structure, 47, 48, 49
Ordering, 45, 46
   effect of, on density, 440
      on electrical properties, 440
      on magnetic properties, 441
      on mechanical properties, 440
      on properties, 439
Ordering reaction, 358
Orientation, preferred, 274, 275, 349
Oriented nucleation, 334
Orowan, E., 197
Overaging, 396
Oxidation, internal, 447

Oxidation, rate of, 444
   thermodynamic considerations, 441
Oxidation of alloys, 445
Oxide film, 136, 161, 162
Oxides, free energy of formation of, 442

Partial dislocation, 103, 104, 105
   Burgers vector for, 103
   Frank, 106
   Shockley, 105
Pasty freezing, 261, 262
Pearlite, 424
   metallography of, 422
   properties of, 420
Pearlite colony, growth of, 417, 418
   nucleus of, 417
Pearlite hardenability, 435
Pearlite lamellae, thickness of, 420, 421
Pearlite transformation, 415
Perfect dislocation, 103
Perfect single crystal, yield stress for, 140
Periodic table, 6, 7, 34
Peritectic, 41
   freezing of, 271
Peritectoid, 41
Peritectoid transformation, 358
Permanent mold casting, 289
Permeability, 223
   initial, 223, 228
Petch, N. J., 197
Pfann, W. G., 256, 257
Phase, 17
Phase diagram, 36, 37, 39, 61, 63
Phase rule, Gibbs, 39
Phases, electron, 67, 68, 69, 76
   Hume-Rothery, 67, 68, 69, 75, 76
Pile-up of dislocations, 151, 155
Pilling-Bedworth ratio, 444, 445
Pipe, shrinkage, 278, 279, 286
Pipe diffusion, 380
Pipe diffusion coefficient, 383
Pit formation, corrosion by, 303
Planck's constant, 2, 8, 65
Plastic deformation, 94, 139, 140
   change of shape of crystals in, 309
   dislocation theory of, 144
   effect of, on diffusionless transformation, 361
   effect of environment on, 162

Plastic deformation, effect of solute on, 163
  effect of radiation on, 318
  effect of temperature on, 162
  fracture by, 195
  polycrystalline aggregates, 177
Plastic properties, effect of crystal size on, 179, 180
Plutonium, allotropic forms of, 25
Point defects, annealing out of, 322
Point imperfections, 92, 93, 96
Pole figure, 311, 333
Polishing, 352
  chemical, 306
Polycrystalline aggregates, elastic properties of, 82
  plastic deformation of, 177
Polygonization, 180, 322, 323, 325
Polymorphism, 25
Polyphase alloys, elastic properties of, 83
Pore, shrinkage, 279
Porosity, 277
  surface, 279
Potential, electrochemical, 298, 299
  single deposition, 291, 292
Potential energy, 15, 16
Pouring basin, 288
Powder metallurgy, 354
Precipitate, dendritic structure of, 391
  kinetics of growth of, 388
  morphology of, 390
  solution of, 405
Precipitation, 358, 385
  crystal structure of, 393
  effect of, on properties, 394
  effect of quenching rate on, 391
  effect of strain on, 393
  irreversible, 385
  nucleation of, 386
  nucleation sites for, 388
  rate of, 389
  reversible, 385
  variation of hardness with time, 394
Precipitation hardening, mechanism of, 395
Precipitation in aluminum-copper system, 393
Precipitation in beryllium-copper system, 398
Precipitation of gases in metals, 403
Precipitation of solute on dislocations, 401
Precipitation temperature, 391
Precision casting, 289
Preferred orientation, 83, 274, 275, 311, 349
Pre-precipitation zones, 391, 395, 441
Pressing, 346
Pressure, 10, 20, 36
Primary creep, 184
Proeutectoid ferrite, 417, 420
Proeutectoid reaction, 358
Proof test, 223
Protection, sacrificial, 302
Protection by insulation, 302
Protective alloying, 445
Proton, 1, 313
p-type conductivity, 67
Pure metals, melting points of, 84
Pyrophoric metals, 444

Quantum numbers, 2
Quantum states, 11
Quasi-viscous creep, 184
Quench aging, 402, 403
Quenching, 432
Quenching cracks, 436
Quenching rate, effect of, on precipitation, 391
Quenching stresses, 391

Radiation, diffraction effects of, 319
  effect of, on elastic properties, 318
  on electrical resistivity, 317
  on plastic deformation, 318
  types of, 313
Radiation damage, 95, 307, 313, 314
Radiation-induced disordering, 439
Radius, critical, 237
Raynor, G. V., 4, 7, 24, 33, 69, 71, 72, 73, 74
Read, W. T., Jr., 106, 117
Recovery, 322
Recovery processes, 307, 321
Recrystallization, 328
  atom movements in, 333
  embryo for, 329
  growth rate, 329
  kinetics of, 328

INDEX

Recrystallization, mechanism of, 330
  nucleation of, 328
  nucleation rate, 329
  secondary, 339
Recrystallization *in situ*, 332
Recrystallization temperature, 330
  effect of deformation on, 331
  effect of impurities on, 331
  effect of time on, 331
  table of, 332
Recrystallization texture, 332
Recrystallization twins, 334
Recrystallized metals, properties of, 336
Redistribution of solute, 250, 258
Reduction of area, 215, 218
Rehbinder effect, 162
Relaxation curve, 184
Relaxation peak, 193
Relaxation time, 193
Residual induction, 224
Residual resistance, 146
Residual stress, 326
Resistance strain gages, 91
Resistivity, 67, 90, 91
Resolved shear stress, critical, 152
Richards, T. Ll., 350
Riser, 288
Rolling, 344
  work hardening by, 313
Runner, 288
Rutter, J., 120, 121

Sacrificial protection, 302
Sand mold casting, 289
Saturation intensity of magnetization, 90
Saturation magnetization, 223
Schmid factor, 150, 151, 152, 153, 154
Screw dislocation, 97
Season cracking of brass, 305
Second phase, hardening by, 165, 172
Secondary creep, 184
Secondary recrystallization, 339
Segregation, 281
  gravity, 283
  interdendritic, 282, 283
  inverse, 283
  normal, 281
Segregation at dislocations, 106, 401
Selective growth, 334

Self-diffusion, 376
  activation energy for, 185
Semicoherent interface, 129
Semiconductor, 66, 67
Sessile dislocation, 106
Shaping of metal, 343
Shear deformation, 140
Shear fracture, 196
Shear modulus, 82
Shear stress, critical, 149
Shear stress for deformation, theoretical, 141
Shell moulding, 289
Shockley partial dislocation, 105
Short range order, 46, 167, 170
Short-range-order parameter, 46, 52
Shrinkage, 277
Shrinkage and gas evolution, interaction of, 280
Shrinkage pipe, 278, 286
Shrinkage pore, 279
Simultaneous shear deformation, 142, 144
Single crystal, 139, 142, 145, 149, 276
Single deposition potential, 291, 292
Single domain particles, 228
Sintering, 354
Size effect on deformation, 161
Skeleton of dendritic crystals, 261
Slip, 101
  inhomogenous, 151
Slip bands, 144
Slip direction, 149
Slip plane, 102, 125, 144, 145, 149, 151
Slip systems, 149
Small particle magnets, 228
Smith, C. S., 130, 131
Smithells, C. J., 53
Smooth freezing, 234
Smooth interface, 259
Snowflake, 404
Soft atoms, 32
Soft testing machine, 216
Softening, geometrical, 151
Soldering, 290
Solid and liquid alloys, equilibrium between, 54
Solid diffusion, 272
Solid-gas reaction, 358
Solid-liquid interface, 112, 130

INDEX

Solid solution, free energy of, 63
  interstitial, 36, 53, 95
  lattice parameter of, 80
  substitutional, 36, 45
Solid-state transformations, 357
Solid-state welding, 352
Solid-vapor interface, 112, 133, 134
Solidification, 231
  applications of, 285
  structure resulting from, 273
Solidus, 37, 43, 85
Solubility, limit of, 53
  unlimited, 53
Solute, distribution of, ahead of interface, 254
  effect of, on freezing and melting, 250
  precipitation on dislocations, 401
  redistribution of, 250, 258
Solution, 295, 385
Solution in aqueous media, 298
Solution in liquid metals, 296
Solution of precipitates, 405
Solution strengthening, 168
Solution temperature, 391
Solvus, 43
Source, Frank-Read, 146, 147
Specific surface energy, crystal size and, 244
Spectroscopic notation, 3
Spheroidal graphite, 267, 268
Spheroidization, 400, 405, 406
Spheroidization of cementite, 438
Spherulitic structure, 267
Spinning, 347
Spiral, growth, 294
Springback, 326
Stability of a structure, 10
Stabilization of austenite, 367
Stabilization of stainless steel, 446
Stacking fault, 104, 106, 112, 127
Stacking sequence, 27
Stagnant layer, 255
Stainless steel, 438
  stabilization of, 446
Steady-state condition for freezing of alloy, 252
Steady-state creep, 184, 189
Steel, air hardening, 436
  annealing of, 438
  austenitic, 437, 438

Steel, eutectoid, 419
  fully annealed, 438
  martensitic transformation in, 361, 362
  normalized, 438
  stainless, 438
Stored energy of deformation, 335, 336
Strain, effect on magnetic properties, 230
  effect on precipitation, 392
Strain aging, 402
Strain anneal method, 341
Strain gage, resistance, 91
Strain-hardening exponent, 179
Stress, engineering, 215
  true, 215
  yield, 215
Stress and temperature, effect of, on creep, 185
Stress concentration, 208
Stress corrosion, 297, 304
Stress raiser, 222
Stress relief, 326
Stress-relieving treatment, 305
Stress-rupture-creep test, 199
Stress-strain curve, 149, 150, 214, 217, 219
Stress-strain-time surface, 183
Structure, atom, 1
  crystalline solids, 21
  deformed metals, 307
  effect of, on creep, 186
  equilibrium, 36
  ingots, 275
  lamellar eutectic, 265, 266
  liquids, 34
  martensite, 362, 363
  metals, 30
  vapors, 34
Structure-insensitive properties, 77, 78
Structure resulting from solidification, 272
Structure-sensitive properties, 139
Subboundary, 112, 125
Substitutional atoms, 93, 95
Substitutional solid solution, 36, 45
Substitutional solutes, diffusion of, 377
Substrate, 242
Substructure hardening, 180
Suits, J. C., 109
Supercooled melt, nucleation in, 261

INDEX

Supercooling, constitutional, 258, 259, 271, 273
Supercooling for homogeneous nucleation, 245
Supersessile dislocation, 106, 155, 162
Surface diffusion, 384
Surface energy, 132, 339
Surface imperfections, 92, 110
Surface nucleation, 246
Surface porosity, 279
Surface roughness, 133
Suzuki mechanism, 168

Taylor, G. I., 177
Taylor dislocation, 97
Temper brittleness, 400
Temperature, effect of, on corrosion, 306
    on fracture, 205
    on plastic deformation, 162
    equilibrium, 19, 56
Temperature inversion, 259
Temperature spike, 314
Tempering of martensite, 399
Tempering temperature, effect on hardness, 400
Tensile test, 214
Ternary diagram, 40, 42, 45
Ternary eutectic point, 45
Ternary system, 42
Tertiary creep, 184
Testing, nondestructive, 223
Testing machine, hard, 216
    soft, 216
Tetrakaidecahedron, 122
Texture, cube, 226
    cube-on-edge, 340
    deformation, 311, 312
    grain-oriented, 225
    recrystallization, 332
Thermal agitation, 15, 16, 17
    amplitude of, 87
Thermal energy, 15, 18, 19
Thermal etching, 134
Thermal expansion, 85, 86
    abnormal, 87
Thermal expansion and melting point, relation between, 88
Thermal expansion of uranium, 89
Thermal spike, 314, 315
Thermocouples, 91

Thermodynamic criterion for equilibrium, 63
Thermodynamically stable imperfections, 92
Thermodynamically unstable imperfections, 92
Thermoelastic effects, 192
Thermoelectric properties, 91
Three-phase equilibrium, 38
Throwing power, 291
Tilt boundary, 115
Time-temperature-transformation diagram, 419, 425, 429
Tin-lead eutectic, 268
Titanium, alpha phase alloys of, 410
    embrittlement of, 199
    interstitial solutes in, 410
Titanium alloys, diffusionless transformation in, 268
    heat treatment and properties of, 410
Titanium alloys containing beta phase, 411
Titchener, A. L., 336
Torsion test, 219
Transcrystalline fracture, 199
Transformation, allotropic, 406
    bainite, 425
    diffusionless, 358, 359
    eutectoid, 358, 415, 417, 419
    heat of, 19
    isothermal, 419
    martensitic, 358, 359
    pearlite, 415, 418
    peritectoid, 358
Transformation temperature, effect of alloying elements on, 431
Transformations, solid-state, 357
Transient creep, 184
Transition lattice theory, 114
Transition metals, 7, 30
Troostite, 424
True stress, 215
T.T.T. curve for eutectoid steel, 420
T.T.T. curve for hypoeutectoid steel, 421
Tube extrusion, 345
Turbulent deformation, 180
Turnbull, D., 132, 245, 382, 383
Twin boundary, 112, 113, 125
    coherent, 127

Twin boundary, energy of, 159
Twinning, 151, 158
  dislocation mechanism for, 158
Twinning plane, 127
Twins, annealing, 335
  recrystallization, 334
Twist boundary, 116, 117
Two-dimensional embryo, 246
Two-phase alloys, freezing of, 262
  grain shapes in, 341

Ultimate tensile strength, 214, 215, 216
Unidirectional freezing, 232
Uniform corrosion, 302
Unit cell, 13, 25
Unlimited solid solubility, 53
Uphill diffusion, 376
Upper bainite, 425
Upper yield point, 216
Uranium, growth of, during temperature cycling, 327
  thermal expansion of, 89

Vacancies, aggregation of, 199
  lattice, 93
Vacancy, formation at interface, 248
Vacancy content, equilibrium, 93, 94
Vacancy disk, 248, 249
  collapse of, 249
Vacancy mechanism of diffusion, 378, 380
Vacuum melting, 288
Van Arkel process, 295
van der Waals forces, 14
van't Hoff's equation, 56
Vapor, growth of crystal from, 292
  structure of, 34
Vaporization, latent heat of, 36, 86, 133
Vegard's law, 80
Vibration, effect of, on grain size, 277
  frequency of thermal, 20
  thermal, 16
Vineyard, G. H., 315, 316, 317, 318

Wagner, R. S., 107, 125
Walker, J. L., 236, 275, 282
Wall, ferromagnetic domain, 136, 138
Walter, J., 135
Walton, D., 260, 275
Weakness, intercrystalline, 213
Wear, 351
Weld, blacksmith's, 355
Welding, cold, 355
  fusion, 290
  solid-state, 352
Whisker, 142, 195, 295
Widmanstatten structure, 391
Wilkinson, W. D., 319
Willsdorf, H. G. F., 143
Wilson, F. H., 333
Wire drawing, 346
Work hardening, 146, 217, 312
Work hardening by rolling, 313
Work-hardening exponent, 218
Wurtzite structure, 71, 72

X-ray diffraction, effect of dislocations on, 109

Yield, suppression by cold work, 401
Yield point, 167, 180, 216, 401
  Cottrell theory of, 107
  upper, 216
Yield stress, 214, 215
Yield stress for perfect single crystals, 140
Young's modulus, 82, 215
Young's modulus of some aluminum alloys, 84

Zener-Hollomon parameter, 186
Zinc blende structure, 71, 72
Zone, Brillouin, 64, 65, 66, 67, 71
  chill, 273
  columnar, 274
  equiaxed, 274
  Guinier Preston, 393
  pre-precipitation, 391, 395, 441
Zone refining, 255, 256, 261

Date Due			
MAR 1 9 62			
MAY 2 5 62			
MAR 2 0 '63			
SEP 6 1963			
MAY 1 1964			
MAY 2 7 1965			
DEC 7 1967			
JAN 4 1968			
APR 1 6 1968			
FEB 6 1969			
NOV 1 7 2002			

Demco 293-5